Outstanding Contributions to Logic

Volume 28

Outstanding Contributions to Logic puts focus on important advances in modern logical research. Each volume is devoted to a major contribution by an eminent logician. The series will cover contributions to logic broadly conceived, including philosophical and mathematical logic, logic in computer science, and the application of logic in linguistics, economics, psychology, and other specialized areas of study.

A typical volume of Outstanding Contributions to Logic contains:

- A short scientific autobiography by the logician to whom the volume is devoted
- The volume editor's introduction. This is a survey that puts the logician's contributions in context, discusses its importance and shows how it connects with related work by other scholars
- The main part of the book will consist of a series of chapters by different scholars that analyze, develop or constructively criticize the logician's work
- Response to the comments, by the logician to whom the volume is devoted
- A bibliography of the logician's publications Outstanding Contributions to Logic is published by Springer as part of the Studia Logica Library.

This book series, is also a sister series to Trends in Logic and Logic in Asia: Studia Logica Library. All books are published simultaneously in print and online. This book series is indexed in SCOPUS.

Proposals for new volumes are welcome. They should be sent to the editor-in-chief sven-ove.hansson@abe.kth.se

Nick Bezhanishvili · Rosalie Iemhoff · Fan Yang
Editors

Dick de Jongh on Intuitionistic and Provability Logics

Springer

Editors
Nick Bezhanishvili
Institute for Logic, Language
and Computation
University of Amsterdam
Amsterdam, The Netherlands

Rosalie Iemhoff
Department of Philosophy and Religious
Studies
Utrecht University
Utrecht, The Netherlands

Fan Yang
Department of Philosophy and Religious
Studies
Utrecht University
Utrecht, The Netherlands

ISSN 2211-2758 ISSN 2211-2766 (electronic)
Outstanding Contributions to Logic
ISBN 978-3-031-47920-5 ISBN 978-3-031-47921-2 (eBook)
https://doi.org/10.1007/978-3-031-47921-2

Cover Photograph by Peter van Emde Boas.

This Springer imprint is published by the registered company Springer Nature Switzerland AG
The registered company address is: Gewerbestrasse 11, 6330 Cham, Switzerland

Paper in this product is recyclable.

Preface

Dick de Jongh has been a central figure in many important subdisciplines of logic and related areas such as intuitionistic and intermediate logics, provability and interpretability logics, theories of well-quasi orders, and formal learning theory. Many groundbreaking results and concepts in these areas bear his name, e.g., de Jongh theorems for intuitionistic logic and for intuitionistic arithmetic, de Jongh formulas for intuitionistic logic, Sambin-de Jongh fixed-point theorem in provability logic, and de Jongh-Parikh theorem on well-quasi orders. Many of these results and their legacy in the retrospective fields are discussed in this volume.

The volume starts with autobiographic notes by Dick de Jongh, which set the course for the whole volume. It also explains in detail the background on how many of the aforementioned results have been obtained, where they were presented first, etc.

The autobiography is followed by a block of four papers on provability and interpretability logics. Visser and Litak's contribution discusses the Sambin-de Jongh theorem in provability logic and the Visser-de Jongh theorem in interpretability logic, and their consequences for the area. Van Benthem's article also reviews the de Jongh-Sambin fixed-point theorem and its generalizations from a model theoretic perspective. These topics are presented in the context of abstract results on generalized well-founded orders. Mojtahedi's paper revisits the Σ_1-provability logic of Heyting arithmetic by providing a translation between two relevant logics. An interesting class of formulas to which de Jongh has turned his attention time and time again, the class of NNIL-formulas, also plays a role in the translation. Joosten, Mikec, and Vukovic's contribution gives an overview of interpretability logic and variants of their Kripke semantics, with a focus on Verbrugge semantics (a.k.a. generalized Veltman semantics).

The next four chapters study modal, intuitionistic, and intuitionistic modal logics. Metcalfe and Tokuda's paper discusses the independence of formulas in the general setting of universal algebra. This notion of intuitionistic logic and intuitionistic arithmetic was first introduced and studied by de Jongh. This paper shows that in many cases (including many well-known modal logics), independence is decidable. For intuitionistic logic this result was obtained by de Jongh and Chagrova (1996).

Balbiani and Gencer's contribution provides a survey of the unification problem in modal logic. The unification problem via Ghilardi's projective formulas is directly linked to exact and extendible formulas studied by de Jongh and Visser (1996). Iemhoff's chapter proves uniform interpolation for the Lax Logic, an interesting and peculiar intuitionistic modal logic. The uniform interpolation property of intuitionistic logic was proved by Pitts (1992) and has played an important role in the work of de Jongh, also in the context of independence of formulas discussed in Chap. 6. Bezhanishvili and Yang overview the concept of de Jongh formulas for intuitionistic propositional logic, their main properties, and also some of their modern applications. Motivated by this, they then developed a new theory of generalized intuitionistic team semantics.

In the next chapter, Weiermann surveys the theory of well partial orders. He also gives a detailed account of de Jongh's contribution to the development of the field. Bolander, Gierasimczuk, and Liberman discuss in their chapter some new developments in formal learning theory—yet another topic of interest of de Jongh to which he made important contributions together with Gierasimczuk.

The second to last chapter contains a paper on origami geometry, in which Beklemishev, Dmitrieva, and Makowsky provide a variant of an axiomatization of origami planes. The contribution can be seen as representing the Master of Logic program of the ILLC in which de Jongh invested a lot of energy. Out of the three authors, Dmitrieva obtained her Master's degree in 2021 at the ILLC with de Jongh being her mentor.

The volume ends with a complete bibliography of Dick de Jongh in the last chapter.

We thank all the authors of this volume for their beautiful contributions. We are also extremely grateful to all the reviewers for their excellent work that improved considerably the quality of this volume.

Amsterdam, The Netherlands — Nick Bezhanishvili
Utrecht, The Netherlands — Rosalie Iemhoff
Utrecht, The Netherlands — Fan Yang
August 2022

Contents

Contributors

Lev Beklemishev Steklov Mathematical Institute of RAS, Moscow, Russia; National Research University Higher School of Economics, Moscow, Russia

Philippe Balbiani Toulouse Institute of Computer Science Research, CNRS — Toulouse University, Toulouse, France

Nick Bezhanishvili Institute for Logic, Language and Computation, University of Amsterdam, Amsterdam, The Netherlands

Thomas Bolander Danmarks Tekniske Universitet, Kongens Lyngby, Denmark

Dick de Jongh Institute for Logic, Language and Computation, University of Amsterdam, Amsterdam, The Netherlands

Anna Dmitrieva University of East Anglia, Norwich, United Kingdom

Çiğdem Gencer Toulouse Institute of Computer Science Research, CNRS — Toulouse University, Toulouse, France;
Faculty of Arts and Sciences, Istanbul Aydın University, Istanbul, Turkey

Nina Gierasimczuk Danmarks Tekniske Universitet, Kongens Lyngby, Denmark

Rosalie Iemhoff Utrecht University, Utrecht, The Netherlands

Joost J. Joosten Department of Philosophy, University of Barcelona, Barcelona, Spain

Tadeusz Litak Informatik 8, FAU Erlangen-Nürnberg, Erlangen, Germany

Johann A. Makowsky Technion – Israel Institute of Technology, Haifa, Israel

George Metcalfe Mathematical Institute, University of Bern, Bern, Switzerland

Luka Mikec Department of Mathematics, University of Zagreb, Zagreb, Croatia

Mojtaba Mojtahedi Department of Mathematics: Analysis, Logic and Discrete Mathematics, Ghent University, Ghent, Belgium;
Department of Mathematics, Statistics and Computer Sciences, University of Tehran, College of Science, Tehran, Iran

Andrés Occhipinti Liberman DTIC, Universitat Pompeu Fabra, Barcelona, Spain

Jan Mas Rovira Heliax, Zug, Switzerland

Naomi Tokuda Mathematical Institute, University of Bern, Bern, Switzerland

Johan van Benthem University of Amsterdam, Amsterdam, The Netherlands;
Stanford University, Stanford, CA, USA;
Tsinghua University, Beijing, China

Albert Visser Philosophy, Faculty of Humanities, Utrecht University, Utrecht, The Netherlands

Mladen Vuković Department of Mathematics, University of Zagreb, Zagreb, Croatia

Andreas Weiermann Vakgroep Wiskunde: Analysis, Logic and Discrete Mathematics, Ghent, Belgium

Fan Yang Department of Philosophy and Religious Studies, Utrecht University, Utrecht, The Netherlands

Chapter 1
Notes on My Scientific Life

Dick de Jongh

I was born in Enschede in the Netherlands on October 19, 1939, just before the start of the Second World War to parents who were both first in their families with a university degree, in medicine, my father, and in pharmacy, my mother. My father, D. K. de Jongh, was at the time a general practitioner. Before the war ended he wrote a dissertation on homeopathy (de Jongh, 1943), a full-fledged history, analysis and rebuttal. We moved to Haarlem near Amsterdam when he became the director of a laboratory for a pharmaceutical company. Still later he became professor in pharmacology in Utrecht and Amsterdam. He was my first teacher with a strong belief in the scientific method and a wide knowledge.

Master study, Amsterdam, 1961–1964. After an uneventful bachelor in Leiden (1957–1961), first chemistry, physics, and later mathematics, I moved for my masters to Amsterdam for a study in foundations with E. W. Beth, with a major in philosophy and minor in mathematics in the same style as the MSc in Logic today. Simultaneously with me, Hans Kamp arrived from Leiden with a physics bachelor from Leiden. We had discussed this move beforehand together.

The period of this masters, 1961–1964, was a very fruitful one. Evert Beth, a man of encyclopedic knowledge, instilled in me from the start the idea of logic not as a subdiscipline of mathematics, but as the center of an interdisciplinary field. He had created a small Institute "Instituut voor Grondslagenonderzoek en Filosofie der Exacte Wetenschappen" (IGLO).[1] In my studies, mathematics, philosophy, computer science avant la lettre, and linguistics, all played a role. I was immediately made a research assistant, funded by the European Atomic Community; Beth was one of the first to recognize the opportunities for external funding even for an ostensibly purely theoretical scientific discipline as logic. The weekly seminar of the research

[1] Institute for Foundational Research and Philosophy of the Exact Sciences.

D. de Jongh (✉)
Institute for Logic, Language and Computation, University of Amsterdam, Amsterdam, The Netherlands
e-mail: d.h.j.dejongh@uva.nl

N. Bezhanishvili et al. (eds.), *Dick de Jongh on Intuitionistic and Provability Logics*, Outstanding Contributions to Logic 28, https://doi.org/10.1007/978-3-031-47921-2_1

group had widely ranging subjects. Later on in the course of this period it was joined by a number of young linguists (among which Pieter Seuren and Rudolf de Rijk) who did not find a sympathetic ear in their own department when they were inclined to study Chomsky's new ideas. For Beth's students Chomsky's "Syntactic Structures" (Chomsky, 1957) was compulsory reading.

Beth assigned to me the study of his relational semantics for intuitionistic logic: a semantics he was developing independently from Kripke (see de Jongh and van Ulsen, 1999). This study became the subject of my master thesis (de Jongh, 1964), but one may say I never really completed his assignment. Beth and his close colleague Arend Heyting regularly had a common seminar. Although not in the least prepossessing, Heyting was an impressive figure. His teaching of intuitionistic mathematics profoundly influenced me and it established in me the core of my view on constructivism. His student Anne Troelstra became a life-long friend. Anne and I have always been able to discuss our studies and research. He joined me during our master studies in working on an issue arising from the work on my master thesis. It introduced the concept of p-morphism and the first results on the duality of algebras and frames, models and, finally, the way in which p-morphisms between finite frames can be thought of as being split into a sequence of α- and β-reductions (Jongh & Troelstra, 1966). This latter splitting was extended to $\mathsf{K4}$-frames in modal logic 56 years later in a master thesis (Carr, 2022) with N. Bezhanishvili and T. Moraschini as supervisors. My master thesis itself, intended as an investigation of the concept of intuitionistic connective in relational models, developed into a partial construction of the universal model of IPC, the intuitionistic propositional calculus (see more on this below in the discussion about my dissertation).

Although Hams Kamp's research later took him in a quite different direction, we had a close relationship during the whole of the master study period in Amsterdam, following the same courses and exploring the philosophical student community. Together, we worked, among other things, on the construction of programs for the intuitionistic propositional calculus in a language that combined ALGOL and LISP on one of the first computers at the Mathematisch Centrum, the predecessor of the CWI. The programs we used were based on semantic tableaux, Beth's main interest at the time of my relationship with him. Although I no longer wrote programs afterwards, this successful project led to much work in later periods, as we will see.

Anne Troelstra and I had been fascinated by Diego's Theorem (Diego, 1966) that the fragment $[\to]$ of IPC, the formulas containing only $\to$, is locally finite, i.e. there are only finitely many equivalence classes of formulas in n variables. From the start I recognized the utility of the computational approach to the study of such fragments.

Beth died in 1963, a year before the end of my studies. Heyting then became the official supervisor of my thesis.

Ph.D. study, University of Wisconsin, 1964–1968. I obtained as Beth's student a scholarship for two years study in the U.S. from the American Council of Learned Societies. S. C. Kleene accepted me as a Ph.D student. I stayed at the University of Wisconsin at Madison for 4 years, the third and fourth year as a teaching assistant. Kleene assigned to me the problem of characterizing IPC. Łukasiewicz had conjectured that IPC among the intermediate logics (between IPC and classical propositional logic) is characterized by its disjunction property (Łukasiewicz, 1952). This was proved wrong by Kreisel and Putnam (1957), who constructed a proper extension of IPC that does have the disjunction property. Kleene conjectured that a more complicated form of the disjunction property based on his slash | would be sufficient for a characterization. My assignment was to prove or disprove this conjecture.

I quickly saw two possible ways to prove the conjecture, an easy one and a very complicated one. I promptly sent him the easy solution, a completely wrong solution as I soon realized. I didn't really want to use the other solution because it was a considerable amount of work based on my master thesis, but I had no choice. I worked very hard to have a sketch ready quickly. Most of the time of writing my thesis consisted of working out this sketch in detail. It involved a proof of a theorem that, unknown to me, as in those years often happened with results from Soviet Union, had been proved a few years before by Jankov (1963). My proof was very different from his and gives completely different formulas axiomatizing the logics involved. His proof was completely algebraic, while mine basically used frames but needed translating forth and back to algebra, using the duality discussed in my work with Troelstra.

It also used the partial construction of the universal model for formulas in n variables of my master's thesis. I further perfected its construction but didn't think of creating one infinite model. I left it as a collection of reduced finite models, i.e. models that do not allow for non-trivial p-morphisms. This made formulations and proofs somewhat awkward because one has to keep mentioning the isomorphisms and embedding functions. The history of the n-universal models of intuitionistic and modal logic is full of names of prominent modal logicians who played a role in their construction or the discovery of its structure: Segerberg (1971), Shehtman (1978), Bellissima (1986), Grigolia (1987), and, finally, Rybakov (1992), who surely made use of it more than anyone else. For more details, see Chagrov and Zakharyaschev (1997, 2006). I will return to the Jankov theorem and the universal models later. In any case, the method of proving the conjecture turned out to be more important than the conjecture itself, not uncommon in mathematics. But the characterization will reappear in this story as well.

In the dissertation (de Jongh, 1968) the concept of connective remained an important theme. One result in that direction was the characterization in Theorem 3.5 of the set connectives generated by the usual ones $\{\wedge, \vee, \rightarrow, \neg\}$ as the set of connectives definable on relational models by using restricted quantifiers ($\forall y \geq x, \exists y \geq x$). This result was published independently by McCullough (1971) in 1971.

In the last period of my time as a Ph.D. student I had also almost finished proving a theorem relating intuitionistic logic to Heyting Arithmetic HA. But Kleene didn't agree that the manuscript was well enough in shape to be part of my doctoral thesis. That was not because of a lack of interest on his part. I still have a manuscript which on many pages has more of his comment than the original text. He was a very meticulous man. The proof uses his concept of functional realizability, a concept developed by him more with intuitionistic real analysis in mind than with arithmetic. And it was the formalization of intuitionistic analysis (Kleene & Vesley, 1965) that occupied his mind and the local seminar at this time. In later terminology, one may call the theorem an arithmetical completeness theorem.

In fact, I was gratified to see that Balbiani and Gencer named it this way in their contribution to this Volume.

The theorem states more or less that, if a formula B is not provable in intuitionistic first order logic IQC, then it has an arithmetical substitution instance which is not provable in Heyting Arithmetic. But the statement I had proved for IQC is not the desirable one. I didn't prove the statement above for B itself but for B with all quantifiers relativized to a predicate letter $A(x)$. Model-theoretically this means that the countermodel for B that is obtained is not defined on the set of natural numbers itself but on a definable subset. The proper form was later proved twice, both times in a dissertation. First by Leivant (1975) in 1975, by a purely proof-theoretical method. Later, in 1991, by Jaap van Oosten in (van Oosten, 1991). For propositional logic IPC the statement was the desirable one. As we will see, for that case Craig Smoryński quickly found a simpler proof. A lot on the content and the later history of this theorem can be found in (de Jongh et al., 2011), where Albert Visser, Rineke Verbrugge and I extended the theorem to many intermediate logics with respect to their corresponding arithmetics. That doesn't cover the most recent progress on the theorem. I will get to that at the end of these notes.

During my studies in Madison I was mostly concerned with mathematical logic, but I did not lose contact with philosophy altogether, I did a minor in philosophy with Julius Weinberg.

Return to Amsterdam, 1968–1970. In 1968 I returned to Amsterdam as an assistant professor accompanying Dana Scott who had accepted a temporary position on Beth's chair. I had met him in the summer of 1968 at the Conference on Constructivism and Proof Theory in Buffalo where many of the significant logicians of the time had gathered.

In Amsterdam I also made my acquaintance with Dov Gabbay who visited there for a while. He proved one theorem after another at that time. While in 1969 we were both at the ASL conference in Manchester we got the idea for our article on intermediate logics for n-ary splitting finite trees on which he immediately wrote a report (Gabbay & Jongh, 1969). The article (Gabbay & de Jongh, 1974), which appeared some time later, was an early paper on specific intermediate logics. Its interest was in introducing this prominent sequence of logics, in showing that the disjunction property is not a rare occurrence, and in making explicit the difference between finite and infinite trees (the full infinite binary tree is characteristic for IPC itself).

Assistant professor at SUNY, Buffalo, NY, 1970–1972. I visited the state University of New York, SUNY at Buffalo from 1970-1972, where a group of people with an interest in intuitionism had gathered around Kleene's student Richard Vesley. Inspiring influences were John Myhill, a wonderfully creative person, and Nicholas Goodman who deepened my insight in formalized arithmetic. This assisted me in developing my work on unary propositional formulas in intuitionistic arithmetic (de Jongh, 1982).

An important person to be mentioned from my Buffalo time is Craig Smoryński whom I met at an ASL conference in Atlantic City. He was a student of Kreisel's who knew about my recent theorem on intuitionistic arithmetic and had found a much easier proof of the theorem for the propositional case using Kripke models for Heyting Arithmetic (Smoryński, 1973). I got so involved in discussions with him that on my way home to Buffalo I realized that I had left Atlantic City after being there many days without seeing the sea.

But, equally important is that I also started to work in Buffalo with Rohit Parikh. First, we found a common interest in the ultra-intuitionistic foundation of mathematics on which I had already done some work (Ehrenfeucht et al., 1971), while he had written his amazing article (Parikh, 1971) which approaches that area from within standard logic. Our discussions got no follow-up when the grant we applied for at NSF was not forthcoming.

But we did connect on another idea by Rohit, continuing after I had returned to Amsterdam in 1972. Wpo's are well-founded partial orderings with no infinite anti-chains. An equivalent formulation is that they are partial orderings which, when extended to a linear ordering, will always become a well-ordering. He hoped that this could throw light on the reason why some hierarchies built on methods that can be seen as having a wpo-basis seem to reach only to low ordinal lengths. We were indeed able to explain this by developing a theory assigning to each wpo an ordinal, namely the limit of the order type of the ordinals of well-orders extending the wpo. The basic theorem of this theory is that for each wpo a specific well-ordering extending it exists that actually reaches this order type (de Jongh & Parikh, 1977).

The basic theorem turned out, also in this case, to be more important than our particular application. As a follow-up I found as a further application the ordinal of the class of all finite binary trees. This is a wpo (actually a wqo, a well-quasi-ordering) by Kruskal's well-known tree theorem (Kruskal, 1960) and I could prove that the ordinal of this class is ε_0. The interest in this is that finite binary trees can be seen coding the application of one binary operation satisfying certain monotonicity condition on one constant to obtain ordinal notations. The theorem shows that with one such binary monotonic operation and one constant one can reach exactly to ε_0. Stunningly, this result filled a gap in the Ph.D. work (Schmidt, 1975) of Diana Schmidt who had given such ordinals for notations using Schütte's Klammer-Symbole (Schütte, 1954) for higher ordinals. On the basis of the theory about wpo's by Rohit and me, she later wrote her Habilitationsschrift (Schmidt, 1979).

> One can find a more extensive overview of this history in a wider context in the contribution of Andreas Weiermann in this Volume. Besides the fact that it is very illuminating, it is fitting that this overview also has much of a personal flavor.

At SUNY at Buffalo, the mathematical logicians did have some contact with philosophy, especially due to the very active John Corcoran at the Philosophy Department.

Back in Amsterdam, the seventies. After these two years in Buffalo I returned to Amsterdam thanks to Anne Troelstra who had secured a position for me. Important work in 1973 was finishing the report on unary propositional formulas in Heyting arithmetic (de Jongh, 1973). Its main theorem about HA stated that, if in HA of a sentence A, neither $\neg\neg A$ nor $\neg\neg A \to A$ is provable, no $F(A)$ for a purely propositional formula $F(p)$ is provable in HA unless $F(p)$ is provable in IPC itself. When I recently looked at Brouwer's (1954), I was pleasantly surprised to see that, for the A he discussed in that article, he was only interested in showing that neither $\neg\neg A$ nor $\neg\neg A \to A$ is provable.

The paper was mainly concerned with the role of IPC in HA. However, it also contained some concepts applicable to the role of IPC in any theory T formalized in IPC and thus specifically to IPC itself, that would keep playing a role in my later work. These were the concepts of exact(ly provable) formula, and the notion of (in)dependence of formulas. Two formulas A, B are independent in T if for no propositional $F(p, q)$, $F(A, B)$ is provable in T unless $F(p, q)$ is provable: there is no non-trivial link between the formulas; similarly for n formulas $A_1, \ldots, A_n$. If A, B are not independent there is a non-trivial $F(p, q)$ such that $F(A, B)$ is provable in T; if this is all that is provable about A, B, in other words $F(p, q)$ axiomatizes the theory of all $G(p, q)$ such that $G(A, B)$ is provable in T, F is called exact in T, and if T is IPC itself, simply exact. Regarding the unary case, there are only 5 exact IPC-formulas, $p, \neg p, \neg\neg p, \neg\neg p \to p, p \to p$. The main result mentioned above states that the same 5 unary IPC-formulas are exact in HA. The intuitive idea is that those are the only five states in which a formula can occur in HA. The paper was published later, in 1982, in the Brouwer Conference Volume (de Jongh, 1982).

We will return to exact and independent formulas later, also in connection with propositional theories (or in a different formulation: Magari and Heyting algebras) that will turn up later. Consider the formula $F(p, q)$ a few lines back that axiomatizes a theory of some A, B. We can similarly consider the (propositional) theory realized by a sequence of formulas $A_0, A_1, \ldots$. Which theories are realized in HA, and for this infinite case, in PA as well?

Besides polishing my work on unary formulas, and on well-partial orderings the period after 1972 was the start of my connection with the logic of provability, as the sadly missed George Boolos coined it in the title of his book (Boolos, 1993).[2] Martin Löb had taken Beth's chair while I was in Buffalo. He had, already in 1955, two years before I started my studies, proved his famous theorem that, if PA proves $Prov(A) \to A$, then it proves A itself (Löb, 1955). It is an extension of Gödel's Second Incompleteness Theorem, but also clearly recognizable in the modal-logical principle $\Box(\Box A \to A) \to \Box A$, now known as GL after Gödel and Löb. Unrelated to this, Johan van Benthem asked me a purely modal question, does GL prove the 4-principle $\Box A \to \Box\Box A$, and I was able to solve it for him.[3]

[2] I will reserve the term provability logic for the logic GL itself.

[3] It turns out I have not been the only one to solve this question. See also (Boolos and Sambin, 1991).

Let me introduce Johan. As a student he finished his masters in philosophy and mathematics during the first year after my arrival, immediately becoming assistant professor as was the habit of the time. He directly initiated serious philosophical-logical courses. He opened my eyes to the new possibilities of applying logic in philosophy, reinforcing in me the vision of logic as the center of an interdisciplinary area. In 1977 he obtained his Ph.D. with Martin Löb. The simultaneous contact with both van Benthem and Löb brought me to the conjecture that Löb's principle axiomatizes the logic of the formalized provability predicate in Peano Arithmetic in the form of the provability logic GL. This was without realizing that in the U.S. several people were already having the same thought. I had a number of discussions with Martin Löb about the possibilities to prove the conjecture but we didn't really get off the ground.

Luck had it that just at that time Craig Smoryński arrived in Amsterdam for a two year stay from 1973 to 1975. He was not only the serious student of intuitionistic logic and its formalized theories that I knew, but just like me had turned to the logic of provability. He acquainted me with his conjecture that in the logic GL each formula $A(p)$ in which p only occurs inside some $\Box$ has a fixed point F, i.e. $\mathsf{GL} \vdash F \leftrightarrow A(F)$. After a while I was able to prove this conjecture. A short time later, Giovanni Sambin, who had also been concerned with the problem, proved it as well by a quite different method. Craig wrote a proof in more or less the style of my proof and published it Smoryński (1985). The theorem became known as the Fixed Point Theorem of de Jongh and Sambin. I am of the opinion now that it would be proper to add Smoryński to these names as well: he clearly saw the potential of the result and inspired us both. In that period Craig and I also wrote another paper together about second order intuitionistic arithmetic showing that simple semantic methods can be quite powerful in that area. Almost simultaneously, in 1976, Robert Solovay proved in an insightful and ingenious paper (Solovay, 1976) that indeed GL is the provability logic of PA, the logic that codifies the principles that are provable in PA about the provability predicate. He proved the necessary modal completeness of GL on the way. His arithmetic completeness result, even more so than the fixed point theorem, put the logic of provability for a while in the limelight. For a fuller history of the initial stages of this subject see Boolos and Sambin (1991).

In this Volume Johan van Benthem wrote a number of observations concerning the fixed point theorem, its proof and modern connections. As always it abounds with ideas that one would like to follow up. Many of his observation have their origin in an extensive correspondence we had in the eighties on the subject. I am grateful he remembered its relevance, for me it was far in the back of my mind. Especially sections 5 and 6 point to a lot of work that can still be done in the area. Good to see is "A simple fixed-point description" in section 6. It describes exactly what one thinks when one is trying to prove a fixed point theorem in a new area in a semantic way. If one succeeds in finding a proof that way one can then easily translate it into a syntactic proof with, for others, a mysterious origin.

Of course, intuitionistic logic always stayed in the back of my mind. When my teacher Kleene was honored with a conference in Madison in 1978, I participated. The main theorem of the paper (de Jongh, 1980) I presented there is that a completely free Heyting Algebra (free also with respect to arbitrarily infinite joins and meets) on two generators does not exist. In other words, this algebra is a proper class. This compares to the same result for Boolean algebras on countably many generators (Kripke, 1967). The method of the proof consists, one may say, of an unbounded iteration of the construction of the 2-universal model.

Developments at the University of Amsterdam, 1973–1986. Except for those articles, the second half of the seventies was research-wise for me a low point. But in other aspects of my scientific existence there were important developments. On the mathematical logic side Henk Barendregt, studying in Utrecht, and I met soon after my return, and together we organized the so-called Intercity Seminar, a country wide mathematical logic seminar, held in turn in Amsterdam and Utrecht. It lasted for many years, certainly until 2003. It was the main mathematical logic seminar where people met, with Troelstra and van Dalen as its senior participants.

It was a loss for Amsterdam in 1977 that Johan van Benthem left for Groningen to take an associate professorship. I took over his tasks as the major logician at the Philosophy Department (at the time: Centrale Interfaculteit). With regard to teaching this meant in any case the teaching and organization of the introduction to logic for philosophy students. I changed the course in a small but for me essential way. It had consisted of a weekly lecture by the main teacher plus practicing and explanation in subgroups led by assistants. I deleted the lecture by the main teacher. All of the lecturing, practicing and explanation was to be done henceforth in the subgroups themselves, one of which was to be lead by the main teacher, who would also coordinate the efforts of the assistants. This has its advantages and disadvantages. For the assistants it becomes more interesting, they have more responsibility. The main teacher no longer reaches all students, a loss for them, and loses some control but obtains more direct access to a smaller number of individual students. Of course, its net success depends much on the capabilities and enthusiasm of the assistants. In this respect the ILLC and its precursors have always been lucky. I, for example, had among my assistants (master's students!) Michiel van Lambalgen and Lex Hendriks. The first later became a distinguished colleague, with the second I would collaborate in many ways later on. In any case, the method did prove to be a success at the time.

I even became chairman of the philosophy department for a year in this turbulent period during which students for a while wielded a considerable amount of power. Somehow they worked on the assumption that I was just as radical as they were, and I am afraid I turned out to be somewhat of a disappointment to them. An important occurrence in this period was the arrival in 1978 of Frank Veltman as a logician in the philosophy department. Really capable logicians driven by purely philosophical-linguistic motives are rare, and his appointment solidified the position of logic inside philosophy, and the coherence of the group of logicians.

Cooperation with the linguistic philosophers (Bartsch, Stokhof, Groenendijk, Dekker and others) flourished. There was a popular bi-weekly Montague Colloquium with Natural Language Semantics as its theme in which many people participated for which this was not their core business. Also, Peter van Emde Boas, a member of the computer science start up at the University of Amsterdam, was included in our unified enterprise, accompanied by his co-worker Theo Janssen, mathematician-linguist.

The cooperation extended to participation in the First International Montague Colloquia (1975, 1977) in which I also figured as one of the many teachers of Montague Grammar. In addition, the first unified introductory teaching of logic and philosophy of language was an exciting common undertaking by Stokhof, Groenendijk, Veltman, me and others. I learned a lot from it. As far as I know it still lasts on under the name *Logic and the Linguistic Turn*.

The ultimate unifying experience though, at least for me, was the writing of *Logic, Language and Meaning* by L. T. F. Gamut (van Benthem et al., 1982a, 1982b). As Wikipedia states: "L. T. F. Gamut was a collective pseudonym for the Dutch logicians Johan van Benthem, Jeroen Groenendijk, Dick de Jongh, Martin Stokhof and Henk Verkuyl. Gamut stands for the Dutch universities of Groningen (G), Amsterdam (am), and Utrecht (ut), then the affiliations of the authors. The initials L.T.F. stand for the discussed topics, respectively, Logic (Dutch: Logica), Language (Dutch: Taal) and Philosophy (Dutch: Filosofie)." It is a two-volume overview of modern logic as used in the analysis of natural language and is still being used for that purpose by students today. The extreme carefulness and common care with which it was written did entail that the writing of both the original Dutch version and the updated English one (van Benthem et al., 1991a, 1991b) took an extended period of time (1972, 1981).

In that period, the position of logic between mathematics and philosophy was somewhat uneasy. The university was slowly changing over from very small faculties to very large ones, and with each step in this change-over the logicians were somehow able to keep some Logic "vakgroep" (subdepartment) intact in various changing forms under different authorities who didn't all equally appreciate our taste for independence. I was always the secretary. It doesn't speak well for my archival capabilities that no solid description of this period is available. Later, when all possibilities to maintain a logic vakgroep finally gave out completely we created with our community a wider imaginary institute, the ITLI (Instituut voor Taal, Logica en Informatie), which functioned for a while quite well because of the clear coherence of the members. It was eventually succeeded by the ILLC.

Around 1984–1985 there was a discussion in the mathematics department about Beth's chair that had been held by Löb. Should it be maintained? It was not unreasonable for the department to consider cutting it. The university was financially at an all-time low. With seven chairs assigned by the university to mathematics, two for logic was a bit much. Maybe against the odds, we were able to save the chair for the university as a whole by an appeal to its historical and interdisciplinary significance. The selection committee chose for Johan van Benthem's return from Groningen to fill this chair. This choice didn't receive unanimous applause, but for me and others with the same ideals it was a very important and positive decision; his return would help pave the way to the realization of Beth's vision in which our growing community wholeheartedly believed.

The eighties, Provability logic, Interpretability logic. The decade of the eighties was a time of high activity for the logic of provability. Albert Visser arrived on the scene with his dissertation (Visser, 1981) in 1981. Like Smoryński and I, he combined an interest in the logic of provability and formalized intuitionistic theories. He was the first to see not only the difficulties in but also the possibilities of a provability logic for Heyting Aritmetic. He established a number of principles for this logic, often using a very surprising version HA^* of HA. Cooperation between the two of us started in this period, and has never stopped.

In the late eighties, I worked a lot with Franco Montagna on the logic of Rosser-type provability predicates initiated by Guaspari and Solovay (1979). I went several times for extended visits to Siena (weeks, months). We wrote five papers together in an intense cooperative effort. The most interesting were the two on so-called provable fixed points (de Jongh & Montagna, 1988, 1989), arithmetical formulas whose fixed points are always provable. The starting point was a theorem in Parikh's (1971) mentioned above. This theorem states that, in Peano arithmetic, theorems exist for which the proof that they are provable is (much) shorter than the shortest proof of the theorem itself. We could now prove this result very easily using the Rosser provability methods: the arithmetical formula "Prov(Prov(x)) has a shorter proof than any of Prov(x)" has only provable fixed points. The two papers investigate the power of the method. In the second paper, the non-linear ordering "much shorter than" (going back quite some time Luce, 1956) plays an interesting role.

Around the same time, Albert Visser had started his work on the logic of interpretability (or Π_1-conservativity). The main new logical idea was the introduction of a binary conditional-like modality to form formulas $A \rhd B$ with arithmetic interpretation: "$T + B$ is interpretable in $T + A$" (T a formalized theory).[4] He immediately obtained important results. He asked Frank Veltman, since his dissertation (Veltman, 1985) the authority on conditionals, whether he could produce a modal semantics for it, handing him the purported system ILM for interpretability over $\mathsf{PA} = T$. Frank quickly succeeded. Together Frank and I produced a modal completeness proof for ILM (de Jongh & Veltman, 1990) with respect to this semantics. My role consisted mainly in adapting Frank's ideas to the peculiar setting of the logic of provability.[5] I presented it at the memorable Heyting Conference in Varna of 1988. Albert and I met there a number of the top Russian logicians for the first time. The use of our modal completeness result for ILM in obtaining an arithmetical completeness proof for ILM with respect to PA, independently by Shavrukov (1988) in Moscow (from Sergei Artemov's school) and Alessandro Berarducci in Berkeley (in his dissertation supervised by, again, Bob Solovay, see the article Berarducci, 1990) quickly followed. Those were exciting times.

The subject led to intensive cooperation between Sergei Artemov's group in Moscow and ours. I visited him and the group inspired by his enthusiastic leadership several times from 1989, the year of the huge Maltsev conference in Novosibirsk, onwards. He visited and taught in Amsterdam, Volodya Shavrukov came as a Ph.D. student, and Lev Beklemishev and others came on a string of visits. Lev taught for quite a number of years in Utrecht.

---

[4] Equivalent to "$T + B$ is interpretable in $T + A$" is, at least in extensions of PA, "$T + B$ is Π_1-conservative over $T + A$", i.e. $T + B$ proves no more Π_1 statements than $T + A$.

[5] Much later we conquered together a tougher logic ILW (de Jongh and Veltman, 1999).

Albert and I supervised two master's theses on interpretability, by Marianne Kalsbeek and by Rineke Verbrugge. In a co-supervised NWO-project Rineke wrote her dissertation "Efficient Metamathematics" (Verbrugge, 1993), mostly on interpretability. It also contained, in a cooperation with Alessandro Berarducci, a partially successful attempt to extend Solovay's theorem to bounded arithmetics, later published as Verbrugge and Visser (1994). The well-known major problems in complexity theory have prevented a full extension of Solovay's statement for PA to those weak arithmetics.

The nineties, PIA Conference, Computation in logic, Dependence. Already in the early nineties the peak activity in the logic of provability as a completely independent subject was reached. For Albert and me this was the organization of the PIA (Provability, Interpretability, Arithmetic) Conference in Utrecht in 1991. Also, I wrote a chapter (de Jongh and Japaridze, 1998) for the Handbook of Proof Theory on the logic of provability with G. Japaridze. Again this was part of the cooperation with Artemov's group.

The logic of provability was slowly integrated into the larger area of proof theory of formalized mathematical theories and combined with the study of truth and provability. This took place under the spiritual leadership of Lev Beklemishev and Albert Visser. This wider area does attract considerable interest, as witnessed by the Wormshops (Workshop on Proof Theory, Modal Logic and Reflection Principles) organized by Joost Joosten and David Fernández-Duque together with Lev.

Joost J. Joosten is a later student of Albert and mine who has stayed closest to the original setting of the logic of provability. He obtained his M.Sc (Joosten, 1998) in Amsterdam under our co-supervision, and his Ph.D. (Joosten, 2004) in Utrecht in 2004, for which Lev joined us in the supervision. Now in Barcelona, he continuously remains productive in the area. Together with Marta Bílková we wrote on the interpretability logic of Primitive Recursive Arithmetic (Bílková et al., 2009). Very recently, Albert Visser and I joined Joost and Ana de Almeida Borges in determining the strictly positive fragment of the quantified provability logic of HA (Almeida Borges et al., 2022).[6] The full quantified provability logic of even PA is known to be not recursively axiomatizable, in fact Π_2-complete, by work of Vardanyan and Artemov.

Very fitting in this Volume is the extensive overview of Generalized Veltman Semantics that Joost gives with a number of co-authors. One of these co-authors, Mladen Vuković, has worked for many years indefatigably on the subject. That in this overview Generalized Veltman Semantics has been renamed Verbrugge Semantics after Rineke, who discovered this semantics in Verbrugge (1992) pleases both Albert and me. The article contains a wealth of facts and insights on the history of interpretability logics. The semantics has become an interesting subject in its own right. One wonders whether the insights gained here can be useful in the wider area of conditional logics.

The detailed investigation into the prerequisites for obtaining explicit fixed points in the area of the logic of provability by Albert with Tadeusz Litak in this Volume shows that my original proofs of the existence of fixed points in the two different situations of provability

[6] The strictly positive fragment consists of implications of formulas containing only $\Diamond$, $\wedge$ and the universal quantifier.

logic and interpretability logic were more different than I realized at the time. It also shows that all the details in the handling of the proof that Albert added and which I thought at the time a kind of exaggerated precision really bore fruit in the present investigation of fixed points.

Albert Visser and I have had a continuous cooperation, both in research and in co-supervising, in the area of intuitionism as well as in the logic of provability. This is a good point to go through this cooperation. In 1991, we wrote on extending the fixed point theorem to interpretability logic (de Jongh & Visser, 1991). On the intuitionistic side we wrote a paper together for the Keele ASL conference (de Jongh & Visser, 1996) on the Heyting Algebra of Heyting Arithmetic. It combined ideas from his thesis and my (de Jongh, 1982). The leading question is which propositional theories (in the terminology described above) are realized in HA and IPC. Notwithstanding the inspiration we got from Shavrukov's theorem about PA (see later), we didn't succeed in finding an answer to that question. A difficulty is that realized theories have to be closed under all admissible rules. But we got a number of interesting results on the way, specifically the characterization of the propositional theories realized in Albert's HA^* as all theories with the disjunction property. That HA^* is so much easier to handle than HA is due to the fact that HA^* has no non-trivial admissible rules. The results of this paper arose during a memorable visit to Prague the two of us made and where we found the time to have discussions in the hotel room, time we didn't find at home in the Netherlands.

After that we wrote the basic article (Visser et al., 1995) on NNIL-formulas together with Johan van Benthem and Gerard Renardel de Lavalette, but under Albert's lead. These are formulas from IPC with no nesting of implications on the left. Albert had introduced the formulas in arithmetic (Visser, 2002), and we had used them in (de Jongh and Visser, 1996) afterwards. In the article the NNIL-formulas are proved to be the Π_1-formulas on models and to be up to equivalence the ones preserved in submodels. A large role in the article was played by subsimulations. In the study of Heyting arithmetic, NNIL-formulas are on the foreground with Σ-sentences substituted for the variables (see Visser, 1985).

One can read about their present day use in Heyting arithmetic in Mojtahedi's contribution to this Volume.

Much later, in 2011, we proved, this time together with Rineke Verbrugge, the arithmetical completeness of intermediate logics with the FMP. Each is complete with respect to the version of HA based on the logic itself (de Jongh et al., 2011). The same statement for arbitrary intermediate logics remains an open problem.

Verbrugge and Joosten were not the only Ph.D. students we co-supervised, sometimes with others. Others were Shavrukov (1994, logic of provability), Zambella (1994, logic of provability, bounded arithmetic), Iemhoff (2001, intuitionistic logic of provability, admissible rules) and Paula Henk (2016, logic of provability). In these supervisions there was always a common goal (generally a shifting goal) for the dissertation. Discussions with the student were generally held by the student and both of us together.

Shavrukov proved in his dissertation (Shavrukov, 1994) (in fact even before) a major theorem on the Magari Algebra (also diagonalizable algebra) of PA (see Shavrukov, 1993). As before I will state it in terms of propositional theories. The theorem states that all r.e. theories T with the disjunction property (if $\Box A \vee \Box B \in T$, then $A \in T$ or $B \in T$) are realized in PA. A second related theorem of his is that the Magari Algebras of PA and ZF are non-isomorphic (see Shavrukov, 1993). Besides work on bounded arithmetic Domenico Zambella gave a very clean and understandable proof of Shavrukov's theorem in his dissertation (Zambella, 1994). In fact, Albert and I heavily leaned on his presentation in writing (de Jongh & Visser, 1996). He did the same with the arithmetical completeness proof for ILM with respect to PA. Both remain excellent introductions to the subject.

The much later dissertation of Henk (2016), this time co-supervised by Volodya Shavrukov, concerned non-standard proof predicates of PA. My supervision concerned mostly her work on the bimodal logic GLT introduced by Lindström in his study of Parikh provability. Her project lead to a renewed fruitful cooperation with Lev Beklemishev's students in the area of the logic of provability, specifically Fedor Pakhomov, when she made an extended visit to Lev.

In the present Volume Lev Beklemishev and his student Anna Dmitrieva showed the reverse direction of this cooperation. As an MSc Logic student in Amsterdam she worked on a logic project mainly under the direction of Lev, but also in cooperation with Makowsky, on Origami Planes. I was her mentor but had little to do with it. In the present article the result is presented. I do admire it of course when such a project in an outlying range is carried out successfully. But she wrote her master's thesis (Dmitrieva, 2021) with Nick Bezhanishvili and Tomasso Moraschini on a typically Amsterdam project, and this thesis won the prize for the best logic thesis of the year in the Netherlands. She is now carrying out a Ph.D. project in the UK.

Otherwise worth mentioning about the nineties is a development that can be seen as a direct sequel to the programs I wrote with Hans Kamp as a student. Several times I had students who improved on the programs, but it was Lex Hendriks who really made it an independent subject. He finished his master's thesis on the subject and aferwards he started his work on his dissertation under the joint supervision of Gerard Renardel de Lavalette and me. In 1996 his dissertation (Hendriks, 1996) on computations in intuitionistic logic was completed. It was concentrated on locally finite fragments of intuitionistic propositional logic IPC in restricted fragments like the fragment $[\rightarrow]$ mentioned before. Already in 1991 the three of us had written an article (de Jongh et al., 1991) on the first results, and we finalized this research much later in (Renardel de Lavalette et al., 2012).

This research was by no means the only thing Lex and I did together. He often assisted me in projects or theses of master's students. Furthermore, our knowledge of intermediate logics, and his of logic programming, led in 2001 to (Hendriks and de Jongh, 2003) on stable model semantics, in which it is shown that the logic KC is the least logic for showing strongly equivalent logic programs to be equivalent. It is worth mentioning that for Lex his work on logic was always a hobby; his main job was a quite different one outside the university.

In the same period, the concept of dependence of formulas made a reappearance on the scene. This was the result of a visit by another Russian visitor, modal and intuitionistic logician Lilia Chagrova. We decided to tackle a problem that had been plaguing me for a long time. I realized that it simply had to be the case that dependence of formulas in IPC is decidable, but I had no clue how to prove it. During her visit it suddenly turned out that it is not difficult to derive this decidability from the uniform interpolation property that Pitts had recently proved for IPC (Pitts, 1992). In the article with this result we also made some progress about characterizing the exact formulas in two variables (Chagrova & de Jongh, 1995). The hope that as in the unary case there are only finitely many such formulas disappeared almost immediately.

Up to the present day nobody except myself and my co-authors wrote on exact and dependent formulas.[7] Of course, this was always somewhat disappointing. However, in their contribution to the present Volume George Metcalfe and Naomi Tokuda have come up with an abundance of results about the concepts. They place the concept of dependence in an algebraic setting and show how uniform interpolation can be used to prove decidability, often via an intermediate property called coherence. They are successful in proving decidability in this manner for a large amount of examples. They also prove for lattices an analogue of the property for single sentences I proved for IPC, which they call minimal provability. I have always had the feeling that using uniform interpolation to show decidability of dependence is a kind of overkill, especially because in many cases where one feels that decidability ought to hold, say for S4, uniform interpolation is not available. But finding a generally available weaker property for this purpose certainly doesn't seem to be an easy task.

The Master of Logic program. Since the late eighties I had led for ITLI the Erasmus and Tempus projects for cooperation and student exchange within the European Community and with Eastern Europe. With the start of the ILLC and Johan van Benthem as its first director I functioned as his unofficial foreign secretary in general. In these ways I got to appreciate the benefits of the way the UvA functions with a scientific director and a manager for more businesslike matters for most of its institutions. Erik Jan van der Linden was the first manager of the ILLC and we worked closely together.

When the university opened up the possibility of starting an international interdisciplinary master program in logic we saw this as an opportunity. The flow of exchange students visiting the ILLC could be turned to our benefit financially, and at the same time give the ILLC a foothold in the area of teaching. It wouldn't need a great organizational effort in the beginning. For the reception of the exchange students in the years before, we had already established the rule that courses should be taught in English as soon as a non-Dutch speaking student joined a course. The logically directed courses across departments were also marked as such. In line with the ITLI philosophy we had obtained for the Dutch students with a specialization in logic in one of the various departments the right to take such courses in other departments.[8] It was easy to give the new master's students the same individual attention

[7] The nice exception is the overview article (Humberstone, 2020) in which various notions of dependence are discussed.

[8] Dutch students couldn't major in logic, they could only major in mathematics, philosophy or computer science with a logic specialization.

that the exchange students and our own major students had always received, helping them find the right courses in our expanding curriculum. This personal attention was later institutionalized in the person of the mentor assigned to the student, an essential factor in the success of the master program. Of course, the most crucial factor in the success of the master program has always been the universal and unrelenting enthusiasm of the teachers of the ILLC for the program.

After a slow start, with one student joining the first year, and two the second, it picked up, four the third year, seven the fourth, stabilizing at around 12 for a number of years. This master program consisted of one (heavy!) year's study.

At the start of the program, I was the director, and Ingrid van Loon became its manager. Again this dual heading developed into a very successful cooperation. When Ingrid became manager of the ILLC, her position in the master program was taken over by Tanja Kassenaar with whom the close cooperation immediately continued. Tanja is still the soul of the program today.

In 2004, right around the time of my retirement, the program was also changing over into an official European master with a two year program. I had been rather anxious about this change, worried that students wouldn't choose to come to Amsterdam for two whole years. This worry turned out to be totally unfounded. This was only partially because Dutch students could now enter the program. The number of students admitted yearly even rose farther and now reaches to around 50. During my last year, I resigned one by one all other responsibilities I had in the ILLC, but remained director of the master program up to the very last day. That was purely because of my strong personal attachment to it, not because I didn't have confidence in the new director. Quite to the contrary. I found Benedikt Löwe the ideal successor. With the rising number of students, a person with more inclination than I have for the administrative side was very welcome. Also with regard to the ideals of the ILLC and its research and teaching he was a perfect fit. He and his later successors Ulle Endriss, Maria Aloni and Paul Dekker have kept the quality, the interdisciplinary, and the individual attention to the students of the program at an unparalleled level. I have remained in contact by still taking a few mentor students each year.

Meanwhile in Barcelona, Joost Joosten has become the director of the MSc Logic program in that city.

Turn of the century, Beth, Rosalie Iemhoff, Projective Formulas. Around the turn of the century my thoughts had a chance to return for a while to Beth. The Ph.D. study of Eva Hoogland, co-supervised by Yde Venema and me, concerned Beth's Definability Theorem and the closely related Craig interpolation theorem in non-classical logics. She finished in 2001 with a very well-written dissertation (Hoogland, 2001). I much enjoyed writing a paper (Areces et al., 2000) with her and Carlos Areces relating Beth's theorem, fixed points and interpolation in the logic of provability, one of the ideas of Craig Smoryński.

I had little involvement with another dissertation, the one of Paul van Ulsen on Beth's scientific work (van Ulsen, 2000), supervised by van Benthem and Troelstra, but I did work with Paul on one special subject, Beth's non-classical valuations (de Jongh & van Ulsen, 1999). Of course this subject was intimately related to the work

I did on relational models. The main direction of the investigation was how Beth's ideas had developed. We concluded that his ideas developed naturally from ideas he had before and that he did not owe the relational models to Saul Kripke.

A major event around the same time was the completion of the dissertation of Rosalie Iemhoff. Her main subject was the simultaneous investigation of the provability logic of HA and the more expressive Σ-preservativity logic of that system. Albert's notion of Σ-preservativity can be seen as a translation of interpretability to the intuitionistic case, much more fitting to intuitionistic logic than interpretability. A theory $\mathsf{HA}+\mathsf{A}$ is Σ-preservative over a theory $\mathsf{HA}+\mathsf{B}$ if any Σ sentence provable in $\mathsf{HA}+\mathsf{B}$ is provable in $\mathsf{HA}+\mathsf{A}$, which is classically equivalent to Π_1-conservativity. My activities in advising her were concentrated in the first part of her Ph.D. research, in which she quickly started to make big strides. She was the first to make real progress on the provability logic of HA since Albert's own dissertation.

A second topic in her dissertation was admissible rules. Propositional rules are an important factor, both in HA and IPC. They struck her imagination, and when Albert and I informed her of the conclusion both of us had come to independently, namely that the obvious rules about the effect of implications in antecedents of implications ought to be sufficient to derive all admissible rules she greatly surprised us by coming up with a proof of this conjecture. Even though it later turned out that part of the result had been reached before in an unnoticed article by Rozière (1993), in practice her result shook up the entire study of admissible rules. Another surprise came after she obtained her Ph.D. In Rosalie's dissertation she had been a semanticist proving results about formal systems as Albert and I had taught her, but afterwards she reformed herself completely into a proof-theorist.

An illustration of Rosalie's work afterwards can be seen in this Volume where she gives a proof of uniform interpolation for Lax Logic. Let me explain about my somewhat uncertain relationship with uniform interpolation. Although I was much in contact with the theorem, it never really clicked and I never gave a full proof of it for any logic myself. When Pitts proved the theorem for IPC this came as a big surprise. But we have seen already that I quickly found a way to apply it in my work with Chagrova. Albert Visser acquainted me with his model-theoretic proof of the theorem for IPC and some modal logics.[9] Together we taught about this at the ESSLLI Summer School in Prague in 1996 (de Jongh & Visser, 1996). Then with Eva Hoogland and Carlos Areces I worked on interpretability logic for Eva's dissertation. This was in the model-theoretic mode. We didn't get to uniform interpolation. Katsumi Sasaki did just that in his dissertation (Sasaki, 2001) under the supervision of Anne Troelstra and me. I learned most from Marta Bílková when she explained to me a lot about the proof-theoretical method during her visits to Utrecht for contact with Albert and me in the time of her Ph.D. studies in Prague. Finally, with Zhiguang Zhao I proved the uniform interpolation for minimal logic. The proof was however not a proof from scratch, we used the theorem for IPC as input. Nevertheless it was a very useful result, the $^+$-operation shows well how the positive fragment fits into IPC as a whole. After all this, I think the proof-theoretic proofs like the one presented by Rosalie in this Volume are clearer.

After her dissertation we remained in contact but only worked together on one paper. It derived from the master's thesis of Chunlai Zhou on intuitionistic

---

[9] Volodya Shavrukov already before gave a proof of the theorem for GL, in his dissertation, but its proof is not easy to grasp. He needed the theorem, and just made sure he got it.

modal logics co-supervised by the two of us. It was published by the three of us together (Iemhoff et al., 2005).

This is a good point to mention that the next big step after Rosalie Iemhoff's work in the search for the provability logic of HA was made rather recently by Mohammad Ardeshir and Mojtaba Mojtahedi in (Ardeshir and Mojtahedi, 2018). After first proving the analogue of my HA-theorem for Basic Arithmetic (arithmetic based on a subintuitionistic logic due to Albert) (Ardeshir & Mojtahedi, 2014) Ardeshir and Mojtahedi turned to HA itself. They characterized the Σ-provability logic of HA. The Σ-provability logic of a theory arises when one allows only Σ_1-substitutions for the propositional variables.

Very recently, Mojtaba announced that he has a proof of the arithmetical completeness of a logical system for the provability logic for HA itself.

In this Volume Mojtaba Mojtahedi returns to the Σ-provability logic of HA (Visser, 1982) and clarifies it by a translation. The Σ-provability logic of PA is GL plus $p \to \Box p$ (the Σ-completeness principle) as Albert proved quite a while ago (Visser, 1982). The Σ-provability logic of HA, not surprisingly, contains many more principles than the intuitionistic version of GL plus $p \to \Box p$, but Mojtaba shows in the present article that the intuitionistic version of GL plus $p \to \Box p$ can nevertheless in some ways be seen as a bound for it. He has succeeded in finding a translation of the complicated system he found with Ardeshir before into the intuitionistic version of GL plus $p \to \Box p$.

Even slightly before Rosalie's dissertation there was an important outside development that had a direct influence on the concept of exact formulas. In our paper on embeddings of Heyting algebras Albert and I had made the conjecture that exact formulas (in IPC) are precisely the ones that in relational models have the descendence property: if they are true in two incomparable worlds then they are true in direct ancestors of both worlds. In fact, Albert had shown already one direction: exact formulas do have the descendence property. In 1999 Silvio Ghilardi proved a major theorem about the unification type[10] of IPC in (Ghilardi, 1999). He applied to formulas a property obtained from algebra called projective, and the crucial part of his proof is showing that projective formulas are exactly the ones with the descendence property. Unintentionally he had shown our conjecture because the circle of the three properties is easy to close. Now the mathematical concept of projectivity is clearly more fruitful and more widely usable than exactness, but the concept of exactness does have a clear logical meaning, and his result makes it easier to discuss the concept. In general, Ghilardi's result had a great impact, especially on the theory of admissible rules, e.g. Rosalie made use of his results in her dissertation.

Identification in the limit. Already in my Buffalo period I had taken up an interest in formal learning theory. There was a lecture, I don't remember when or by whom, about identification in the limit, a learning paradigm stemming from (Gold, Gold (1967)). As happens to others as well I was sold immediately. I read Gold's magnificent article, the book by Osherson et al. (1990), and a series of articles written by the first and third of those authors about identifiability in the limit of classes of r.e. languages. And then some of Dana Angluin's articles, specifically (Angluin, 1980)

[10] See the article by Balbiani and Gencer in this Volume.

with her characterization of identifiability in the limit of uniformly recursive classes of languages. For this characterization she introduced the concept of telltale set of a language. This is a finite set that, once it has been observed as occurring in the input, makes it safe to conjecture the language it is related to. In my last longer visit to Siena I found myself working with Franco Montagna, not on provability logic, but on different problems in formal learning theory (I had temporarily infected Franco with my enthusiasm). He worked more in the direction of probability, which never attracted me much, while I was going for an extension of Angluin's basic characterization theorem to classes of r.e. languages. Also, I did some refereeing in the area, not generally something I am very fond of.

In the meantime, at Stanford, Johan van Benthem supervised a Ph.D. student Makoto Kanazawa, also a champion of identification in the limit, with a dissertation on applications of Angluin's theorems to classes of languages generated by categorial grammars. Makoto came as a visitor to the ILLC for a year. We found each other and agreed to work together on my project, taught a course at the 1995 ESSLLI in Barcelona, and finished an article with an Angluin type characterization theorem for classes of r.e. languages plus extensions to conservative learning (de Jongh & Kanazawa, 1996). By this time I had somehow become recognized as the expert in the Netherlands on the subject, mainly by the lack of other converts here. When the Handbook of Logic and Language was edited (van Benthem & Termeulen, 1997) by van Benthem and ter Meulen I was assigned to supervise the chapter on formal learning theory. Its authors Daniel Osherson, Eric Martin and Scott Weinstein thought it a better solution to integrate me as a co-author with a minor task (Osherson et al., 1997).

Let me continue on my later encounters with the subject of identifiability in the limit and return to the flow of time later. Two years after my official retirement, in 2006, Nina Gierasimczuk arrived in Amsterdam as a student. Since she arrived here as an already dedicated champion of identification in the limit, I was assigned to her as her mentor. Her persistence and great communicative abilities convinced me to take her on as a Ph.D. student. We found Johan van Benthem ready to be her co-supervisor for her Ph.D. thesis (Gierasimczuk, 2010) as well, which for both Johan and me turned out to be a fruitful decision. For the part of her thesis that was supervised by me she came up with the theme of finite identifiability where the object of the learning process is not reached in the limit by a stabilizing hypothesis on the inputs of information but by an instant decision after a finite amount of time. This area had been put on the map in the nineties by (Lange and Zeugmann, 1992; Mukouchi 1992), but it had not attracted further interest. Under my supervision she made it a respectable subject with its own character, and we published the main results on finite identifiability in (Gierasimczuk and de Jongh, 2013). I will just mention here the most interesting concept of the second article, which is *fastest learning*. This amounts to finite identification with the crucial decision being taken at the objectively first possible moment. Our theorem about fastest learning states that if one expects the decision to be taken effectively this type of learning is really a stronger, even more restricted form of learning than finite identifiability proper. It shows to my mind clearly that instant updating cannot always be an effective procedure. We expressed our views on logic and learning more generally together with Vincent Hendricks in (Gierasimczuk et al., 2014).

> In her contribution to this Volume Nina applies formal learning theory together with Thomas Bolander and Andrés Occhipinti Liberman to action learning of more complex objects and events in partially observable domains. The setting seems very realistic and there are a number of solid results. Good to see a new interpretation of behaviourally correct learning and the difference in power between different types of learning, especially between finite identification as studied by Nina and me, and identification in the limit.

Nina also made herself a convert to identification in the limit in Ana Lucia Vargas Sandoval. I already knew Ana Lucia as one of my mentor students in the MSc program and by a common article connected to a project she did with me. After a master's thesis on the subject with Nina and Jakub Szymanik Ana wrote a Ph.D. thesis (Vargas-Sandoval, 2020) under the supervision of Alexandru Baltag, Aybüke Özgün and me. The part I mainly supervised was a continuation of the research on finite identification. We concentrated on the difference between learning with only positive information (only examples of the concept are presented), and complete information (also counter-examples are presented), which had already occupied Ana Lucia in her master's thesis. A first paper was presented in one of the Tbilisi workshops (de Jongh and Vargas-Sandoval, 2019). The study of finite learning with this difference in mind turned out to split into two very distinct parts. When effectiveness (recursiveness) is disregarded the investigation into what classes of languages can be identified becomes a very combinatorial affair. When recursiveness is taken into consideration the investigation is more similar to standard research on identification in the limit. In both cases complete information is distinctly more powerful than purely positive information. This is less so when one restricts to fastest learning. In this second part the research was in the initial stages helped to a large extent by Sebastiaan Terwijn, an ex-Ph.D. student of mine as we will see.

Ph.D.'s with subjects outside my usual range. This is just as good a time as any to mention the few dissertations I supervised at different times on more outlying subjects. A first, in 1992, was the one of Michael Morreau whom I supervised from the first day he started his philosophical studies, including his master's thesis. I co-supervised his dissertation in 1992 with Hans Kamp. It concerned conditionals (Morreau, 1992). After his Ph.D. he has had an academic career in pure philosophy.

The next philosophical dissertation I supervised was in 1998 on special relativity by Nataša Rakić. This was a co-supervision with Frank Veltman and Johan van Benthem. It was not an easy subject for us but we were happy with the result.

A third one was a really linguistic dissertation by Yoav Seginer in 2007 (Seginer, 2007). It did concern language learning, not in the context of formal leaning theory, but by using empirical computational models. To uncover syntactical structure it constructs an incremental unsupervised parser. For that purpose a completely new representation of syntactic structure is developed: common cover links. He has certainly been one of my strongest students, and worked very independently.

One student who fits in this category as well is Sebastiaan Terwijn. I co-supervised him with Leen Torenvliet for his master's thesis and dissertation (Terwijn, 1998) (for which Leen and I were joined by Klaus Ambos Spies). He was a student who found his own area in the theory of computability and looked for and solved his own problems. Most of the time the only thing I felt I could do was to try to create opportunities for him to do this. Later we found we could discuss formal learning theory together. See more under the caption "Identification in the limit".

The first decade of the two thousands, Nick Bezhanishvili, renewed interest in my dissertation. The first decade of the new century brought me Nick Bezhanishvili as a Ph.D. student (in a co-supervision with Yde Venema). As Leo Esakia's student he came with a lot of knowledge, but also with an interest in the results and methods of my own dissertation. He saw the importance of universal models and the connection with the Jankov Theorem. In his dissertation (Bezhanishvili, 2006) he did many things, but an important part was devoted to a proper construction of the n-universal model. He came up with the concept of frame-based formulas that define frame classes like the Jankov-deJongh formulas, as he named those formulas, and subframe formulas in the form of NNIL-formulas as introduced in his dissertation. He isolated their important properties, showed they cannot axiomatize all intermediate logics, and characterized them (Bezhanishvili, 2008).

That Nick came to Amsterdam at all can be seen as a consequence of the fact that the ILLC has invested a lot in a cooperation with Georgia. The biennial TbiLLC workshops started in 1995. Let us review their history. No ILLC members participated in the first workshop, but at the Second Workshop in 1997, there were Paul Dekker, Ingrid van Loon and myself.[11] From that time onwards the ILLC has taken a large part of the responsibility for the workshops, always supporting them financially. I felt this responsibility myself as well and I have acted upon it. I did so for several reasons. In the first place the workshops covered from the start a vast range including linguistics and computation, and have kept that up against the spirit of the times that leads to more and more specialized meetings. In the second place, regarding logic itself, the small but very strong Georgian logic group led by Leo Esakia has always had an interest in modal and intuitionistic logic and the accompanying algebras both also prevalent in the ILLC. The leader of the group after Leo's untimely death is David Gabelaia who holds a Master of Logic of the ILLC. All this combined with the hospitality of the Georgian people leads to a very relaxed atmosphere so that not only have all managers of the ILLC supported the workshops, but a whole series of them, Erik Jan van der Linden, Peter Blok, Ingrid van Loon and Peter van Ormondt participated in person and felt very much at home! I myself was always there from 1997 onwards. My consistent presence and attention earned me an honorary doctorate of Tbilisi State University in 2015. Naturally, I now share my concerns with TbiLLC with Nick, and leave most of the practical issues to him.

In any case the friendship has led to a number of visitors, master's students from Georgia visiting and studying at the ILLC, and, importantly, Nick as a Ph.D. student. After he did his Ph.D., he left for a number of years as a postdoc, but he returned to the ILLC in 2014 as an assistant (now associate professor). The closeness of our

[11] Robert van Rooij, the present director of the ILLC, was there as well, but it would take years before he would be a member of the ILLC.

interests and our friendship has assisted me in remaining linked to the ILLC the last few years.

Already during his Ph.D. studies Nick and I taught a course in intuitionistic logic at the 2005 ESSLLI in Edinborough. The course notes were extended to lecture notes (Bezhanishvili & Jongh, 2006), which I used several time for courses in intuitionistic logic, the last time with Ivano Ciardelli in 2011. In the period after his Ph.D. we wrote a number of papers together. I will discuss the more important ones. A first one (Bezhanishvili et al., 2008), written together with his brother Guram, concerned the logic of the Rieger-Nishimura lattice (for both of us a favorite structure, it was presented on the cover of my farewell lecture in 2004 (de Jongh, 2004)).

The next one in 2012 (Bezhanishvili & de Jongh, 2012) extended my work with Lilia Chagrova by fully characterizing the IPC-formulas in two variables. The title "Extendible Formulas in two variables in Intuitionistic Logic" shows that we heavily leaned on Ghilardi's results (exact = projective = having the descendence property). It relied much on Nick's insights on the relationship between the 2-universal model and the free Heyting algebra on 2 generators. The paper unfortunately undercut my conviction that universal models will always lead to very elegant proofs. In this case the proofs involved much painstaking case distinctions. It is no surprise, therefore, that neither we, nor anybody else has ever attempted to generalize our results from 2 to n (but one never knows as the story of dependence shows).

The next paper with Nick (Bezhanishvili & de Jongh, 2018), in 2018, meant a return to NNIL-formulas. And to morphisms as well, not to the original p-morphisms, but to simple monotonic maps. The original NNIL article concentrated on the behavior of NNIL on models. This time, the behavior on frames was central. NNIL-formulas are shown to axiomatize exactly the subframe logics, i.e. logics with frame classes closed under subframes. On models they are (up to equivalence) the formulas conversely preserved by color-preserving monotonic maps. Stable logics, on the other hand, are the ones with frame classes closed under straightforward monotonic maps. We were able to construct a syntactic class of ONNILLI-formulas (Only NNIL to the left of Implications) which are preserved under monotonic maps and axiomatize all stable logics.

Let us return once more to my dissertation. In 2004 Nick also proved its characterization of IPC in the originally intended manner without a detour through algebra, in a paper (Bezhanishvili, 2004) for my sixtieth birthday. Its idea goes in the direction of my later paper with Fan Yang (de Jongh & Yang, 2009) on other aspects of the dissertation. Already three years before, Rosalie Iemhoff had found another characterization of IPC by showing that it is the only intermediate logic with the disjunction property and all its admissible rules (Iemhoff, 2001). In her proof the Gabbay-deJongh logics function as a kind of limit sequence to IPC. She relates her characterization to mine.

Foreign visits, new students, later years. In the nineties and later I traveled a lot outside of Amsterdam, to Georgia, Russia, Turkey, China, Japan, Iran, Portugal and, as before, to Italy and Spain. This gave me not only new friends, but also new co-authors and students. Istanbul became a favorite city. Hostess Çiğdem Gencer introduced me to her work on unifiability with Rybakov. Rybakov was the first to notice its relevance to admissible rules. He and Çiğdem wrote (Rybakov et al., 1999) on the subject with M. Terziler. Çiğdem and I showed in a follow-up (Gençer & de Jongh, 2009) that interesting technical results can be obtained in that area. Putting universal models to act in that rather distant area was for me a special attraction.

In their article in this Volume, she and Philippe Balbiani put that work in perspective by writing a very clear and useful overview of unification.

At one of the short courses I gave in Istanbul at the Bogaziçi University, I met bachelor student Aybüke Özgün. She came to Amsterdam as a master of logic student the year after. Except for mentoring her the first year of her master studies I had little or nothing to do with her very speedy and successful career at the ILLC that followed.

From Beijing Johan van Benthem directed student Fenrong Liu to the ILLC as a master of Logic student. I became her mentor and was a spectator in seeing her develop from a beginning student to an excellent researcher. I was co-supervisor for her Ph.D. with Johan van Benthem. Her dissertation (Liu, 2008) was about preference and preference change. The part I supervised concerned the way preference can be based on one's priorities, and how changes in the priorities and changes in one's belief about the application of the priorities can lead to changes in preference. Our article (de Jongh & Liu, 2009) appeared a little later. When she returned to Beijing I visited her to give some courses. It was the year of the Gödel centennial, so I gave a lecture on the incompleteness theorems (de Jongh, 2006).

Fenrong also introduced me to Sujata Ghosh who was working on a distantly related problem, namely believing one thing more than another. We were able to construct some clear systems for this extending the usual belief system KD45 (Ghosh & de Jongh, 2012; de Jongh & Ghosh, 2009). I especially liked the logic of plausibility arising from equating A to be plausible to: believing more in A than in $\neg A$. Both the paper with Fenrong and the one with Sujata ventured into the, for me, new area of Dynamic Epistemic Logic. Both concerned, unsurprisingly for me, orderings and stuck to qualitative methods, no probabilities. The area is fascinating but it was not for me to keep working in it. It moves too rapidly, with so many young people coming in.

While teaching in Guangzhou yet another beginning master's student, Fan Yang, caught my attention. An MSc Logic study in Amsterdam followed also for her. Nick's arrival had revived the ideas of my thesis, also in myself. I found in Fan Yang an ideal companion, first in her master's thesis (Yang, 2008) and later in (de Jongh and Yang, 2009) to produce the construction of the universal model for IPC and now prove the Jankov theorems in a frame setting as I had originally envisioned it, which had not been possible at the time without an algebraic detour. In that setting we also gave a new proof of the so-called second Jankov Theorem that states that KC is the (unique) strongest intermediate logic that has the same positive fragment as IPC. Fan also started already in her thesis the construction of the more complex universal model for NNIL-formulas, a large amount of work in a short time. This construction was finished much later in a cooperation with Ilin et al. (2021), first in the latter's dissertation (Ilin, 2018). That article also contains a new proof of the finite model property for logics axiomatized by NNIL-formulas, i.e. the subframe logics. In our article, my running argument with Albert Visser about the effectivity of simulations vs morphisms was continued. In the original NNIL article subsimulations dominated, here monotonic maps struck back.

> Of course, in the meantime Fan had found her own main area of interest in her Ph.D. (Yang, 2004) and afterwards: dependence logic. But that area has given her ample opportunity to keep deepening her knowledge of intuitionistic logic, and with Nick she used the team semantics that is so effective in dependence logic in a joint effort to give a wider view on intermediate logics in this Volume. Differently from some other extensions like bi-intuitionistic logic, Jankov-de Jongh formulas play a role in this setting of intuitionistic logic.

In Beijing I met bachelor student Zhiguang Zhao. When he came to Amsterdam the year after I became his mentor. I did some interesting work with him when he was still a student. It can be seen as the start of one of the three research lines that have kept me busy the last few years. One other line is the one arising from identification in the limit, which I already discussed above, and where my work with Ana Lucia still awaits publication. A third is the area of subintuitionistic logics with Fatemeh Shirmohammadzadeh to which I will turn after this.

The research line started with Zhiguang has its origin in Jankov's second theorem as it was proved in Fan Yang's thesis using the n-universal models. The second Jankov Theorem states that KC is the (unique) strongest intermediate logic that has the same positive fragment $[\to, \wedge, \vee]$, as IPC. The proof in Fan Yang's thesis was adequate but not really perspicuous. The positive fragment by itself deserved closer a look. Moreover, it is very close to minimal logic MPC, i.e. IPC without the Ex Falso axiom, because an equivalent formulation of MPC consists of the positive fragment of IPC plus a propositional constant f without any special properties. Zhiguang and I defined an operation $^+$ which turns any propositional formula φ into a formula φ^+ that in many ways expresses its positive content (de Jongh & Zhao, 2017). We related this operation to the notion of top model, a model with a top point: the behavior of φ and φ^+ on top models is identical. Among other things, it enabled us to prove uniform interpolation for both the positive fragment, and for minimal logic using the property for IPC itself. A striking phenomenon is that the right uniform interpolants of a positive formula in MPC and in IPC may well be distinct: in the case of IPC this interpolant of a positive formula may necessarily need to contain negation. Afterwards, Zhiguang and I discussed, together with N. Bezhanishvili and A. Tzimoulis, in (Bezhanishvili et al., 2017) the universal model of the positive fragment and of minimal logic MPC. Among other things, the proof of the second Jankov theorem of (Bezhanishvili et al., 2017) was improved.

This line of research was continued by equipping the positive fragment with several weaker versions of the minimal negation. This work was started by Ana Lucia Vargas Sandoval, and continued by Almudena Colacito. A first publication (Colacito et al., 2017) was completed, quickly followed by an extensive master's thesis by Almudena Colacito (2016). The semantic part relied on neighborhood models. This led to a number of characteristic systems, most notably the logic of Contraposition CoPC axiomatized by $(A \to B) \to (\neg B \to \neg A)$ over the positive fragment. The thesis contained a proof-theoretic part supervised by Marta Bílková. The subminimal logics turned out to be nicely behaved proof-theoretically. Interpolation was proved for a number of them (see Bílková and Colacito, 2020). Uniform interpolation for CoPC remains an unproved conjecture. In another extension of the thesis Almudena and I proved together with Nick (Bezhanishvili et al., 2019) that there are uncountably many subminimal logics, even between two logics that are very close. Also, we provided the subminimal logics with modal companions with two modal operators.

Next in this project came the master's thesis by Heerkens (2018). This thesis deals with the use of minimal logic in formalized intuitionistic theories. It became clear that if one uses the most natural way to approach negation in HA and formalized analysis, namely interpreting the negation of A as $A \to 1 = 0$, then one gets the intuitionistic negation including Ex Falso for free. In set theory one can use $\forall x, y(x \in y)$ for this purpose. Ex Falso is provable in these systems from basic principles. So, if one proceeds in this manner there is no place for a weaker minimal negation. Alternatively, one can add a minimal negation by including a constant f for the contradiction in the language. Then weaker systems are generated. This is also the case for theories like the ones of linear and partial order which are introduced axiomatically and are not based in one of the usual mathematical systems like HA.

A third line of research is the area of subintuitionistic logics. Fatemeh Shirmohammadzadeh Maleki visited Amsterdam as an Iranian Ph.D. student in 2015. As a topic she brought neighborhood models for intuitionistic logic with her. We started a project in this area on which we wrote six papers (de Jongh & Shirmohammadzadeh, 2017, 2018, 2019, 2021, 2023; Shirmohammadzadeh & Jongh, 2017). We first studied Corsi's logics (Corsi, 1987) which have Kripke models with fewer requirements than IPC, and for which Corsi had already constructed modal companions. We then defined weaker logics, and constructed two kinds of neighborhood semantics for these logics. One kind is close to the ordinary neighborhood models for modal logics. They give rise to modal companions in the ordinary manner, down to Classical Modal Logic with the necessity rule. For our weaker logics this kind of model didn't suffice. We needed a neighborhood semantics with pairs of sets instead of single sets as elements. We studied the relationship between the two kinds of models, which is kind of intricate. Also, we started a study of the strength of the various logics by studying which IPC-provable implications are provable in which systems. Finally, for the weakest logics, which do not have ordinary modal companions, we found modal companions with a binary operator. This gave rise to our latest project to study the relations between standard modal logics and logics with a binary operator (de Jongh & Shirmohammadzadeh, 2021, 2023).

Let me finally return to the history of what I would like to call the arithmetical completeness theorem relating IQC and HA. For many years an important extension of the theorem remained an open problem: to prove the corresponding theorem for intuitionistic set theory. Only very recently, Robert Paßmann, Ph.D. student at the ILLC, gave a proof for the most important system CZF, forthcoming in the JSL (Paßmann, 2022): its logic is IQC. He used set register machines and basically the method of proof originally due to Jaap van Oosten. He has been working on related problems since his master thesis (Paßmann, 2018) under supervision of Benedikt Löwe. Before the result on CZF he had already proved a similar theorem for Kripke-

Platek set theory together with Rosalie Iemhoff (2021). It is remarkable that for a competing intuitionistic set theory, IZF, Friedman and Scedrov (1986) proved quite a while ago a negative result, the logic of IZF properly contains IQC.

These notes show how closely interrelated with others the life of a scientist nowadays is. You have seen how lucky I have been with the people I have met on my way, and with the structures I have dwelt in. Of course I am getting into quieter waters now, but surprises may still lie ahead.

The story above describes my scientific life. And life, certainly in the later part, has much more to offer than science alone. Here I haven't got into this part of my life at all, and the people I have been closest to have hardly appeared in these pages.

That this book appeared is very much due to the continuous efforts and great patience of the editors Nick Bezhanishvili, Rosalie Iemhoff and Fan Yang. I can only be thankful. I feel really honored by the eminently suitable articles presented by the contributors to this book. Finally, I want to express my thanks to a number of people who read earlier versions of the Autobiographical Notes, Frank Veltman, Henk Verkuyl, Chase Ford, Hans Kamp and Rineke Verbrugge. They helped me a great deal. Valuable also was the assistance by Swapnil Ghosh with the bibliography.

References

de Almeida Borges, A., de Jongh, D., Joosten, J. J., & Visser, A. (2022). The strictly positive fragment of provability logic for Heyting arithmetic.

Angluin, D. (1980). Inductive inference of formal languages from positive data. *Information and Control, 45*, 117–135.

Ardeshir, M., & Mojtahedi, M. (2014). The de Jongh property for Basic Arithmetic. *Archive for Mathematical Logic, 53*, 881–895.

Ardeshir, M., & Mojtahedi, M. (2018). The Σ_1-provability logic of HA. *Annals of Pure and Applied Logic, 169*(10), 997–1043.

Areces, C., Hoogland, E., & de Jongh, D. (2000). Interpolation, definability and fixed points in interpretability logics. In M. Zakharyaschev, K. Segerberg, M. de Rijke & H. Wansing (Eds.), *Advances in modal logic* (Vol. 2, pp. 53–76). CSLI Publications.

Bellissima, F. (1986). Finitely generated free Heyting algebras. *Journal of Symbolic Logic, 51*(1), 152–165.

van Benthem, J., de Jongh, D., Groenendijk, J., Stokhof, M., & Verkuyl, H. (1982). (under the name: Gamut). *Logica, Taal en Betekenis, Boek 1*. Inleiding in de Logica, Aula Paperback no. 76, Spectrum, Utrecht.

van Benthem, J., de Jongh, D., Groenendijk, J., Stokhof, M., & Verkuyl, H. (1982). (under the name: Gamut). *Logica, Taal en Betekenis, Boek 2*. Intensionele Logica en Logische Grammatica, Aula Paperbacks no. 77, Spectrum, Utrecht.

van Benthem, J., de Jongh, D., Groenendijk, J., Stokhof, M., & Verkuyl, H. J. (1991). (under the name: Gamut). *Logic, language and meaning, book 1, introduction to logic*. The University of Chicago Press.

van Benthem, J., de Jongh, D., Groenendijk, J., Stokhof, M., Verkuyl, H. J. (1991). (under the name: Gamut). *Logic, language and meaning, book 2, intensional logic and logical grammar*. The University of Chicago Press.

van Benthem, J., & Stokhof, M. (2021). *A brief history of the institute for logic, language and computation. People, ideas and practices*. Universiteit van Amsterdam, Institute for Logic, Language and Computation.
van Benthem, J., & Termeulen, A. (Eds.). (1997). *Handbook of logic and language*. Elsevier Science Publishers.
Berarducci, A. (1990). The interpretability logic of Peano arithmetic. *Journal of Symbolic Logic, 55*, 1058–1089.
Bezhanishvili, G., Bezhanishvili, N., & de Jongh, D. (2008). The Kuznetsov-Gerciu and Rieger-Nishimura logics: The boundaries of the finite model property. *Logic and Logical Philosophy, 17*, 73–110.
Bezhanishvili, N. (2004). *De Jongh's*. In Liber Amicorum Dick de Jongh: Characterization of intuitionistic propositional calculus.
Bezhanishvili, N. (2006). *Lattices of intermediate and cylindric modal logics*. Ph.D. thesis, ILLC, Universiteit van Amsterdam.
Bezhanishvili, N. (2008). Frame based formulas for intermediate logics. *Studia Logica, 90*(2), 139–159.
Bezhanishvili, N., Colacito, A., & de Jongh, D. (2019). A study of subminimal logics of negation and their modal companions. In *Post-proceedings of TbiLLC 2017*, LNCS 11456 (pp. 21–41). Springer.
Bezhanishvili, N., & de Jongh, D. (2018). Stable formulas in intuitionistic logic. *Notre Dame Journal of Formal Logic, 59*(3), 307–324.
Bezhanishvili, N., & de Jongh, D. (2006). Intuitionistic Logic. Technical Report PP-2006-26, ILLC, Universiteit van Amsterdam.
Bezhanishvili, N., & de Jongh, D. (2012). Extendible formulas in two variables in intuitionistic logic. *Studia Logica, 100*(1), 61–89.
Bezhanishvili, N., de Jongh, D., Tzimoulis, A., & Zhao, Z. (2017). Universal models for the positive fragment of intuitionistic logic. In H. H. Hansen (Ed.), *11th international tbilisi symposium on logic and language, and computation, TbiLLC 2015 Revised Selected Papers*. Lecture notes in computer science (Vol. 10148, pp. 229–250).
Bílková, M., de Jongh, D., & Joosten, J. J. (2009). Interpretability in PRA. *Annals of Pure and Applied Logic, 161*(2), 128–138.
Boolos, G. (1993). *The logic of provability*. Cambridge University Press.
Boolos, G., & Sambin, G. (1991). Provability: The emergence of a mathematical modality. *Studia Logica, 50*, 1–23.
Brouwer, L. E. J. (1954). Points and spaces. *Canadian Journal of Mathematics, 6*(1), 1–17 (Also in *Complete Works*, Vol. 1).
Bílková, M., & Colacito, A. (2020). Proof theory for positive logic with weak negation. *Studia Logica, 108*(4), 649–686.
Carr, J. (2022). Hereditary structural completeness over K4: Rybakov's Theorem revisited. Master's thesis, ILLC, University of Amsterdam.
Chagrov, A., & Zakharyaschev, M. (1997). *Modal logic*, volume 35 of *Oxford logic guides*. Oxford University Press.
Chagrova, L., & de Jongh, D. (1995). The decidability of dependency in intuitionistic propositional logic. *Journal of Symbolic Logic, 60*, 498–504.
Chomsky, N. (1957). *Syntactic Structures*. Mouton.
Colacito, A. (2016). Minimal and subminimal logic of negation. Master's thesis, ILLC, Universiteit van Amsterdam.
Colacito, A., de Jongh, D., & Sandoval, A. L. (2017). *Subminimal negation. Soft computing, 21*, 165–174.
Corsi, G. (1987). Weak Logics with Strict Implication. *Zeitschrift fur Mathematische Logik und Grundlagen der Mathematik, 33*, 389–406.
Diego, A. (1966). *Sur les algèbres de Hilbert*. Collection de Logique Mathématique, Série A 21, Gauthier-Villars.

Dmitrieva, A. (2021). Positive modal logic beyond distributivity: duality, preservation and completeness. Master's thesis, ILLC, Universiteit van Amsterdam.

Ehrenfeucht, D. A., Geiser, J., Gordon, C. E., & de Jongh, D. (1971). Semantics for non-iterated local observation. In *Proceedings of Symposia in Pure Mathematics, Tarski Symposium* (Vol. 25).

Friedman, H., & Ščedrov, A. (1986). On the quantificational logic of intuitionistic set theory. *Mathematical Proceedings of the Cambridge Philosophical Society, 99*(1), 5–10.

Gabbay, D. M., & de Jongh, D. (1969). A sequence of decidable finitely axiomatizable intermediate logics with the disjunction property. volume No. 35 of *Technical Reports, U.S. Office of Naval Research Information Systems Branch.*

Gabbay, D. M., & de Jongh, D. (1974). A Sequence of Decidable Intermediate Logics with the Disjunction Property. *The Journal of Symbolic Logic, 39*, 67–78.

Gençer, Ç., & de Jongh, D. (2009). Unifiability in extensions of K4. *Logic Journal of IGPL, 17*(2), 159–172.

Ghilardi, S. (1999). Unification in intuitionistic logic. *The Journal of Symbolic Logic, 64*(2), 859–880.

Ghosh, S., & de Jongh, D. (2012). Comparing strengths of beliefs explicitly. *Logic Journal of the IGPL, 21*, 488–514.

Gierasimczuk, N. (2010). *Knowing one's limits. Logical analysis of inductive inference*. Ph.D. thesis, ILLC, Universiteit van Amsterdam, DS-2010-11.

Gierasimczuk, N., & de Jongh, D. (2013). On the complexity of conclusive update. *The Computer Journal, 56*(3), 365–377.

Gierasimczuk, N., Hendricks, V. F., & de Jongh, D. (2014). Logic and learning. In A. Baltag & S. Smets (Eds.), *Johan van Benthem on Logic and Information Dynamics, volume 5 of Outstanding Contributions to Logic* (pp. 267–288). Springer.

Gold, M. (1967). Language identification in the limit. *Information and Control, 10*(5), 447–474.

Grigolia, R. (1987). *Free algebras of non-classical logics*. Metsniereba Press. (In Russian).

Guaspari, D., & Solovay, R. M. (1979). Rosser sentences. *The Annals of Mathematical Logic, 16*, 81–99.

Heerkens, N. (2018). Studies in minimal mathematics. Master's thesis, ILLC, Universiteit van Amsterdam.

Hendriks, A. (1996). *Computations in propositional logic*. Ph.D. thesis, ILLC, Universiteit van Amsterdam, DS-1996-01.

Hendriks, A., & de Jongh, D. (2003). Characterization of strongly equivalent logic programs in intermediate logics. *The Journal for the Theory and Practice of Logic programming (TPLP), 3*, 259–270.

Henk, P. (2016). *Nonstandard provability for Peano arithmetic. A modal perspective*. Ph.D. thesis, ILLC, University of Amsterdam, DS-2016-10.

Hoogland, E. (2001). *Definability and Interpolation: Model-theoretic investigations*. Ph.D. thesis, ILLC, University of Amsterdam, DS-2001-05.

Humberstone, L. (2020). Explicating logical independence. *Journal of Philosophical Logic, 49*(2020), 135–218.

Iemhoff, R. (2001). A(nother) characterization of intuitionistic propositional logic. *Annals of Pure and Applied Logic, 113*(1–3), 161–173.

Iemhoff, R., de Jongh, D., & Zhou, C. (2005). Properties of intuitionistic provability and preservativity. *Logic Journal of the IGPL, 13*(6), 615–636.

Iemhoff, R., & Paßmann, R. (2021). Logics of intuitionistic Kripke-Platek set theory. *Annals of Pure and Applied Logic, 172*, 103014.

Ilin, J. (2018). *Filtration revisited: Lattices of stable non-classical logics*. Ph.D. thesis, ILLC, Universiteit van Amsterdam.

Ilin, J., de Jongh, D., & Yang, F. (2021). NNIL formulas revisited: Universal models and finite model property. *Journal of Logic and Computation, 31*(2), 573–596.

Jankov, V. A. (1963). On the relation between deducibility in intuitionistic propositional calculus and finite implicative structures. *Doll. Akad. Nauk SSSR, 151*, 1293–1294. (In Russian).

de Jongh, D. (2006). The incompleteness theorems, their content and their meaning. *Philosophical Trends*.

de Jongh, D. (1964). Onderzoekingen over de intuitionistische propositielogica. Master's thesis, Universiteit van Amsterdam. https://drive.google.com/file/d/14drwxJqi31FXHEfmuXAZb5QhpAcKEr7t/view?usp=sharing

de Jongh, D. (1968). *Investigations on the intuitionistic propositional calculcus*. Ph.D. thesis, University of Wisconsin. https://eprints.illc.uva.nl/id/eprint/1837/2/HDS-05-Dick-de-Jongh.text.pdf

de Jongh, D. (1973). Formulas of one propositional variable in intuitionistic arithmetic. In *Reports 73-03*. Mathematisch Instituut, Roetersstraat 15, Amsterdam.

de Jongh, D. (1980). A class of intuitionistic connectives. In H. J. Keisler Barwise & K. Kunen (Eds.), *The kleene symposium* (pp. 103–111). North Holland.

de Jongh, D. (1982). Formulas of one propositional variable in intuitionistic arithmetic. In Troelstra & van Dalen (Eds.), *Brouwer centenary symposium* (pp. 51–64). North-Holland.

de Jongh, D. (2004). Volledigheid en onvolledigheid. *ILLC-prepublications PP-2004-30*, Universiteit van Amsterdam.

de Jongh, D., & Ghosh, S. (2009). Comparing strengths of beliefs explicitly. In Maldoni et al. (Ed.), *Proceedings of the 2nd Multi-agent Logics, Languages, and Organizations Federated, MALLOW 2009. CEUR Workshop Proceedings* (Vol. 494).

de Jongh, D., Hendriks, A., & Renardel de Lavalette, G. R. (1991). Computations in fragments of intuitionistic propositional logic. *Journal of Automated Reasoning, 7*, 537–561.

de Jongh, D., & Japaridze, G. (1998). The logic of provability. In S. Buss (Ed.), *Handbook of proof theory* (pp. 475–546). North-Holland/Elsevier.

de Jongh, D., & Kanazawa, M. (1996). Angluin's theorem for indexed families of r.e. sets and applications. In *Proceedings of the Nineth Annual Conference on Computational Learning Theory* (pp. 193–204). The Association for Computing Machinery.

de Jongh, D., & Liu, F. (2009). Preference, priorities and belief. In T. Grüne-Yanoff & S. O. Hansson (Eds.), *Preference change: Approaches from philosophy, economics and psychology* (pp. 85–107).

de Jongh, D., & Montagna, F. (1988). Provable fixed points. Zeitschr. f. math Logik und Grundlagen d. *Math, 34*, 229–250.

de Jongh, D., & Montagna, F. (1989). Much shorter proofs. Zeitschr. f. math Logik und Grundlagen d. *Math, 35*, 247–260.

de Jongh, D., & Parikh, R. (1977). Well-partial Orderings and Hierarchies. In *Proceedings of the Koninklijke Akademie van Wetenschappen, series A* (Vol. 80)

de Jongh, D., & Shirmohammadzadeh Maleki, F. (2017). Subintuitionistic logics with Kripke semantics. In *11th International tbilisi symposium on logic, language, and computation, TbiLLC 2015, LNCS, Volume 10148* (pp. 333–354). Springer.

de Jongh, D., & Shirmohammadzadeh Maleki, F. (2018). Subintuitionistic logics and the implications they prove. *Indagationes Mathematicae, 6*(3), 1525–1545.

de Jongh, D., & Shirmohammadzadeh Maleki, F. (2019). Two neighborhood semantics for Subintuitionistic logics. In *12th international tbilisi symposium on logic, language, and computation, TbiLLC 2018, LNCS, Volume 11456* (pp. 64–85). Springer.

de Jongh, D., & Shirmohammadzadeh Maleki, F. (2021). Binary modal companions for subintuitionistic logics. *Mathematics, logic, and their philosophies: Essays in honour of Mohammad Ardeshir* (pp. 35–52).

de Jongh, D., & Shirmohammadzadeh Maleki, F. (2023). Binary modal logic and unary modal logic. *Logic Journal of the IGPL*. To be published.

de Jongh, D., & Troelstra, A. S. (1966). On the connection of partially ordered sets with some pseudo-Boolean Algebras. *Indagationes Mathematicae,* (3).

de Jongh, D., & van Ulsen, P. (1999). Beth's non-classical valuations, in: Un Logicien consciencieux, La Philosophie de Evert Willem Beth. *Philosophiae Scientiae, Éditions Kimé, 3*, cahier 4, 279–302.

de Jongh, D., & Vargas-Sandoval, A. L. (2019). Finite identification with positive and with complete data. In Bezhanishivili, N. et al. (Ed.), *Post-proceedings of the twelfth tbilisi symposium on language, logic and computation, TbiLLC 2017, Lecture notes in computer science* (pp. 42–63). Springer.

de Jongh, D., & Veltman, F. (1990). Provability logics for relative interpretability. In Petkov (Ed.), *Mathematical logic, proceedings of the heyting 1988 summer school in Varna* (pp. 31–42). Plenum Press.

de Jongh, D., & Veltman, F. (1999). Modal completeness of ILW. In J. Gerbrandy, M. Marx, M. de Rijke & Y. Venema (Eds.), *Essays dedicated to J. van Benthem on the occasion of his 50th birthday*. Amsterdam University Press.

de Jongh, D., Verbrugge, R., & Visser, A. (2011). Intermediate logics and the de Jongh property. *Archive for Mathematical Logic, 50*, 197–213.

de Jongh, D., & Visser, A. (1991). Explicit fixed points in interpretability logic. *Studia Logica, 50*, 39–50.

de Jongh, D., & Visser, A. (1996). Embeddings of Heyting algebras. European logic colloquium. In W. Hodges, M. Hyland, C. Steinhorn, & J. Truss (Eds.), *Logic: from foundations to applications* (pp. 187–214). Clarendon Press.

de Jongh, D., & Visser, A. (1996). Bisimulations, model descriptions and propositional quantifiers (1996)

de Jongh, D., & Yang, F. (2009). Jankov's Theorems for intermediate logics in the setting of Universal models. In N. Bezhanishvili, S. Loebner, K. Schwabe & L. Spada (Eds.), *Post-proceedings of the Eighth Tbilisi Symposium on language, logic and computation, TbiLLC 2009, Lecture notes in computer science* (Vol. 6618, pp. 53–76). Springer.

de Jongh, D., & Zhao, Z. (2017). Positive formulas in intuitionistic and minimal logic. In *10th international tbilisi symposium on logic, language, and computation, TbiLLC, Gudauri 2015*, volume 10148 of *Lecture notes in computer science* (pp. 229–250). Springer.

de Jongh, D. K. (1943). *Critische Beschouwingen Over de Homeopathie*. Noord-Hollandsche Uitgevers Maatschappij

Joosten, J. J. (1998). Towards the interpretability logic of all reasonable arithmetical theories. Master's thesis, Universiteit van Amsterdam.

Joosten, J. J. (2004). *Interpretability formalized*. Ph.D. thesis, Utrecht University.

Kleene, S. C., & Vesley, R. E. (1965). *The foundations of intuitionistic mathematics*. North-Holland Publishing Company.

Kreisel, G., & Putnam, H. (1957). Eine Unableitbarkeitsbeweismethode für den Intuitionistischen Aussagenkalkül. *Zeitschrift für Mathematische Logik and Grundlagen der Mathematik, 3*, 74–78.

Kripke, S. A. (1967). An extension of a theorem of Gaifman-Hales-Solovay. *Fundamenta Mathematicae, 61*, 29–32.

Kruskal, J. B. (1960). Well-quasi-ordering, the tree theorem, and Vazsonyi's conjecture. *Transactions of the American Mathematical Society, American Mathematical Society, 95*(2), 210–225.

Lange, S., & Zeugmann, T. (1992). Types of monotonic language learning and their characterization. In *COLT'92: Proceedings of the 5th annual ACM conference on computational learning theory* (pp. 377–390). ACM.

Leivant, D. (1975). *Absoluteness of intuitionistic logic*. Ph.D thesis, Mathematisch Centrum. https://eprints.illc.uva.nl/id/eprint/1845/1/HDS-13-Daniel-Leivant.text.pdf

Liu, F. (2008). *Changing for the better: Preference dynamics and agent diversity*. Ph.D thesis, ILLC, Universiteit van Amsterdam.

Löb, M. H. (1955). Solution of a problem of Leon Henkin. *The Journal of Symbolic Logic, 20*(2), 115–118.

Luce, R. D. (1956). Semi-orders and a theory of utility discrimination. *Econometrica, 24*, 178–191.

Łukasiewicz, J. (1952). On the intuitionistic theory of deduction. *Indagationes Mathematicae, 14*, 202–212.

McCullough, D. P. (1971). Logical connectives for intuitionistic propositional logic. *The Journal of Symbolic Logic, 36*, 15–20.

Morreau, M. (1992). *Conditionals in philosophy and artificial intelligence*. Ph.D thesis, Universiteit van Amsterdam.
Mukouchi, Y. (1992). Characterization of finite identification. In K. P Jantke (Ed.), *Analogical and inductive inference. AII 1992. Lecture notes in computer science* (Vol. 642)
van Oosten, J. (1991). *Exercises in realizability*. Ph.D. thesis, ILLC, Universiteit van Amsterdam, 1991. https://eprints.illc.uva.nl/id/eprint/1858/3/HDS-26-Jaap-van-Oosten.text.pdf
Osherson, D., de Jongh, D., Martin, E., & Weinstein, S. (1997). Formal learning theory. In J. van Benthem & A. Termeulen (Eds.), *Handbook of logic and language* (pp. 737–775). Elsevier Science Publishers.
Osherson, D., Stob, M., & Weinstein, S. (1990). *Systems that learn: An introduction to learning theory for cognitive and computer scientists*. MIT Press.
Parikh, R. (1971). Existence and feasibility in arithmetic. *The Journal of Symbolic Logic, 36*(3), 494–508.
Paßmann, R. (2022). The first-order logic of CZF is intuitionistic first-order logic. *Journal of Symbolic Logic* (upcoming). https://www.doi.org/10.1017/jsl.2022.51
Paßmann, R. (2018). Loyalty and faithfulness of model constructions for constructive set theory. Master's thesis, ILLC, Universiteit van Amsterdam.
Pitts, A. (1992). On an interpretation of second order quantification in first order intuitionistic propositional logic. *Journal of Symbolic Logic, 57*(1), 33–52.
Renardel de Lavalette, G. R., Hendriks, A., & de Jongh, D. (2012). Intuitionistic implication without disjunction. *Journal of Logic and Computation, 22*(3), 375–404.
Rozière, P. (1993). Admissible and derivable rules in intuitionistic logic. *Mathematical Structures in Computer Science, 3*, 129–136.
Rybakov, V. (1992). Rules of inference with parameters for intuitionistic logic. *The Journal of Symbolic Logic, 57*(3), 912–923.
Rybakov, V. V., Terziler, M., & Gençer, Ç. (1999). An essay on unification and inference rules for modal logics. *Bulletin of the Section of Logic, 28*(3), 145–157.
Sasaki, K. (2001). *Logics and provability*. Ph.D thesis, ILLC, Universiteit van Amsterdam.
Schmidt, D. (1975). Bounds for the closure ordinals of replete monotonic increasing functions. *The Journal of Symbolic Logic, 40*(3), 305–316.
Schmidt, D. (1979). *Well-partial orderings and their maximal order types*. Habilitationsschrift.
Schütte, K. (1954). Kennzeichnung von Ordinalzahlen durch rekursiv erklärte funktionen. *Mathematische Annalen, 127*, 15–32.
Segerberg, K. (1971). *An essay in classical modal logic, volume 13 of Philosophical studies.*
Seginer, Y. (2007). *Learning syntactic structure*. Ph.D. thesis, ILLC, University of Amsterdam, DS-2007-05.
Yu Shavrukov, V. (1993). A note on the diagonalizable algebras of PA and ZF. *Annals of Pure and Applied Logic, 61*(1–2), 161–173.
Shavrukov, V. Y. (1993). Subalgebras of diagonalizable algebras of theories containing arithmetic. *Dissertationes Mathematicae, CCCXXIII*, 1–82.
Shavrukov, V. Y. (1994). *Adventures in diagonalizable algebras*. Ph.D. thesis, ILLC, University of Amsterdam, DS-1994-07.
Shavrukov, V. Y. (1988). The logic of relative interpretability over Peano arithmetic (in Russian). *Technical Report 5*, Stekhlov Mathematical Institute.
Shehtman, V. (1978). *Rieger-Nishimura lattices. Soviet Mathematics Doklady, 19*, 1014–1018.
Shirmohammadzadeh Maleki, F., & de Jongh, D. (2017). Weak subintuitionistic logics. *Logic Journal of the IGPL, 25*(2), 214–231.
Smoryński, C. (1973). Applications of Kripke models. In A. S. Troelstra (Ed.), *Metamathematical investigation of intuitionistic arithmetic and analysis*. Springer.
Smoryński, C. (1985). *Self-reference and modal logic*. Springer.
Solovay, R. M. (1976). Provability interpretations of modal logic. *Israel Journal of Mathematics, 25*(3–4), 287–304.
Terwijn, S. (1998). *Computability and measure*. Ph.D. thesis, ILLC, Universiteit van Amsterdam.

van Ulsen, P. (2000). *E. W. Beth als logicus*. Ph.D. thesis, ILLC, Universiteit van Amsterdam.
Vargas-Sandoval, A. L. (2020). *On the path to the truth*. Ph.D. thesis, ILLC, University of Amsterdam, DS-2020-07.
Veltman, F. (1985). *Logics for conditionals*. Ph.D. thesis, Universiteit van Amsterdam. https://eprints.illc.uva.nl/id/eprint/1834/3/HDS-02-Frank-Veltman.text.pdf
Verbrugge, R. (1992). Verzamelingen-Veltman frames en modellen (set Veltman frames and models). Unpublished manuscript, Amsterdam.
Verbrugge, R. (1993). *Efficient metamathematics*. Ph.D. thesis, Universiteit van Amsterdam. https://eprints.illc.uva.nl/id/eprint/1960/3/DS-1993-03.text.pdf
Verbrugge, R., & Visser, A. (1994). A small reflection principle for bounded arithmetic. *Journal of Symbolic Logic, 59*(3), 785–812.
Visser, A. (1981). *Aspects of diagonalization and provability*. Ph.D. thesis, Utrecht University. https://eprints.illc.uva.nl/id/eprint/1854/3/HDS-22-Albert_Visser.text.pdf
Visser, A. (1982). On the completeness principle: A study of provability in Heyting's arithmetic and extensions. *Annals of Mathematical, 22*(3), 263–295.
Visser, A. (1985). Evaluation, provably deductive equivalence in Heyting's Arithmetic of substitution instances of propositional formulas. In *Technical Report 4*. Dept. of Philosophy, University of Utrecht, Heidelberglaan 8, 3584 CS Utrecht.
Visser, A. (2002). Substitutions of Σ^0_1 sentences: Explorations between intuitionistic propositional logic and intuitionistic arithmetic. *Annals of Pure and Applied Logic, 114*(1–3), 227–271.
Visser, A., van Benthem, J., de Jongh, D., & Renardel de Lavallette, G. R. (1995). NNIL, a Study in Intuitionistic Propositional Logic. In A. Ponse, M. de Rijke, & Y. Venema (Eds.), *Modal logic and process algebra, a bisimulation perspective, CSLI lecture notes no* (pp. 289–326). CSLI Stanford.
Yang, F. (2004). *On extensions and variants of dependence logic*. Ph.D. thesis, University of Helsinki.
Yang, F. (2008). Intuitionistic Subframe formulas, NNIL-formulas and n-universal models. Master's thesis, ILLC, Universiteit van Amsterdam.
Zambella, D. (1994). *Chapters on bounded arithmetic and provability logic*. Ph.D. thesis, ILLC, University of Amsterdam, DS-1994-08.

Chapter 2
Lewisian Fixed Points I: Two Incomparable Constructions

Tadeusz Litak and Albert Visser

Abstract Our paper studies what one might call "the reverse mathematics of explicit fixed points". We discuss two methods of constructing such fixed points for formulas whose principal connective is the intuitionistic Lewis arrow $\dashv$. Our main motivation comes from metatheory of constructive arithmetic, but the systems in question allow several natural kinds of semantics. The first of these methods, inspired by de Jongh and Visser, turns out to yield a modal system $\mathsf{L}^{\flat}_{\mathsf{a}}$, extending the "gathering" axiom 4_{a} with the standard ("box") version of the Löb axiom. The second one, inspired by de Jongh and Sambin, seemingly simpler, leads to a modal theory $\mathsf{JS}^{\flat}$, which proves harder to axiomatize in an elegant way. Apart from showing that both theories are mutually incomparable, we axiomatize their join and investigate several subtheories, whose axioms are obtained as fixed points of simple formulas. We also show that both $\mathsf{L}^{\flat}_{\mathsf{a}}$ and $\mathsf{JS}^{\flat}$ are *extension stable*, that is, their validity in the corresponding preservativity logic of a given arithmetical theory transfers to its finite extensions.

Keywords Provability logic · Fixed points · Lewis arrow · Constructivism

2.1 Introduction

In Sect. 2.1.1, we recall Dick de Jongh's contributions to the theory of propositional fixpoints on which our paper builds. As the details are not sufficiently well-known by the broader public, we provide a more detailed introduction in Sect. 2.1.2. Then Sect 2.1.3 presents our own contributions.

T. Litak
Informatik 8, FAU Erlangen-Nürnberg, Erlangen, Germany
e-mail: tadeusz.litak@fau.de

A. Visser (✉)
Philosophy, Faculty of Humanities, Utrecht University, Janskerkhof 13, 3512BL Utrecht, The Netherlands
e-mail: a.visser@uu.nl

N. Bezhanishvili et al. (eds.), *Dick de Jongh on Intuitionistic and Provability Logics*, Outstanding Contributions to Logic 28, https://doi.org/10.1007/978-3-031-47921-2_2

2.1.1 Dick de Jongh and Explicit Fixed Points

In 1974, Dick de Jongh gave both a semantical proof and a syntactical proof of the existence of explicit fixed points in Löb's Logic. The same insight was independently discovered by Sambin (1976), who explicitly states that an intuitionistic propositional base is sufficient. The result is a remarkable generalization of the already known facts that $\top$ is equivalent to the Henkin Fixed Point and that the consistency statement is equivalent to the Gödel Fixed Point.

The de Jongh-Sambin theorem is one of the three great results of classical provability logic. The second one is the uniqueness of fixed points modulo provable equivalence, due to Dick de Jongh (unpublished), Giovanni Sambin in the above-quoted paper (Sambin, 1976), and Claudio Bernardi in 1974 (published in Bernardi, 1976). The third one is the arithmetical completeness theorem of Solovay (1976).

In 1991, Dick de Jongh, in collaboration with Albert Visser, generalized the de Jongh-Sambin theorem to the interpretability logic IL. Dick first invented a semantical proof which was subsequently translated to a syntactical one by Albert. See de Jongh and Visser (1991). The de Jongh-Visser computation employs, as we will see, a step that is genuinely not present in the de Jongh-Sambin calculation. This observation was also made, for the classical case, in Iwata et al. (2022).

In 2005, Rosalie Iemhoff, Dick de Jongh, and Chunlai Zhou gave a fixed point calculation for the constructive Lewis systems denoted here as $\mathsf{L}^{\sharp}_{\Box} \oplus 4_{\mathsf{a}} = \mathsf{L}^{\sharp}_{\mathsf{a}}$ and $\mathsf{W}^{\sharp}$.[1] See Iemhoff et al. (2005a).

Tohei Iwata, Taishi Kurahashi, and Yuya Okawa studied the fixed point property for sub-logics of IL with considerable depth. See Iwata et al. (2020), Iwata et al. (2022), Okawa (2022).[2]

In this tribute to Dick, we take up the line of fixed point research from the point where it was brought by Iemhoff, de Jongh, and Zhou in their 2005 paper. One of the insights of our paper is that the de Jongh-Visser computation really differs from the de Jongh-Sambin one. In a sense that will be explicated these computations are mutually incomparable. In the remainder of this section, we provide a more general introduction for a broader audience.

2.1.2 Explicit Fixed Points in More Detail

The definability of explicit fixed points for the case of Löb's Logic can be described as follows. Given any formula understood as a polynomial $\varphi(p)$ where all occurrences of p are guarded by $\Box$, one can use the de Jongh-Sambin algorithm to compute a formula χ (not involving p) such that $\mathsf{GL} \vdash \chi \leftrightarrow \varphi(\chi)$. Actually, GL is obtained

---

[1] They call these systems iPL, respectively iPW, and our earlier paper (Litak & Visser, 2018) denotes these systems as $\mathsf{i\text{-}GL_a}$, respectively $\mathsf{i\text{-}GW_a}$.

[2] We found these papers when most of the present paper was written. This explains the absence of a more careful comparison between the results of these papers and ours.

precisely as the smallest extension of $\mathsf{K4}$ (i.e., the logic of the transitivity axiom $\Box\varphi \to \Box\Box\varphi$) in which guarded fixed points are definable. This follows immediately from the fact that, by Löb's argument, Löb's principle is entailed by the presence of guarded fixed points.

This fixed point result encodes the algebraic content of Löb's Theorem and Gödel's Second Incompleteness Theorem. The modal analysis gives us the conceptual resources to say that the consistency statement is the explicit form of the Gödel sentence: we need the simple modal language to say that *w.r.t. to this language* the fixed points have explicit forms. More mundanely, the de Jongh-Sambin Theorem can be seen as an elimination result for the fixed point operator. The original statement is restricted to guarded fixed points, but one can indeed extend this classical result to elimination of the positive fixed points of the ordinary μ-calculus (van Benthem, 2006; Visser, 2005) and further beyond (see Sect. 2.12). Given the Kripkean meaning of GL as the logic of Noetherian (a.k.a. conversely well-founded) transitive frames, such results in turn have found applications, e.g., in characterizing the expressivity of XPath fragments (Ten Cate et al. 2010, Sect. 3.1). The de Jongh-Sambin result has inspired Nakano's seminal work on *modality for recursion* (Nakano 2000, Sect. 7), (Nakano 2001, Sect. 5); see Visser and Litak (202x) for a detailed discussion of this point and related references. Last, but definitely not least, the result can be used to prove other metaresults about GL, such as the Beth definability property, as observed first by Maksimova (1989) (cf. also Areces et al., 1998; Hoogland, 2001; Iemhoff, 2005a, and Sect. 2.11 below).

What happens when we broaden the investigation beyond the classical base and unary provability $\Box$? Regarding the former restriction, already (Sambin, 1976) paper noted that the fixed point theorem works over the intuitionistic propositional calculus (IPC). Unfortunately, despite decades of efforts (Ardeshir & Mojtahedi, 1994; Iemhoff, 2001a, 2001b; Visser, 2008, 2018; Visser & Zoethout 2019), there is no known axiomatization of the provability logic of Heyting Arithmetic (HA) and related systems; it is a system much stronger than intuitionistic GL, including principles such as

$$\Box(\neg\neg\Box\varphi \to \Box\varphi) \;\to\; \Box\Box\varphi$$

underivable even in classical GL, as classically it implies $\Box\Box\bot$ (cf. also Litak and Visser, 2018, Sect. 5.3 for more examples).[3] The algebraic core of a weak theory can include powerful schemes refutable in a stronger theory enjoying a different provability predicate. This phenomenon is often caused, e.g., by the fact that the weaker theory is closed under some translation method, whereas the stronger one is not.

Allowing non-unary connectives opens up vast new landscapes, especially in the constructive setting. In our earlier paper, apart from providing a general framework of *schematic logics* (Litak & Visser 2018, Sect. 5.1), we have made the case for the

---

[3] By an amazing coincidence, as we were belatedly submitting the ready-for-print version of this paper, Mojtahedi (2022b, 2022a) announced an axiomatization of the provability logic of HA. The proof is presently being checked by other experts in the area.

constructive strict implication $\dashv$, also called the *Lewis arrow*. It allows defining $\Box\,\varphi$ as $\top \dashv \varphi$. We list its arithmetical interpretations in Sect. 2.4. The most important one, in the study of metatheory of HA, is provided by Γ-preservativity for a theory T, where Γ is a class of sentences, most commonly taken to be Σ^0_1 (Iemhoff, 2003; Iemhoff et al., 2005a; Visser, 1985, 1994, 2002). Other ones include contraposed conservativity/interpretability (Artemov & Beklemishev, 2004; Berarducci, 1990; Japaridze & de Jongh, 1998; Shavrukov, 1988; Visser, 1998), the logic of admissible schemes or the $\boxdot$-provability interpretation (Litak & Visser 2018, Sect. C.3). See Sect. 2.4 for details. Almost needless to say, constructive arithmetic can be replaced by any other foundational theory rich enough for standard encodings of syntactic notions. There are also computational interpretations originating elsewhere (Litak & Visser 2018, Sect. 7.1), such as Hughes' "classical" arrows in functional programming (de Groot et al., 2021; Hughes, 2000; Lindley et al., 2011). All these interpretations extend our base system $\mathsf{HLC}^\flat$. Interestingly enough, not all of them validate principles like Di (Sect. 2.2) which hold in the standard Kripke semantics of $\dashv$ (Sect. 2.3.2). But what axioms do we need to ensure fixed point results, and how can we compute these explicit fixed points?

2.1.3 Our Contributions

The classical construction of explicit guarded fixed points proceeds in two stages. One first proves the result for formulas where the main connective is the modal operator and then one shows how to extend the result to all modalized formulas (and possibly beyond that). Our results directly concern the first step. See Sect. 2.12 for a brief discussion of the second one.

We have two known paradigms for the construction of fixed points of formula with as main operator a modal one. First, there is the original de Jongh-Sambin construction (see Sect. 2.7) as generalized by Smoryński (1985, Chap. 4). Secondly, there is the construction given by de Jongh and Visser (1991) for the interpretability logic IL (see Sect. 2.6). As it simplifies to the de Jongh-Sambin construction when one adds the principle W (see Fig. 2.1), it seemed the *master* construction. Our results show that de Jongh-Visser construction and the de Jongh-Sambin construction are mutually incomparable. The incomparability result also holds in the classical case. This result for the classical case was also discovered by Sohei Iwata, Taishi Kurahashi and Yuya Okawa. See Iwata et al. (2022).

Our paper is the first study of what one might call "the reverse mathematics of explicit fixed points". After discussing algebraic and Kripke semantics for extensions of $\mathsf{HLC}^\flat$ (Sect. 2.3), we investigate the effect of adding explicit schemes stating that a given method (de Jongh-Visser or de Jongh-Sambin) indeed yields fixed points of formulas whose principal connective is $\dashv$.

We show that these extensions with fixed point-equations are good in a very specific sense: they are *extension stable*. In order to explain extension stabiity, we direct the reader's attention to arithmetical interpretations. As we noted above, the validity of a scheme in the logic (algebraic core) of a given arithmetical theory (Sect. 2.4) does not need to transfer to the logic of some given finite extension. However, if the provability logic of a theory is preserved under finite extensions, then this is reflected in a purely modal property of the logic, to wit *extension stability*. Similarly, the provability logic of all finite extensions of a given theory will be extension stable. Löb's logic in the classical case is the prime example of an extension stable theory. Not coincidentally, it is also the prime example of a (non-elementary) *subframe logic*. We give more examples of extension stable and extension unstable theories in Sect. 2.5. We also clarify that in the classical unimodal setting, a logic is extension stable iff it is subframe.

After recalling the information on extension stability (Sect. 2.5), we will show that the minimal theory in which the de Jongh-Visser construction works is the theory $\mathsf{L}_{\mathsf{a}}^{\flat}$(Sect. 2.6), which is extension stable. Thus, for the de Jongh-Visser construction we have a precise analogue of Löb's Logic.[4] In Sect. 2.7 we show that the case of the de Jongh-Sambin construction is more complex. The theory of this construction, to wit $\mathsf{JS}^{\flat}$, is incomparable to $\mathsf{L}_{\mathsf{a}}^{\flat}$. We show that $\mathsf{JS}^{\flat}$ is extension stable.

In Sect. 2.8 we axiomatize the join of both theories $\mathsf{L}_{\mathsf{a}}^{\flat}$ and $\mathsf{JS}^{\flat}$. In Sect. 2.9 we investigate several subtheories of $\mathsf{JS}^{\flat}$, whose axioms are obtained as de Jongh-Sambin fixed points of simple formulas. A large part of our results on axiomatizing explicit fixed points is concisely summarized by Fig. 2.2 therein. In Sect. 2.10 we present Kripke semantics for some principles investigated in earlier sections and uncover a simple nonconservativity phenomenon.

2.2 Basics

Our basic system is (the "flat" version of) *Heyting-Lewis Calculus* ($\mathsf{HLC}^{\flat}$) (de Groot et al., 2021; Litak & Visser, 2018) in the language of intuitionistic propositional calculus (IPC) extended with a binary connective $\dashv$. We use infix notation for $\dashv$ and follow the usual convention of omitting outer brackets. The system is given by the following axioms:

prop axioms and rules for IPC

Tr $((\varphi \dashv \psi) \wedge (\psi \dashv \chi)) \to (\varphi \dashv \chi)$

$\mathsf{K_a}$ $((\varphi \dashv \psi) \wedge (\varphi \dashv \chi)) \to (\varphi \dashv (\psi \wedge \chi))$

$\mathsf{N_a}$ $\vdash \varphi \to \psi \quad \Rightarrow \quad \vdash \varphi \dashv \psi$

We take $\Box\varphi := \top \dashv \varphi$ and $\boxdot\varphi$ for $\varphi \wedge \Box\varphi$. One can easily derive the intuitionistic version of the classical system K (without $\Diamond$) for the $\Box$-language from $\mathsf{HLC}^{\flat}$. A basic insight is contained in the following lemma.

Lemma 2.1 $\mathsf{HLC}^{\flat} \vdash \Box(\varphi \to \psi) \to (\varphi \dashv \psi)$.

The "sharp" system $\mathsf{HLC}^{\sharp}$ extends $\mathsf{HLC}^{\flat}$ with

---

[4] We do not have an example of an arithmetical theory for which this logic is arithmetically complete.

Di $((\varphi \dashv\!\!3\; \chi) \wedge (\psi \dashv\!\!3\; \chi)) \to ((\varphi \vee \psi) \dashv\!\!3\; \chi)$

A(n) ($\mathsf{HLC}^\flat$-)*logic* Λ is an extension of $\mathsf{HLC}^\flat$ that is closed under modus ponens, necessitation and substitution. Let X be a set of formulas (or a formula scheme, or a single axiom). We write $\Lambda \oplus X$ for the closure of $\Lambda \cup X$ under modus ponens and necessitation. Note that $\Lambda \oplus X$ is not automatically a logic. On the other hand, if X is closed under substitution, then so is $\Lambda \oplus X$.

Remark 2.1 All theorems we claim for $\mathsf{HLC}^\flat$ also hold when we omit disjunction from the language, in the sense that we still have all schemes, where the interpretations of the schematic letters are restricted to disjunction-free formulas. Our "flat" proofs also work in the disjunction-free setting.

Theorems 2.31 and 2.32 illustrate that Di is sometimes needed to derive principles not involving $\vee$. A similar example is provided by the trivialization of $\dashv\!\!3$ in $\mathsf{CPC}^\sharp$ (Litak & Visser 2018, Lemma 4.6). In the latter case, we know we need Di to make the argument work since the classical interpretability logic IL does not trivialize. In other words, some derivations are only possible if one sharpens up.[5]

Taking our notational experiment further, we abbreviate $\mathsf{HLC}^\flat \oplus X$ as $X^\flat$ and $\mathsf{HLC}^\sharp \oplus X$ as $X^\sharp$. For example, $\mathsf{HLC}^\sharp = \mathsf{Di}^\flat$ and for X containing (all instances of) Di, $X^\flat = X^\sharp$.

At some points, we will use a convenient notation for substitution. Suppose a variable, say r, of substitution is given in the context. We will write $\varphi\psi$ for $\varphi[r := \psi]$. We note that $(\varphi\psi)\chi$ is equal to $\varphi(\psi\chi)$. So we may write $\varphi\psi\chi$.

We remind the reader of the traditional principles $4_\Box$ and $\mathsf{L}_\Box$.

$4_\Box$ $\Box\varphi \to \Box\Box\varphi$
$\mathsf{L}_\Box$ $\Box(\Box\varphi \to \varphi) \to \Box\varphi$

Lemma 2.2 *Let a designated variable of substitution r be given. We have:*

Sub1 $4_\Box^\flat \vdash \boxdot(\varphi \leftrightarrow \psi) \to (\chi\varphi \leftrightarrow \chi\psi)$.
Sub2 *Suppose every occurrence of r is in the scope of an occurrence* $\dashv\!\!3$ *in* χ. *We have:*

$$4_\Box^\flat \vdash \Box(\varphi \leftrightarrow \psi) \to (\chi\varphi \leftrightarrow \chi\psi).$$

Proof We prove Sub1 by induction on χ. The only interesting case is where $\chi = \nu_0 \dashv\!\!3\; \nu_1$. By the Induction Hypothesis, for $i = 0, 1$, we have:

$$4_\Box^\flat \vdash \boxdot(\varphi \leftrightarrow \psi) \to (\nu_i\varphi \leftrightarrow \nu_i\psi).$$

[5] The "sharp" versus "flat" convention is introduced in the final version of the present paper. Previously, we used minus as a superscript for flat variants and no specific notation for sharp ones. There are also other changes, for example we used to denote $\mathsf{HLC}^\sharp$ as iA (for *intuitionistic arrows*) and earlier references such as Iemhoff et al. (2005a) were using iP, presumably due to connections with preservativity.

It follows that:

$$4^{\flat}_{\Box} \vdash \boxdot(\varphi \leftrightarrow \psi) \to \Box(\nu_i\varphi \leftrightarrow \nu_i\psi).$$

By Lemma 2.1, we find:

$$4^{\flat}_{\Box} \vdash \boxdot(\varphi \leftrightarrow \psi) \to (\nu_i\varphi \strictiff \nu_i\psi).$$

The desired result is now immediate.

We prove Sub2. Suppose $\chi = \nu_0 \strictif \nu_1$. We have, by Sub1,

$$4^{\flat}_{\Box} \vdash \boxdot(\varphi \leftrightarrow \psi) \to (\nu_i\varphi \leftrightarrow \nu_i\psi).$$

Hence,

$$4^{\flat}_{\Box} \vdash \Box(\boxdot(\varphi \leftrightarrow \psi) \to (\nu_i\varphi \leftrightarrow \nu_i\psi)).$$

So,

$$4^{\flat}_{\Box} \vdash \Box\boxdot(\varphi \leftrightarrow \psi) \to \Box(\nu_i\varphi \leftrightarrow \nu_i\psi).$$

Ergo, since $4^{\flat}_{\Box} \vdash \Box\,\varphi \to \Box\boxdot\,\varphi$, we find:

$$4^{\flat}_{\Box} \vdash \Box(\varphi \leftrightarrow \psi) \to (\nu_i\varphi \strictiff \nu_i\psi).$$

Thus the desired result follows for the case that $\chi = \nu_0 \strictif \nu_1$. The remaining cases are by induction on outer non-modal propositional connectives. ⊣

Lemma 2.3 (Uniqueness of fixed points) *Fix a designated variable of substitution r. Suppose every occurrence of r is in the scope of an occurrence $\strictif$ in χ. Suppose q does not occur in χ. We have:*

a. $\mathsf{L}^{\flat}_{\Box} \vdash (\boxdot(r \leftrightarrow \chi r) \wedge \boxdot(q \leftrightarrow \chi q)) \to (r \leftrightarrow q)$.
b. $\mathsf{L}^{\flat}_{\Box} \vdash (\boxdot(r \leftrightarrow \chi r) \wedge \boxdot(q \leftrightarrow \chi q)) \to \boxdot(r \leftrightarrow q)$.
c. $\mathsf{L}^{\flat}_{\Box} \vdash (\Box(r \leftrightarrow \chi r) \wedge \Box(q \leftrightarrow \chi q)) \to \Box(r \leftrightarrow q)$.

Proof We use the familiar fact that we have $4_{\Box}$ in $\mathsf{L}^{\flat}_{\Box}$. It is easy to see that we have the strengthened Löb's Rule in $\mathsf{L}^{\flat}_{\Box}$, to wit:

$$\text{if } \mathsf{L}^{\flat}_{\Box} \vdash (\textstyle\bigwedge_{i<n} \boxdot\,\varphi_i \wedge \Box\,\psi) \to \psi, \text{ then } \mathsf{L}^{\flat}_{\Box} \vdash \textstyle\bigwedge_{i<n} \boxdot\,\varphi_i \to \psi.$$

We use this rule in our proof. We only prove the first item, as the other two follow then immediately. We reason in $\mathsf{L}^{\flat}_{\Box}$. Suppose $\boxdot(r \leftrightarrow \chi r)$ and $\boxdot(q \leftrightarrow \chi q)$ and $\Box(r \leftrightarrow q)$. Since we have $4_{\Box}$, we also have Sub2 and, hence, $\chi r \leftrightarrow \chi q$. Thus, we find $r \leftrightarrow q$. By the strengthened Löb's Rule, we may conclude $r \leftrightarrow q$ without the assumption of $\Box(r \leftrightarrow q)$. ⊣

Remark 2.2 We find a closely related development in Craig Smoryński's paper (1985, Chap. 4). We note that our development is constructive and studies a binary operator, where Smoryński works classically and studies a unary operator. Moreover, since Smoryński aims at a version of the de Jongh-Sambin Theorem, he imposes an additional axiom written (in an adjusted notation) as $\vdash \triangle\varphi \to \Box \triangle\varphi$.

2.3 Semantics

Most proofs in this paper are of purely syntactic nature. Nevertheless, we occasionally still need semantics, e.g., for non-derivability results and for better understanding of syntactic systems and notions studied below. For our purposes, it is enough to consider algebraic semantics (Sect. 2.3.1) and Kripke semantics (Sect. 2.3.2).

2.3.1 Algebraic Semantics

We briefly recapitulate algebraic semantics. A generic algebraic completeness result after the manner of Lindenbaum and Tarski is obtainable for almost any "natural" logic and extensions of $\mathsf{HLC}^\flat$ are no exception. We can in fact put it in a general setting: The "global consequence relation" of any extension of $\mathsf{HLC}^\flat$ including both Modus Ponens and $\mathsf{N_a}$ is easily seen to be *algebraizable* (Blok & Pigozzi, 1989; Font et al., 2003), in fact an instance of what Rasiowa calls an *implicative logic* (Font, 2006; Rasiowa, 1974). More explicitly, the algebraic semantics looks as follows:

Definition 2.1 A *Lewisian Heyting Algebra Expansion*, or L-HAE,[6] is a tuple $\mathcal{A} := \langle A, \wedge, \vee, \dashv, \to, \bot, \top \rangle$, whose *intuitionistic/Heyting reduct* (i.e., the reduct without $\dashv$) is a Heyting algebra and the following laws are satisfied:

CK $(a \dashv b) \wedge (a \dashv c) = a \dashv (b \wedge c)$,
CT $(a \dashv b) \wedge (b \dashv c) \leq a \dashv c$,
CI $a \dashv a = \top$.

Moreover, $\mathcal{A}$ is a *Lewisian Heyting Algebra with Operators* (L-HAO) if its *strict reduct* (the reduct without $\to$) is a weak Heyting algebra (Celani & Jansana, 2005), i.e., in addition to CK, CT, and CI it satisfies:

CD $(a \dashv c) \wedge (b \dashv c) = (a \vee b) \dashv c$.

The principles CK, CD, CT, and CI are called by Celani and Jansana (2005) C1 – C4, respectively.

Remark 2.3 The notion of an operator is well-known in the boolean context. Since the groundbreaking work of Jónsson and Tarski (1951, 1952), an operator on a

[6] We write L-HAEs for the class of all Lewisian Heyting Algebra Expansions.

boolean algebra is supposed to distribute over finite joins. In the presence of involutive negation, the choice between preservation or reverse preservation of meets or joins is largely a question of personal preference due to mutual term-definability. Things are obviously more complicated in the Heyting/positive setting and existing literature seems to be liberal regarding terminology. This is why we allowed ourselves to use this name for an operation turning joins into meets in the first co-ordinate.

A valuation v in $\mathcal{A}$, as usual, maps propositional atoms to elements of A and is inductively extended to $\hat{v}$ defined on all formulas in the obvious way. Given any K that is a subset of the class of L-HAEs. and any set of formulas Λ, we define:

- $\mathcal{A}, v \Vdash \varphi$ iff $\hat{v}(\varphi) = \top$,
- $\mathcal{A} \Vdash \varphi$ iff $\mathcal{A}, v \Vdash \varphi$, for all v,
- $\mathcal{A} \Vdash \Lambda$ iff $\mathcal{A} \Vdash \varphi$, for every $\varphi \in \Lambda$,
- $K \Vdash \varphi$ iff $\mathcal{A} \Vdash \varphi$, for every $\mathcal{A} \in K$.
- $\mathsf{Mod}(\Lambda) := \{\mathcal{A} | \mathcal{A} \Vdash \Lambda\}$,
- $\mathsf{Th}(K) := \{\varphi | K \Vdash \varphi\}$.

Theorem 2.1 (Algebraic Completeness)

- *For any $K \subseteq$ L-HAEs, we have $\mathsf{Th}(K)$ is a logic.*
- *For any logic Λ, we have $\Lambda = \mathsf{Th}(\mathsf{Mod}(\Lambda))$.*

Proof One can apply techniques of abstract algebraic logic (AAL) as discussed in the standard references (Blok & Pigozzi, 1989; Font, 2006; Font et al., 2003; Rasiowa, 1974). However, we can also provide a detailed proof.

For the first part (soundness): closure under substitution is trivial. Modus Ponens and IPC axioms are standard using the fact that our algebras have Heyting reducts. We get Tr via the validity of CT and $\mathsf{K_a}$ via the validity of CK. For $\mathsf{N_a}$, assume $\mathcal{A}, v \Vdash \varphi \to \psi$. By standard facts regarding Heyting algebras, $\hat{v}(\varphi) \leq \hat{v}(\psi)$, hence $\hat{v}(\varphi) = \hat{v}(\varphi \wedge \psi)$. By CI, we get $\hat{v}(\varphi \dashv (\varphi \wedge \psi)) = \top$. Now use CK to derive that $\hat{v}(\varphi \dashv \psi) = \top$.

For the second part (completeness): $\mathsf{Th}(\mathsf{Mod}(\cdot))$ is clearly a closure operator, we only need to show that whenever $\Lambda \nvdash \varphi$, there is $\mathcal{A} \in \mathsf{Mod}(\Lambda)$ and a valuation v s.t. $\hat{v}(\varphi) \neq \top$. Simply pick $\mathcal{A}$ to be the Lindenbaum-Tarski algebra of formulas quotiented by Λ-provable equivalence. We only need to show that this yields a $\dashv$-algebra. The Heyting part is standard. For one half of CK, we use $\mathsf{K_a}$. For the other, we use Tr and $\mathsf{N_a}$. CT directly follows by Tr and CI directly follows by $\mathsf{N_a}$. ⊣

2.3.2 *Kripke Semantics*

We briefly recapitulate basic information from our overview (Litak & Visser, 2018) and refer interested readers to a much more detailed treatment in de Groot et al.

Box	$\varphi \mathrel{-\!\!3} \psi \to \Box(\varphi \to \psi)$ brilliant	$k \sqsubset \ell \preceq m \Rightarrow k \sqsubset m$	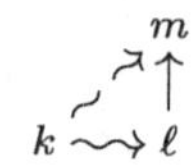
$4_\Box$	$\Box\varphi \to \Box\Box\varphi$ semi-transitive	$k \sqsubset \ell \sqsubset m \Rightarrow$ $\exists x. k \sqsubset x \preceq m$	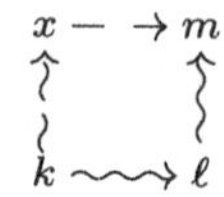
4_a	$\varphi \mathrel{-\!\!3} \Box\varphi$ gathering	$k \sqsubset \ell \sqsubset m \Rightarrow \ell \preceq m$	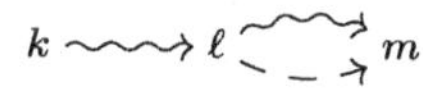
$L_\Box$	$\Box(\Box\varphi \to \varphi) \to \Box\varphi$	Löb: $\sqsubset$-Noetherian (conversely well-founded) and semi-transitive	
W	$((\varphi \wedge \Box\psi) \mathrel{-\!\!3} \psi) \to \varphi \mathrel{-\!\!3} \psi$ supergathering	on finite frames: $k \sqsubset \ell \sqsubset m \Rightarrow \exists x \sqsupset k.(\ell \prec x \preceq m)$	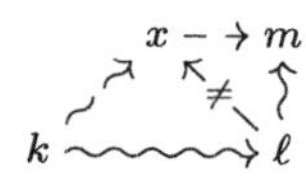
S	$\varphi \to \Box\varphi$ strong	$k \sqsubset \ell \Rightarrow k \preceq \ell$	

Fig. 2.1 Basic axioms and their correspondence conditions (Litak & Visser 2018, Sect. 6). Further correspondence conditions are provided in Sect. 2.10

(2021). A (*Lewisian*) *Kripke frame* is a triple $\mathfrak{F} := \langle W, \preceq, \sqsubset\rangle$, where $\preceq$ is a partial order on W, $\sqsubset \subseteq W \times W$ and furthermore

$\mathrel{-\!\!3}$-p if $k \preceq \ell \sqsubset m$, then $k \sqsubset m$, that is, $(\preceq \cdot \sqsubset) \subseteq \sqsubset$.[7]

Admissible valuations map propositional atoms to upward $\preceq$-closed sets and intuitionistic connectives are interpreted as usual using $\preceq$. The clause for $\mathrel{-\!\!3}$ in a model is

$$k \Vdash \varphi \mathrel{-\!\!3} \psi \text{ if, for all } \ell \sqsupset k, \text{ if } \ell \Vdash \varphi, \text{ then } \ell \Vdash \psi. \tag{2.1}$$

Given such a Kripke frame, it is an easy exercise to define its dual algebra whose Heyting reduct is given as usual by $\mathrm{Up}^{\preceq}(W)$, i.e., upsets of W, and the strict reduct (interpretation of $\mathrel{-\!\!3}$) is induced by (2.1). The dual algebra is a L-HAO, i.e., the equality CD corresponding to Di holds. In other words, extensions of $\mathsf{HLC}^\flat$ which are not extensions of $\mathsf{HLC}^\sharp$ can only be Kripke sound, but not Kripke complete. Definitions of notions such as the *finite model property* (fmp), i.e., completeness w.r.t. a class of finite frames are standard.

Some modal axioms that we need together with their correspondence conditions are given in Fig. 2.1. Its fuller version can be found in Litak and Visser (2018, Sect. 6), along with an extended version of the following summary of existing results.[8]

[7] We write "·" for composition in the order of reading.

[8] Note that (Litak & Visser, 2018) refers to our unpublished (and unfinished) note "Lewis arrow fell off the wall: decompositions of constructive strict implication". Presently, the relevant results are available in de Groot et al. (2021). Furthermore, some results summarized here in Theorem 2.2

Theorem 2.2 *We have:*

(a) $\mathsf{HLC}^\sharp$ *has the finite model property (Iemhoff 2001b, Proposition 4.1.1; Iemhoff 2003, Proposition 7; Zhou 2003, Theorem 2.1.10).*
(b) $4^\sharp_{\mathsf{a}}$ *corresponds to the class of gathering frames and has the fmp (Iemhoff 2001b, Proposition 4.2.1; Iemhoff 2003, Proposition 8).*
(c) $4^\sharp_{\Box}$ *corresponds to semi-transitivity (Iemhoff 2001b, Proposition 4.2.2).*
(d) $\mathsf{L}^\sharp_{\Box}$ *corresponds to Noetherian semi-transitivity (Iemhoff 2001b, Proposition 4.3.2; Zhou 2003, Theorem 2.2.7).*
(e) $\text{i-}\mathsf{GL}_{\mathsf{a}} = \mathsf{L}^\sharp_{\Box} \oplus 4_{\mathsf{a}} = \mathsf{L}^\sharp_{\mathsf{a}}$ *corresponds*[9] *to the class of Noetherian gathering frames (Iemhoff 2003, Lemma 9; Iemhoff et al. 2005a, Lemma 3.10).*
(f) $\text{i-}\mathsf{GW} = \mathsf{W}^\sharp$ *corresponds to the "supergathering" property of Fig. 2.1 on the class of finite frames (Zhou 2003, Lemma 3.5.1; Iemhoff et al. 2005a, Theorem 3.31).*

The reader may have noted that in the above summary of existing results, clauses (c), (d), (e), and (f) do not mention the finite model property for the full language with $\dashv$; indeed, we are not aware whether such results have been obtained (cf. Footnote 8). For the four principles in question, the above references show the finite model property for the corresponding logics restricted to the language with $\Box$ (i.e., formulas not involving binary $\dashv$).

It should be mentioned that (de Groot et al., 2021) provide an alternative route towards correspondence, strong completeness/canonicity and the fmp via a Gödel-McKinsey-Tarski embedding into classical logics with two unary modalities, extending earlier results for intutionistic modal logics by Wolter and Zakharyaschev (1997, 1998). In particular, these results automatically yield correspondence and strong completeness for canonical axioms such as 4_{a}, $4_{\Box}$, Box and S, cf. de Groot et al. (2021, Theorem 4.23 and Example 4.24). These results also allow transfer of the fmp, but the most useful criterion presently available (de Groot et al. 2021, Corollary 5.5 and Example 5.6) applies to (cofinal) subframe logics including the scheme P introduced in Sect. 2.7 below, which ensures transitivity of the modal accessibility relation (Theorem 2.29). This covers systems such as $\text{i-}\mathsf{SA} = \mathsf{S}^\sharp$ and $\text{i-}\mathsf{SL}_{\mathsf{a}} = \mathsf{L}^\sharp_{\Box} \oplus \mathsf{S}$, but not the systems mentioned in the preceding paragraph.

Question 2.1 Do $4^\sharp_{\Box}$, $\mathsf{L}^\sharp_{\Box}$, $\text{i-}\mathsf{GL}_{\mathsf{a}} = \mathsf{L}^\sharp_{\Box} \oplus 4_{\mathsf{a}}$ and $\text{i-}\mathsf{GW} = \mathsf{W}^\sharp$ have the finite model property?

Question 2.2 Can the techniques of de Groot et al. (2021) be used also to transfer fixpoint calculations between classical bimodal setting and constructive Lewisian setting?

---

are misrepresented in Litak and Visser (2018). In particular, while results of (Iemhoff, 2001b) and subsequent references do cover the finite model property of fragments of $4^\sharp_{\Box}$, $\mathsf{L}^\sharp_{\Box}$ and $\mathsf{L}^\sharp_{\Box} \oplus 4_{\mathsf{a}} = \text{i-}\mathsf{GL}_{\mathsf{a}}$ restricted to the language where the only modality is $\Box$ (as stated in Theorem 2.2 here), we were unable to find an explicit statement of the fmp for the corresponding full systems. We extended Question 2.1, originally concerned only with the fmp of $\mathsf{W}^\sharp = \text{i-}\mathsf{GW}$, to these systems as well.

[9] See Sect. 2.6 for a discussion of L_{a}.

Remark 2.4 There are other semantics for $\mathsf{HLC}^\flat$ and $\mathsf{HLC}^\sharp$, which are intermediate between algebraic and Kripke ones. We do not treat them here, as our paper is primarily syntactic in nature, but in subsequent work we are going to discuss, e.g., suitable generalizations of Veltman semantics, or semantics merging the Kripkean interpretation of intuitionistic connectives with the neighbourhood interpretation of $\dashv$.

2.4 Arithmetical Interpretations

In our earlier paper, we proposed a general framework of *schematic logics* (Litak & Visser 2018, Sect. 5.1) for arithmetical interpretations of logical systems, in particular extensions of $\mathsf{HLC}^\flat$. It can be also seen as semantics of propositional modal logics and this aspect is our main focus here. Let us present the framework restricted to the $\dashv$-signature.

We restrict ourselves in our presentation to theories extending the intuitionistic version $\mathsf{i\text{-}EA}$ of Elementary Arithmetic in the same language. Intuitionistic Elementary Arithmetic consists, over intuitionistic predicate logic, of the basic axioms for successor, addition and muliplication plus Δ_0-induction, plus an axiom expressing that exponentiation is total.[10] We assume that a $\Delta_0(\mathsf{exp})$ formula (a.k.a. an *elementary* formula) α_T representing the given axiom set of T is part of the data for T. NB: we thus treat arithmetical theories differently from propositional logics that are given as sets of sentences.

Consider a function F that assigns to T an arithmetical formula $A_T(v_0, v_1)$ as interpretation of $\dashv$. We note that, by our convention, F operates intensionally on α_T, in other words, it operates not on T qua set of theorems or even T qua set of axioms, but on T as given by the formula α_T that represents its axiom set. We write $B_0 \dashv_{F,T} B_1$ for $F(T)(\ulcorner B_0 \urcorner, \ulcorner B_1 \urcorner)$. Here $\ulcorner C \urcorner$ is the numeral of the Gödel number of C. Suppose f is a mapping from the propositional atoms to arithmetical sentences. We define $(\varphi)^f_{F,T}$ as the translation that uses f to interpret atoms, $\dashv_{F,T}$ to interpret $\dashv$, and that commutes with the propositional connectives.

We say that φ is T*-valid* w.r.t. F if, for all assignments f of arithmetical sentences to the propositional atoms, we have $T \vdash (\varphi)^f_{F,T}$. We write $\Lambda_F(T)$ for the set of modal formulas that are T-valid w.r.t. F. Here are several examples of interpretations fitting in this framework, each of which yields a logic extending $\mathsf{HLC}^\flat$.

- The most important one, for our purposes, is provided by $\Lambda^\circ_\Gamma(T)$, i.e., the logic of Γ*-preservativity* for a theory T, where Γ is a class of sentences. We assume that

[10] There is a Δ_0-definition of the graph of exponentiation. See, e.g., Hájek and Pudlák (1993, Sect. 5.3.c), where this is proved for the classical case. It is easy to see the considerations involved are constructive. Moreover, we may conservatively add a term exp for exponentiation to the language. If we do this, we can prove $\Delta_0(\mathsf{exp})$-induction in $\mathsf{i\text{-}EA}$. See Gaifman and Dimitracopoulos (1982), where these facts are proved in the classical case. Again it is easy to see that everything needed here is constructive.

Γ is given by an elementary formula, say γ, that arithmetically represents it. We define:

- $A \dashv_{\Gamma,T} B$ if, for all Γ-sentences C, if $T \vdash C \to A$, then $T \vdash C \to B$.[11]

A minimal assumption on Γ is that it includes a sentence (equivalent to) $\top$. Let us call such Γ *preservation-suitable*. This ensures, in particular, that $\Box\varphi$ defined as $\top \dashv \varphi$ still encodes provability. It is most common to fix Γ as Σ_1^0 (Iemhoff, 2003; Iemhoff et al., 2005a; Visser, 1985, 1994, 2002), in which case we drop the subscript Γ (cf. Litak and Visser, 2018, Sect. 5 for the notation). The relation of Σ_1^0-preservativity is an important tool in the study of metatheory of HA. In particular, constructive preservativity logic is conceivably easier to analyze than its provability fragment. See Litak and Visser (2018) for further motivation.

- Classically, Σ_1^0-preservativity can be seen as contraposed form of Π_1^0-*conservativity*, which in turn, over theories with full induction, is classically equivalent to *interpretability*. See Artemov and Beklemishev 1990, Shavrukov (1988), Japaridze & de Jongh (1998), Visser (1998), Artemov & Beklemishev (2004), and also Litak and Visser (2018, C.3). It is also possible to investigate constructive interpretability logic, although it appears somewhat less well-motivated (Litak & Visser, 2018, C.4). We note that Σ_1^0-preservativity, Π_1^0-conservativity and interpretability yield pairwise different logics over HA.
- Yet another example of a schematic interpretation is provided by the $\boxdot$-translation. That is, $\dashv$ is interpreted as $\Box_T((A \wedge \Box_T A) \to (B \wedge \Box_T B))$, where $\Box_T$ is provability in T.

2.5 Extension Stability

Our paper focuses on axiomatizing principles yielding explicit fixed points in the logic $\Lambda_F(T)$ of an arithmetical theory T (relative to chosen F). Nevertheless, we have already mentioned that regardless of the interpretation of $\dashv$, $\Lambda_F(\cdot)$ is not necessarily monotone: $T \subseteq T'$ does not entail that $\Lambda_F(T) \subseteq \Lambda_F(T')$.

This makes the following question interesting: in what cases is $\Lambda_F(T)$ preserved under finite extensions? In other words, given an arithmetical theory T, when do we have $\Lambda_F(T) \subseteq \Lambda_F(T + A)$, for all A? It also motivates an interest in the study of

- $\widetilde{\Lambda}_F(T) := \bigcap\{\Lambda_F(T + A) | A \text{ is a sentence of the arithmetical language}\}$.

The consideration of finite extensions of the base theory leads to a corresponding *modal* notion, to wit *extension stability*, which will be explained below. Apart from playing an important role in our arithmetical (reverse) correspondence theory, this notion seems to be of interest for ordinary modal model theory; it turns out to be

[11] We note that $\dashv_{\Gamma,T}$ is a rather uniform interpretation. There is an arithmetical formula $P(v_0, v_1, X, Y)$ with two free second-order variables, such that $B_0 \dashv_{\Delta,T} B_1 = P(\ulcorner B_0 \urcorner, \ulcorner B_1 \urcorner, \alpha_T, \gamma)$.

an overlooked generalization of the notion of a *subframe logic* (Fine, 1985; Wolter, 1993). We provide a detailed discussion in Appendix 3. Recall that above classical K4, subframeness ensures nice completeness properties. This is important for Löb-like logics: they are non-elementary and hence cannot be proved complete via Sahlqvist-style arguments. However, we postpone a detailed model-theoretic discussion to a later work.

Given a formula φ and a fresh propositional variable p, we define the translation $\varphi^{\ulcorner p \urcorner}$ inductively as commuting with the propositional variables and the connectives of IPC, with the $\dashv$ clause being

- $(\psi \dashv \chi)^{\ulcorner p \urcorner} := ((p \to \psi^{\ulcorner p \urcorner}) \dashv (p \to \chi^{\ulcorner p \urcorner}))$.

As $\Box\,\varphi$ is $\top \dashv \varphi$, we get $\mathsf{HLC}^\flat \vdash (\Box\,\varphi)^{\ulcorner p \urcorner} \leftrightarrow \Box(p \to \varphi^{\ulcorner p \urcorner})$.

We note that for any logic Λ and any φ, if $\Lambda \vdash \varphi^{\ulcorner p \urcorner}$, then $\Lambda \vdash \varphi$.

A logic Λ is *extension stable* if, whenever $\Lambda \vdash \varphi$ and p not in φ, we have $\Lambda \vdash p \to \varphi^{\ulcorner p \urcorner}$. We can give a slightly different formulation by using the notion of *stabilizer* $\mathfrak{S}\Lambda$ of Λ.

- $\mathfrak{S}\Lambda := \{\varphi \mid \text{for some } p \text{ not in } \varphi,\ \Lambda \vdash p \to \varphi^{\ulcorner p \urcorner}\}$.

Theorem 2.3 *$\mathfrak{S}\Lambda$ is a logic.*

Proof is delegated to Appendix 1.

We note that Λ is extension stable iff $\Lambda = \mathfrak{S}\Lambda$. As noted in Remark 2.5 below, there is a dual notion of a *minimal stable extension* $\mathfrak{Q}\Lambda$, with which we could reformulate all results in this section. While $\mathfrak{Q}$ is a (not necessarily topological) closure operator, $\mathfrak{S}$ is of interior character:

Theorem 2.4 *For any logic Λ:*

(i) *if $\Lambda \subseteq \Lambda'$, then $\mathfrak{S}\Lambda \subseteq \mathfrak{S}\Lambda'$.*
(ii) *$\mathfrak{S}\Lambda \subseteq \Lambda$.*
(iii) *$\mathfrak{S}\mathfrak{S}\Lambda = \mathfrak{S}\Lambda$.*

Proofs is delegated to Appendix 2.

Theorem 2.5 *Suppose the logic Λ is axiomatized over $\mathsf{HLC}^\flat$ by X (w.r.t. the $\mathsf{HLC}^\flat$-rules Modus Ponens and Necessitation) and suppose that, for every $\varphi \in X$ and p not in φ, we have $\Lambda \vdash p \to \varphi^{\ulcorner p \urcorner}$. Then Λ is extension stable.*

Proof Straightforward induction on proofs.

Definition 2.2 We say that F is an *elegant interpretation* if for any recursively axiomatizable arithmetical theory U and any arithmetical sentences A, B, C, we have that $U \vdash B \dashv_{F,U+A} C \leftrightarrow (A \to B) \dashv_{F,U} (A \to C)$.

Theorem 2.6 *For any preservation-suitable Γ, Γ-preservativity is an elegant interpretation.*

Proof
$B \dashv_{\Gamma, U+A} C$

$$\begin{aligned}
&\leftrightarrow \forall D \in \Gamma\, (\Box_{U+A}(D \to B) \to \Box_{U+A}(D \to C)) \\
&\leftrightarrow \forall D \in \Gamma\, (\Box_U(D \to (A \to B)) \to \Box_U(D \to (A \to C))) \\
&\leftrightarrow (A \to B) \dashv (A \to C).
\end{aligned}$$

Theorem 2.7 *If F is an elegant interpretation, then*

$$\mathfrak{S}\Lambda_F(T) = \widetilde{\Lambda}_F(T).$$

Consequently, the logic of T is preserved under finite extensions iff it is extension stable.

Proof Let $p \notin \mathsf{var}(\varphi)$. We find by induction on φ that

$$U \vdash (\varphi)^{\sigma}_{F,U+A} \leftrightarrow (\varphi^{\ulcorner p \urcorner})^{\sigma[p:=A]}_{F,U}.$$

Hence,

$$\begin{aligned}
\forall A \;\; \Lambda_F(U+A) \vdash \varphi &\Leftrightarrow \forall A\, \forall \sigma \;\; U + A \vdash (\varphi)^{\sigma}_{F,U+A} \\
&\Leftrightarrow \forall A\, \forall \sigma \;\; U + A \vdash (\varphi^{\ulcorner p \urcorner})^{\sigma[p:=A]}_{F,U} \\
&\Leftrightarrow \forall \tau \;\; U \vdash (p \to \varphi^{\ulcorner p \urcorner})^{\tau}_{F,U} \\
&\Leftrightarrow \Lambda_F(U) \vdash p \to \varphi^{\ulcorner p \urcorner} \\
&\Leftrightarrow \mathfrak{S}(\Lambda_F(U)) \vdash \varphi
\end{aligned}$$

Remark 2.5 We can also build a closure operator $\mathfrak{Q}$ mapping a logic to its minimal stable extension: consider the set of $\varphi^{\ulcorner p \urcorner}$ for $\Lambda \vdash \varphi$, where $p \notin \mathsf{var}(\varphi)$, and close it off to make it a logic. This gives the desired extension of Λ.

Interestingly, this construction yields the well-known Kripke proof[12] of Löb's Theorem from Gödel's Theorem in the classical case: if we start with $\mathsf{HLC}^\flat$ plus the excluded middle plus $\Box\Diamond\top \to \Box\bot$, we get $\mathsf{L}_\Box$. In a similar way, we get $(\Box\varphi \to \varphi) \dashv \varphi$ from $\Diamond\top \dashv \bot$. Regrettably, we need the law of excluded middle here. In the constructive context we only get Löb's Theorem for negated formulas. For a truly constructive version of Kripke's argument, we need to employ (an iterated version of) the Friedman translation on the arithmetical side (Friedman, 1978; Troelstra, 1973). However, it is not clear to us how to employ that method, e.g., in the context of set theory.

Example 2.1 Here are some positive and negative examples.

[12] The argument is credited to Kripke in folklore.

+ Positive examples of extension stability: the interpretability logic ILM of classical sequential essentially reflexive theories. See Shavrukov (1988), Berarducci (1990), Japaridze and de Jongh (1998), or Artemov and Beklemishev (2004). The logic ILP of finitely axiomatized classical arithmetical theories extending $\mathrm{I}\Delta_0 + \mathsf{supexp}$. See Visser (1990).
– Failure of extension stability: the interpretability logic of Elementary Arithmetic EA. See Kalsbeek (1991). The provability logic of HA and, *ipso facto*, the Σ_1-preservativity logic of HA, since primitive recursive Markov's Principle can be axiomatized by a single sentence, to wit:

$$\mathsf{M} := \forall S \in \Sigma_1^0\text{-}\mathsf{sent}\,(\neg\neg\,\mathsf{true}_{\Sigma_1}(S) \to \mathsf{true}_{\Sigma_1}(S)).$$

In $\Lambda^{\circ}(\mathsf{HA})$ we have, e.g., the principle

$$\varphi_* := (\neg\neg\Box q \prec \Box q) \to \Box\Box q.$$

We find that:

$$\varphi_*^{\ulcorner p \urcorner} = ((p \to \neg\neg\Box(p \to q)) \prec (p \to \Box(p \to q))) \to \Box(p \to \Box(p \to q)).$$

If interpret p as Markov's Principle M and q as $\bot$ and $\prec$ as Σ_1-preservativity over HA, we find that the antecedent is $\varphi_*^{\ulcorner p \urcorner}$ is valid. Thus, if $p \to \varphi_*^{\ulcorner p \urcorner}$ were arithmetically valid, then we would have $\mathsf{HA} + \mathsf{M} \vdash \Box_{\mathsf{HA+M}}\Box_{\mathsf{HA+M}}\bot$. *Quod non.*

Here are four formulas in the provability logic of HA that fail in the provability logic of a finite extension of HA. The first column gives a principle φ that is in the provabiliy logic of HA and the second column gives a sentence A such that φ is not in the provability logic of $\mathsf{HA} + A$.

$\Box(\Box\bot \vee \neg\Box\bot) \to \Box\Box\bot$	$\Box_{\mathsf{HA}}\bot \vee \neg\Box_{\mathsf{HA}}\bot$
$\Box\neg\neg\Box\bot \to \Box\Box\bot$	$\neg\neg\Box_{\mathsf{HA}}\bot$
$\Box(\neg\neg\Box\bot \to \Box\bot) \to \Box\Box\bot$	$\neg\neg\Box_{\mathsf{HA}}\bot \to \Box_{\mathsf{HA}}\bot$
$\Box(\neg p \to (q \vee r)) \to \Box((\neg p \to q) \vee (\neg p \to r))$	$\neg\neg\Box_{\mathsf{HA}}\bot \to \Box_{\mathsf{HA}}\bot$

Question 2.3 Here are some questions that arise about the provability logic of all extensions of a given theory.

I. Is the provability logic of all finite extensions of T always the provability logic of some finite extension of T?
II. Is the provability logic of all T-verifiable extensions of T always the same as the provability logic of all finite extensions of T?
III. Is the provability logic of all finite extensions of HA the logic $\mathsf{iGL}_{\Box}$? If not can we axiomatize the provability logic of all finite extensions of HA?

We note that the methods of Mojtahedi (2022a, 2022b) do not work for these questions since e.g. extension with the finite statement of Primitive Recursive Markov's Principle already blocks some of the steps.

2.6 Löb-Lewis Logic Meets de Jongh and Visser

In this section we study a fixed point calculation that is due to de Jongh and Visser (1991), originally in the context of the classical interpretability logic IL. It transfers naturally to the context of the Lewis arrow in the presence of axiom $\mathsf{L_a}$, as observed first by Iemhoff et al. (2005a). We show that the scheme JV corresponding to the de Jongh-Visser calculation allows deriving $\mathsf{L_a}$ (hence $\mathsf{L}^\flat_\mathsf{a}$ is precisely the logic of this construction) and that the resulting logic is extension stable.

We define *Löb-Lewis Logic* as $\mathsf{L}^\flat_\mathsf{a}$, where:

$\mathsf{L_a}$ $(\Box\,\varphi \to \varphi) \strictif \varphi$.

One may show that $\mathsf{L}^\flat_\Box \oplus 4_\mathsf{a}$ coincides with Löb-Lewis Logic (Litak & Visser, 2018, Lemma 4.11).

Remark 2.6 If we add classical logic to Löb-Lewis Logic and replace $\strictif$ by the contraposed operator $\rhd$, we obtain the interpretability logic IL. We note that Di is not valid in contraposed IL. See de Jongh and Visser (1990), Visser (1991).

The symmetric Löb rule is given by:

$\mathsf{sLR_a}$ $\Box\,\varphi \vdash \varphi \leftrightarrow \psi \quad \Rightarrow \quad \vdash \varphi \strictiff \psi$.

Lemma 2.4 *The rule* $\mathsf{sLR_a}$ *is admissible for Löb-Lewis Logic.*

Proof Suppose $\Box\,\varphi \vdash_{\mathsf{L}^\flat_\mathsf{a}} \varphi \leftrightarrow \psi$. It follows that $\psi \vdash_{\mathsf{L}^\flat_\mathsf{a}} \Box\,\varphi \to \varphi$. Hence, by $\mathsf{N_a}$, we find $\vdash_{\mathsf{L}^\flat_\mathsf{a}} \psi \strictif (\Box\,\varphi \to \varphi)$ and, thus, by $\mathsf{L_a}$, $\vdash_{\mathsf{L}^\flat_\mathsf{a}} \psi \strictif \varphi$.

For the converse direction, we have $\boxdot\,\varphi \vdash_{\mathsf{L}^\flat_\mathsf{a}} \psi$. Hence, $\vdash_{\mathsf{L}^\flat_\mathsf{a}} \boxdot\,\varphi \strictif \psi$. Since, by 4_a, we have $\vdash_{\mathsf{L}^\flat_\mathsf{a}} \varphi \strictif \boxdot\,\varphi$, by Tr, we find $\vdash_{\mathsf{L}^\flat_\mathsf{a}} \varphi \strictif \psi$. $\dashv$

It was noted by Iemhoff et al. (2005a) that the following result, first obtained in the classical setting (de Jongh & Visser, 1991), holds constructively as well.

Theorem 2.8 *We fix a variable of substitution* r*. Let* $\varphi := \psi \strictif \chi$ *and let* $\theta := (\psi\,\Box\,\chi\top \strictif \chi\top)$*. Then, we have* $\mathsf{L}^\flat_\mathsf{a} \vdash \theta \leftrightarrow \varphi\theta$*.*

We note that θ can also be written as $(\psi\,\Box\, r \strictif r)\chi\top$ or as $(\psi\,\Box\,\chi \strictif \chi)\top$.

Proof We have $\Box\,\chi\top \vdash_{\mathsf{L}^\flat_\mathsf{a}} \theta$. Hence, $\Box\,\chi\top \vdash_{\mathsf{L}^\flat_\mathsf{a}} \boxdot(\top \leftrightarrow \theta)$. So, it follows that $\Box\,\chi\top \vdash_{\mathsf{L}^\flat_\mathsf{a}} \chi\top \leftrightarrow \chi\theta$. Hence, by $\mathsf{sLR_a}$, (a) $\vdash_{\mathsf{L}^\flat_\mathsf{a}} \chi\top \strictiff \chi\theta$.

We also have $\Box\,\psi\,\Box\,\chi\top \vdash_{\mathsf{L}_\mathsf{a}^\flat} \Box\,\chi\top \leftrightarrow \theta$. (We note that this step already works in $\mathsf{HLC}^\flat$.) So, $\Box\,\psi\,\Box\,\chi\top \vdash_{\mathsf{L}_\mathsf{a}^\flat} \psi\,\Box\,\chi\top \leftrightarrow \psi\theta$. Thus, by $\mathsf{sLR_a}$, we find (b) $\vdash_{\mathsf{L}_\mathsf{a}^\flat} \psi\,\Box\,\chi\top \strictiff \psi\theta$.

Combining (a) and (b), we now have:

$$\begin{aligned}\mathsf{L}_\mathsf{a}^\flat \vdash \theta &\leftrightarrow (\psi\,\Box\,\chi\top \strictif \chi\top)\\ &\leftrightarrow (\psi\theta \strictif \chi\theta)\\ &\leftrightarrow \varphi\theta\end{aligned}$$

⊣

We define the principle JV as follows:

$\mathsf{JV}\quad (\psi\,\Box\,\chi\top \strictif \chi\top) \quad \leftrightarrow (\psi \strictif \chi)(\psi\,\Box\,\chi\top \strictif \chi\top)$

We have seen that $\mathsf{L}_\mathsf{a}^\flat \vdash \mathsf{JV}$. We now show that, conversely, $\mathsf{JV}^\flat \vdash \mathsf{L_a}$, and, hence, $\mathsf{JV}^\flat$ coincides with $\mathsf{L}_\mathsf{a}^\flat$.

Theorem 2.9 $\mathsf{JV}^\flat \vdash \mathsf{L_a}$.

Proof The JV-fixed point of $\Box(p \to \varphi)$, where p is fresh for φ, is $\Box(\top \to \varphi)$. This tells us that, $\mathsf{JV}^\flat$-verifiably, we have $\mathsf{L}_\Box$. It follows, by Lemma 2.3, that we have uniqueness of fixed points in $\mathsf{JV}^\flat$. Now consider the formula

$$(p \to \varphi) \strictif \varphi,$$

where p does not occur in φ. On the one hand, $\top$ is a fixed point of this formula even over $\mathsf{HLC}^\flat$. On the other hand, JV gives us $(\Box\,\varphi \to \varphi) \strictif \varphi$ as a fixed point. By uniqueness, we find that $(\Box\,\varphi \to \varphi) \strictif \varphi$ is $\mathsf{JV}^\flat$-provable. ⊣

We note that, in the proof of Theorem 2.9, we used that we are allowed to choose p locally. We summarize Theorems 2.8 and 2.9 in one statement.

Corollary 2.1 *Löb-Lewis Logic coincides with* $\mathsf{JV}^\flat$.

It follows from that $\mathsf{JV}^\flat$ is indeed a logic in that it is closed under substitution. This could have been proven, of course, without the detour over the characterization.

Theorem 2.10 $\mathsf{JV}^\flat$ *is extension stable.*

Proof It is sufficient to show that Löb-Lewis Logic is extension stable. We leave the verification that $\mathsf{HLC}^\flat$ is extension stable to the reader. We have:

$$((\Box\,\varphi \to \varphi) \strictif \varphi)^{\ulcorner p \urcorner} \quad = \quad (p \to (\Box(p \to \varphi^{\ulcorner p \urcorner}) \to \varphi^{\ulcorner p \urcorner})) \strictif (p \to \varphi^{\ulcorner p \urcorner}).$$

Clearly,

$$(\dagger) \qquad \mathsf{HLC}^\flat \vdash \quad ((\Box\,\varphi \to \varphi) \strictif \varphi)^{\ulcorner p \urcorner} \quad \leftrightarrow \quad (\Box(p \to \varphi^{\ulcorner p \urcorner}) \to (p \to \varphi^{\ulcorner p \urcorner})) \strictif (p \to \varphi^{\ulcorner p \urcorner}).$$

The right-hand-side of (†) is an instance of $\mathsf{L_a}$. ⊣

We end this section by comparing $\mathsf{L}_{\mathsf{a}}^{\flat}$ to the stronger theory $\mathsf{W}^{\flat}$. We remind the reader that the principle W can be equivalently written as

$$\mathsf{W'}\quad (\varphi \mathrel{-\!\!3} \psi) \to ((\Box\, \psi \to \varphi) \mathrel{-\!\!3} \psi).$$

Lemma 2.5 (Iemhoff et al. 2005b, Corollary 2.7) W *and* $\mathsf{W'}$ *axiomatize the same logics over* $\mathsf{HLC}^{\flat}$.

Proof In one direction:

$$\begin{aligned} \varphi \mathrel{-\!\!3} \psi &\to_{\mathsf{HLC}^{\flat}} ((\Box\, \psi \to \varphi) \wedge \Box\, \psi) \mathrel{-\!\!3} \psi \\ &\to_{\mathsf{W}^{\flat}} (\Box\, \psi \to \varphi) \mathrel{-\!\!3} \psi. \end{aligned}$$

For the converse,

$$\begin{aligned} (\varphi \wedge \Box\, \psi) \mathrel{-\!\!3} \psi &\to_{\mathsf{W'}^{\flat}} (\Box\, \psi \to (\varphi \wedge \Box\, \psi)) \mathrel{-\!\!3} \psi \\ &\to_{\mathsf{HLC}^{\flat}} (\Box\, \psi \to \varphi) \mathrel{-\!\!3} \psi \\ &\to_{\mathsf{HLC}^{\flat}} \varphi \mathrel{-\!\!3} \psi. \end{aligned}$$ ⊣

In what follows, we are going to use W and W' as equivalent schemes. To separate $\mathsf{L}_{\mathsf{a}}^{\flat}$ from $\mathsf{W}^{\flat}$, we use here a lemma that is an adaptation of an example from de Jongh and Visser (1991, p. 48).

Lemma 2.6 $\mathsf{L}_{\mathsf{a}}^{\flat} \nvdash (\Box \bot \mathrel{-\!\!3} \bot) \to \Box \bot$.

Proof We can show a stronger statement $\mathsf{L}_{\mathsf{a}}^{\sharp} \nvdash (\Box \bot \mathrel{-\!\!3} \bot) \to \Box \bot$ on a three-point Noetherian gathering model (Litak & Visser, 2018, Example 6.7):

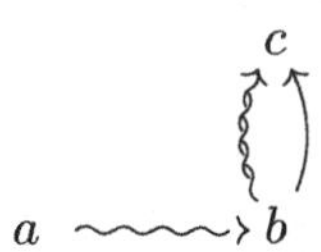

All points are $\sqsubset$-irreflexive. We set $b \preceq c$, $a \sqsubset b \sqsubset c$ and the valuation is empty, i.e., $V(a) = V(b) = V(c) = \emptyset$.

Clearly, $a \Vdash \Box\bot \mathrel{-\!\!3} \bot$, but $a \nVdash \Box\bot$. ⊣

Theorem 2.11 $\mathsf{W}^{\flat}$ *strictly extends Löb-Lewis Logic.*

Proof To see that $\mathsf{W}^{\flat}$ extends Löb-Lewis Logic, we note that specializing W by taking $\psi := \varphi$ gives us L_{a} over $\mathsf{HLC}^{\flat}$.

We take $\varphi := \Box \bot$ and $\psi := \bot$ in W. This yields $\mathsf{W}^{\flat} \vdash (\Box \bot \mathrel{-\!\!3} \bot) \to \Box \bot$. By Lemma 2.6, $\mathsf{L}_{\mathsf{a}}^{\flat} \nvdash (\Box \bot \mathrel{-\!\!3} \bot) \to \Box \bot$. ⊣

2.7 Löb-Lewis Logic Meets de Jongh and Sambin

We proceed with the study of explicit fixed points in the style of de Jongh and Sambin. We define JS as follows.

JS $(\psi \dashv\!\!3\, \chi)\top \leftrightarrow (\psi \dashv\!\!3\, \chi)(\psi \dashv\!\!3\, \chi)\top$.
Written differently:
$(\psi\top \dashv\!\!3\, \chi\top) \leftrightarrow (\psi \dashv\!\!3\, \chi)(\psi\top \dashv\!\!3\, \chi\top)$.

The principle JS is *prima facie* simpler than JV, however, the impression is somewhat misleading. As we will see, over $\mathsf{HLC}^\flat$, neither principle implies the other.

We begin with a natural preliminary result:

Theorem 2.12 $\mathsf{JS}^\flat$ *is a logic.*

We remind the reader that the results of applying $\oplus$ are not automatically closed under substitution. As before, we use $\chi\rho$ for substitution w.r.t. a designated substitution variable r. We use $\theta[q : \eta]$ for q not equal to r.

Proof We only have to check closure under substitution for JS. Consider formulas $\psi \dashv\!\!3\, \chi$ and ν and a variable q. We fix a substitution variable r different from q. By possibly renaming r in $\psi \dashv\!\!3\, \chi$, we can arrange that it does not occur in ν. We have:

$$((\psi \dashv\!\!3\, \chi)\top \leftrightarrow (\psi \dashv\!\!3\, \chi)(\psi \dashv\!\!3\, \chi)\top)[q : \nu] = \\ (\psi[q : \nu] \dashv\!\!3\, \chi[q : \nu])\top \leftrightarrow \\ ((\psi[q : \nu] \dashv\!\!3\, \chi[q : \nu])(\psi[q : \nu] \dashv\!\!3\, \chi[q : \nu])\top).$$

Thus, the result of substitution is again an instance of JS. ⊣

Theorem 2.13 $\mathsf{JS}^\flat$ *is extension stable.*

Proof Let q be distinct from the variables in ψ and χ and from the substitution variable p.
We have:

$$((\psi \dashv\!\!3\, \chi)\top \leftrightarrow (\psi \dashv\!\!3\, \chi)(\psi \dashv\!\!3\, \chi)\top)^{\ulcorner q \urcorner} = \\ (((q \to \psi^{\ulcorner q \urcorner}) \dashv\!\!3\, (q \to \chi^{\ulcorner q \urcorner}))\top \leftrightarrow ((q \to \psi^{\ulcorner q \urcorner}) \dashv\!\!3 \\ (q \to \chi^{\ulcorner q \urcorner}))((q \to \psi^{\ulcorner q \urcorner}) \dashv\!\!3\, (q \to \chi^{\ulcorner q \urcorner}))\top).$$

The resulting formula is again an instance of JS. ⊣

The proof of the following theorem is essentially due to de Jongh and Visser (1991). In the intuitionistic setting, it has been also noted by Iemhoff et al. (2005a).

Theorem 2.14 $\mathsf{W}^\flat \vdash \mathsf{JS}$.

Proof Since $\mathsf{W}^\flat$ extends Löb-Lewis Logic, we have JV in $\mathsf{W}^\flat$. Moreover:

$$\begin{aligned}\mathsf{W}^\flat \vdash (\psi \Box \chi \top \mathrel{-\!\!3} \chi \top) &\leftrightarrow ((\Box \chi \top \to \psi \Box \chi \top) \mathrel{-\!\!3} \chi \top)\\ &\leftrightarrow ((\Box \chi \top \to \psi \top) \mathrel{-\!\!3} \chi \top)\\ &\leftrightarrow (\psi \top \mathrel{-\!\!3} \chi \top)\end{aligned}$$

Here the first and third equivalences are instances of W and the second equivalence is an instance of the substitution principle $\mathsf{Su1}$, noting that $\Box \chi \top$ is equivalent to $\boxdot(\Box \chi \top \leftrightarrow \top)$. ⊣

Now consider the principle:

P $(\varphi \mathrel{-\!\!3} \psi) \to \Box(\varphi \mathrel{-\!\!3} \psi)$.

We prove JS in $\mathsf{L}_\Box^\flat \oplus \mathsf{P}$. The development here is essentially the one from Smoryński (1985, Chap. 4).

Theorem 2.15 $\mathsf{L}_\Box^\flat \oplus \mathsf{P} \vdash \mathsf{JS}$.

Proof We reason in $\mathsf{L}_\Box^\flat \oplus \mathsf{P}$.

Suppose $(\psi \mathrel{-\!\!3} \chi)\top$. Then, $\boxdot((\psi \mathrel{-\!\!3} \chi)\top \leftrightarrow \top)$. Hence, $(\psi \mathrel{-\!\!3} \chi)(\psi \mathrel{-\!\!3} \chi)\top$.

Conversely, suppose $(\psi \mathrel{-\!\!3} \chi)(\psi \mathrel{-\!\!3} \chi)\top$ and $\Box(\psi \mathrel{-\!\!3} \chi)\top$. From the latter we can derive

$$\Box((\psi \mathrel{-\!\!3} \chi)\top \leftrightarrow \top).$$

Since in $(\psi \mathrel{-\!\!3} \chi)(\psi \mathrel{-\!\!3} \chi)\top$ the formula $(\psi \mathrel{-\!\!3} \chi)\top$ occurs only modalized, we find: $(\psi \mathrel{-\!\!3} \chi)\top$. By Löb's Rule, we may drop the assumption $\Box(\psi \mathrel{-\!\!3} \chi)\top$. ⊣

We end this section by showing some incomparability results.

Lemma 2.7 $\mathsf{L}_\Box^\sharp \oplus \mathsf{P} \nvdash 4_\mathsf{a}$. *Consequently, Löb-Lewis Logic is not a subtheory of* $\mathsf{L}_\Box^\sharp \oplus \mathsf{P}$.

Proof If, over the minimal intuitionistic provability logic $\mathsf{iGL}_\Box$, we define $\varphi \mathrel{-\!\!3} \psi$ as $\Box(\varphi \to \psi)$, this validates an extension of $\mathsf{L}_\Box^\sharp \oplus \mathsf{P}$. However, 4_a would translate to the principle $\Box(\varphi \to \Box \varphi)$, which is obviously not in $\mathsf{L}_\Box^\sharp \oplus \mathsf{P}$, or even in the classical version theoreof.

Alternatively, we can consider the Kripke model on three different points a, b, c. We take $\sqsubset := \{\langle a, b\rangle, \langle a, c\rangle, \langle b, c\rangle\}$ and $\preceq = \{\langle a, a\rangle, \langle b, b\rangle, \langle c, c\rangle\}$ (that is, a classical model). Let $b \Vdash p$ and $c \nVdash p$. Then, the model satisfies $\mathsf{L}_\Box^\sharp \oplus \mathsf{P}$ but $a \nVdash p \mathrel{-\!\!3} \Box p$. ⊣

In fact, the non-containment goes both ways, i.e., extends to incomparability.

Theorem 2.16 $\mathsf{L}^{\flat}_{\Box} \oplus \mathsf{P}$ *is not contained in* $\mathsf{W}^{\sharp}$ *(and a fortiori not in* $\mathsf{L}^{\sharp}_{\mathsf{a}}$*).*

Proof One possible argument is via interpretability logic, however this only works for non-containment in $\mathsf{W}^{\flat}$.[13] Thus, consider the following Kripke frame:

$$a \sqsubset b \prec c \prec d, \quad a \sqsubset c, \qquad b \sqsubset d, \quad b \prec d.$$

We set that $d \Vdash p$ and p does not hold anywhere else. This model is easily seen to be supergathering, but a does not validate $(p \dashv\!\!3 \bot) \to \Box(p \dashv\!\!3 \bot)$. ⊣

Let us note here the following equivalence between combinations of principles in question.

Theorem 2.17 $\mathsf{L}^{\flat}_{\Box} \oplus \mathsf{P} \oplus 4_{\mathsf{a}} = \mathsf{W}^{\flat} \oplus \mathsf{P}$.

In particular, the right-to-left inclusion, noted already as Litak and Visser (2018, Lemma 4.14c), is the constructive version of the fact that ILP proves W. The classical case was first proved in Visser (1990). We are not sure who first noted the result in the constructive case.

Proof We first show that $\mathsf{L}^{\flat}_{\Box} \oplus \mathsf{P} \oplus 4_{\mathsf{a}} \vdash \mathsf{W}$. By Lemma 2.5[14], it is enough to derive W':

$$\begin{aligned}
\varphi \dashv\!\!3\, \psi &\to_{\mathsf{P}} && \Box(\varphi \dashv\!\!3\, \psi) \\
&\to_{\mathsf{HLC}^{\flat}} && \Box(\Box\varphi \to \Box\psi) \\
&\to_{\mathsf{HLC}^{\flat}} && \Box((\Box\psi \to \varphi) \to (\Box\varphi \to \varphi)) \\
&\to_{\mathsf{HLC}^{\flat}} && (\Box\psi \to \varphi) \dashv\!\!3\, (\Box\varphi \to \varphi) \\
&\to_{\mathsf{L}^{\flat}_{\Box}\oplus 4_{\mathsf{a}}} && (\Box\psi \to \varphi) \dashv\!\!3\, \varphi.
\end{aligned}$$

For the other direction it is sufficient to show that $\mathsf{W}^{\flat}$ proves $\mathsf{L}_{\Box}$ and 4_{a}. We reason in $\mathsf{W}^{\flat}$. We first prove L_{a}. Since, $((\Box\varphi \to \varphi) \wedge \Box\varphi) \dashv\!\!3\, \varphi$, we find $(\Box\varphi \to \varphi) \dashv\!\!3\, \varphi$.

[13] Suppose we extend $\mathsf{W}^{\flat}$ and $\mathsf{L}^{\flat}_{\Box} \oplus \mathsf{P}$ with classical logic, and we add 4_{a}, and we switch to the contraposed reading $\rhd$. We thus obtain, respectively, the interpretability logics ILW and ILP. It is well known that ILP strictly extends ILW. So it follows that $\mathsf{L}^{\flat}_{\Box} \oplus \mathsf{P}$ is not a subtheory of $\mathsf{W}^{\flat}$. Hence, a fortiori, it is not a subtheory of Löb-Lewis Logic.

[14] Here is an alternative deduction directly to W:

$$\begin{aligned}
(\varphi \wedge \Box\psi) \dashv\!\!3\, \psi &\to \Box((\varphi \wedge \Box\psi) \dashv\!\!3\, \psi) \wedge ((\varphi \wedge \Box\psi) \dashv\!\!3\, \psi) \\
&\to \Box(\Box(\varphi \wedge \Box\psi) \to \Box\psi) \wedge ((\varphi \wedge \Box\psi) \dashv\!\!3\, \psi) \\
&\to \Box((\Box\varphi \wedge \Box\Box\psi) \to \Box\psi) \wedge ((\varphi \wedge \Box\psi) \dashv\!\!3\, \psi) \\
&\to \Box(\Box\varphi \to \Box\psi) \wedge ((\varphi \wedge \Box\psi) \dashv\!\!3\, \psi) \\
&\to (\varphi \wedge \Box\varphi) \dashv\!\!3\, \psi \\
&\to \varphi \dashv\!\!3\, \psi
\end{aligned}$$

From $\mathsf{L_a}$ we immediately have $\mathsf{L}_\Box$. We prove 4_a. We have:

$$\varphi \strictif (\Box(\varphi \wedge \Box\,\varphi) \rightarrow (\varphi \wedge \Box\,\varphi)).$$

So, by $\mathsf{L_a}$, we find $\varphi \strictif (\varphi \wedge \Box\,\varphi)$ and, hence, $\varphi \strictif \Box\,\varphi$. ⊣

There are more such equivalences, some of them involving novel principles not discussed in this paper. In order not to distract the reader, another example of this kind has been delegated to Appendix 4.

Since $\mathsf{JS}^\flat$ is contained in the incomparable theories $\mathsf{W}^\flat$ and $\mathsf{L}_\Box^\flat \oplus \mathsf{P}$, it is strictly contained in both.

Theorem 2.18 *The theory* $\mathsf{JS}^\flat$ *is strictly contained in* $\mathsf{W}^\flat$ *and* $\mathsf{L}_\Box^\flat \oplus \mathsf{P}$.

Theorem 2.19 *Löb-Lewis Logic is incomparable to* $\mathsf{JS}^\flat$.

Proof Löb-Lewis Logic proves 4_a, but, by Lemma 2.7, $\mathsf{L}_\Box^\flat \oplus \mathsf{P}$ does not. So, a fortiori, $\mathsf{JS}^\flat$ does not prove 4_a.

By considering the JS-fixed point of $p \strictif \bot$, we see that $\mathsf{JS}^\flat$ proves

$$(\Box\bot \strictif \bot) \rightarrow \Box\bot.$$

However, by Lemma 2.6, Löb-Lewis Logic does not. ⊣

Question 2.4 Can we give an interesting characterization of $\mathsf{JS}^\flat$? E.g. can we derive it from finitely many schematic fixed point equations? Or can we refute that this is possible? In the rest of this paper we will make the question a bit more specific.

2.8 The Join of JV and JS

What happens if we add an axiom that says that the JV-fixed point of $\psi \strictif \chi$ is equal to the JS-fixed point?

Consider the scheme:

X $(\psi \Box \chi\top \strictif \chi\top) \leftrightarrow (\psi \strictif \chi)\top$.

Theorem 2.20 $\mathsf{X}^\flat \vdash \mathsf{W}$.

Proof Consider the formula $(p \rightarrow \varphi) \strictif \psi$, where p does not occur in φ and ψ. The JV-fixed point of $(p \rightarrow \varphi) \strictif \psi$ is $(\Box\,\psi \rightarrow \varphi) \strictif \psi$ and the JS-fixed point is modulo $\mathsf{HLC}^\flat$-provable equivalence $\varphi \strictif \psi$. ⊣

Corollary 2.2 *The following theories are equal:* $\mathsf{JV}^\flat \oplus \mathsf{JS}$, $\mathsf{X}^\flat$ *and* $\mathsf{W}^\flat$.

Proof Since $\mathsf{JV}^\flat \oplus \mathsf{JS}$ extends $\mathsf{L}_\Box^\flat$, we have uniqueness of fixed points (Lemma 2.3) and hence X. Moreover, $\mathsf{X}^\flat \vdash \mathsf{W}$. Finally, we have already seen that $\mathsf{W}^\flat$ proves JS and JV. ⊣

Corollary 2.3 $\mathsf{L}_\Box^\flat \oplus \mathsf{P}$ *does not prove* JV *and Löb-Lewis Logic does not prove* JS.

Question 2.5 Can we find a logic below Löb-Lewis Logic and $\mathsf{L}_\Box^\flat \oplus \mathsf{P}$ that extends $\mathsf{iGL}_\Box$ with explicit fixed points? We note that these fixed points would collapse into the JV-fixed points on extension to Löb-Lewis Logic and into the JS-fixed points upon extension to $\mathsf{L}_\Box^\flat \oplus \mathsf{P}$ or $\mathsf{W}^\flat$.

2.9 Four Salient Fixed Points: Subtheories of $\mathsf{JS}^\flat$

There are four approaches to finding a nice axiomatization of $\mathsf{JS}^\flat$: (i) a brilliant flash of insight, (ii) trying out some salient fixed points and see whether they already give us the desired scheme, (iii) analyzing what is shared between the $\mathsf{W}^\flat$-proof of JS and the $\mathsf{L}_\Box^\flat \oplus \mathsf{P}$-proof and (iv) investigating what JS means in Kripke models.

In this section we contribute to approach (ii), by computing four salient JS-fixed points.

2.9.1 *The Principle* W^*

The JS-fixed point of $(p \to \varphi) \dashv\!\!\! 3\, \psi$ is modulo $\mathsf{HLC}^\flat$-provable equivalence $\varphi \dashv\!\!\! 3\, \psi$. The fixed point equation is $(\varphi \dashv\!\!\! 3\, \psi) \leftrightarrow (((\varphi \dashv\!\!\! 3\, \psi) \to \varphi) \dashv\!\!\! 3\, \psi)$. Modulo $\mathsf{HLC}^\flat$-provable equivalence this equation simplifies to the principle W^*:

W^* $(\varphi \dashv\!\!\! 3\, \psi) \to (((\varphi \dashv\!\!\! 3\, \psi) \to \varphi) \dashv\!\!\! 3\, \psi)$

Here is a first insight about W^*.

Theorem 2.21 *Over* $\mathsf{CPC}^\sharp$, *the principle* W^* *axiomatizes the same logic as* $4_\Box$.

Proof Over $\mathsf{CPC}^\sharp$, $\theta \dashv\!\!\! 3\, \rho$ collapses to $\Box(\theta \to \rho)$. See Litak and Visser (2018, Lemma 4.6). This tells us that, over $\mathsf{CPC}^\sharp$, the principle W^* reduces to:

$$(\dagger) \quad \Box(\varphi \to \psi) \to \Box((\Box(\varphi \to \psi) \to \varphi) \to \psi).$$

Putting $\varphi := \neg\chi$ and $\psi := \bot$, $(\dagger)$ gives us classically $4_\Box$, i.e., $\Box\chi \to \Box\Box\chi$. Conversely, we easily get $(\dagger)$ from $4_\Box$. ⊣

Corollary 2.4 $\mathsf{W}^{*\flat}$ *does not contain* $\mathsf{L}_\Box$.

JV	$(\psi\Box\chi\top \mathrel{-\!\!3} \chi\top) \leftrightarrow (\psi \mathrel{-\!\!3} \chi)(\psi\Box\chi\top \mathrel{-\!\!3} \chi\top)$	the de Jongh-Visser scheme of explicit fixed points
$\mathsf{L_a}$	$(\Box\varphi \to \varphi) \mathrel{-\!\!3} \varphi$	its direct axiomatization
JS	$(\psi\top \mathrel{-\!\!3} \chi\top) \leftrightarrow (\psi \mathrel{-\!\!3} \chi)(\psi\top \mathrel{-\!\!3} \chi\top)$	the de Jongh-Sambin scheme of explicit fixed points
P	$(\varphi \mathrel{-\!\!3} \psi) \to \Box(\varphi \mathrel{-\!\!3} \psi)$	one of principles ensuring JS holds over $\mathsf{L}_\Box$
$\mathsf{W}_\circ$	$((\varphi \wedge (\varphi \mathrel{-\!\!3} \psi)) \mathrel{-\!\!3} \psi) \to (\varphi \mathrel{-\!\!3} \psi)$	the most powerful consequence of JS we know
W^*	$(\varphi \mathrel{-\!\!3} \psi) \to (((\varphi \mathrel{-\!\!3} \psi) \to \varphi) \mathrel{-\!\!3} \psi)$	an alternative axiomatization for $\mathsf{W}_\circ$ over $\mathsf{L}_\Box$
X	$(\psi\Box\chi\top \mathrel{-\!\!3} \chi\top) \leftrightarrow (\psi \mathrel{-\!\!3} \chi)\top$	the scheme collapsing JV and JS
W	$(\varphi \mathrel{-\!\!3} \psi) \to ((\Box\psi \to \varphi) \mathrel{-\!\!3} \psi)$	its direct axiomatization (here as W′, cf. Lemma 5)
$\mathsf{L}_\circ$	$(\varphi \mathrel{-\!\!3} ((\varphi \mathrel{-\!\!3} \psi) \to \psi)) \mathrel{-\!\!3} (\varphi \mathrel{-\!\!3} \psi)$	a principle holding under both JS and JV
$4_\circ$	$(\varphi \mathrel{-\!\!3} \psi) \to (\varphi \mathrel{-\!\!3} (\varphi \mathrel{-\!\!3} \psi))$	a principle deriving $\mathsf{L}_\circ$ over $\mathsf{L}_\Box$

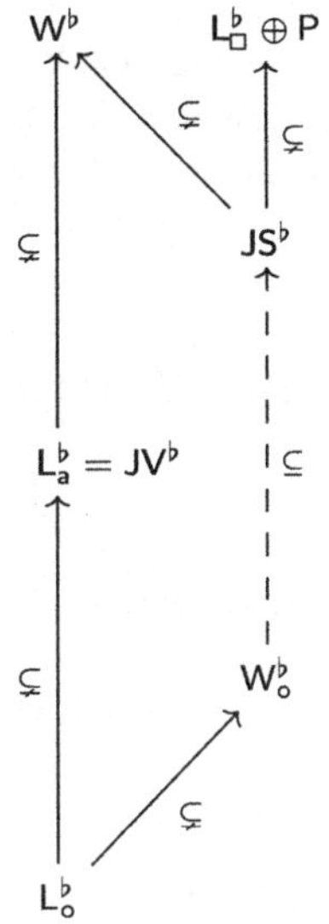

Fig. 2.2 Relationships between logics axiomatized by either fixed point principles or axioms obtained as fixed points of simple formulas

2.9.2 *The Principle* $\mathsf{W}_\circ$

The JS-fixed point of $(p \wedge \varphi) \mathrel{-\!\!3} \psi$ is $\mathsf{HLC}^\flat$-provably equivalent to $\varphi \mathrel{-\!\!3} \psi$. The fixed point equation is $(\varphi \mathrel{-\!\!3} \psi) \leftrightarrow ((\varphi \wedge (\varphi \mathrel{-\!\!3} \psi)) \mathrel{-\!\!3} \psi)$. Modulo $\mathsf{HLC}^\flat$-provable equivalence this equation simplifies to the principle $\mathsf{W}_\circ$:

$$\mathsf{W}_\circ \quad ((\varphi \wedge (\varphi \mathrel{-\!\!3} \psi)) \mathrel{-\!\!3} \psi) \to (\varphi \mathrel{-\!\!3} \psi)$$

We note that, if we put $\varphi := \top$ in $\mathsf{W}_\circ$, we get $(\Box\psi \mathrel{-\!\!3} \psi) \to \Box\psi$. This last, by $\mathsf{HLC}^\flat$-reasoning, implies $\mathsf{L}_\Box$.

Theorem 2.22 *Over* $\mathsf{CPC}^\sharp$, $\mathsf{W}_\circ$ *axiomatizes the same logic as* $\mathsf{L}_\Box$.

Proof By Lemma 4.6 in Litak and Visser (2018), over $\mathsf{CPC}^\sharp$ $\mathsf{W}_\circ$ reduces to:

$$(\dagger) \quad \Box((\varphi \wedge \Box(\varphi \to \psi)) \to \psi) \to \Box(\varphi \to \psi)$$

By putting $\varphi := \top$ we find $\mathsf{L}_\Box$ modulo $\mathsf{HLC}^\sharp$-provable equivalence. Conversely, it is easily seen that $(\dagger)$ is $\mathsf{HLC}^\sharp$-provably equivalent to an instance of $\mathsf{L}_\Box$. ⊣

Theorem 2.23 *We have:* $\mathsf{W}^{*\flat} \subsetneq \mathsf{W}^{*\flat} \oplus \mathsf{L}_{\Box} = \mathsf{W}^{\flat}_{\circ} \subseteq \mathsf{JS}^{\flat}$.

Proof The fact that $\mathsf{W}^{*\flat} \neq \mathsf{W}^{*\flat} \oplus \mathsf{L}_{\Box}$ is Corollary 2.4.

We show that $\mathsf{W}^{*\flat} \oplus \mathsf{L}_{\Box} \vdash \mathsf{W}_{\circ}$. We work in $\mathsf{W}^{*\flat} \oplus \mathsf{L}_{\Box}$. Let $\beta := (\varphi \wedge (\varphi \dashv\!\!3\, \psi))$. We want to show that we have $(\beta \dashv\!\!3\, \psi) \to (\varphi \dashv\!\!3\, \psi)$. We assume:

$$\text{(a)}\quad \Box\,((\beta \dashv\!\!3\, \psi) \to (\varphi \dashv\!\!3\, \psi)).$$

We apply W^{*} with β in the role of φ and ψ in the role of ψ obtaining:

$$\text{(b)}\quad (\beta \dashv\!\!3\, \psi) \to (((\beta \dashv\!\!3\, \psi) \to \beta) \dashv\!\!3\, \psi).$$

By (a), we have:

$$\begin{aligned}((\beta \dashv\!\!3\, \psi) \to \beta) &\;\varepsilon\!\!\dashv\!\!3\; ((\varphi \dashv\!\!3\, \psi) \to (\varphi \wedge (\varphi \dashv\!\!3\, \psi)))\\ &\;\varepsilon\!\!\dashv\!\!3\; ((\varphi \dashv\!\!3\, \psi) \to \varphi)\end{aligned}$$

So, (b) gives us:

$$\text{(c)}\quad (\beta \dashv\!\!3\, \psi) \to (((\varphi \dashv\!\!3\, \psi) \to \varphi) \dashv\!\!3\, \psi).$$

From (c) it is immediate that $(\beta \dashv\!\!3\, \psi) \to (\varphi \dashv\!\!3\, \psi)$. Finally, we apply Löb's Rule and we are done.

We have already seen that $\mathsf{W}^{\flat}_{\circ} \vdash \mathsf{L}_{\Box}$. We show that $\mathsf{W}^{\flat}_{\circ} \vdash \mathsf{W}^{*}$. We work in $\mathsf{W}^{\flat}_{\circ}$. We will use that we have $\mathsf{L}_{\Box}$ in this theory. Let $\alpha := (\varphi \dashv\!\!3\, \psi) \to \varphi$. We want to show that $(\varphi \dashv\!\!3\, \psi) \to (\alpha \dashv\!\!3\, \psi)$. We assume:

$$\text{(a)}\quad \Box\,((\varphi \dashv\!\!3\, \psi) \to (\alpha \dashv\!\!3\, \psi)).$$

We apply $\mathsf{W}_{\circ}$ with α in the role of φ and ψ in the role of ψ obtaining:

$$\text{(b)}\quad ((\alpha \wedge (\alpha \dashv\!\!3\, \psi)) \dashv\!\!3\, \psi) \to (\alpha \dashv\!\!3\, \psi).$$

By (a), we have:

$$\begin{aligned}(\alpha \wedge (\alpha \dashv\!\!3\, \psi)) &\;\varepsilon\!\!\dashv\!\!3\; (((\varphi \dashv\!\!3\, \psi) \to \varphi) \wedge (\varphi \dashv\!\!3\, \psi))\\ &\;\varepsilon\!\!\dashv\!\!3\; (\varphi \wedge (\varphi \dashv\!\!3\, \psi)).\end{aligned}$$

So, (b) gives us:

$$\text{(c)}\quad ((\varphi \wedge (\varphi \dashv\!\!3\, \psi)) \dashv\!\!3\, \psi) \to (\alpha \dashv\!\!3\, \psi).$$

From (c), it is immediate that $(\varphi \dashv\!\!3\, \psi) \to (\alpha \dashv\!\!3\, \psi)$. Finally, we apply Löb's Rule and we are done.

Finally, we have already shown the inclusion of $\mathsf{W}^{\flat}_{\circ}$ in $\mathsf{JS}^{\flat}$. $\dashv$

Here is a further insight.

Theorem 2.24 $\mathsf{W}^{\flat}_{\circ}$ *is incomparable to Löb-Lewis Logic.*

Proof Since $\mathsf{W}^{\flat}_{\circ}$ is contained in $\mathsf{L}^{\flat}_{\Box} \oplus \mathsf{P}$, it cannot contain Löb-Lewis Logic. Conversely, putting $\varphi := \top$ and $\psi := \bot$, we see that

$$\mathsf{W}^{\flat}_{\circ} \vdash (\Box\bot \dashv\!\!\!3\ \bot) \to \Box\bot.$$

But, by Lemma 2.6, Löb-Lewis Logic does not prove this. ⊣

2.9.3 *The Principle* $\mathsf{L}_{\circ}$

Suppose p does not occur in φ and ψ. The JS-fixed point of $\varphi \dashv\!\!\!3\ (p \to \psi)$ is $\varphi \dashv\!\!\!3\ \psi$, modulo $\mathsf{HLC}^{\flat}$-provable equivalence. The fixed point equation is: $(\varphi \dashv\!\!\!3\ \psi) \leftrightarrow (\varphi \dashv\!\!\!3\ ((\varphi \dashv\!\!\!3\ \psi) \to \psi))$. Modulo $\mathsf{HLC}^{\flat}$-provable equivalence this reduces to:

$\mathsf{L}_{\circ}$ $(\varphi \dashv\!\!\!3\ ((\varphi \dashv\!\!\!3\ \psi) \to \psi)) \to (\varphi \dashv\!\!\!3\ \psi)$.

We note that putting $\varphi := \top$ in $\mathsf{L}_{\circ}$ gives us precisely $\mathsf{L}_{\Box}$.

Theorem 2.25 $\mathsf{W}^{\flat}_{\circ} \vdash \mathsf{L}_{\circ}$. *In other words,* $\mathsf{W}^{\flat}_{\circ}$ *extends* $\mathsf{L}^{\flat}_{\circ}$.

Proof In $\mathsf{W}^{\flat}_{\circ}$ we have:

$$\begin{aligned}(\varphi \dashv\!\!\!3\ ((\varphi \dashv\!\!\!3\ \psi) \to \psi)) &\to ((\varphi \wedge (\varphi \dashv\!\!\!3\ \psi)) \dashv\!\!\!3\ \psi) \\ &\to \varphi \dashv\!\!\!3\ \psi\end{aligned}$$

⊣

2.9.4 *The Principle* $\mathsf{4}_{\circ}$

The JS-fixed point of $\varphi \dashv\!\!\!3\ (p \wedge \psi)$, where p does not occur in φ and ψ is, modulo $\mathsf{HLC}^{\flat}$-provable equivalence, $\varphi \dashv\!\!\!3\ \psi$. The fixed point equation is

$$(\varphi \dashv\!\!\!3\ \psi) \leftrightarrow (\varphi \dashv\!\!\!3\ ((\varphi \dashv\!\!\!3\ \psi) \wedge \psi)).$$

Over $\mathsf{HLC}^{\flat}$ this simplifies to the following principle:

$4_{\circ}$ $(\varphi \dashv\!\!\!3\ \psi) \to (\varphi \dashv\!\!\!3\ (\varphi \dashv\!\!\!3\ \psi))$

Putting $\varphi := \top$ in $4_{\circ}$ gives us precisely $4_{\Box}$.

Theorem 2.26 $\mathsf{L}^\flat_\circ \vdash 4_\circ$.

Proof We reason in $\mathsf{L}^\flat_\circ$.

$$\begin{aligned}\varphi \dashv \psi &\to \varphi \dashv ((\varphi \dashv (\psi \wedge (\varphi \dashv \psi))) \to (\psi \wedge (\varphi \dashv \psi)))\\ &\to \varphi \dashv (\psi \wedge (\varphi \dashv \psi))\\ &\dashv \varphi \dashv (\varphi \dashv \psi)\end{aligned}$$

⊣

Theorem 2.27 $\mathsf{L}^\flat_\Box \oplus 4_\circ \vdash \mathsf{L}_\circ$ *and hence* $\mathsf{L}^\flat_\Box \oplus 4_\circ$ *coincides with* $\mathsf{L}^\flat_\circ$.

Proof Let $\gamma := ((\varphi \dashv \psi) \to \psi)$. We have over $\mathsf{L}^\flat_\Box \oplus 4_\circ$:

$$\begin{aligned}&(\varphi \dashv \gamma) \wedge \Box((\varphi \dashv \gamma) \to (\varphi \dashv \psi)) \quad \to\\ &\to (\varphi \dashv \gamma) \wedge \Box((\varphi \dashv \gamma) \to (\varphi \dashv \psi)) \wedge (\varphi \dashv (\varphi \dashv \gamma))\\ &\to (\varphi \dashv \gamma) \wedge (\varphi \dashv (\varphi \dashv \psi))\\ &\to (\varphi \dashv \psi)\end{aligned}$$

We find the desired result by applying Löb's Rule. ⊣

We note that we can interpret the unimodal Löb system in $\mathsf{L}^\flat_\circ$ by reading $\varphi \dashv (\cdot)$ for the box.

Theorem 2.28 $\mathsf{L}^\flat_\mathsf{a} \vdash \mathsf{L}_\circ$.

In other words, Löb-Lewis Logic extends $\mathsf{L}^\flat_\circ$.

Proof We reason in Löb-Lewis Logic.

$$\begin{aligned}(\varphi \dashv ((\varphi \dashv \psi) \to \psi)) &\to (\varphi \dashv (\Box\psi \to \psi))\\ &\to (\varphi \dashv \psi)\end{aligned}$$

⊣

Corollary 2.5 $\mathsf{L}^\flat_\circ$ *is strictly contained in* $\mathsf{W}^\flat_\circ$ *and in Löb-Lewis Logic.*

Proof The strictness follows from the fact that $\mathsf{L}^\flat_\circ$ is contained in the incomparable theories $\mathsf{W}^\flat_\circ$ and Löb-Lewis Logic. ⊣

Figure 2.2 summarizes the relationships between the theories studied. Theories that are incomparable in the picture are really incomparable.

Question 2.6 Is $\mathsf{W}^\flat_\circ$ equal to $\mathsf{JS}^\flat$ and do systems such as $\mathsf{W}^\flat_\circ$ or, perhaps, even $\mathsf{L}^\flat_\circ$ have explicit fixed points at all?

2.10 Correspondence and Non-conservativity

We can now extend the list of known correspondences given in Theorem 2.2 and Fig. 2.1 to several additional principles, whose importance has been highlighted by our study.

Theorem 2.29 *In Kripke semantics,* P *corresponds to the transitivity of* $\sqsubset$.

Proof Clearly, transitivity implies P. Conversely, consider a frame $\mathcal{F}$ with nodes $a \sqsubset b \sqsubset c$ and $a \not\sqsubset c$. We define $x \Vdash p$ iff $c \preceq x$, and $x \Vdash q$ iff $x \not\preceq c$. Clearly, $c \Vdash p$ and $c \nVdash q$, so $a \nVdash \Box(p \prec\!\!\!\!\!\!\;\!\;\!\!\:\!\dashv q)$. On the other hand, consider any z with $a \sqsubset z \Vdash p$. It follows that $c \preceq z$. By our assumption $z \neq c$, so $z \not\preceq c$. Ergo $z \Vdash q$. Thus, $a \Vdash p \dashv q$. $\dashv$

Let us remind the reader that P plays a central rôle in the fmp proofs obtained by de Groot et al. (2021, Sect. 5).

Theorem 2.30 *In Kripke semantics,* $4_\circ$ *corresponds to gather-transitivity: if* $x \sqsubset y \sqsubset z$, *then* $x \sqsubset z$ *or* $y \preceq z$.

Proof Suppose $\mathcal{F}$ is gather-transitive. Consider any model on $\mathcal{F}$. Suppose we have $a \Vdash \varphi \dashv \psi$. Suppose we have $a \sqsubset b \sqsubset c$ and $b \Vdash \varphi$ and $c \Vdash \varphi$. In case $a \sqsubset c$, we get $c \Vdash \psi$ and in case $b \preceq c$ we also get $c \Vdash \psi$. Done.

Conversely, consider a frame $\mathcal{F}$ with $a \sqsubset b \sqsubset c$ and (i) $a \not\sqsubset c$ and (ii) $b \not\preceq c$. We define $x \Vdash p$ iff $b \preceq x$ or $c \preceq x$, and $x \Vdash q$ iff $x \not\preceq c$. Clearly, $b \Vdash p$ and $c \Vdash p$ and $c \nVdash q$, so $a \nVdash \Box(p \dashv q)$. On the other hand, consider any z with $a \sqsubset z \Vdash p$. It follows that $b \preceq z$ or $c \preceq z$. In case we would have $z \preceq c$, it follows that either $b \preceq c$ or $z = c$. The first possibility is excluded by (ii). The second possibility is excluded since it would follow that $a \sqsubset c$ contradicting (i). Ergo, $z \not\preceq c$ and, so, $z \Vdash q$. Thus, $a \Vdash p \dashv q$. $\dashv$

In the presence of Di, we also have:

$44_\circ$ $(\varphi \dashv (\psi \dashv \chi)) \to (\varphi \dashv (\psi \dashv (\varphi \dashv (\psi \dashv \chi))))$.

Theorem 2.31 $4^\sharp_\circ \vdash 44_\circ$.

Proof We reason in $4^\sharp_\circ$. First, $4_\circ$ gives us that

$$\varphi \dashv (\psi \dashv \chi) \;\to\; \varphi \dashv (\varphi \dashv (\psi \dashv \chi)).$$

On the other hand, it also yields that

$$\varphi \dashv (\psi \dashv \chi) \;\to\; \varphi \dashv (\psi \dashv (\psi \dashv \chi)).$$

We use Di to derive:

$$\varphi \strictif (\psi \strictif \chi) \;\to\; \varphi \strictif ((\varphi \vee \psi) \strictif (\psi \strictif \chi)).$$

... and, now, we use $4_\circ$ again:

$$\begin{aligned}\varphi \strictif (\psi \strictif \chi) \;&\to\; \varphi \strictif ((\varphi \vee \psi) \strictif (\psi \strictif \chi))\\ &\to \varphi \strictif ((\varphi \vee \psi) \strictif ((\varphi \vee \psi) \strictif (\psi \strictif \chi)))\\ &\to \varphi \strictif (\psi \strictif (\varphi \strictif (\psi \strictif \chi))).\end{aligned}$$

⊣

Nevertheless, the above derivation crucially uses Di. It is a cautionary tale that the Kripke correspondence conditions lose some of their importance for extensions of $\mathsf{HLC}^\flat$ which do not extend $\mathsf{HLC}^\sharp$.

Theorem 2.32 $4^\flat_\circ \nvdash 44_\circ$.

Proof We use algebraic semantics for this purpose. A smallest possible, 6-element countermodel was found with the help of the program Mace4 (McCune, 2003). Its Heyting reduct looks as follows:

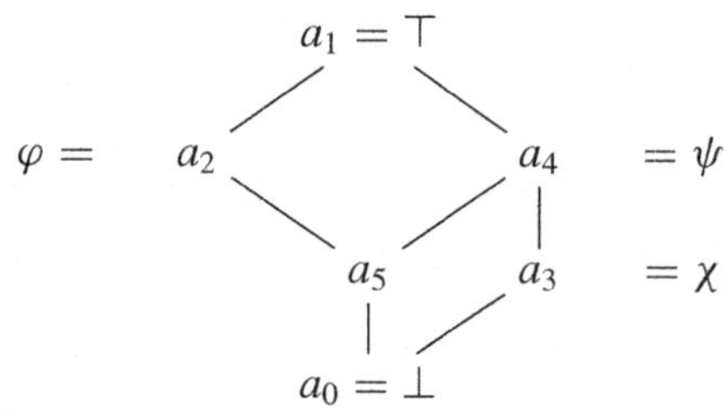

Some key fact about the definition of $\strictif$ are:

- For every $a \neq \top$, $\top \strictif a = \bot$;
- $a_3 \strictif a_2 = \bot$;
- $a_5 \strictif \bot = a_5 \strictif a_3 = a_4 \strictif a_3 = a_4$;
- $a_2 \strictif \bot = a_2 \strictif a_3 = a_5$;
- $a_2 \strictif a_4 = a_2 \strictif a_5 = a_2$.

Note that via $a_4 \strictif a_3$ and $a_2 \strictif a_3$ one can disprove Di. The full truth table looks as follows:

$\strictif$	$\bot$	$\top$	a_2	a_3	a_4	a_5
$\bot$	$\top$	$\top$	$\top$	$\top$	$\top$	$\top$
$\top$	$\bot$	$\top$	$\bot$	$\bot$	$\bot$	$\bot$
a_2	a_5	$\top$	$\top$	a_5	a_2	a_2
a_3	$\bot$	$\top$	$\bot$	$\top$	$\top$	$\bot$
a_4	$\bot$	$\top$	$\bot$	a_4	$\top$	$\bot$
a_5	a_4	$\top$	$\top$	a_4	$\top$	$\top$

⊣

2.11 Explicit Fixed Points and the Beth Property

It has been shown first by Maksimova (1989) that explicit fixed points yield the Beth definability property and subsequent literature provides further investigation of this phenomenon (Areces et al., 1998; Hoogland, 2001). For constructive $\dashv$-logics, this argument has already been made by Iemhoff et al. (2005a, Sect. 4.3). Hence, we simply state as an observation that any logic which extends either $\mathsf{JS}^\flat$ or i-$\mathsf{GL_a} = \mathsf{JV}^\flat$ enjoys this property.

2.12 Conclusions and Future Work

In this paper, we have studied the surprisingly complex first stage of the construction of guarded fixed points, that is the construction for formulas whose principal connective is $\dashv$ itself. Even its restricted scope has left some questions unanswered.

In follow-up work, we will investigate the question of how to extend the definability result from a restricted class of formulas to a wider one, including the computational cost of such extensions; the syntactic setup necessary for that purpose (generalizing that of de Jongh and Visser, 1991; the issue whether the resulting fixed point-operators commute with substitutions has intriguing Beck-Chevalley aspects) would distract too much from the main goal of the present paper. In particular, we have mentioned that classically, it is possible to use elimination of guarded fixed points for elimination of positive fixed points of ordinary μ-calculus (van Benthem, 2006; Visser, 2005). Furthermore, one can define a portmanteau notion of *semipositive* formulas (where every occurrence of p is either guarded or positive), which in general allow explicit *locally minimal* fixed points (Visser, 2005). It is possible to generalize such results to the intuitionistic setting, but the techniques and results involved are more demanding.

In the present paper, we have provided a few footnotes to Dick's work on fixed points. We have seen that, at each stage of the development of the fixed point-calculation, Dick was actively and decisively involved: in the heroic past of the original calculation, in the second stage of interpretability logic, and in the third stage of generalization to the constructive case. We hope to have offered the reader a glimpse of the rich landscape of research questions that opens up when one follows the paths on which Dick led the way.

Appendix 1: Proof of Theorem 2.3

Proof We choose p distinct from $\mathsf{var}(\varphi) \cup \mathsf{var}(\psi) \cup \mathsf{var}(\chi)$.

We note that:

$$(((\varphi \dashv \psi) \wedge (\psi \dashv \chi)) \to (\varphi \dashv \chi))^{\ulcorner p \urcorner} =$$
$$((((p \to \varphi^{\ulcorner p \urcorner}) \dashv (p \to \psi^{\ulcorner p \urcorner})) \wedge$$
$$((p \to \psi^{\ulcorner p \urcorner}) \dashv (p \to \chi^{\ulcorner p \urcorner}))) \to$$
$$((p \to \varphi^{\ulcorner p \urcorner}) \dashv (p \to \chi^{\ulcorner p \urcorner}))).$$

So the translation of an instance of Tr is itself an instance of Tr. It follows that

$$\Lambda \vdash p \to (((\varphi \dashv \psi) \wedge (\psi \dashv \chi)) \to (\varphi \dashv \chi))^{\ulcorner p \urcorner}.$$

We note that:

$$\begin{aligned}(((\varphi \dashv \psi) \wedge (\varphi \dashv \chi)) \to (\varphi \dashv (\psi \wedge \chi)))^{\ulcorner p \urcorner} = \\ ((((p \to \varphi^{\ulcorner p \urcorner}) \dashv (p \to \psi^{\ulcorner p \urcorner})) \wedge \\ ((p \to \varphi^{\ulcorner p \urcorner}) \dashv (p \to \chi^{\ulcorner p \urcorner}))) \to \\ ((p \to \varphi^{\ulcorner p \urcorner}) \dashv (p \to (\psi \wedge \chi)^{\ulcorner p \urcorner}))).\end{aligned}$$

Over $\mathsf{HLC}^\flat$, $(p \to (\psi \wedge \chi)^{\ulcorner p \urcorner})$ is equivalent to

$$(p \to \psi^{\ulcorner p \urcorner}) \wedge (p \to \chi^{\ulcorner p \urcorner}).$$

So the translation of an instance of $\mathsf{K_a}$ is, modulo $\mathsf{HLC}^\flat$-provability, itself an instance of $\mathsf{K_a}$. It follows that

$$\Lambda \vdash p \to (((\varphi \dashv \psi) \wedge (\varphi \dashv \chi)) \to (\varphi \dashv (\psi \wedge \chi)))^{\ulcorner p \urcorner}.$$

Suppose φ and $\varphi \to \psi$ are in $\mathfrak{S}\Lambda$. Then, both we have $\Lambda \vdash p \to \varphi^{\ulcorner p \urcorner}$ and $\Lambda \vdash p \to (\varphi \to \psi)^{\ulcorner p \urcorner}$. Since $(\cdot)^{\ulcorner p \urcorner}$ commutes with the propositional connectives, we find $\Lambda \vdash p \to \psi^{\ulcorner p \urcorner}$.

Suppose $(\varphi \to \psi) \in \mathfrak{S}\Lambda$. Then we have $\Lambda \vdash p \to (\varphi \to \psi)^{\ulcorner p \urcorner}$. It follows that

$$\Lambda \vdash (p \to \varphi^{\ulcorner p \urcorner}) \to (p \to \psi^{\ulcorner p \urcorner}).$$

Hence, $\Lambda \vdash (p \to \varphi^{\ulcorner p \urcorner}) \dashv (p \to \psi^{\ulcorner p \urcorner})$. Ergo, $\Lambda \vdash (\varphi \dashv \psi)^{\ulcorner p \urcorner}$, and, *a fortiori*, we find $\Lambda \vdash p \to (\varphi \dashv \psi)^{\ulcorner p \urcorner}$. Hence, $(\varphi \dashv \psi) \in \mathfrak{S}\Lambda$.

Suppose $\varphi \in \mathfrak{S}\Lambda$. We want to show that $\varphi[q := \psi]$ in Λ. We can arrange that p is distinct from q. We now prove by induction on p-free formulas ν that

$$(\nu[q := \psi])^{\ulcorner p \urcorner} = (\nu^{\ulcorner p \urcorner})[q := \psi^{\ulcorner p \urcorner}].$$

Since $\varphi \in \mathfrak{S}\Lambda$, we have $\Lambda \vdash p \to \varphi^{\ulcorner p \urcorner}$. Hence,

$$\Lambda \vdash (p \to \varphi^{\ulcorner p \urcorner})[q := \psi^{\ulcorner p \urcorner}],$$

and so $\Lambda \vdash p \to (\varphi^{\ulcorner p \urcorner}[q := \psi^{\ulcorner p \urcorner}])$. It follows that

$$\Lambda \vdash p \to (\varphi[q := \psi])^{\ulcorner p \urcorner}.$$

Thus, $(\varphi[q := \psi]) \in \mathfrak{S}\Lambda$.

Appendix 2: Proof of Theorem 2.4

Proof (i) is trivial.

We treat (ii). Suppose p does not occur in φ. Suppose $\varphi \in \mathfrak{S}\Lambda$. Then, we have $\Lambda \vdash p \to \varphi^{\ulcorner p \urcorner}$. So, $\Lambda \vdash (p \to \varphi^{\ulcorner p \urcorner})[p := \top]$. Clearly, $(p \to \varphi^{\ulcorner p \urcorner})[p := \top]$ is equivalent to φ over $\mathsf{HLC}^\flat$.

We treat (iii). By (ii) it is sufficient to show that $\mathfrak{S}\mathfrak{S}\Lambda \supseteq \mathfrak{S}\Lambda$. Let p and p' be distinct variables not in φ. Suppose $\varphi \in \mathfrak{S}\Lambda$. Then, $\Lambda \vdash p \to \varphi^{\ulcorner p \urcorner}$. It follows that $\Lambda \vdash (p \to \varphi^{\ulcorner p \urcorner})[p := (p \wedge p')]$. Clearly, over $\mathsf{HLC}^\flat$, $(p \to \varphi^{\ulcorner p \urcorner})[p := (p \wedge p')]$ is equivalent to $p' \to (p \to \varphi^{\ulcorner p \urcorner})^{\ulcorner p' \urcorner}$.

Appendix 3: An Algebraic Reformulation of Extension Stability

In this appendix, we are going to clarify the algebraic meaning of extensions stability. In particular, we are going to see that in the classical unimodal setting, a logic is extension stable iff it is a subframe logic (Fine, 1985; Wolter, 1993). The algebraic construction involved turns out to fit into the following frameworks:

- elementary J-operators of Fourman and Scott (1979) generalized to arbitrary HAEs) as the *open quotient*;
- *modal operators on Heyting algebras* of Macnab (1981) (generalized to arbitrary HAEs). These days, these operators are known as *nuclei* and (Bezhanishvili & Ghilardi, 2007) have shown that in the intuitionistic setting, they allow a neat algebraic definition of subframeness;
- describable operations of Wolter (1993) (generalized from BAOs to HAEs).

Given any L-HAE $\mathcal{A} := \langle A, \wedge, \vee, \prec\!\!\!\!\!-\!\!\!3, \to, \bot, \top \rangle$ and any $a \in A$, define

$$A^{\ulcorner a \urcorner} := \{a \to b | b \in A\}.$$

Furthermore, for any n-ary operation f in the signature of L-HAE and any vector $b_1, \ldots, b_n \in A^{\ulcorner a \urcorner}$, let

$$f^{\ulcorner a \urcorner}(b_1, \ldots, b_n) := a \to f(b_1, \ldots, b_n).$$

Now we can define

$$\mathcal{A}^{\ulcorner a \urcorner} := \langle A^{\ulcorner a \urcorner}, \wedge, \vee^{\ulcorner a \urcorner}, \prec\!\!\!\!\!-\!\!\!3^{\ulcorner a \urcorner}, \to, \bot^{\ulcorner a \urcorner}, \top \rangle.$$

Lemma 2.8 *For arbitrary* $b_1 = (a \to b_1') \in A^{\ulcorner a \urcorner}$ *and* $b_2 = (a \to b_2') \in A^{\ulcorner a \urcorner}$

$$b_1 \wedge^{\ulcorner a \urcorner} b_2 = b_1 \wedge b_2, \qquad b_1 \to^{\ulcorner a \urcorner} b_2 = b_1 \to b_2, \qquad \top^{\ulcorner a \urcorner} = \top.$$

Proof Straightforward. ⊣

Theorem 2.33 *For any* L-HAE $\mathcal{A}$ *and any* $a \in A$, $\mathcal{A}^{\ulcorner a \urcorner}$ *is a* L-HAE.

Proof The Heyting laws have been verified in references such as Fourman and Scott (1979), Macnab (1981). The remaining ones are essentially an algebraic translation of Theorem 2.3. ⊣

Let us now define another translation $(\cdot)^{\ulcorner p \urcorner}$, at first sight complementary to the one from Sect. 2.5. Given a formula φ and a fresh propositional variable p, we define the $\varphi^{\ulcorner p \urcorner}$ inductively as commuting with $\wedge$, $\to$ and $\top$, with other clauses being $f^{\ulcorner p \urcorner}(\varphi_1, \ldots, \varphi_n) := p \to f(\varphi_1^{\ulcorner p \urcorner}, \ldots, \varphi_n^{\ulcorner p \urcorner})$ for remaining n-ary operations f in the signature of L-HAE (that is, $\bot$, $\vee$, $\dashv\!\!\!3$) and $q^{\ulcorner p \urcorner} := p \to q$.

Lemma 2.9 *Let* $a \in A$, p *be an atom fresh for* φ, v *a valuation of atoms in* $\mathcal{A}^{\ulcorner a \urcorner}$, *and* w *a valuation in* $\mathcal{A}$ *s.t.* $w(p) = a$ *and for every atom* q *occurring in* φ, $v(q) = \hat{w}(p \to q)$. *Then* $\hat{v}(\varphi) = \hat{w}(\varphi^{\ulcorner p \urcorner})$.

Proof Using induction and Lemma 2.8. ⊣

For any L-HAE $\mathcal{A} := \langle A, \wedge, \vee, \dashv\!\!\!3, \to, \bot, \top \rangle$, denote $\mathfrak{S}\mathcal{A} := \{\mathcal{A}^{\ulcorner a \urcorner} | a \in A\}$.

Theorem 2.34 *For any* $\mathcal{A}$, φ *and* p *fresh for* φ, *we have that*

$$\mathcal{A} \Vdash \varphi^{\ulcorner p \urcorner} \quad \textit{iff} \quad \mathfrak{S}\mathcal{A} \Vdash \varphi$$

Proof From left to right: assume there is $a \in A$ and a valuation v in $\mathcal{A}^{\ulcorner a \urcorner}$ s.t. $\hat{v}(\varphi) \neq \top$. Define $w(p) := a$ and $w(q) := v(q)$ (with other atoms set arbitrarily, for example to $\bot$). Note that, as $v(q)$ as an element of A is of the form $a \to^{\mathcal{A}} b$ for some b, we have that

$$v(q) = a \to^{\mathcal{A}} b = a \to^{\mathcal{A}} (a \to^{\mathcal{A}} b) = w(p) \to^{\mathcal{A}} w(q) = \hat{w}(p \to q).$$

Using Lemma 2.9, we show that $\varphi^{\ulcorner p \urcorner}$ is refuted in $\mathcal{A}$.

From right to left: assume that a valuation w in $\mathcal{A}$ refuting $\varphi^{\ulcorner p \urcorner}$ exists and consider $\mathcal{A}^{\ulcorner w(p) \urcorner}$. Define $v(q) := \hat{w}(p \to q)$. Lemma 2.9 immediately implies that φ is refuted in $\mathcal{A}^{\ulcorner w(p) \urcorner}$. ⊣

We could develop this into a proof that $\mathfrak{S}$ is a *describable operation* in the sense of Wolter (1993) (suitably generalized from BAOs to HAEs). Observe that there is nothing in this development that specifically relies on axioms for $\dashv\!\!\!3$. However, at present our main goal is to relate the present framework to that of Sect. 2.5.

Theorem 2.35 *For any p fresh for φ, $p \to \varphi^{\lceil p \rceil}$ is $\mathsf{HLC}^\flat$-equivalent to $\varphi^{\llbracket p \rrbracket}$.*

We could dramatically weaken the $\mathsf{HLC}^\flat$ base in the statement of this result: the only relevant property of $\mathbin{-\!\!\!3}$ is its congruentiality wrt logical equivalence.

Proof By induction on φ. The base cases of atoms, $\bot$ and $\top$ are immediate. For conjunction,

$$\begin{aligned}(\varphi \wedge \psi)^{\llbracket p \rrbracket} &\leftrightarrow (\varphi^{\llbracket p \rrbracket} \wedge \psi^{\llbracket p \rrbracket}) \\ &\leftrightarrow ((p \to \varphi^{\lceil p \rceil}) \wedge (p \to \psi^{\lceil p \rceil})) \\ &\leftrightarrow p \to (\varphi^{\lceil p \rceil} \wedge \psi^{\lceil p \rceil}) \\ &\leftrightarrow p \to (\varphi \wedge \psi)^{\lceil p \rceil}.\end{aligned}$$

For implication,

$$\begin{aligned}(\varphi \to \psi)^{\llbracket p \rrbracket} &\leftrightarrow (\varphi^{\llbracket p \rrbracket} \to \psi^{\llbracket p \rrbracket}) \\ &\leftrightarrow ((p \to \varphi^{\lceil p \rceil}) \to (p \to \psi^{\lceil p \rceil})) \\ &\leftrightarrow ((p \to \varphi^{\lceil p \rceil}) \wedge p \to \psi^{\lceil p \rceil}) \\ &\leftrightarrow (\varphi^{\lceil p \rceil} \wedge p \to \psi^{\lceil p \rceil}) \\ &\leftrightarrow p \to (\varphi^{\lceil p \rceil} \to \psi^{\lceil p \rceil}) \\ &\leftrightarrow p \to (\varphi \to \psi)^{\lceil p \rceil}.\end{aligned}$$

For disjunction,

$$\begin{aligned}(\varphi \vee \psi)^{\llbracket p \rrbracket} &\leftrightarrow p \to (\varphi^{\llbracket p \rrbracket} \vee \psi^{\llbracket p \rrbracket}) \\ &\leftrightarrow p \to (p \wedge ((p \to \varphi^{\lceil p \rceil}) \vee (p \to \psi^{\lceil p \rceil}))) \\ &\leftrightarrow p \to ((p \wedge (p \to \varphi^{\lceil p \rceil})) \vee (p \wedge (p \to \psi^{\lceil p \rceil}))) \\ &\leftrightarrow p \to ((p \wedge \varphi^{\lceil p \rceil}) \vee (p \wedge \psi^{\lceil p \rceil})) \\ &\leftrightarrow p \to (p \wedge (\varphi^{\lceil p \rceil} \vee \psi^{\lceil p \rceil})) \\ &\leftrightarrow p \to (\varphi^{\lceil p \rceil} \vee \psi^{\lceil p \rceil}) \\ &\leftrightarrow p \to (\varphi \vee \psi)^{\lceil p \rceil}.\end{aligned}$$

For strict implication,

$$\begin{aligned}(\varphi \mathbin{-\!\!\!3} \psi)^{\llbracket p \rrbracket} &\leftrightarrow p \to (\varphi^{\llbracket p \rrbracket} \mathbin{-\!\!\!3} \psi^{\llbracket p \rrbracket}) \\ &\leftrightarrow p \to ((p \to \varphi^{\lceil p \rceil}) \mathbin{-\!\!\!3} (p \to \psi^{\lceil p \rceil})) \\ &\leftrightarrow p \to (\varphi \mathbin{-\!\!\!3} \psi)^{\lceil p \rceil}.\end{aligned}$$

$\dashv$

We can now express ourselves in terms familiar to anybody who studied negative translations: while $(\cdot)^{\ulcorner p \urcorner}$ is a suitable generalization of the Gödel-Gentzen translation, $p \to (\cdot)^{\ulcorner p \urcorner}$ is the obvious generalization of the Kuroda translation.[15] Furthermore, we can prove that using the same notation for the stabilizer of a logic and for the above-introduced operation on (classes of) algebras was neither a mistake, nor a coincidence.

Theorem 2.36 *Assume that a logic Λ is the (logical counterpart of) the equational theory of a class of algebras K, that is, $\Lambda = \mathsf{Th}(K)$ (in the notation of Sect. 2.3.1). Then $\mathfrak{S}\Lambda = \mathsf{Th}(\mathfrak{S}K)$.*

Proof By definition, $\varphi \in \mathfrak{S}\Lambda$ iff for some fresh p, $\Lambda \vdash p \to \varphi^{\ulcorner p \urcorner}$. By Theorem 2.35, this is equivalent to $\Lambda \vdash \varphi^{\ulcorner p \urcorner}$. By assumption, this is equivalent to $K \Vdash \varphi^{\ulcorner p \urcorner}$. By Theorem 2.34, this is equivalent to $\mathfrak{S}K \Vdash \varphi$. ⊣

Correspondingly, we can call a variety[16] V *extension stable* if $\mathfrak{S}V \subseteq V$.

Our final task is to show that in the case of normal unimodal logics over CPC, corresponding to modal algebras, extension stability is equivalent to being a *subframe logic* (Wolter, 1993). Wolter defines the following self-translation[17] of the unimodal language with $\Box$:

$$\bot \downarrow p := \bot, \quad \top \downarrow p = p, \quad q \downarrow p = p \wedge q, \quad (\psi \wedge \chi) \downarrow p := \psi \downarrow p \wedge \chi \downarrow p$$
$$(\psi \vee \chi) \downarrow p := p \wedge (\psi \downarrow p \vee \chi \downarrow p), \quad (\psi \to \chi) \downarrow p := p \wedge (\psi \downarrow p \to \chi \downarrow p),$$
$$(\Box \psi) \downarrow p := p \wedge \Box(p \to \psi \downarrow p).$$

Consider now $\mathsf{Box}^\flat$, i.e., the minimal extension of $\mathsf{HLC}^\flat$, in which $\prec\!\!3$ is definable in terms of $\Box$; recall that $\mathsf{Box}^\flat = \mathsf{Box}^\sharp$. We can extend the Wolter translation to such systems by treating $\prec\!\!3$ as a defined connective. The reader can easily verify that $(\Box(\psi \to \chi)) \downarrow p$ is equivalent to

$$p \wedge \Box(p \to (\psi \downarrow p \to \chi \downarrow p)).$$

We can prove an analogue of Theorem 2.35.

Theorem 2.37 *For any p fresh for φ,*

$$p \to \varphi^{\ulcorner p \urcorner} \text{ is } \mathsf{Box}^\flat\text{-equivalent to } p \to \varphi \downarrow p.$$

[15] See Litak et al. (2017) for a more detailed discussion of negative translations in the modal context. These translations and results corresponding to Theorem 2.35 can be generalized to other nuclei, but we postpone this discussion to a separate paper.

[16] Recall that a *variety* is a class of algebras of the form $V = \mathsf{Mod}(\Lambda)$ for some logic Λ, or alternatively (over ZFC) an HSP-closed class of algebras.

[17] Strictly speaking, as Wolter is assuming classical connectives throughout, his primitives do not include $\vee$ or $\to$ and $\bot$ is replaced by negation. The reader can easily check our translation is equivalent to his.

Proof Base cases and that of conjunction are left for the reader. For disjunction:

$$\begin{aligned}
p \to (\psi \vee \chi)^{\ulcorner p \urcorner} &\leftrightarrow_{\text{def}} p \to (\psi^{\ulcorner p \urcorner} \vee \chi^{\ulcorner p \urcorner}) \\
&\leftrightarrow_{\mathsf{IPC}} p \to ((p \to \psi^{\ulcorner p \urcorner}) \vee (p \to \chi^{\ulcorner p \urcorner})) \\
&\leftrightarrow_{\text{IH}} p \to ((p \to \psi \downarrow p) \vee (p \to \chi \downarrow p)) \\
&\leftrightarrow_{\mathsf{IPC}} p \to (\psi \downarrow p \vee \chi \downarrow p) \\
&\leftrightarrow_{\mathsf{IPC}} p \to (p \wedge (\psi \downarrow p \vee \chi \downarrow p)) \\
&\leftrightarrow_{\text{def}} p \to (\psi \vee \chi) \downarrow p.
\end{aligned}$$

For implication:

$$\begin{aligned}
p \to (\psi \to \chi)^{\ulcorner p \urcorner} &\leftrightarrow_{\text{def}} p \to (\psi^{\ulcorner p \urcorner} \to \chi^{\ulcorner p \urcorner}) \\
&\leftrightarrow_{\mathsf{IPC}} p \to (p \to (\psi^{\ulcorner p \urcorner} \to \chi^{\ulcorner p \urcorner})) \\
&\leftrightarrow_{\mathsf{IPC}} p \to ((p \to \psi^{\ulcorner p \urcorner}) \to (p \to \chi^{\ulcorner p \urcorner})) \\
&\leftrightarrow_{\text{IH}} p \to ((p \to \psi \downarrow p) \to (p \to \chi \downarrow p)) \\
&\leftrightarrow_{\mathsf{IPC}} p \to (p \to (\psi \downarrow p \to \chi \downarrow p)) \\
&\leftrightarrow_{\mathsf{IPC}} p \to (\psi \downarrow p \to \chi \downarrow p) \\
&\leftrightarrow_{\mathsf{IPC}} p \to (p \wedge (\psi \downarrow p \to \chi \downarrow p)) \\
&\leftrightarrow_{\text{def}} p \to (\psi \to \chi) \downarrow p.
\end{aligned}$$

For the strict implication:

$$\begin{aligned}
p \to (\psi \dashv\!3\, \chi)^{\ulcorner p \urcorner} &\leftrightarrow_{\text{def}} p \to (p \to \psi^{\ulcorner p \urcorner}) \dashv\!3\, (p \to \chi^{\ulcorner p \urcorner}) \\
&\leftrightarrow_{\mathsf{Box}^\flat} p \to \Box((p \to \psi^{\ulcorner p \urcorner}) \to (p \to \chi^{\ulcorner p \urcorner})) \\
&\leftrightarrow_{\mathsf{Box}^\flat} p \to \Box(p \to ((p \to \psi^{\ulcorner p \urcorner}) \to (p \to \chi^{\ulcorner p \urcorner}))) \\
&\leftrightarrow_{\text{IH}} p \to \Box(p \to (\psi \downarrow p \to \chi \downarrow p)) \\
&\leftrightarrow_{\mathsf{Box}^\flat} p \to (\psi \dashv\!3\, \chi) \downarrow p.
\end{aligned}$$

⊣

The proof above highlights the fact that for any $a \in A$, the algebra $\mathcal{A}^{\ulcorner a \urcorner}$ of fixpoints of the open nucleus induced by a is isomorphic to the ideal of elements below a. Theorems 2.35, 2.36, 2.37 in conjuction with Wolter (1993, Theorem 2.2.2) entail that classical normal logics are subframe iff they are extension stable.[18]

---

[18] The reason why Wolter's translation matches ours in the sense specified by Theorem 2.37, even above intutionistic propositional base, is that it employs (syntactically) the *open* perspective rather than the *closed* perspective on classical subframes. Had Wolter used $\vee$ as the principal connective in his translation, we would been forced to limit the statement of Theorem 2.37 to classical normal logics. On the other hand, a subframe logic in the intuitionistic sense has to be closed under ***all*** nuclei (in particular both open and closed ones) (Bezhanishvili & Ghilardi, 2007) and thus one should not directly apply the syntactic definition employed by Wolter anyway. In fact, for superintuitionistic logics without additional operators, their corresponding varieties are always closed under *open* quotients: in proofs above, we have seen that the open nucleus is in fact a Heyting homomorphism. We will discuss this more extensively in a separate paper.

Appendix 4: A Further Principle

We are building here on the thread started by Theorem 2.17. Let wW be the principle $(\varphi \wedge \Box\, \psi) \strictif \psi \to (\varphi \wedge \Box\, \varphi) \strictif \psi$. We have:

Theorem 2.38 $\mathsf{L}^{\flat}_{\Box} \oplus \mathsf{P} = \mathsf{wW}^{\flat} \oplus \mathsf{P}$.

Proof We first show that $\mathsf{L}^{\flat}_{\Box} \oplus \mathsf{P} \vdash \mathsf{wW}$. We reason in $\mathsf{L}^{\flat}_{\Box} \oplus \mathsf{P}$. We have:

$$\begin{aligned}
(\varphi \wedge \Box\, \psi) \strictif \psi &\to \Box((\varphi \wedge \Box\, \psi) \strictif \psi) \wedge ((\varphi \wedge \Box\, \psi) \strictif \psi) \\
&\to \Box(\Box(\varphi \wedge \Box\, \psi) \to \Box\, \psi) \wedge ((\varphi \wedge \Box\, \psi) \strictif \psi) \\
&\to \Box((\Box\, \varphi \wedge \Box\Box\, \psi) \to \Box\, \psi) \wedge ((\varphi \wedge \Box\, \psi) \strictif \psi) \\
&\to \Box(\Box\, \varphi \to \Box\, \psi) \wedge ((\varphi \wedge \Box\, \psi) \strictif \psi) \\
&\to (\varphi \wedge \Box\, \varphi) \strictif \psi
\end{aligned}$$

For the other direction it is sufficient to show that $\mathsf{wW}^{\flat}$ proves $\mathsf{L}_{\Box}$. We reason in $\mathsf{wW}^{\flat}$. Since, $((\Box\, \varphi \to \varphi) \wedge \Box\, \varphi) \strictif \varphi$, we find $((\Box\, \varphi \to \varphi) \wedge \Box(\Box\, \varphi \to \varphi)) \strictif \varphi$. It follows that $\Box((\Box\, \varphi \to \varphi) \wedge \Box(\Box\, \varphi \to \varphi)) \to \Box\, \varphi$. Since P implies $4_{\Box}$, we immediately have $\mathsf{L}_{\Box}$. $\dashv$

We are abstaining here from further investigation of wW in its own right. We simply hope that the reader has been convinced that the space of potential Löb-like principles (and their mutual interactions) in the constructive Lewis setting is rich, under-explored and requires further study.

References

Ardeshir, M., & Mojtahedi, M. (2018). The Σ_1-provability logic of HA. *Annals of Pure and Applied Logic, 169*(10), 997–1043.

Areces, C., Hoogland, E., & de Jongh, D. (1998). Interpolation, definability and fixed points in interpretability logics. *Advances in Modal Logic, 2*, 35–58.

Artemov, S., & Beklemishev, L. (2004). Provability logic. In D. Gabbay & F. Guenthner (Eds.), *Handbook of philosophical logic* (2nd ed., Vol. 13, pp. 229–403). Springer.

Berarducci, A. (1990). The interpretability logic of Peano arithmetic. *The Journal of Symbolic Logic, 55*, 1059–1089.

Bernardi, C. (1976). The uniqueness of the fixed-point in every diagonalizable algebra. *Studia Logica, 35*(4), 335–343.

Bezhanishvili, G., & Ghilardi, S. (2007). An algebraic approach to subframe logics. Intuitionistic case. *Annals of Pure and Applied Logic, 147*(1–2), 84–100

Blok, W.J., Pigozzi, D. (1989). Algebraizable logics. *Memoirs AMS, 77*(396).

Celani, S., & Jansana, R. (2005). Bounded distributive lattices with strict implication. *Mathematical Logic Quarterly, 51*(3), 219–246.

de Jongh, D., & Visser, A. (1991). Explicit fixed points in interpretability logic. *Studia Logica, 50*, 39–50.

de Groot, J., Litak, T., & Pattinson, D. (2021). Gödel-McKinsey-Tarski and Blok-Esakia for Heyting-Lewis implication. In: *Proceedings of the 36th Annual ACM/IEEE Symposium on Logic in Computer Science (LICS)* (pp. 1–15). https://doi.org/10.1109/LICS52264.2021.9470508

Fine, K. (1985). Logics containing K4. part II. *The Journal of Symbolic Logic, 50*(3), 619–651.

Font, J. M. (2006). Beyond Rasiowa's algebraic approach to non-classical logics. *Studia Logica, 82*(2), 179–209.

Font, J.M., Jansana, R., Pigozzi, D. (2003). A survey of abstract algebraic logic. *Studia Logica, 74*(1–2), 13–97. http://dx.doi.org/10.1023/A:1024621922509

Fourman, M.P., & Scott, D.S. (1979). Sheaves and logic. In: M. Fourman, C. Mulvey & D. Scott (Eds.), *Applications of Sheaves: Proceedings of the Research Symposium on Applications of Sheaf Theory to Logic, Algebra, and Analysis Durham* (pp. 302–401). Springer. https://doi.org/10.1007/BFb0061824

Friedman, H. (1978). Classically and intuitionistically provably recursive functions. In G. Müller & D. Scott (Eds.), *Higher Set Theory* (pp. 21–27). Springer.

Gaifman, H., & Dimitracopoulos, C. (1982). Fragments of Peano's arithmetic and the MDRP theorem. *Logic and Algorithmic, L'Enseignement Mathématique, monographie nr.,30*, 319–329.

Hájek, P., & Pudlák, P. (1993). *Metamathematics of First-Order Arithmetic. Perspectives in Mathematical Logic*. Springer.

Hoogland, E. (2001). *Definability and interpolation: Model-theoretic investigations*. ILLC, University of Amsterdam. https://www.illc.uva.nl/Research/Publications/Dissertations/DS/

Hughes, J. (2000). Generalising monads to arrows. *Science of Computer Programming, 37*(1–3), 67–111.

Iemhoff, R. (2001a). A modal analysis of some principles of the provability logic of Heyting Arithmetic. In *Proceedings of AiML'98* (Vol. 2). Uppsala.

Iemhoff, R. (2001b). *Provability logic and admissible rules*. Ph.D. thesis, ILLC, University of Amsterdam. https://www.illc.uva.nl/Research/Publications/Dissertations/DS/

Iemhoff, R. (2003). Preservativity logic: An analogue of interpretability logic for constructive theories. *Mathematical Logic Quarterly, 49*(3), 230–249.

Iemhoff, R., de Jongh, D., & Zhou, C. (2005). Properties of intuitionistic provability and preservativity logics. *Logic Journal of IGPL, 13*(6), 615–636.

Iemhoff, R., De Jongh, D., & Zhou, C. (2005). Properties of intuitionistic provability and preservativity logics. *Logic Journal of the IGPL, 13*(6), 615–636.

Iwata, S., Kurahashi, T., & Okawa, Y. (2020). *The fixed point and the craig interpolation properties for sublogics of IL*. https://doi.org/10.48550/ARXIV.2007.05427

Iwata, S., Kurahashi, T., & Okawa, Y. (2022). The persistence principle over the logic $\mathbf{IL}^-$. https://doi.org/10.48550/ARXIV.2203.02183

Japaridze, G., & de Jongh, D. (1998). The logic of provability. In S. Buss (Ed.), *Handbook of proof theory* (pp. 475–546). North-Holland Publishing Co.

Jónsson, B., & Tarski, A. (1951). Boolean algebras with operators Part I. *American Journal of Mathematics, 73*(4), 891–939.

Jónsson, B., & Tarski, A. (1952). Boolean algebras with operators Part II. *American Journal of Mathematics, 74*(1), 127–162.

Kalsbeek, M. (1991). Towards the interpretability logic of $textI\Delta_0 + \mathsf{EX}P$. *Logic Group Preprint Series 61, Faculty of Humanities, Philosophy*. Utrecht University. https://dspace.library.uu.nl/handle/1874/26517

Lindley, S., Wadler, P., & Yallop, J.: Idioms are oblivious, arrows are meticulous, monads are promiscuous. *ENTCS, 229*(5), 97–117 (2011). Proceedings of MSFP

Litak, T., Polzer, M., & Rabenstein, U. (2017). Negative translations and normal modality. In D. Miller (Ed.), *2nd International Conference on Formal Structures for Computation and Deduction (FSCD 2017), Leibniz International Proceedings in Informatics (LIPIcs)* (Vol. 84, pp. 27:1–27:18). http://drops.dagstuhl.de/opus/volltexte/2017/7741

Litak, T., Visser, A. (2018). Lewis meets Brouwer: constructive strict implication. *Indagationes Mathematicae, 29*, 36–90. https://arxiv.org/abs/1708.02143, a special issue L.E.J. Brouwer, fifty years later

Macnab, D.S. (1981). Modal operators on Heyting algebras. *Algebra Universalis, 12*(1), 5–29. https://doi.org/10.1007/BF02483860

Maksimova, L. L. (1989). Definability theorems in normal extensions of the probability logic. *Studia Logica,48*(4), 495–507.

McCune, W. (2003). Mace4 reference manual and guide. CoRR cs.SC/0310055. http://arxiv.org/abs/cs.SC/0310055, tool URL: https://www.cs.unm.edu/~mccune/prover9/

Mojtahedi, M. (2022a). On provability logic of HA. https://doi.org/10.48550/ARXIV.2206.00445

Mojtahedi, M. (2022b). Relative unification in intuitionistic logic: Towards provability logic of HA. https://doi.org/10.48550/ARXIV.2206.00446

Nakano, H. (2000). A modality for recursion. In *Proceedings of LiCS* (pp. 255–266). IEEE. http://www.computer.org/proceedings/lics/0725/07250255abs.htm

Nakano, H. (2001). Fixed-point logic with the approximation modality and its Kripke completeness. In N. Kobayashi & B. C. Pierce (Eds.), *Proceedings of TACS, LNCS* (Vol. 2215, pp. 165–182). Springer. http://link.springer.de/link/service/series/0558/bibs/2215/22150165.htm

Okawa, Y. (2022). Countably many sublogics of the interpretability logic IL having fixed point properties. *Journal of Logic and Computation.*

Rasiowa, H. (1974). *An Algebraic approach to non-classical logics*. North Holland.

Sambin, G. (1976). An effective fixed-point theorem in intuitionistic diagonalizable algebras. *Studia Logica, 35*, 345–361.

Shavrukov, V. (1988). The logic of relative interpretability over Peano arithmetic (in Russian). Technical Report No.5, Stekhlov Mathematical Institute

Smoryński, C. (1985). *Self-reference and modal logic*. Universitext.

Solovay, R. (1976). Provability interpretations of modal logic. *Israel Journal of Mathematics, 25*, 287–304.

Ten Cate, B., Fontaine, G., & Litak, T. (2010) Some modal aspects of XPath. *Journal of Applied Nonclassical Logics, 20*(3), 139–171. http://jancl.e-revues.com/article.jsp?articleId=15670

Troelstra, A. (1973). Metamathematical investigations of intuitionistic arithmetic and analysis. In *Springer Lecture Notes* (Vol. 344). Springer.

van Benthem, J. (2006). Modal frame correspondences and fixed-points. *Studia Logica, 83*(1–3), 133–155.

Visser, A. (1985). Evaluation, provably deductive equivalence in Heyting's Arithmetic of substitution instances of propositional formulas. In *Logic group preprint series 4, Faculty of humanities, philosophy*. Utrecht University. https://dspace.library.uu.nl/handle/1874/26232

Visser, A. (1990). Interpretability logic. In P. Petkov (Ed.), *Mathematical logic, Proceedings of the Heyting 1988 summer school in Varna* (pp. 175–209). Plenum Press.

Visser, A. (1994). Propositional combinations of Σ-sentences. In *Heyting's arithmetic. logic group preprint series 117, Faculty of humanities, philosophy*. Utrecht University. https://dspace.library.uu.nl/handle/1874/27700

Visser, A. (1998). An Overview of Interpretability Logic. In M. Kracht, M. de Rijke, H. Wansing, & M. Zakharyaschev (Eds.), *Advances in modal logic, CSLI lecture notes, vol 1, 87* (pp. 307–359). Center for the Study of Language and Information.

Visser, A. (2002). Substitutions of Σ^0_1-sentences: explorations between intuitionistic propositional logic and intuitionistic arithmetic. *Annals of Pure and Applied Logic, 114*, 227–271.

Visser, A. (2005). Löb's Logic Meets the μ-Calculus. In A. Middeldorp, V. van Oostrom, F. van Raamsdonk, & R. de Vrijer (Eds.), *Processes, terms and cycles, steps on the road to infinity, essays dedicated to Jan Willem Klop on the occasion of His 60th Birthday, LNCS 3838* (pp. 14–25). Springer.

Visser, A. (2008). Closed fragments of provability logics of constructive theories. *Journal of Symbolic Logic, 73*(3), 1081–1096.

Visser, A., & Zoethout, J. (2019). Provability logic and the completeness principle. *Annals of Pure and Applied Logic, 170*(6), 718–753.
Visser, A., Litak, T. (202x). *Lewis and Brouwer meet strong Löb*. In preparation
Wolter, F. (1993). Lattices of modal logics. Ph.D. thesis, Fachbereich Mathematik, Freien Universität.
Wolter, F., & Zakharyaschev, M. (1997). On the relation between intuitionistic and classical modal logics. *Algebra and Logic, 36*, 121–125.
Wolter, F., & Zakharyaschev, M. (1998). Intuitionistic modal logics as fragments of classical bimodal logics. In E. Orlowska (Ed.), *Logic at Work* (pp. 168–186). Essays in honour of Helena Rasiowa. Springer.
Zhou, C. (2003). Some intuitionistic provability and preservativity logics (and their interrelations). Master's thesis, ILLC, University of Amsterdam. https://msclogic.illc.uva.nl/theses/archive/

Chapter 3
An Abstract Look at the Fixed-Point Theorem for Provability Logic

Johan van Benthem

Abstract In this brief note, I discuss some general backgrounds of the well-known Fixed-Point Theorem for Provability Logic, taking my cues from an elegant abstract argument by Dick de Jongh in the 1980s.

Keywords Modal provability logic · Well-founded orderings · Definability

3.1 Introduction

In the mid 1980s, I was into 'semantic automata', (van Benthem, 1986), classifying linguistic quantifiers in terms of the complexity of their verification procedures on Venn diagrams. The next step in developing this 'procedural semantics' was an analysis of linguistic expressions that depend on the underlying structure of the object domain, and so, I developed an interest in tree automata whose computation rule is recursive in a given tree ordering. This led to an intensive and fruitful correspondence with Dick de Jongh about connections with provability logic, where such recursive definitions can be made explicit. In this correspondence, Dick came up with an elegant generalization of the key step in the Fixed-Point Theorem which applied far beyond the modalities, namely, to arbitrary generalized quantifiers satisfying suitable abstract conditions. Dick's result was included in my somewhat long and meandering paper 'Toward a Computational Semantics', (van Benthem, 1987) where it has remained hidden since.

The purpose of this brief note is twofold. I want to advertize Dick's result by itself, and its elegant level of abstraction. After that, I go further in this spirit and add some simple observations showing how the Fixed-Point Theorem can be seen as an instance of a family of abstract results on generalized well-founded orders. Much of

J. van Benthem (✉)
University of Amsterdam, Amsterdam, The Netherlands
e-mail: j.vanbenthem@uva.nl

Stanford University, Stanford, CA, USA

Tsinghua University, Beijing, China

N. Bezhanishvili et al. (eds.), *Dick de Jongh on Intuitionistic and Provability Logics*, Outstanding Contributions to Logic 28, https://doi.org/10.1007/978-3-031-47921-2_3

what follows may be present in the folklore or the expert literature (more on this in Sect. 3.3), but a compact story may be useful.[1]

3.2 The Fixed-Point Theorem for Provability Logic

Consider the language of basic modal logic interpreted over relational models where the accessibility relation is transitive and upward well-founded. The standard modalities $\Box$, $\Diamond$ then range over all relational successors in this ordering. Call a modal formula $\varphi = \varphi(p, \mathbf{q})$ *p-modalized* if all occurrences of the proposition letter p in φ are in the scope of at least one modal operator $\Box$, while there is no syntactic constraint on the occurrences of the proposition letters in the sequence $\mathbf{q}$. We state the following result in its semantic version; a proof-theoretic version can be stated using the completeness theorem for Gödel-Löb logic.

Theorem 3.1 (Fixed-Point Theorem, De Jongh-Sambin-Bernardi) *For any p-modalized formula $\varphi(p, \mathbf{q})$, there exists a formula $\alpha = \alpha(\mathbf{q})$ such that $\varphi(\alpha(\mathbf{q}), \mathbf{q}) \leftrightarrow \alpha(\mathbf{q})$ is valid. This fixed-point formula α can be found by an effective procedure. Moreover, the fixed-point is unique up to logical equivalence.*

There are many proofs of the existence of fixed-points in provability logic, both model-theoretic and proof-theoretic, including a pleasantly non-constructive one via Beth's Theorem, while Maksimova also derived Beth's theorem from the fixed-point theorem, (Hoogland, 2001). A compact informative survey of provability logic can be found in Verbrugge (2005). The analysis of fixed-points in this paper is also indebted to a perspicuous proof in Reidhaar-Olson (1989).

Two major features. All these proofs make recurrent use of two general features. One is the semantic import of the restriction to p-modalized formulas $\varphi(p)$. In any model $\mathbf{M}$, their truth value at a point s is only sensitive to the intersection of the valuation set $V(p)$ of the model with the set of successors of s in $\mathbf{M}$. In *generalized quantifier* terminology [see below], the set $\{t \mid Rst\}$ is a *live-on set* for the quantifier defined by $\varphi(p)$ at s. The other feature is the availability of *induction principles* over well-founded orderings, in the form of Löb's Axiom or suitable variants. In fact, well-foundedness just *is* an induction principle.

The Recursion Theorem in set theory. A common feeling is that something general is going on in the above Fixed-Point Theorem beyond the specifics of a modal language. To see this, recall the Recursion Theorem from your introductory course in set theory, (Hrbacek & Jech, 1999), or perhaps you encountered an arithmetical version in recursion theory. One of its many versions says that functions defined by recursion on the transitive closure of their arguments in the epsilon relation exist

[1] This paper offers a perspective that is complementary to (Litak and Visser, this volume), though the latter paper goes into far more proof-theoretic depth and width, and also covers different fixed-point constructions arising with binary modalities.

and are unique. The proof shows striking similarities with the provability logic case: e.g., one first proves uniqueness by epsilon-induction, and then uses this to prove existence, again by epsilon induction. But the content of the Recursion Theorem is more sweeping: the language is not a simple modal one, but that of set theory, and also, the well-founded relation is not transitive.

One can reproduce the proof of the Recursion Theorem in more restricted formal languages. Still close to set theory is *monadic second-order logic* over well-founded orderings, but in what follows, we will also think of the system SOML: *second-order modal logic* with quantification over proposition letters. Fixed-points arguments in the setting of SOML specialize to simpler modal languages, and yet bring out the essence of proofs in richer formalisms.

But this is not yet the whole story. The striking fact about the Fixed-Point Theorem for provability logic is that the fixed-points exist *inside the first-order modal language* used to state the fixed-point equation. This brings us to the definability of fixed-points, and to De Jongh's style of analysis.

3.3 De Jongh's Generalized Quantifier Analysis

A parametrized generalized quantifier (henceforth called a 'quantifier', for short) is a relation between points and sets of points. In any model **M**, each modal formula $\varphi(p)$ defines a modal generalized quantifier $Q_{\varphi,p}$ as follows:

$$Q_{\varphi,p}sX \text{ iff } \mathbf{M}[p := X], s \models \varphi(p)$$

Alternatively, one can think of these point-dependent generalized quantifiers as functions from sets to sets, as with the next-approximation maps used in the semantics of the modal μ-calculus, (Venema, 2020). We now turn to two special properties for such quantifiers that will be central to what follows.

As is well-known, all modal formulas are *invariant for generated submodels*, (Blackburn et al., 2001), and thus, the generalized quantifiers $Q_{\varphi,p}sX$ defined in the above way only depend on the intersection of the set X with the set $\{s\} \cup \{t \mid Rst\}$. When $\varphi(p)$ is p-modalized, this strengthens to dependence only on the set of accessible points $X \cap \{t \mid Rst\}$. Let us call a quantifier satisfying the latter property *future-oriented*.

Next, here is a special property that is crucial to getting explicit fixed-points. We call a quantifier Q *persistent* if QsX implies QtX for all t with Rst: in modal terms, for persistent quantifiers, $Q\varphi \to \Box Q\varphi$ is valid.

With all this in place, here is Dick's result.

Fact 3.1 All future-oriented persistent generalized quantifiers Q have the fixed-point $Q\top$, i.e., the equivalence $Q\top \leftrightarrow QQ\top$ is valid on well-founded models.

We committed some harmless abuse of notation here, using 'Q' both in syntax and semantics. The proof to follow is the essence of a first step in a standard proof of the Fixed-Point Theorem for special formulas $\Box\varphi(p)$, (Smoryński, 1984).

Proof We reason inside models $\mathbf{M}, s$ at some specific point s.

(i) $Q\top \rightarrow QQ\top$. Let $\mathbf{M}, s \models Q\top$. By persistence, we also have $\Box Q\top$, and therefore $\Box(Q\top \leftrightarrow \top)$. Using future-orientation, we get that $\mathbf{M}, s \models QQ\top$.
(ii) $QQ\top \rightarrow Q\top$ is proved using well-foundedness. Assume that $\mathbf{M}, s \models \neg Q\top$, then $\neg Q\top \wedge \Box Q\top$ is true at s or at some t with Rst. Using future-orientation as before, $\neg QQ\top$ is then true at t, and hence it is also true at s: either directly, or using persistence once more. (Note: The contraposition style in this second leg is just for convenience; one can also run it positively via well-founded induction.)

Remark on proof style. A word about this semantic style of reasoning, that will be used throughout. Uniqueness and existence hold locally at points s in models based only on what is true in the generated submodel at s. In a proof-theoretic version of our semantic arguments, this additional information would translate into explicit modal assumptions such as $\Box(p \leftrightarrow q)$ that keep track of exactly what is used. The proofs to follow should be understood with this in mind.

Fact 3.1 yields just one fixed-point, perhaps a trivial-looking one. But in fact, it is the only fixed-point up to logical equivalence. A standard argument for the following fact depends only on well-foundedness, and does not need persistence.

Fact 3.2 If a future-oriented quantifier has a fixed-point, that fixed-point is unique.

What Dick's analysis shows is how a well-chosen notation identifies abstract assumptions that go into more complex syntactic arguments. One is reminded of how provability logic itself clarifies key patterns in complex metamathematical proofs with well-chosen simple notation. There is also the beauty of abstraction, and the suggestion of more general perspectives.

In particular, there is nothing sacrosanct about the basic modal setting. The above result also applies, say, to *graded modal logic*, since the modality

$[\leq \mathrm{k}]\varphi$ 'φ is true in all except k successors'

is future-oriented and persistent. As an example of a de Jongh fixed-point in graded modal logic, consider the formula $p \leftrightarrow [\leq 1]p$, which might be read as 'up to at most one exception, p is refutable'. The fixed-point $[\leq 1]\bot$ is only true in endpoints or points that see only one endpoint. But in fact, Dick's result applies to any quantifier $\Box Q$, whether Q is modal, first-order, or not: say,

in *most* successors of each successor, or
for each successor, in an *even number* of its successors

In finite at most binary trees, the de Jongh fixed-point for the latter quantifier defines the 'well-balanced trees' where no node has just one successor.

Languages with such extended generalized quantifiers go far beyond the original arithmetical motivations for provability logic, and have their uses elsewhere. But still, one might be allowed to wonder whether, e.g., the two just-mentioned modalities correspond to weaker notions of arithmetical provability, say, allowing small sets of exceptions in a numerical or measure-theoretic sense.

Still, we are not done yet, as we want fixed-point theorems for complex formulas, not just for single generalized quantifiers. I believe this can be done in a suitably generalized quantifier perspective, but in this paper, I follow a more standard route, borrowing known ideas from provability logic.

Historical comment In response to a first draft of this paper, Albert Visser pointed me to (Smoryński, 1985) which has Dick's idea in a proof-theoretic setting. In Chapter 4, Smoryński introduces a theory SR with two basic axioms:

(a) $\Delta\varphi \to \Box\Delta\varphi$
(b) $\Box(\varphi \leftrightarrow \psi) \to (\Delta\varphi \leftrightarrow \Delta\psi)$,

where Δ is a modal operator with an arbitrary interpretation, for which the book gives concrete examples in the realm of arithmetical provability. Smoryński proves various results, and claims a full fixed-point theorem. Given this, it seems that the ideas reported here may well have independent double origins.

For now, we first generalize in another direction, showing that fixed-points exist just as well with not necessarily transitive well-founded relations. This will yield a general fixed-point definition of which Dick's $Q\top$ is a special case.

3.4 Fixed-Points on Generalized Well-Founded Orders

Dropping the transitivity assumption on relational models makes sense in many settings, for instance, with tree automata for procedural semantics that work in layers, (van Benthem, 1987). Recursion in non-transitive settings also drives the fixed-point logic analysis of Zermelo's Theorem and Backward Induction solutions for games in van Benthem and Gheerbrant (2006).

In this setting, we need to assume *well-foundedness* in the following sense: there are no infinite upward sequences $s_1 R s_2 R s_3, \ldots$ in the ordering R (this terminology is for brevity: the modal literature usually calls this 'converse well-foundedness'). This still gives the following induction principle to work with: each non-empty set X has a maximal element $s \in X$ in the sense that no t with R^*st is in X, where, in this paper, R^* denotes the transitive closure of R. Restated in positive terms, this is an abstract form of course-of-values induction.

We now also generalize the notion of *future-orientation* for quantifiers Qp to dependence on the intersection of $V(p)$ with $\{t \mid R^*st\}$. The following assertion is easy to prove by the new form of induction, in a way that resembles, as announced earlier, the proof of the Recursion Theorem in set theory.

Fact 3.3 If Q is future-oriented, then it has a unique fixed-point.

The following gives more information about the shape of this fixed-point.

Proof (i) Uniqueness follows by a standard well-foundedness argument. Beyond a last point s where two putative fixed-points p, q differ, they coincide on the points reachable from s in the transitive closure R^*. But at that s, by future-orientation, Qp, Qq have the same truth value, and hence so do p, q. (ii) As for existence, the following second-order formula in fact defines the fixed-point:

$$\exists p\,(Qp \wedge [R^*](Qp \leftrightarrow p))$$

Call this formula α: we will prove that $Q\alpha \leftrightarrow \alpha$ by the above form of induction. Assume that $[R^*](Q\alpha \leftrightarrow \alpha)$ is true. We show that $Q\alpha \leftrightarrow \alpha$ is true as well.

(a) Let $Q\alpha$ be true. Then α describes a unary predicate (or set) p such that $Qp \wedge [R^*](Qp \leftrightarrow p)$, and the latter fact means by definition that α holds.
(b) Let α be true. That is, for some p we have that $Qp \wedge [R^*](Qp \leftrightarrow p)$ is true. We also have $[R^*](Q\alpha \leftrightarrow \alpha)$, and so, by unicity of fixed-points, we obtain that $[R^*](\alpha \leftrightarrow p)$. Using future-orientation w.r.t. Qp, we then get $Q\alpha$.

Now Dick's result follows as a consequence of this general analysis.

Corollary 3.1 *The preceding result implies Fact 3.1.*

Proof We show that, if the quantifier Q is persistent, then the above formula $\alpha = \exists p\,(Qp \wedge [R^*](Qp \leftrightarrow p))$ is in fact equivalent with $Q\top$.

(i) Suppose that $Q\top$ holds at s, then it holds in the whole generated submodel at s, by repeated applications of persistence. Therefore, $Q\top$ is equivalent to $\top$ in the whole generated submodel, and α is true by taking p to be $\top$.
(ii) Now let α hold at s. Then, for some p, Qp is true, and so by persistence, Qp is equivalent to $\top$ in all reachable successors. Also, we have $[R^*](Qp \leftrightarrow p)$, and so $[R^*](\top \leftrightarrow p)$. By the future-orientation of Qp, $Q\top$ then holds at s.

3.5 Fixed-Point Theorems for Generalized Modal Languages

Next, consider a propositional modal language with many different modalities [first-order or higher-order], viewed as generalized quantifiers $Q_1, \ldots, Q_k$ that are all future-oriented in the sense of the preceding section. Fixed-point equations are formulas $p \leftrightarrow \psi(p)$ for possibly complex p-modalized formulas ψ in this language. Each such formula can be viewed as being of the form

$$\varphi[Q_1\alpha_1(p), \ldots, Q_k\alpha_k(p)]$$

where the displayed subformulas list all occurrences of p in the p-free 'skeleton formula' φ. (Incidentally, while such decompositions always exist, they need not be unique.) Clearly, such a formula can also be seen as defining one generalized quantifier. w.r.t. the propositional variable p. Moreover, given the invariance for generated submodels and the future-orientation of all the quantifiers occurring inside the formula, this overall quantifier is future-oriented.

Let us spell out this fact, as it also underlies later proofs. Let Q be future-oriented and α invariant for generated submodels. Consider a point s in a model $\mathbf{M}$ where (i) $[R^*](p \leftrightarrow q)$ and (ii) $Q\alpha(p)$ are true. By (ii), $Qs\,\{t \mid \mathbf{M} \models \alpha(p)\}$ which holds iff $Qs\,\{t \mid \mathbf{M_s} \models \alpha(p)\}$, with $\mathbf{M}_s$ the generated submodel of $\mathbf{M}$ at point s. Next, by the future-orientation of Q, (i) implies that $Qs\,\{t \mid \mathbf{M_s} \models \alpha(q)\}$, and by invariance for generated submodels once more, we get $Qs\,\{t \mid \mathbf{M} \models \alpha(q)\}$. [Actually, less is needed than invariance for generated submodels to make this argument work, but we are not aiming for a minimal version in this paper.]

Now, given the future-orientation, by Fact 3.3, there is a unique fixed-point for $\psi(p)$ which can be defined explicitly by one unary existential second-order quantifier $\exists p$ over a formula of the given quantifier language (at least, if we also have a transitive-closure modality for the well-founded order in the language). This analysis leads to the following result for second-order modal languages.

Fact 3.4 The Fixed-Point Theorem holds for second-order modal logic over a well-founded order with arbitrary added future-oriented modalities.

Proof The above analysis applies to the extended modal language of SOML, where in particular, each formula is still invariant for generated submodels. Therefore, our abstract fixed-point of Fact 3 is available. Moreover, we need no separate modality for the operator $[R^*]$ in the definition, since this is known to be definable as a unary fixed-point which can itself be defined in SOML.

An alternative analysis of Fact 3.3 shows that the Fixed-Point Theorem holds for extended modal logics that allow for elimination of 'bisimulation quantifiers', (Hollenberg, 1996), such as suitable fragments of the modal μ-calculus.

Now let us return to basic modal languages without second-order quantifiers. Even here, lifting the transitivity constraint is not unnatural. Van Benthem (2006) studies variants of Löb's Axiom using transitive closure modalities which separate out its well-foundedness content from its transitivity content.

The following extension of Dick's result generalizes the Fixed-Point Theorem for provability logic, but cf. our remark about (Smoryński, 1985).

Theorem 3.2 *For any modal language over a well-founded order whose formulas are invariant for generated submodels and whose vocabulary contains arbitrary future-oriented persistent modalities, each fixed-point equation has a unique solution definable inside the language.*

Proof We follow the proof in Reidhaar-Olson (1989), but cutting out the proof-theoretic excursions, and taking a direct route to existence. Moreover, unlike in that proof, we use the earlier abstract solution of Fact 3.3 at a crucial stage.

Also as before, we reason locally at some arbitrary point s in a model. Consider a formula $\psi(p) = \varphi[Q_1\alpha_1(p), \ldots, Q_k\alpha_k(p)]$ as described above. This describes a unique fixed-point α with a second-order definition as stated earlier in the proof of Fact 3.3, so the following formula is true throughout our model:

$$\alpha \leftrightarrow \varphi[Q_1\alpha_1(\alpha), \ldots, Q_k\alpha_k(\alpha)]$$

Here, since combinations $Q_i\alpha_i(p)$ with formulas α_i that are themselves invariant under generated submodels still define future-oriented invariant quantifiers, we can simplify notation in what follows without loss of generality to

$$\psi(p) = \varphi[Q_1 p, \ldots, Q_k p]$$

where the unique fixed-point simplifies to:

$$\alpha = \varphi[Q_1(\alpha), \ldots, Q_k(\alpha)]$$

Now we prove the definability of this fixed-point inside our language. The base case with $k = 0$ is obvious, as the formula ψ itself is then a fixed-point.

Next, let $k > 0$, and assume that we have the result already for all formulas with fewer than k occurrences of p in the display. Define ψ_i as the formula

$$\varphi[Q_1 p, \ldots, \top, \ldots, Q_k p]$$

where the ith occurrence $Q_i p$ has been replaced by $\top$. By the inductive assumptions, this has a fixed-point δ_i satisfying

$$\delta_i \leftrightarrow \varphi[Q_1(\delta_i), \ldots, \top, \ldots, Q_k(\delta_i)]$$

Lemma 3.1 *The following equivalence holds everywhere:* $Q_i(\delta_i) \leftrightarrow Q_i(\alpha)$.

Proof (i) First assume that $Q_i(\delta_i)$ is true at a point s. By the persistence of Q_i, we have as in earlier arguments that $(Q_i(\delta_i) \leftrightarrow \top) \wedge [R^*](Q_i(\delta_i) \leftrightarrow \top)$ is true at s. By invariance of φ for generated submodels and replacement of equivalents in the generated submodel at s, this implies that we have the following equivalence true throughout this generated submodel:

$$\delta_i \leftrightarrow \varphi[Q_1(\delta_i), \ldots, Q_i(\delta_i), \ldots, Q_k(\delta_i)]$$

But then δ_i is a fixed-point for the original formula ψ, and as fixed-points are unique, we have $\delta_i \leftrightarrow \alpha$ true in the generated submodel, with α the earlier fixed-point. It then follows from the truth of $Q_i(\delta_i)$ that $Q_i(\alpha)$ is true as well.

(ii) Next let $\neg Q_i(\delta_i)$ be true at a point s. Then by well-foundedness, there is a point t R-reachable from s in 0 or more steps with $\neg Q_i(\delta_i) \wedge [R^*]Q_i(\delta_i)$ true. In this point, we also have $[R^*](Q_i(\delta_i) \leftrightarrow \top)$ true. As before, it follows that

$$[R^*](\delta_i \leftrightarrow \varphi[Q_1(\delta_i), \ldots, Q_i(\delta_i), \ldots, Q_k(\delta_i)])$$

and hence, by uniqueness of fixed-points again, $[R^*](\delta_i \leftrightarrow \alpha)$ is true at t. Then, by the future-orientation of Q_i, $\neg Q_i(\delta_i)$ implies that $\neg Q_i(\alpha)$ is true at t: and, either directly, or using persistence of Q_i as often as necessary on the finite path from s to t, we have that $\neg Q_i(\alpha)$ is true at s.

The rest of the argument is now easy. Using the lemma repeatedly, together with semantic replacement of equivalents, we can replace the successive occurrences of α in $\varphi[Q_1(\alpha), \ldots, Q_k(\alpha)]$ to obtain the equivalence

$$\alpha \leftrightarrow \varphi[Q_1(\delta_1), \ldots, Q_k(\delta_k)]$$

where the formula to the right-hand side is inside our modal language.

I believe that the preceding proof identifies the essentials of the Fixed-Point Theorem in its original setting. But in line with what was said earlier about Dick's result, it applies much more widely, to graded modal logic, modal logics with persistent future-oriented second-order quantifiers, and so on. As a simple example, it is easy to compute a counterpart for the Löb fixed-point in terms of graded 'almost provability': $[\leq 1]\neg p \rightarrow q$ has the fixed-point $[\leq 1]\neg q \rightarrow q$. More exciting examples can be found with iterated persistent graded modalities.

All this expressive power still leaves an issue that may hamper the generality. Normally, in extended modal languages, modalities can refer to *many relations*, whereas here, all modalities refer to one fixed well-founded order R. Can this be lifted? A few thoughts on this will be found in Sect. 3.7.

3.6 Fixed-Point Theorems for Non-persistent Modalities

Persistence is a sufficient condition on future-oriented quantifiers to guarantee a fixed-point theorem. But is it necessary? As reported in van Benthem (1987), De Jongh also had a nice example showing that things at least become more difficult without persistence. Consider the simple fixed-point equation

$$p \leftrightarrow \text{MOST}\neg p$$

where 'MOST' is the future-oriented but not persistent modality 'in most successors'. When computing this fixed-point on the natural numbers $\{0, 1, 2, \ldots\}$ with $>$, p turns out to be true in all *odd numbers*. But the property of being odd is not definable in the pure modal language with MOST.

The fixed-point in the preceding example is still quite regular. Indeed, in the model of the preceding example, the odd positions are definable as follows in the language of *propositional dynamic logic* PDL over the immediate predecessor relation R_a.

This widely used formalism shifts our earlier notation slightly, and in what follows, Kleene star will denote *reflexive-transitive closure*:

$$\langle a; (a; a)^* \rangle [a] \bot$$

Here is a general fact behind the preceding observation. Consider upward well-founded deterministic relations R_a, i.e., taking the transitive closure: finite linear orders, or in terms of formal languages and automata theory, *finite words*. Take a propositional modal language of '*bare* PDL' with propositional constants $\top, \bot$, one propositional variable p, Boolean operations, and modalities $\langle \pi \rangle$ where π is a program expression constructed from a using just the regular operations ; $\cup$, $*$. In this setting, a formula $\varphi(p)$ is called *p-modalized* if each occurrence of p in $\varphi(p)$ occurs in the scope of at least one modality $\langle a \rangle$. (This syntactic format can be generalized, since in the proof to follow, we could also allow modalities for complex PDL-programs that are equivalent to a ; π for some program π.)

Fact 3.5 Fixed-point equations $p \leftrightarrow \varphi(p)$ with p-modalized $\varphi(p)$ over finite words in an alphabet $\{a\}$ are uniquely solvable with a fixed-point defined in bare PDL.

Proof Consider the natural numbers with the immediate predecessor relation R_a. Computing the fixed-point upward from the endpoint 0 yields a unique pattern of truth values for p, since $\varphi(p)$ does not refer to any structure except the bare order. In this structure, any modal formula defines a set of points, or equivalently, because of the invariance of modal formulas for generated submodels, a set of linear orders. Also, given the finite modal depth k of $\varphi(p)$, the truth value of p at a point only depends on the truth values of p in the k preceding positions. Call any occurring distribution of truth values for p over k successive positions a *p-segment*. We now analyze what this means combinatorially.

Looking at successive intervals of length k, there must be a p-segment S that occurs infinitely often, possibly separated by other intervals. Moreover, given the functional nature of the computation for $\varphi(p)$, the truth values for p to the right of occurrences of S are all in the same order. Therefore, there is in fact a unique interval T separating the occurrences of S. Thus, from some finite arbitrary interval of truth values for p onward, the whole ordering consist of repetitions of the interval $S \cup T$. But then we can define p as a disjunction of

(a) Explicit modal definitions for the p-positions in the arbitrary initial interval, say, $\langle a ; a \rangle \neg \langle a \rangle \top$ for distance 2 to the endpoint, (b) For each p-position in $S \cup T$: a modal formula describing the distance of that position to the end of $S \cup T$ followed by $\langle (a^m)^* \rangle \langle a^n \rangle \neg \langle a \rangle \top$, where m is the length of $S \cup T$, and n is the length of the arbitrary initial interval.

In terms of natural numbers, this defines *semi-linear sets* with just one period.

One would like to generalize to full PDL with more atoms and tests inside programs, over branching well-founded orders. However, this seems infeasible.

Counter-example: solving games. Zermelo's Theorem says that all finite-horizon two-player zero-sum games of perfect information are determined: one of the players

i, j has a winning strategy. The key to its proof is the following recursive definition of the positions in the game tree where player i has a winning strategy. Here move stands for the union of all moves, win_i marks winning end positions for player i, and turn_i marks turns for i:

$$\text{WIN}_i \leftrightarrow$$
$$((\neg\langle \texttt{move}\rangle\top \wedge \text{win}_i) \vee (\text{turn}_i \wedge \langle \texttt{move}\rangle \text{WIN}_i) \vee (\text{turn}_j \wedge [\texttt{move}]\text{WIN}_i))$$

This is a fixed-point equation of the form

$$p \leftrightarrow (q \vee (r \wedge \Diamond p) \vee (s \wedge \Box p))$$

which does not seem to yield an explicit definition for p inside the language of PDL. The intuitive reason is that the solution pattern $(\Diamond\Box)^*q$ over finite trees demands the existence of a subtree encoding the winning strategy, as opposed to a mere path, (van Benthem, 2006).

Still, at present, this is just a conjecture: it seems an open problem whether Zermelo-style fixed-points can be defined in PDL. Is propositional dynamic logic closed under fixed-points on finite trees, or on well-founded orders in general?

A simple fixed-point description. Consider a basic modal fixed-point formula $\varphi(p)$ with p modalized, on finite trees which has a fixed-point α by our general results. Since subformulas $\Box\psi$ are true in endpoints, evaluating $\varphi(p)$ on endpoints makes this formula equivalent with a Boolean combination of proposition letters other than p, say, φ^0. Next, for points at level 1 (where all successors are endpoints), we substitute φ^0 for p to obtain a formula $\varphi^1 = \varphi(\varphi^0/p)$ which is easily seen to be true at the same points at levels 0, 1 as the fixed-point α. Working upward inductively with successive substitutions, we obtain formulas φ^k for each natural number k. Given this construction, it is easy to prove the following fact:

On points up to level k, φ^k is equivalent with the fixed-point formula α.

Let us define φ^* as the countable disjunction of all formulas φ^k. It follows from the preceding observation that φ^* defines the fixed-point of $\varphi(p)$.

Modal substitution logic. This suggests an extension of the basic modal language to a system MSL with a minimum needed for the preceding to work. We introduce explicit notation for *substitutions* $p := \psi$ that are interpreted semantically as transforming models $\mathbf{M}$ by resetting the valuation for p to the truth set $[[\psi]]^{\mathbf{M}}$. Moreover, we add *regular operations* over these transformations, in particular, iterations $(p := \psi)^*$. Thus, the syntax of MSL is that of PDL, now with formulas plus expressions for model-transforming programs, but it is more powerful.

For instance, the above pattern $(\Diamond\Box)^*\varphi$ is definable as

$$\langle p := \varphi;\ (p := \Diamond\Box p)^*\rangle\top$$

where p is a fresh proposition letter not occurring in φ. The above fixed-points are all definable in this formalism, using finitary iterations of substitutions.

The language of the system MSL may look unusual, but it is finitary, and it is semantically modal in that its formulas are invariant for bisimulation. One can also show that MSL extends PDL, while, on finite trees, it is contained in the modal μ-calculus. The general case may have higher complexity, like the logic of iterated relativization modalities in Miller and Moss (1960). A further investigation of modal substitution logic is left to another occasion.

The use of dynamic modalities for model transformation in MSL brings iterated manipulations into the syntax of a logic in a generic manner. An alternative format inspired by fixed-point analysis is the graph-based ‘cyclic Henkin syntax’ of Visser (2027), and it would be of interest to compare the two approaches.

3.7 Further Directions

At this point, many directions open up. One can extend our analysis to multiple fixed-points for *simultaneous fixed-point equations* and to generalized quantifiers with more arguments. One can move from model theory to *proof theory*, and determine the precise content of the preceding semantic arguments and the logical strength they presuppose (classical, intuitionistic, or still less). Also, the generalized quantifier perspective employed here suggests analogies with *neighborhood semantics* for modal logic, (Pacuit, 2017), since the usual modal neighborhood frames are nothing but point-dependent generalized quantifiers in our sense. Instead of pursuing these issues, we mention three other points of interest.

Other fixed-point logics. Provability logic is one modal fixed-point logic, more widely used is the *modal μ-calculus*. The two systems are related: (van Benthem, 2006) embedded provability logic faithfully into the μ-calculus, (Visser, 2005) has a partial converse. Also, the mechanics of the two systems are close: computing fixed-points for modalized formulas on well-founded orders in μ-calculus approximation style eventually stabilizes to the provability logic fixed-points, despite negative occurrences of the running variable p. Taking our generalized quantifier perspective to the μ-calculus is natural, since that system also works for *arbitrary monotonic modalities*, not just standard ones or first-order quantifiers. One can even move the analysis to first-order fixed-point logic LFP(FO), (Ebbinghaus & Flum, 1995) or its fragment of ‘guarded fixed-point logic’, (Grädel, 2002).

Connections with automata theory and tree logics. The analysis in Sect. 3.6 employs essentially weak monadic second-order logic of one successor for finite strings, a special case of monadic second-order logic MSOL over strings which defines the regular languages, (Büchi, 1960). We believe that the classical results by Büchi and Rabin connecting MSOL over strings and trees with second-order languages, (Grädel et al., 2002), have a deeper bearing on the fixed-point theorems discussed in this paper. In fact, automata provide an alternative syntax format whose state graphs are more tolerant of defining fixed-points. However, the precise connection remains to be understood

Extending the languages sideways. Modal languages are often about many orderings connected in certain ways. Also, general fixed-point logics can have a lot of vocabulary referring to complex relational similarity types for their models. Of course, there is a distinguished well-founded order R in the background: often the order of the ordinals, that can be added to models explicitly as a further component when proving basic results such as the Downward Löwenheim-Skolem theorem for LFP(FO), (Flum, 1995). What happens to the fixed-point theorem in the presence of other relations with matching modalities?

Some extensions are straightforward. For a formula $p \leftrightarrow (\langle S\rangle q \vee \neg[R]p)$, with R a well-founded relation and S an arbitrary further relation, a fixed-point can be computed as usual, treating the disjunct $\langle S\rangle q$ as a proposition letter. For more general cases, modalized occurrences of p must occur under modalities for relations that are well-founded (these can be basic, but combinations $R\,;\,S$ can be well-founded too). Also acceptable are combinations like $\langle R \cap S\rangle p$, as this just states R-reachability by a special route. If we do not require all this, fixed-points may still exist, but easy examples show that they need not be unique. (We could even allow occurrences of p that are not guarded in this manner as long as they are positive, and thus support a μ-calculus style recursion.)

A good test case for lifting the results of this paper to the broader poly-modal settings suggested here might be the *Bounded Fragment* of first-order logic, where all formulas are invariant for generated submodels, (Feferman & Kreisel, 1995; Areces et al., 2001).

3.8 Conclusion

The fixed-point theorem for provability logic and what makes it tick remains forever intriguing, and we owe this wide world to Dick de Jongh and his friends.

Acknowledgements I thank the referees for this volume, as well as Thomas Icard, Wolfgang Thomas, Yde Venema, Rineke Verbrugge, and especially, Albert Visser for their comments on this paper when still under construction.

References

Areces, C., Blackburn, P., & Marx, M. (2001). Hybrid logics: Characterization, interpolation and complexity. *Journal of Symbolic Logic, 66*(3), 977–1010.

Blackburn, P., de Rijke, M., & Venema, Y. (2001). *Modal logic*. Cambridge University Press.

Büchi, J. (1960). Weak second-order logic and finite automata. *Zeitschrift für Mathematische Logik und Grundlagen der Mathematik., 6*, 66–92.

Ebbinghaus, H.-D., & Flum, J. (1995). *Finite model theory*. Springer Science Publishers.

Feferman, S., & Kreisel, G. (1966). Persistent and invariant formulas relative to theories of higher order. *Bulletin of the American Mathematical Society, 72*, 480–485.

Flum, J. (1995). *On the (Infinite) model theory of fixed-point languages, mathematical institute*. University of Freiburg.

Grädel, E. (2002). Guarded fixed-point logics and the monadic theory of countable trees. *Theoretical Computer Science, 288*, 129–152.

Grädel, E., Thomas, W., & Wilke, T. (Eds.). (2002). *Automata, logics, and infinite games. Lecture notes in computer science* (Vol. 2500). Springer.

Hollenberg, M. (1996). *Logic and bisimulation*, Ph.D. Thesis, Department of Philosophy, University of Utrecht.

Hoogland, E. (2001). *Definability and interpolation: Model-theoretic investigations*, Ph.D/ Thesis DS-2001-05, ILLC, University of Amsterdam.

Hrbacek, K., & Jech, Th. (1999). *Introduction to set theory*. Revised and Expanded. CRS Press.

Litak, T., & Visser, A. (2021). *Lewisian fixed points I: Two incomparable constructions*, this volume.

Miller, J., & Moss, L. (2005). The undecidability of iterated modal relativization. *Studia Logica, 79*(373), 4–7.

Pacuit, E. (2017). *Neighborhood semantics for modal logic*. Springer Science Publishers.

Reidhaar-Olson, L. (1989). A new proof of the fixed-point theorem of provability logic. *Notre Dame Journal of Formal Logic, 31*(1), 37–43.

Smoryński, C. (1984). Modal logic and self-reference. In D. Gabbay & F. Guenthner (Eds.), *Handbook of philosophical logic* (pp. 441–495). Reidel.

Smoryński, C. (1985). *Self-reference and modal logic*. Springer Science Publishers.

van Benthem, J. (1986). *Essays in logical semantics*. Reidel.

van Benthem, J. (1987). Towards a computational semantics. In P. Gärdenfors (Ed.), *Generalized quantifiers: Linguistic and logical approaches* (pp. 31–71). Reidel.

van Benthem, J. (2006). Modal Frame Correspondences and Fixed-Points. *Studia Logica, 83*(1), 133–155.

van Benthem, J., & Gheerbrant, A. (2010). Game Solution, Epistemic Dynamics, and Fixed-Point Logics. *Fundamenta Informaticae, 100*, 19–41.

Venema, Y. (2020). *Lectures on the modal μ -Calculus*. University of Amsterdam.

Verbrugge, R. (2017). *Provability logic*. Stanford Encyclopedia of Philosophy.

Visser, A. (2005). Löb's logic meets the μ-calculus, processes, terms and cycles: Steps on the road to infinity. In *Lecture notes in computer science* (Vol. 3838, pp. 14–25). Springer

Visser, A. (2020). *Provability logic and modalized fixed points, lecture slides*. Online Seminar on Foundations of Mathematics. Wuhan University.

Visser, A. (2021). Cyclic Henkin logic, manuscript, Department of Philosophy, University of Utrecht. https://arxiv.org/abs/2101.11462

Chapter 4
The Σ_1-Provability Logic of HA Revisited

Mojtaba Mojtahedi

Abstract The Σ_1-provability logic of Peano Arithmetic PA, is characterized by (Visser, 1982) as $\mathsf{GLC_a}$, the Gödel-Löb logic GL plus the completeness principle for atomic variables. Also the Σ_1-provability logic of the Heyting Arithmetic HA, is characterized by (Ardeshir & Mojtahedi, 2018) as iH'_σ (for definition of iH'_σ, see Sect. 4.7). In this paper, we find some translation $(.)^{\boxed{h}}$, which embeds iH'_σ in $\mathsf{iGLC_a}$, the intuitionistic counterpart of $\mathsf{GLC_a}$.

Keywords Heyting arithmetic · Provability logic · Intuitionistic modal logic

4.1 Preface

The first arithmetical completeness result, is due to Dick de Jongh in 1969 where he proves in an unpublished paper that the intuitionistic logic is complete for arithmetical interpretations in the Heyting's Arithmetic HA. This result appears even before the Solovay's arithmetical completeness results for the provability logic of PA (Solovay, 1976). de Jongh's proof uses an ingenious combination of Kleene's realizability with Kripke semantics. An extended abstract of this result appears in de Jongh (1970). For a short survey on *de Jongh's theorem*, we refer the reader to de Jongh et al. (2011). The author of this paper, first entered the fascinating world of the arithmetical completenesses, through *de Jongh's theorem*, where together with Mohammad Ardeshir, author of this paper proved de Jongh theorem for the Basic Arithmetic (Ardeshir & Mojtahedi, 2014) and also for $\mathsf{HA} + \neg\mathsf{Con}(\mathsf{HA})$ (Ardeshir & Mojtahedi, 2018).

This chapter serves as a humble tribute to Dick de Jongh, a figure of immense respect and renown within the global logician community. Having had the privilege to meet Dick on two significant occasions, once in Tehran and another in ILLC,

M. Mojtahedi (✉)
Department of Mathematics: Analysis, Logic and Discrete Mathematics, Ghent University, Krijgslaan 281 S8, 9000 Ghent, Belgium
e-mail: mojtahedy@gmail.com

Department of Mathematics, Statistics and Computer Sciences, University of Tehran, College of Science, Tehran, Iran

N. Bezhanishvili et al. (eds.), *Dick de Jongh on Intuitionistic and Provability Logics*, Outstanding Contributions to Logic 28, https://doi.org/10.1007/978-3-031-47921-2_4

Amsterdam, the author was deeply inspired by his encouraging demeanor. These interactions not only left a lasting impression but also greatly motivated the author's continued exploration and contributions to the field.

4.2 Introduction

There are at least three reliable surveys on the subject of provability logic (Artemov & Beklemishev, 2004; Beklemishev & Visser, 2006; Japardize & de Jongh, 1997). Here for the sake of completeness we introduce very shortly the subject and try to cover the recent results related to intuitionistic provability logics. Historically, the first provability interpretation for modal operator $\Box$, considered by Gödel in Gödel (1933). Then the very natural question raised: what remains of a first-order theory T in the modal language, when we interpret $\Box$ as provability in T? The answer to this question for strong theories is stable: all strong enough theories like PA and ZF share the same provability logic GL (Solovay, 1976). Then (Visser, 1982) characterizes the Σ_1-provability logic of PA as $\mathsf{GLC_a} := \mathsf{GL}$ plus the completeness principle $p \to \Box p$ for atomic variables p. (Ardeshir & Mojtahedi, 2018; Visser & Zoethout, 2019) characterize the Σ_1-provability logic of HA (the intuitionistic counterpart of PA) as iH'_σ. For definition of iH'_σ, see Sect. 4.7. In this paper, we show that iH'_σ is embeddable in $\mathsf{iGLC_a}$. Before we explain more about this embedding, let us see an example of such embeddings. The Gödel's negative translation $A^{\neg}$, puts double negation behind every subformula of A. It is well-known that this translation, embeds classical logic into the intuitionistic logic. In this paper (Theorem 4.5), we define a translation $(.)^{\boxed{h}}$ such that

- $\mathsf{iH}'_\sigma \vdash A \leftrightarrow A^{\boxed{h}}$,
- $\mathsf{iH}'_\sigma \vdash A$ iff $\mathsf{iGLC_a} \vdash A^{\boxed{h}}$,

This result is interesting because, it shows that while the Σ_1-provability logic of HA has some additional axioms over $\mathsf{iGLC_a}$ (see Sect. 4.7), it can be embedded in $\mathsf{iGLC_a}$, which is the intuitionistic counterpart of the Σ_1-provability logic of PA.

4.3 Basic Definitions

The propositional modal language is $\mathcal{L}_\Box := \{\to, \vee, \wedge, \bot, \Box\}$. The propositional non-modal language is indicated by $\mathcal{L}_0 := \{\to, \vee, \wedge, \bot\}$. We use $\boxdot A$ as a shorthand for $A \wedge \Box A$. IPC is the intuitionistic propositional non-modal logic over usual propositional non-modal language. $\mathsf{IPC}_\Box$ is the same theory IPC in the extended language of propositional modal language, i.e. its language is propositional modal language and its axioms and rules are the same as in IPC. Because we have no axioms for $\Box$ in $\mathsf{IPC}_\Box$, it is obvious that $\Box A$ for each A, behave exactly like an atomic variable inside $\mathsf{IPC}_\Box$. Note that nothing more than the symbol of A plays a role in $\Box A$. The first-order intuitionistic theory is denoted with IQC and CQC is its classical closure, i.e. IQC plus the principle of excluded middle. We have the usual first-order language of arithmetic which has a primitive recursive function symbol for each primitive recursive function. We use the same notations and definitions for Heyting's arithmetic HA as in Troelstra and van Dalen (1988), and Peano Arithmetic PA is HA plus the principle of excluded middle. For a set of sentences and rules $\Gamma \cup \{A\}$ in propositional non-modal, propositional modal or first-order language, $\Gamma \vdash A$ means that A is derivable from Γ in the system IPC, $\mathsf{IPC}_\Box$, IQC, respectively. For an arithmetical formula A, $\ulcorner A \urcorner$ represents the Gödel number of A. For an arbitrary arithmetical theory T with a set of Δ_0-axioms, we have the Δ_0-predicate $\mathsf{Prov}_T(x, \ulcorner A \urcorner)$, that is a formalization of "x is code of a proof for A in T". We also have the provability predicate $\mathsf{Pr}_T(\ulcorner A \urcorner) := \exists x\ \mathsf{Prov}_T(x, \ulcorner A \urcorner)$.

Definition 4.1 Suppose T is an r.e. arithmetical theory and σ is a function from atomic variables to arithmetical sentences. We extend σ to all modal propositions A, inductively:

- $\sigma_T(A) := \sigma(A)$ for atomic A,
- σ_T distributes over $\wedge, \vee, \rightarrow$,
- $\sigma_T(\Box A) := Pr_T(\ulcorner \sigma_T(A) \urcorner)$, in which $Pr_T(x)$ is the Σ_1-predicate that formalizes T-provability of a sentence with Gödel number x.

We call σ to be a Σ_1-substitution, if for every atomic p, $\sigma(p)$ be a Σ_1-formula.

Definition 4.2 Provability logic of a sufficiently strong theory T is defined as

$$\mathsf{PL}(T) := \{A \in \mathcal{L}_\Box : \forall \sigma\ T \vdash \sigma_T(A)\}.$$

Also the Σ_1-provability logic of T is defined as

$$\mathsf{PL}(T) := \{A \in \mathcal{L}_\Box : \text{for all } \Sigma_1\text{-substitution } \sigma \text{ we have } T \vdash \sigma_T(A)\}.$$

4.4 NNIL Propositions

The class of *No Nested Implications to the Left*, NNIL formulae in the nonmodal language $\mathcal{L}_0$, was introduced in Visser et al. (1995), and more explored in Visser (2002). The crucial result of Visser (2002) is to provide an algorithm (Sect. 4.4.1) that takes $A \in \mathcal{L}_0$ and returns its best NNIL approximation A^* from below, i.e., $\mathsf{IPC} \vdash A^* \rightarrow A$ and for all NNIL formulae B such that $\mathsf{IPC} \vdash B \rightarrow A$, we have $\mathsf{IPC} \vdash B \rightarrow A^*$. Also for all Σ_1-substitutions σ, we have $\mathsf{HA} \vdash \sigma_{\mathsf{HA}}(\Box A \leftrightarrow \Box A^*)$ (Visser, 2002). The class NNIL of propositions in $\mathcal{L}_\Box$ is defined inductively:

- NNIL contains all atomic and boxed propositions,

- if $A, B \in \mathsf{NNIL}$, then $A \vee B, A \wedge B, \Box A \in \mathsf{NNIL}$,
- if all $\to$ occurred in A are contained in the scope of a $\Box$ and $B \in \mathsf{NNIL}$, then $A \to B \in \mathsf{NNIL}$.

Before we continue with the NNIL algorithm, let us present a crucial definition from Visser (2002):

Definition 4.3 For any two modal propositions $A, B \in \mathcal{L}_\Box$, define $[A]B$ by induction on the complexity of B:

- $[A]p = p$, for atomic p, $\top$ and $\bot$.
- $[A](B_1 \circ B_2) = [A](B_1) \circ [A](B_2)$, for $\circ \in \{\vee, \wedge\}$,
- $[A](B_1 \to B_2) = A \to (B_1 \to B_2)$,
- $[A](\Box B) = \Box B$.

For a set Γ of modal propositions we define $[A]\Gamma := \bigvee_{B \in \Gamma}[A]B$.

4.4.1 The NNIL-*Algorithm*

For a modal proposition A, the proposition A^* is defined inductively as follows (Visser, 2002):

1. A is atomic or boxed, take $A^* := A$.
2. $A = B \wedge C$, take $A^* := B^* \wedge C^*$.
3. $A = B \vee C$, take $A^* := B^* \vee C^*$.
4. $A = B \to C$, we have several sub-cases. In the following, an occurrence of E in D is called an *outer occurrence*, if E is neither in the scope of an implication nor in the scope of a boxed formula. Also an outer occurrence of a conjunction is leftmost, if it is the leftmost outer conjunction.

(a) C contains an outer occurrence of a conjunction. To make the algorithm deterministic, let $D \wedge E$ be the leftmost outer conjunction in C. In this case, there is some formula $J(q)$ such that

- q is a propositional variable not occurring in A.
- q is outer in J and occurs exactly once.
- $C = J[q|(D \wedge E)]$.

Now set $C_1 := J[q|D]$, $C_2 := J[q|E]$ and $A_1 := B \to C_1$, $A_2 := B \to C_2$ and finally, define $A^* := A_1^* \wedge A_2^*$.

(b) B contains an outer occurrence of a disjunction. To make the algorithm deterministic, let $D \vee E$ be the leftmost outer disjunction in B. In this case, there is some formula $J(q)$ such that

- q is a propositional variable not occurring in A.

- q is outer in J and occurs exactly once.
- $B = J[q|(D \vee E)]$.

Now set $B_1 := J[q|D]$, $B_2 := J[q|E]$ and $A_1 := B_1 \to C$, $A_2 := B_2 \to C$ and finally, define $A^* := A_1^* \wedge A_2^*$.

(c) $B = \bigwedge X$ and $C = \bigvee Y$ and X, Y are sets of implications, atomics or boxed formulas. We have several sub-cases:

(i) X contains an atomic variable or a boxed formula E. To make the algorithm deterministic, let E be the leftmost atomic or boxed in B. We set $D := \bigwedge(X \setminus \{E\})$ and take $A^* := E^* \to (D \to C)^*$.
(ii) X contains $\top$. Define $D := \bigwedge(X \setminus \{\top\})$ and take $A^* := (D \to C)^*$.
(iii) X contains $\bot$. Take $A^* := \top$.
(iv) X contains only implications. For any $D = E \to F \in X$, define

$$B \downarrow D := \bigwedge((X \setminus \{D\}) \cup \{F\}).$$

Let $Z := \{E \mid E \to F \in X\} \cup \{C\}$ and define:

$$A^* := \bigwedge\{((B \downarrow D) \to C)^* | D \in X\} \wedge \bigvee\{([B]E)^* \mid E \in Z\}$$

4.4.2 The Binary Relation $\rhd$

Visser (2002) axiomatizes the NNIL-algorithm via a binary relation $\rhd_\sigma$. This binary relation is tightly related to the interpretability logic (de Jongh & Veltman, 1990; Berarducci, 1990; Shavrukov, 1988; Visser, 1990) and preservativity logic (Iemhoff, 2003, 2001). We use some modal variant of Visser's $\rhd_\sigma$. The relation $\rhd_\sigma$ is defined to be the smallest relation $\rhd$ on modal propositions in $\mathcal{L}_\Box$ satisfying A1-A4, B1-B3: (Ardeshir & Mojtahedi, 2018)

A1. If $\mathsf{iK4} \vdash A \to B$, then $A \rhd B$.
A2. If $A \rhd B$ and $B \rhd C$, then $A \rhd C$.
A3. If $C \rhd A$ and $C \rhd B$, then $C \rhd A \wedge B$.
A4. If $A \rhd C$ and $B \rhd C$, then $A \vee B \rhd C$.
B1. If $A \rhd A$, then $\Box A \rhd \Box B$.
B2. Let X be a set of implications, $B := \bigwedge X$ and $A := B \to C$. Also assume that $Z := \{E | E \to F \in X\} \cup \{C\}$. Then $A \rhd [B]Z$,
B3. If $A \rhd B$, then $(C \to A) \rhd (C \to B)$, for atomic or boxed C.

The following theorem, states that the NNIL-algorithm only uses the axioms and rules presented in the above inductive definition of $\rhd_\sigma$:

Theorem 4.1 *For every $A \in \mathcal{L}_\Box$ we have $A^* \in$ NNIL, $\mathsf{IPC}_\Box \vdash A^* \to A$ and $A \rhd_\sigma A^*$. Moreover, if $B \in$ NNIL is such that $\mathsf{IPC}_\Box \vdash B \to A$, then $\mathsf{IPC}_\Box \vdash B \to A^*$.*

Proof See Visser (2002, Sect. 7) or Ardeshir and Mojtahedi (2018, Theorem 4.5). $\Box$

Corollary 4.1 *For every* $A \in \mathsf{NNIL}$, *we have* $\mathsf{IPC}_\Box \vdash A \leftrightarrow A^*$.

Proof By Theorem 4.1 we have $\mathsf{IPC}_\Box \vdash A^* \to A$. On the other hand, since $A \in \mathsf{NNIL}$ and $\mathsf{IPC}_\Box \vdash A \to A$, Theorem 4.1 implies $\mathsf{IPC}_\Box \vdash A \to A^*$, as desired. $\Box$

One interesting consequence of Theorem 4.1 follows by taking $B = \top$: $\mathsf{IPC}_\Box \vdash A$ implies $\mathsf{IPC}_\Box \vdash A^*$. However we may strengthen this to the following stronger form:

Corollary 4.2 *For every* $A, B \in \mathcal{L}_\Box$, *if* $\mathsf{IPC}_\Box \vdash A \to B$, *then* $\mathsf{IPC}_\Box \vdash A^* \to B^*$.

Proof Let $\mathsf{IPC}_\Box \vdash A \to B$. By Theorem 4.1 we have $\mathsf{IPC}_\Box \vdash A^* \to A$ and hence $\mathsf{IPC}_\Box \vdash A^* \to B$. Again by Theorm 4.1 we have $\mathsf{IPC}_\Box \vdash A^* \to B^*$. $\Box$

The following lemma states that the NNIL-algorithm does not change the sign of occurrences of boxed propositions. We will use this fact later in Sects. 4.8 and 4.4.3.

Lemma 4.1 *Let* $A, B \in \mathcal{L}_\Box$ *and* $\Box B$ *be a subformula of* A^*. *Then* $\Box B$ *is also a subformula of* A. *Moreover if* $\Box B$ *occurs negatively (positively) in* A^*, *then it also occurs negatively (positively) in* A.

Proof The proof is by induction on the complexity of A. One must consider all cases separately as explained in the above algorithm, and we leave this long and tedious observation to the reader. $\Box$

4.4.3 $\mathsf{NNIL}^\Box$ *Propositions*

A proposition $A \in \mathcal{L}_\Box$ belongs to $\mathsf{NNIL}^{\overrightarrow{\Box}}$, if for any positive occurrence of a boxed proposition $\Box B$ in A, we have $B \in \mathsf{NNIL}$. We say that $A \in \mathcal{L}_\Box$ belongs to $\mathsf{NNIL}^{\overleftarrow{\Box}}$, if for any negative occurrence of a boxed proposition $\Box B$ in A, we have $B \in \mathsf{NNIL}$. Finally we say that A belongs to $\mathsf{NNIL}^\Box$, if it belongs to both $\mathsf{NNIL}^{\overrightarrow{\Box}}$ and $\mathsf{NNIL}^{\overleftarrow{\Box}}$.

Then define $A^{\overrightarrow{\boxed{*}}}$, $A^{\overleftarrow{\boxed{*}}}$ and $A^{\boxed{*}}$ inductively as follows. These translations will nest the translation $(.)^*$ inside positive, negative and all occurrences of boxes, respectively.

- $p^{\overrightarrow{\boxed{*}}} := p^{\overleftarrow{\boxed{*}}} := p^{\boxed{*}} := p$ for atomic p,
- $(\Box B)^{\overrightarrow{\boxed{*}}} := \Box((B^{\overrightarrow{\boxed{*}}})^*)$, and $(\Box B)^{\overleftarrow{\boxed{*}}} := \Box(B^{\overleftarrow{\boxed{*}}})$ and $(\Box B)^{\boxed{*}} := \Box((B^{\boxed{*}})^*)$,
- $(B \circ C)^{\overrightarrow{\boxed{*}}} := B^{\overrightarrow{\boxed{*}}} \circ C^{\overrightarrow{\boxed{*}}}$, $(B \circ C)^{\overleftarrow{\boxed{*}}} := B^{\overleftarrow{\boxed{*}}} \circ C^{\overleftarrow{\boxed{*}}}$ and $(B \circ C)^{\boxed{*}} := B^{\boxed{*}} \circ C^{\boxed{*}}$, for $\circ \in \{\vee, \wedge\}$,
- $(B \to C)^{\overrightarrow{\boxed{*}}} := B^{\overleftarrow{\boxed{*}}} \to C^{\overrightarrow{\boxed{*}}}$, $(B \to C)^{\overleftarrow{\boxed{*}}} := B^{\overrightarrow{\boxed{*}}} \to C^{\overleftarrow{\boxed{*}}}$ and $(B \to C)^{\boxed{*}} := B^{\boxed{*}} \to C^{\boxed{*}}$.

Roughly speaking, one must replace every negatively boxed subformulas of A with its $(.)^*$ translation to calculate $A^{\overleftarrow{\boxed{*}}}$. We have a similar description for $A^{\overrightarrow{\boxed{*}}}$ and $A^{\boxed{*}}$.

Remark 4.1 By induction on the complexity of A, one may observe that

$$A^{\boxed{*}} = \left(A^{\overrightarrow{\boxed{*}}}\right)^{\overleftarrow{\boxed{*}}} = \left(A^{\overleftarrow{\boxed{*}}}\right)^{\overrightarrow{\boxed{*}}}$$

Remark 4.2 If $A \in \mathsf{NNIL}$ then $A^{\overrightarrow{\boxed{*}}} \in \mathsf{NNIL}$ $A^{\overleftarrow{\boxed{*}}} \in \mathsf{NNIL}$ and $A^{\boxed{*}} \in \mathsf{NNIL}$.

The following two lemmas will be used in Theorm 4.5 where we prove the essential properties of the embedding $(.)^{\boxed{h}}$ from iH_σ to $\mathsf{iGLC_a}$.

Lemma 4.2 *For every $A \in \mathcal{L}_\Box$, we have $A^{\overrightarrow{\boxed{*}}} \in \mathsf{NNIL}^{\overrightarrow{\Box}}$ and $A^{\overleftarrow{\boxed{*}}} \in \mathsf{NNIL}^{\overleftarrow{\Box}}$.*

Proof We prove both statements simultaneously by induction on the complexity of A. □

- A is atomic: Trivial.
- $A = \Box B$: Then $A^{\overleftarrow{\boxed{*}}} = \Box B^{\overleftarrow{\boxed{*}}}$. Consider some negative occurrence of $\Box C$ in $A^{\overleftarrow{\boxed{*}}}$. Then $\Box C$ occurs negatively in $B^{\overleftarrow{\boxed{*}}}$ and hence by induction hypothesis, $C \in \mathsf{NNIL}$. This argument shows that $A^{\overleftarrow{\boxed{*}}} \in \mathsf{NNIL}^{\overleftarrow{\Box}}$, as desired. Next we show that $A^{\overrightarrow{\boxed{*}}} = \Box(B^{\overrightarrow{\boxed{*}}})^* \in \mathsf{NNIL}^{\overrightarrow{\Box}}$. Consider some positive occurrence of a boxed formula $\Box C$ in $A^{\overrightarrow{\boxed{*}}}$. If $C = (B^{\overrightarrow{\boxed{*}}})^*$, by Theorem 4.1 we have $C \in \mathsf{NNIL}$. Otherwise, $\Box C$ positively occurs in $(B^{\overrightarrow{\boxed{*}}})^*$. Lemma 4.1 implies that $\Box C$ occurs positively in $B^{\overrightarrow{\boxed{*}}}$ and hence by induction hypothesis $C \in \mathsf{NNIL}$.
- $A = B \circ C$ for $\circ \in \{\vee, \wedge\}$. In this case, every positive (negative) occurrence of $\Box E$ in $A^{\overrightarrow{\boxed{*}}}$ ($A^{\overleftarrow{\boxed{*}}}$), either positively (negatively) occurs in $B^{\overrightarrow{\boxed{*}}}$ ($B^{\overleftarrow{\boxed{*}}}$) or in $C^{\overrightarrow{\boxed{*}}}$ ($C^{\overleftarrow{\boxed{*}}}$). Hence by induction hypothesis we have $E \mathsf{NNIL}$, as desired.
- $A = B \to C$. If $\Box E$ occurs positively in $A^{\overrightarrow{\boxed{*}}}$, then either it negatively occurs in $B^{\overleftarrow{\boxed{*}}}$ or positively occurs in $C^{\overrightarrow{\boxed{*}}}$. Hence induction hypothesis implies $E \in \mathsf{NNIL}$, as desired. With a similar argument, one may show that $E \in \mathsf{NNIL}$ when $\Box E$ negatively occurs in $A^{\overleftarrow{\boxed{*}}}$. □

For arbitrary $A \in \mathcal{L}_\Box$ we have:

- $A^* \in \mathsf{NNIL}$, (Theoem 4.1)
- $A^{\overrightarrow{\boxed{*}}} \in \mathsf{NNIL}^{\overrightarrow{\Box}}$, (Lemma 4.2)
- $A^{\overleftarrow{\boxed{*}}} \in \mathsf{NNIL}^{\overleftarrow{\Box}}$, (Lemma 4.2)
- $A^{\boxed{*}} \in NNIL^{\Box}$, (Lemma 4.2 and Remark 4.1)

Lemma 4.3 *For every $A \in \mathcal{L}_\Box$ we have $(A^{\boxed{*}})^* = (A^*)^{\boxed{*}}$.*

Proof The precise proof, calls for a long and tedious checking of the algorithm of $(.)^*$. We only give some informal argument here and leave details to the reader. One may observe that the algorithm for the computation of $(.)^*$, treats boxed formulas as atomic variables. Also the translation and $(.)^{\boxed{*}}$ only replaces boxed formulas with some other boxed formulas, so its replacements does not affect the computation of $(.)^*$. □

4.5 Leivant's Translation

It is well-known that HA has disjunction property, i.e. HA $\vdash A \vee B$ implies HA $\vdash A$ or HA $\vdash B$. For a proof of this fact, see (Troelstra and van Dale, 1988, Theorem 3.5.10). However HA can not prove the formalization of this fact, i.e. HA $\nvdash \Box(A \vee B) \to (\Box A \vee \Box B)$, in which $\Box$ must be interpreted as "provability in HA" (see Myhill, 1973; Friedman, 1975). Daniel Leivant then shows that a weaker version of the disjunction property is valid in HA. Leivant (1979) shows HA $\vdash \Box(A \vee B) \to \Box(\boxdot A \vee \boxdot B)$, in which again $\Box$ must be interpreted as "provability in HA". With almost the same proof, one may extend the Leivant's principle $\Box(A \vee B) \to \Box(\boxdot A \vee \boxdot B)$. To be able to state an extension of Leivant's principle (that is adequate to axiomatize Σ_1-provability logic of HA Ardeshir and Mojtahedi, 2018) we need a translation $(.)^l$ on modal language which we name it Leivant's translation and recursively defined as follows.

- $A^l := A$ for atomic A, boxed A or $A = \bot$.
- $(A \wedge B)^l := A^l \wedge B^l$.
- $(A \vee B)^l := \boxdot A^l \vee \boxdot B^l$.
- $(A \to B)^l$ is defined by cases: If $A \in$ NOI, i.e. A has no implication outside the scope of $\Box$, define $(A \to B)^l := A \to B^l$, else define $(A \to B)^l := A \to B$.

The following two lemmas will be used later in Sect. 4.7 to show that two different axiomatizations for the Σ_1-provability logic of HA are equivalent (Theorem 4.4).

Lemma 4.4 *For* $A \in$ NNIL *we have* $\mathsf{IPC}_\Box \vdash A^l \to A$.

Proof Use induction on the complexity of A. All cases are straightforward and left to the reader. $\Box$

Lemma 4.5 iK4 $\vdash (A^*)^l \to (A^l)^*$.

Proof Use induction on the complexity of A. All cases are trivial except for $\vee$ and $\to$: $\Box$

- $A = B \vee C$. Then $(A^*)^l = \boxdot(B^*)^l \vee \boxdot(C^*)^l$. Induction hypothesis implies

$$\mathsf{iK4} \vdash [\boxdot(B^*)^l \vee \boxdot(C^*)^l] \to [\boxdot(B^l)^* \vee \boxdot(C^l)^*].$$

  Since $\mathsf{IPC}_\Box \vdash E^* \to E$, we have

$$\mathsf{iK4} \vdash [\boxdot(B^l)^* \vee \boxdot(C^l)^*] \to [((B^l)^* \wedge \Box B^l) \vee ((C^l)^* \wedge \Box C^l)].$$

  The right hand side of the above equation is equal to $(A^l)^*$. Hence iK4 $\vdash (A^*)^l \to (A^l)^*$.
- $A = B \to C$ and $B \in$ NOI. Then it is not difficult to observe that $\mathsf{IPC}_\Box \vdash (A^*)^l \leftrightarrow (B \to (C^*)^l)$. Then by induction hypothesis, iK4 $\vdash (B \to (C^*)^l) \to (B \to (C^l)^*)$. Also the right hand side is $\mathsf{IPC}_\Box$-equivalent to $(A^l)^*$.
- $A = B \to C$ and $B \notin$ NOI. By definition, $A^l = A$. Also Lemma 4.4 implies $\mathsf{IPC}_\Box \vdash (A^*)^l \to A^*$. Hence $\mathsf{IPC}_\Box \vdash (A^*)^l \to (A^*)^l$, as desired. $\Box$

4.6 Gödel's Translation and Heyting's Normal Form

The following translation, is some variant of the Gödel's celebrated translation for embedding IPC in $\mathsf{S4}$ (Gödel, 1933). The Gödel's translation is considered as first provability interpretation for the modal operator. $\square$

Definition 4.4 For every proposition A in modal propositional language, we define A^g inductively as follows:

- $A^g := A$, for atomic or boxed A.
- $(A \circ B)^g := A^g \circ B^g$. for $\circ \in \{\vee, \wedge\}$.
- $(A \to B)^g := \boxdot(A^g \to B^g)$.

Then define the nested Gödel's translations $(.)^{\overleftarrow{\boxed{g}}}$, $(.)^{\overrightarrow{\boxed{g}}}$ and $(.)^{\boxed{g}}$ as follows.

- $p^{\overrightarrow{\boxed{g}}} := p^{\overleftarrow{\boxed{g}}} := p^{\boxed{g}} := p$ for atomic p,
- $(\Box B)^{\overrightarrow{\boxed{g}}} := \Box((B^{\overrightarrow{\boxed{g}}})^g)$, $(\Box B)^{\overleftarrow{\boxed{g}}} := \Box(B^{\overleftarrow{\boxed{g}}})$ and $(\Box B)^{\boxed{g}} := \Box((B^{\boxed{g}})^g)$,
- $(B \circ C)^{\overrightarrow{\boxed{g}}} := B^{\overrightarrow{\boxed{g}}} \circ C^{\overrightarrow{\boxed{g}}}$, $(B \circ C)^{\overleftarrow{\boxed{g}}} := B^{\overleftarrow{\boxed{g}}} \circ C^{\overleftarrow{\boxed{g}}}$ and $(B \circ C)^{\boxed{g}} := B^{\boxed{g}} \circ C^{\boxed{g}}$, for $\circ \in \{\vee, \wedge\}$,
- $(B \to C)^{\overrightarrow{\boxed{g}}} := B^{\overleftarrow{\boxed{g}}} \to C^{\overrightarrow{\boxed{g}}}$, $(B \to C)^{\overleftarrow{\boxed{g}}} := B^{\overrightarrow{\boxed{g}}} \to C^{\overleftarrow{\boxed{g}}}$ and $(B \to C)^{\boxed{g}} := B^{\boxed{g}} \to C^{\boxed{g}}$.

Also define $A^h := (A^*)^g$ and $A^{\boxed{h}} := (A^{\boxed{*}})^{\boxed{g}}$. We say that A is in Heyting's normal form ($A \in \mathsf{HNF}_\sigma$), if either A is not an implication and $A = B^g$ for some $B \in \mathsf{NNIL}$ or $A = B^g \to C^g$ for some $(B \to C) \in \mathsf{NNIL}$. Also define $A \in \mathsf{HNF}_\sigma^{\overrightarrow{\Box}}$ ($A \in \mathsf{HNF}_\sigma^{\overleftarrow{\Box}}$), if for every positive (negative) occurrence of $\Box B$ in A we have $B \in \mathsf{HNF}_\sigma$. Finally let $\mathsf{HNF}_\sigma^{\Box} := \mathsf{HNF}_\sigma^{\overrightarrow{\Box}} \cap \mathsf{HNF}_\sigma^{\overleftarrow{\Box}}$.

Remark 4.3 If $A \in \mathsf{NNIL}$ then $A^g \in \mathsf{NNIL}$. Also $B \to C \in \mathsf{NNIL}$ implies $B^g \to C^g \in \mathsf{NNIL}$.

Remark 4.4 By induction on the complexity of A, one may observe that

$$A^{\boxed{g}} = \left(A^{\overleftarrow{\boxed{g}}}\right)^{\overrightarrow{\boxed{g}}} = \left(A^{\overrightarrow{\boxed{g}}}\right)^{\overleftarrow{\boxed{g}}}$$

Remark 4.5 If $A \in \mathsf{NNIL}$ then $A^{\overleftarrow{\boxed{g}}}, A^{\overrightarrow{\boxed{g}}}, A^{\boxed{g}} \in \mathsf{NNIL}$.

The following lemma expresses the main characteristic of a proposition in the range of $(.)^g$.

Lemma 4.6 $\mathsf{iK4C_a} \vdash A^g \to \Box A^g$.

Proof Use induction on the complexity of A. $\Box$

The rest of this subsection is devoted to some properties of the Gödel's translation and its nested forms. All these properties will be used later in the paper. The reader may simply skip them and read them whenever it is referenced.
The original definition of the Gödel's translation, puts a $\boxdot$ behind every subformula of A. However we considered a variant of Gödel's translation, for technical reasons. The following property does not hold for the original definition of the translation.

Lemma 4.7 *If* $A \in \mathsf{NOI}$, *then* $A^g = A$.

Proof Use induction on the complexity of A. $\Box$

One may easily observe from the definition of $(.)^g$ that every positive occurrence of $\Box B$ in A^g, either positively occurs in A or $B = B_0^g \to B_1^g$ and $B_0 \to B_1$ positively occurs in A. The following lemma is this fact when additionally we have $A \in \mathsf{NNIL}$.

Lemma 4.8 *Let* $A \in \mathsf{NNIL}$ *and* $\Box B$ *positively occurs in* A^g. *Then either* $\Box B$ *positively occurs in* A, *or* $B = B_0^g \to B_1^g$ *for some* $(B_0 \to B_1) \in \mathsf{NNIL}$.

Proof Use induction on the complexity of A. $\Box$

- A is atomic, boxed, conjunction or disjunction. These cases are easy and left to the reader.
- $A = E \to F$. Then $A^g = \boxdot(E^g \to F^g)$. Consider a positive occurrence of $\Box B$ in $\boxdot(E^g \to F^g)$. One of the following cases hold:
- $\Box B = \Box(E^g \to F^g)$. Then $B = E^g \to F^g$ and moreover, since $A \in \mathsf{NNIL}$, we have $E \to F \in \mathsf{NNIL}$.
- $\Box B$ negatively occurs in E^g. Since $A \in \mathsf{NNIL}$, we have $E \in \mathsf{NOI}$ and hence by Lemma 4.7 we have $E^g = E$. Hence $\Box B$ negatively occurs in E, which implies that $\Box B$ positively occurs in A, as desired.
- $\Box B$ positively occurs in F^g. Then induction hypothesis for F, implies the desired result. $\Box$

Lemma 4.9 *For every* $A \in \mathsf{NNIL}^{\Box}$, *we have* $A^{\overrightarrow{\boxed{g}}} \in \mathsf{HNF}_\sigma^{\overrightarrow{\Box}}$ *and* $A^{\overleftarrow{\boxed{g}}} \in \mathsf{HNF}_\sigma^{\overleftarrow{\Box}}$.

Proof We prove both statements by induction on the complexity of A. $\Box$

- A is atomic: trivial.
- $A = \Box B$: Then $A^{\overleftarrow{\boxed{g}}} = \Box(B^{\overleftarrow{\boxed{g}}})$. Consider some negative occurrence of $\Box E$ in $A^{\overleftarrow{\boxed{g}}}$. Then $\Box E$ also negatively occurs in $B^{\overleftarrow{\boxed{g}}}$. Hence induction hypothesis implies the desired result. For the proof of $A^{\overrightarrow{\boxed{g}}} \in \mathsf{HNF}_\sigma^{\overrightarrow{\Box}}$, consider some positive occurrence of $\Box E$ in $A^{\overrightarrow{\boxed{g}}} = \Box((B^{\overrightarrow{\boxed{g}}})^g)$. If $E = (B^{\overrightarrow{\boxed{g}}})^g$, since $B \in \mathsf{NNIL}$ and hence $B^{\overrightarrow{\boxed{g}}} \in \mathsf{NNIL}$, we are done. Otherwise, $\Box E$ positively occurs in $(B^{\overrightarrow{\boxed{g}}})^g$. Lemma 4.8 implies either $\Box E$ positively occurs in $B^{\overrightarrow{\boxed{g}}}$, or $E = E_0^g \to E_1^g$ for some $(E_0 \to E_1) \in \mathsf{NNIL}$. In the first case by induction hypothesis we have the desired result. In the second case, by definition we have $E \in \mathsf{HNF}_\sigma$.

- $A = B \circ C$ for $\circ \in \{\vee, \wedge, \rightarrow\}$. Easy by induction hypothesis and left to the reader. $\square$

Lemma 4.10 *For every* $A \in \mathsf{NNIL}$, *we have* $\mathsf{iK4} \vdash A^g \rightarrow A$.

Proof The proof can be carried out by induction on the complexity of A. We only treat the case for implication. Let $A = B \rightarrow C$. Since $A \in \mathsf{NNIL}$, we have $B \in \mathsf{NOI}$. Lemma 4.7 implies that $A^g = \boxdot(B \rightarrow C^g)$. Hence by induction hypothesis for C, we have $\mathsf{iK4} \vdash A^g \rightarrow (B \rightarrow C)$. $\square$

Lemma 4.11 *For* $A \in \mathcal{L}_\Box$ *we have*

- If $A \in \mathsf{HNF}_\sigma$, then $A \in \mathsf{NNIL}$.
- If $A \in \mathsf{HNF}_\sigma^{\overrightarrow{\Box}}$, then $A \in \mathsf{NNIL}^{\overrightarrow{\Box}}$.
- If $A \in \mathsf{HNF}_\sigma^{\overleftarrow{\Box}}$, then $A \in \mathsf{NNIL}^{\overleftarrow{\Box}}$.
- If $A \in \mathsf{HNF}_\sigma^{\Box}$, then $A \in \mathsf{NNIL}^{\Box}$.

Proof If $A \in \mathsf{HNF}_\sigma$, then by definition, $A = B^g$ for some $B \in \mathsf{NNIL}$. Hence by Lemma 4.10 we have $A \in \mathsf{NNIL}$. The other statements, can be proved by induction on the complexity of A, and the first item. $\square$

The following theorem, is the key to state that two different axiomatizations iH_σ and iH'_σ for the Σ_1-provability logic of HA are equivalent (see Lemma 4.4).

Theorem 4.2 *For every* $A \in \mathsf{NNIL}$, *we have* $\mathsf{iK4C_a} \vdash A^g \leftrightarrow \boxdot A^l$.

Proof Use induction on the complexity of A: $\square$

- A is atomic. We have $A^g = A^l = A$ and since $\mathsf{iK4C_a} \vdash \boxdot A \leftrightarrow A$, we are done.
- $A = \Box B$. Then $A^g = A^l = A$ and since $\mathsf{iK4} \vdash \boxdot\Box B \leftrightarrow \Box B$, we are done.
- $A = B \wedge C$. Use induction hypothesis and $\mathsf{iK4} \vdash \boxdot(E \wedge F) \leftrightarrow (\boxdot E \wedge \boxdot F)$.
- $A = B \vee C$. First note that Lemma 4.6 implies $\mathsf{iK4C_a} \vdash \boxdot E^g \leftrightarrow E^g$, for every $E \in \mathcal{L}_\Box$. Hence in $\mathsf{iK4C_a}$ we have $A^g \leftrightarrow \boxdot A^g$. Then by definition of $(.)^g$, we have $\mathsf{iK4C_a} \vdash A^g \leftrightarrow \boxdot(B^g \vee C^g)$. By induction hypothesis $\mathsf{iK4C_a} \vdash \boxdot(B^g \vee C^g) \leftrightarrow \boxdot(\boxdot B^l \vee \boxdot C^l)$. Hence by definition of $(.)^l$ we have $\mathsf{iK4C_a} \vdash \boxdot(\boxdot B^l \vee \boxdot C^l) \leftrightarrow \boxdot(A^l)$.
- $A = B \rightarrow C$. Then by definition of $(.)^g$ we have $A^g = \boxdot(B^g \rightarrow C^g)$. By induction hypothesis $\mathsf{iK4C_a} \vdash \boxdot(B^g \rightarrow C^g) \leftrightarrow \boxdot(\boxdot B^l \rightarrow \boxdot C^l)$. By a simple reasoning in $\mathsf{iK4}$, we have $\mathsf{iK4} \vdash \boxdot(\boxdot B^l \rightarrow \boxdot C^l) \leftrightarrow \boxdot(\boxdot B^l \rightarrow C^l)$. Hence again by induction hypothesis, $\mathsf{iK4C_a} \vdash \boxdot(\boxdot B^l \rightarrow C^l) \leftrightarrow \boxdot(B^g \rightarrow C^l)$. Since $A \in \mathsf{NNIL}$, we have $B \in \mathsf{NOI}$ and Lemma 4.7 implies $\boxdot(B^g \rightarrow C^l) = \boxdot(B \rightarrow C^l)$. Finally by definition of $(.)^l$ we have $\boxdot(B \rightarrow C^l) = \boxdot A^l$. $\square$

Lemma 4.12 *For $A \in \mathsf{NNIL}$ we have* $\mathsf{iK4} \vdash (A^{\boxast})^g \leftrightarrow (A^g)^{\boxast}$.

Proof We prove this by induction on the complexity of A. All cases are trivial except for $\to$. Let $A = B \to C \in \mathsf{NNIL}$. By definition we have

$$(A^{\boxast})^g = \boxdot((B^{\boxast})^g \to (C^{\boxast})^g) \quad \text{and} \quad (A^g)^{\boxast} = ((B^g)^{\boxast} \to (C^g)^{\boxast}) \wedge \Box((B^g)^{\boxast} \to (C^g)^{\boxast})^*$$

Since $A \in \mathsf{NNIL}$, by Remarks 4.2 and 4.3 we have $(A^g)^{\boxast} \in \mathsf{NNIL}$ and hence $(B^g)^{\boxast} \to (C^g)^{\boxast} \in \mathsf{NNIL}$. Hence by Corollary 4.1 we have $\mathsf{IPC}_\Box \vdash ((B^g)^{\boxast} \to (C^g)^{\boxast})^* \leftrightarrow ((B^g)^{\boxast} \to (C^g)^{\boxast})$. This implies $\mathsf{iK4} \vdash (A^g)^{\boxast} \leftrightarrow \boxdot((B^g)^{\boxast} \to (C^g)^{\boxast})$. Hence the induction hypothesis implies the desired result. $\Box$

Lemma 4.13 *For every $A \in \mathcal{L}_\Box$, we have* $\mathsf{iK4} \vdash A^g \leftrightarrow (A^g)^g$.

Proof Use induction on the complexity of A. The only nontrivial case is $A = B \to C$. Then $A^g = \boxdot(B^g \to C^g)$ and $(A^g)^g = \boxdot((B^g)^g \to (C^g)^g) \wedge \Box(B^g \to C^g)$. By induction hypothesis, $\mathsf{iK4} \vdash (A^g)^g \leftrightarrow \boxdot(B^g \to C^g) \wedge \Box(B^g \to C^g)$. Hence $\mathsf{iK4} \vdash (A^g)^g \leftrightarrow A^g$, as desired. $\Box$

4.7 Axiom Schemas and Theories

In this section, we gather all definitions for the axiom-schemata and theories and logics which we will need later on. Firs axiom-schemata:

- $\underline{\mathsf{i}}$: all axiom schemas of IPC in the Hilbert style proof system, i.e. the following axiom schemas: (1) $A \to (B \to A)$, (2) $(A \to (B \to C)) \to ((A \to B) \to (A \to C))$, (3) $A \to (B \to (A \wedge B))$, (4) $(A \wedge B) \to A$, (5) $(A \wedge B) \to B$, (6) $(A \to C) \to ((B \to C) \to ((A \vee B) \to C))$, (7) $A \to (A \vee B)$, (8) $B \to (A \vee B)$.
- $\underline{\mathsf{K}}$: $\Box(A \to B) \to (\Box A \to \Box B)$,
- $\underline{\mathsf{4}}$: $\Box A \to \Box\Box A$,
- The Löb's axiom, $\underline{\mathsf{Löb}}$ or $\underline{\mathsf{L}}$: $\Box(\Box A \to A) \to \Box A$. (Löb, 1955).
- Restriction of Completeness Principle to atomic variables, $\underline{\mathsf{CP_a}}$ or $\underline{\mathsf{C_a}}$: $p \to \Box p$, for atomic p. (Visser, 1982).
- Leivant's Principle, $\underline{\mathsf{Le}}$: $\Box(B \vee C) \to \Box(\Box B \vee C)$. (Leivant, 1975)
- Extended Σ_1-Leivant's Principle, $\underline{\mathsf{Le}_\sigma}$: $\Box A \to \Box A^l$. (Ardeshir & Mojtahedi, 2018).
- Extended Σ_1-Leivant's Principle, (the weak version) $\underline{\mathsf{Le}'_\sigma}$: $\Box A \to \Box A^g$, for every $A \in \mathsf{NNIL}$.
- The Visser's axiom schema $\underline{\mathsf{V}}$: $\Box A \to \Box A^*$.
- For an axiom schema $\underline{\mathsf{A}}$, the axiom schema $\overline{\mathsf{A}}$ indicates the box of every axiom instance of $\underline{\mathsf{A}}$. Also A indicates $\underline{\mathsf{A}} \wedge \overline{\mathsf{A}}$.

All modal systems which will be defined here, only have one inference rule: modus ponens $\frac{B \quad B \to A}{A}$. Also, celebrated modal logics, like K4, which have the necessitation rule of inference, $\frac{A}{\Box A}$, by abuse of notation, are considered here with the same name and with the same set of theorems, however without the necessitation rule. The reason for this alternate definition of systems, is quite technical. Of course one may define them with the necessitation rule, but at the cost of losing the uniformity of definitions. So in the rest of this paper, all modal systems, are considered with the modus ponens rule of inference. Note that in the presence of the axiom schema 4, one may finitely axiomatize logics such as iK4 and extensions, without necessitation rule.

Consider a list $\mathsf{A}_1, \ldots, \mathsf{A}_n$ of axiom schemas and also L is a modal logic. The notation $\mathsf{A}_1\mathsf{A}_2 \ldots \mathsf{A}_n$ ($L\mathsf{A}_1\mathsf{A}_2 \ldots \mathsf{A}_n$) will be used in this paper for a modal system containing all axiom instances of all axiom schemas A_i (and all axioms of L), and is closed under modus ponens. This general notation makes things uniform and easy to remember for later usage. However, we make the following exceptions:

- $\mathsf{iGL} := \mathsf{iKL}$,
- $\mathsf{GL} := \mathsf{iKLP}$,
- $\mathsf{iH}_\sigma := \mathsf{iGL} + \mathsf{Le}_\sigma + \mathsf{CP_a} + \mathsf{V}$.
- $\mathsf{iH}'_\sigma := \mathsf{iGL} + \mathsf{Le}'_\sigma + \mathsf{CP_a} + \mathsf{V}$.

The following theorem is the main result in Ardeshir and Mojtahedi (2018):

Theorem 4.3 *The Σ_1-Provability logic of* HA *is* iH_σ, *i.e.* $\mathsf{PL}_\Sigma(\mathsf{HA}) = \mathsf{iH}_\sigma$.

Proof For the proof, see Ardeshir and Mojtahedi (2018), Theorems 8.1 and 8.2. □

Theorem 4.4 $\mathsf{iH}_\sigma = \mathsf{iH}'_\sigma$.

Proof Theorem 4.2 implies that iH'_σ is included in iH_σ. To show iH_σ is included in iH'_σ, it is enough to show that $\mathsf{iH}'_\sigma \vdash \Box A \to \Box A^l$ for every $A \in \mathcal{L}_\Box$. By Le'_σ, we have $\mathsf{iH}'_\sigma \vdash \Box A^* \to \Box (A^*)^g$. Then Theorem 4.2 implies $\mathsf{iH}'_\sigma \vdash \Box A^* \to \Box (A^*)^l$. Lemma 4.5 implies $\mathsf{iH}'_\sigma \vdash \Box A^* \to \Box (A^l)^*$. Since by Theorem 4.1 we have $\mathsf{iH}'_\sigma \vdash \Box E \leftrightarrow \Box E^*$, we may conclude $\mathsf{iH}'_\sigma \vdash \Box A \to \Box A^l$, as desired. □

4.8 Embedding of iH'_σ in $\mathsf{iGLC_a}$

In this section, we show (Theorem 4.5) that $(.)^{\boxed{h}}$ embeds iH_σ, the Σ_1-provability logic of HA, in $\mathsf{iGLC_a}$, the intuitionistic counterpart of the Σ_1-provability logic of PA. Moreover we show (Theorem 4.6 and Corollary 4.6) that iH_σ is $\mathsf{HNF}^{\Box}_\sigma$-conservative and also $\mathsf{HNF}^{\overleftarrow{\Box}}_\sigma$-conservative extension of $\mathsf{iGLC_a}$. (For definitions of $\mathsf{HNF}^{\overleftarrow{\Box}}_\sigma$ and $\mathsf{HNF}^{\Box}_\sigma$, see Definition 4.4.)

The following lemma is a first step toward Theorem 4.5, the embedding theorem. It states that $(.)^{\boxed{*}}$ preserves proofs in iK4 and some extensions of it.

Lemma 4.14 iK4X *for any* $\mathsf{X} \subseteq \{\mathsf{CP_a}, \mathsf{Löb}, \mathsf{Le}'_\sigma\}$ *is closed under* $(.)^{\boxed{*}}$.

Proof Use induction on the complexity of the proof $\mathsf{iK4} + \mathsf{X} \vdash \varphi$ and prove $\mathsf{iK4} + \mathsf{X} \vdash \varphi^{\boxed{*}}$. Note that in this paper, the necessitation rule of inference, is missing. The only inference rule is the modus ponens. Nevertheless, necessitation is admissible in the presence of 4 and modus ponens. Having this in our mind, we have only one inductive step: φ is derived by modus ponens from ψ and $\psi \to \varphi$. Then by induction hypothesis, we have $\psi^{\boxed{*}}$ and $\psi^{\boxed{*}} \to \varphi^{\boxed{*}}$. Hence again by modus ponens, we have $\varphi^{\boxed{*}}$, as desired. □

$\underline{\mathsf{i}}$: Note that for every axiom instance φ for $\mathsf{IPC}_\Box$ in the Hilbert style system, $\varphi^{\boxed{*}}$ is also an instance for the same axiom schema.
$\overline{\mathsf{i}}$: By Theorem 4.1, for every axiom instance φ of $\mathsf{IPC}_\Box$, we have $\mathsf{IPC}_\Box \vdash \varphi^*$. Since $\varphi^{\boxed{*}}$ is an instance for the same axiom schema which φ belongs, we have $\mathsf{IPC}_\Box \vdash (\varphi^{\boxed{*}})^*$. Hence $\mathsf{iK4} \vdash (\Box\varphi)^{\boxed{*}}$, as desired.
$\underline{\mathsf{K}}$: We have $\mathsf{IPC}_\Box \vdash [(A^{\boxed{*}} \to B^{\boxed{*}}) \wedge A^{\boxed{*}}] \to B^{\boxed{*}}$. Hence by Corollary 4.2, we have

$$\mathsf{IPC}_\Box \vdash [(A^{\boxed{*}} \to B^{\boxed{*}})^* \wedge (A^{\boxed{*}})^*] \to (B^{\boxed{*}})^*.$$

This implies $\mathsf{iK4} \vdash [\Box(A \to B) \to (\Box A \to \Box B)]^{\boxed{*}}$, as desired.
$\overline{\mathsf{K}}$: Note that

$$\left((\Box(A \to B) \to (\Box A \to \Box B))^{\boxed{*}}\right)^* = (\Box(A \to B) \to (\Box A \to \Box B))^{\boxed{*}}$$

Then by previous item and necessitation we have

$$\mathsf{iK4} \vdash \Box\left((\Box(A \to B) \to (\Box A \to \Box B))^{\boxed{*}}\right)^*.$$

4: Easy and left to the reader.
$\mathsf{CP_a}$: Easy and left to the reader.
$\underline{\mathsf{Löb}}$: Since $(\Box E \to F)^* = \Box E \to F^*$, we have

$$[\Box(\Box A \to A) \to \Box A]^{\boxed{*}} = \Box\left(\Box\left(A^{\boxed{*}}\right)^* \to \left(A^{\boxed{*}}\right)^*\right) \to \Box\left(A^{\boxed{*}}\right)^*$$

The right hand side is an instance of the axiom schema $\underline{\mathsf{Löb}}$.
$\overline{\mathsf{Löb}}$: Easy and left to the reader.
$\underline{\mathsf{Le}'_\sigma}$: Let $A \in \mathsf{NNIL}$. Then by Remark 4.2 $A^{\boxed{*}} \in \mathsf{NNIL}$. Hence by $\underline{\mathsf{Le}'_\sigma}$ we have $\mathsf{iK4Le}'_\sigma \vdash \Box A^{\boxed{*}} \to \Box(A^{\boxed{*}})^g$. Lemma 4.12 implies $\mathsf{iK4Le}'_\sigma \vdash \Box A^{\boxed{*}} \to \Box(A^g)^{\boxed{*}}$. Since $A \in \mathsf{NNIL}$, by Remarks 4.3 and 4.2 we have $(A^g)^{\boxed{*}}, A^{\boxed{*}} \in \mathsf{NNIL}$. Hence by Corollary 4.1 we have $\mathsf{IPC}_\Box \vdash ((A^g)^{\boxed{*}})^* \leftrightarrow ((A^g)^{\boxed{*}})$ and $\mathsf{IPC}_\Box \vdash (A^{\boxed{*}})^* \leftrightarrow A^{\boxed{*}}$ which implies $\mathsf{iK4Le}'_\sigma \vdash \Box(A^{\boxed{*}})^* \to \Box((A^g)^{\boxed{*}})^*$.

Since $\Box(A^{\boxed{*}})^* \to \Box((A^g)^{\boxed{*}})^* = (\Box A \to \Box A^g)^{\boxed{*}}$, we have $\mathsf{iK4Le}'_\sigma \vdash (\Box A \to \Box A^g)^{\boxed{*}}$.

$\overline{\mathsf{Le}'_\sigma}$: Let $A \in \mathsf{NNIL}$. By previous item we have $\mathsf{iK4Le}'_\sigma \vdash \Box(\Box A \to \Box A^g)^{\boxed{*}}$. Since $(\Box A \to \Box A^g)^{\boxed{*}} \in \mathsf{NNIL}$, we have $\mathsf{iK4Le}'_\sigma \vdash \Box[(\Box A \to \Box A^g)^{\boxed{*}}]^*$. Since $\Box[(\Box A \to \Box A^g)^{\boxed{*}}]^* = [\Box(\Box A \to \Box A^g)]^{\boxed{*}}$, we have the desired result. □

In the second step toward Theorem 4.5, we show that the translation $(.)^{\boxed{g}}$ respects the proofs in $\mathsf{iK4C_a}$ and $\mathsf{iGLC_a}$.

Lemma 4.15 $\mathsf{iK4C_a}$ *and* $\mathsf{iGLC_a}$ *are closed under the Gödel's nested translation* $(.)^{\boxed{g}}$.

Proof Use induction on the complexity of the proof $\mathsf{iK4C_a} \vdash \varphi$. We have only one inductive step: φ is derived by modus ponens from ψ and $\psi \to \varphi$. Then by induction hypothesis, we have $\psi^{\boxed{g}}$ and $\psi^{\boxed{g}} \to \varphi^{\boxed{g}}$. Hence again by modus ponens, we have $\varphi^{\boxed{g}}$, as desired. It remains to show here that $\varphi^{\boxed{g}}$ is provable in $\mathsf{iGLC_a}$ if φ is an axiom schema of $\mathsf{iGLC_a}$. □

$\underline{\mathsf{i}}$: If φ is an axiom instance of IPC, observe that $\varphi^{\boxed{g}}$ is an axiom instance of the same schema, and hence provable in $\mathsf{IPC}_\Box$.

$\overline{\mathsf{i}}$: Let $\varphi = \Box A$ for some axiom instance A of IPC. Then also $A^{\boxed{g}}$ is an instance of the same axiom schema. Hence by fact 1, we have $\mathsf{iK4C_a} \vdash (A^{\boxed{g}})^g$ and then $\mathsf{iK4C_a} \vdash \Box(A^{\boxed{g}})^g$. Since by definition, $\varphi^{\boxed{g}} = \Box(A^{\boxed{g}})^g$, we have the desired result.

$\underline{\mathsf{K}}$: Let $\varphi = \Box(A \to B) \to (\Box A \to \Box B)$. We have $\varphi^{\boxed{g}} = \Box((A^{\boxed{g}} \to B^{\boxed{g}})^g) \to (\Box(A^{\boxed{g}})^g \to \Box(B^{\boxed{g}})^g)$. Since in $\mathsf{iK4}$ we have $\Box = \Box\boxdot$, we have

$$\mathsf{iK4} \vdash \varphi^{\boxed{g}} \leftrightarrow \Box((A^{\boxed{g}})^g \to (B^{\boxed{g}})^g) \to (\Box(A^{\boxed{g}})^g \to \Box(B^{\boxed{g}})^g).$$

The right hand side is again an instance of the axiom schema $\underline{\mathsf{K}}$. Hence $\mathsf{iK4} \vdash \varphi^{\boxed{g}}$.

$\overline{\mathsf{K}}$: Observe that $\mathsf{iK4} \vdash (\Box\varphi)^{\boxed{g}} \leftrightarrow \Box(\varphi^{\boxed{g}})$, in which φ is as in the previous item. Then use the previous item and necessitation.

$\underline{\mathsf{L\ddot{o}b}}$: Let $\varphi = \Box(\Box A \to A) \to \Box A$. Observe that $\mathsf{iK4} \vdash \varphi^{\boxed{g}} \leftrightarrow [\Box(\Box(A^{\boxed{g}})^g \to (A^{\boxed{g}})^g) \to \Box(A^{\boxed{g}})^g]$. Since the right hand side is also an instance of the axiom schema Löb, we have $\mathsf{iGL} \vdash \varphi^{\boxed{g}}$, as desired.

$\overline{\mathsf{L\ddot{o}b}}$: Observe that $\mathsf{iK4} \vdash (\Box\varphi)^{\boxed{g}} \leftrightarrow \Box\varphi^{\boxed{g}}$ and use necessitation together with previous item. (φ is as defined in previous item.) □

Since $A^{\boxed{h}} = (A^{\boxed{*}})^{\boxed{g}}$, with Lemmas 4.15 and 4.14, we almost have all required preliminaries to prove Theorem 4.5. However we also need some technical lemmas which comes afterwards.

Lemma 4.16 $\mathsf{iK4C_aX}$ *for* $\mathsf{X} \subseteq \{\mathsf{Löb}, \mathsf{Le}'_\sigma, \mathsf{V}\}$ *is closed under the translation* $(.)^g$.

Proof Use induction on the complexity of the proof $\mathsf{iK4C_a} \vdash \varphi$. We have only one inductive step: φ is derived by modus ponens from ψ and $\psi \to \varphi$. Then by induction hypothesis, we have ψ^g and $\psi^g \to \varphi^g$. Hence again by modus ponens, we have φ^g, as desired. It remains to show here that φ^g is provable in $\mathsf{iK4C_aX}$ for every axiom instance φ from $\mathsf{iK4C_aX}$. All cases are trivial except for when φ is an axiom instance of IPC: (1) $A \to (B \to A)$, (2) $(A \to (B \to C)) \to ((A \to B) \to (A \to C))$, (3) $A \to (B \to (A \wedge B))$, (4) $(A \wedge B) \to A$, (5) $(A \wedge B) \to B$, (6) $(A \to C) \to ((B \to C) \to ((A \vee B) \to C))$, (7) $A \to (A \vee B)$, (8) $B \to (A \vee B)$. We only treat the first axiom here. We have $\mathsf{IPC}_\Box \vdash A^g \to (B^g \to A^g)$. By necessitation $\mathsf{iK4} \vdash \Box A^g \to \Box(B^g \to A^g)$. Then Lemma 4.6 implies $\mathsf{iK4C_a} \vdash A^g \to \boxdot(B^g \to A^g)$. Hence $\mathsf{iK4C_a} \vdash (A \to (B \to A))^g$, as desired. □

Lemma 4.17 *For every* $A \in \mathcal{L}_\Box$ *we have* $\mathsf{iK4C_a} \vdash (A^g)^{\boxed{g}} \leftrightarrow (A^{\boxed{g}})^g$.

Proof Use induction on the complexity of A. The only nontrivial case is $A = B \to C$. Then $\left(A^{\boxed{g}}\right)^g = \boxdot\left(\left(B^{\boxed{g}}\right)^g \to \left(C^{\boxed{g}}\right)^g\right)$ and

$$(A^g)^{\boxed{g}} = \left((B^g)^{\boxed{g}} \to (C^g)^{\boxed{g}}\right) \wedge \Box\boxdot\left(\left((B^g)^{\boxed{g}}\right)^g \to \left((C^g)^{\boxed{g}}\right)^g\right).$$

By induction hypothesis we have

$$\mathsf{iK4C_a} \vdash \left((B^g)^{\boxed{g}} \to (C^g)^{\boxed{g}}\right) \leftrightarrow \left(\left(B^{\boxed{g}}\right)^g \to \left(C^{\boxed{g}}\right)^g\right).$$

Lemma 4.16 implies

$$\mathsf{iK4C_a} \vdash \boxdot\left(\left((B^g)^{\boxed{g}}\right)^g \to \left((C^g)^{\boxed{g}}\right)^g\right) \leftrightarrow \boxdot\left(\left(\left(B^{\boxed{g}}\right)^g\right)^g \to \left(\left(C^{\boxed{g}}\right)^g\right)^g\right).$$

Also by Lemma 4.13 we have

$$\mathsf{iK4} \vdash \boxdot\left(\left(\left(B^{\boxed{g}}\right)^g\right)^g \to \left(\left(C^{\boxed{g}}\right)^g\right)^g\right) \leftrightarrow \boxdot\left(\left(B^{\boxed{g}}\right)^g \to \left(C^{\boxed{g}}\right)^g\right).$$

Hence

$$\mathsf{iK4C_a} \vdash (A^g)^{\boxed{g}} \leftrightarrow \left[\left(\left(B^{\boxed{g}}\right)^g \to \left(C^{\boxed{g}}\right)^g\right) \wedge \Box\boxdot\left(\left(B^{\boxed{g}}\right)^g \to \left(C^{\boxed{g}}\right)^g\right)\right].$$

Since in $\mathsf{iK4}$, $\Box\boxdot$ is equivalent to $\Box$, the right hand side of the above formula, is equivalent to $(A^{\boxed{g}})^g$, as desired. □

Corollary 4.3 *For every* $A \in \mathcal{L}_\Box$ *we have* $\mathsf{iK4C_a} \vdash \left((A^g)^{\boxed{g}}\right)^g \leftrightarrow \left(A^{\boxed{g}}\right)^g$ *and* $\mathsf{iK4C_a} \vdash \left((A^g)^{\boxed{g}}\right)^g \leftrightarrow (A^g)^{\boxed{g}}$.

Proof Lemma 4.17 implies $\mathsf{iK4C_a} \vdash (A^g)^{\boxed{g}} \leftrightarrow (A^{\boxed{g}})^g$ and then by Lemma 4.16 we have $\mathsf{iK4C_a} \vdash \left((A^g)^{\boxed{g}}\right)^g \leftrightarrow \left((A^{\boxed{g}})^g\right)^g$. Hence by Lemma 4.13 we have $\mathsf{iK4C_a} \vdash \left((A^{\boxed{g}})^g\right)^g \leftrightarrow (A^{\boxed{g}})^g$.

For proof of the other statement, by Lemma 4.13 we have $\mathsf{iK4} \vdash A^g \leftrightarrow (A^g)^g$. Hence Lemma 4.15 implies $\mathsf{iK4C_a} \vdash (A^g)^{\boxed{g}} \leftrightarrow ((A^g)^g)^{\boxed{g}}$. Then Lemma 4.17 implies $\mathsf{iK4C_a} \vdash ((A^g)^g)^{\boxed{g}} \leftrightarrow \left((A^g)^{\boxed{g}}\right)^g$. $\square$

Lemma 4.18 *Let $A \in \mathcal{L}_\Box$. Every negative (positive) occurrence of $\Box B$ in $A^{\overrightarrow{\boxed{*}}}$ (in $A^{\overleftarrow{\boxed{*}}}$) also negatively (positively) occurs in A. Moreover, if $A \in \mathsf{NNIL}$, every negative (positive) occurrence of $\Box B$ in $A^{\overrightarrow{\boxed{g}}}$ (in $A^{\overleftarrow{\boxed{g}}}$) also negatively (positively) occurs in A.*

Proof The proof can be carried out easily by induction on the complexity of A. All cases are trivial except for $A = \Box B$, for which we have the desired result by Lemma 4.1. $\square$

Corollary 4.4 *For every $A \in \mathcal{L}_\Box$, if $A \in \mathsf{NNIL}^{\overleftarrow{\Box}}$ ($A \in \mathsf{NNIL}^{\overrightarrow{\Box}}$) then $A^{\overrightarrow{\boxed{*}}} \in \mathsf{NNIL}^{\overleftarrow{\Box}}$ ($A^{\overleftarrow{\boxed{*}}} \in \mathsf{NNIL}^{\overrightarrow{\Box}}$).*

Proof Use Lemma 4.18. $\square$

Corollary 4.5 *For every $A \in \mathcal{L}_\Box$, if $A \in \mathsf{HNF}_\sigma^{\overleftarrow{\Box}}$ ($A \in \mathsf{HNF}_\sigma^{\overrightarrow{\Box}}$) then $A^{\overrightarrow{\boxed{*}}}, A^{\overrightarrow{\boxed{g}}} \in \mathsf{HNF}_\sigma^{\overleftarrow{\Box}}$ ($A^{\overleftarrow{\boxed{*}}}, A^{\overleftarrow{\boxed{g}}} \in \mathsf{HNF}_\sigma^{\overrightarrow{\Box}}$).*

Proof Use Lemma 4.18. $\square$

Lemma 4.19 *For every $A \in \mathcal{L}_\Box$, $\mathsf{iH}_\sigma \vdash A$ implies $\mathsf{iGLC_a} \vdash A^{\boxed{h}}$.*

Proof Let $\mathsf{iH}_\sigma \vdash A$. Hence by Theorem 4.4 $\mathsf{iH}'_\sigma \vdash A$. Then $\mathsf{iGLC_aLe}'_\sigma \vdash \bigwedge X \to A$ for some finite set X of instances of the axiom schema V. Then by Lemma 4.14 we have $\mathsf{iGLC_aLe}'_\sigma \vdash \bigwedge X^{\boxed{*}} \to A^{\boxed{*}}$, in which $X^{\boxed{*}} := \{B^{\boxed{*}} : B \in X\}$. For every $B^{\boxed{*}} \in X^{\boxed{*}}$ we show $\mathsf{iK4} \vdash B^{\boxed{*}}$. We have the following cases:

- $B = \Box E \to \Box E^*$: then $B^{\boxed{*}} = \Box(E^{\boxed{*}})^* \to \Box((E^*)^{\boxed{*}})^*$. Theorem 4.1 implies $E^* \in \mathsf{NNIL}$ and Remark 4.2 implies $(E^*)^{\boxed{*}} \in \mathsf{NNIL}$. Then Corollary 4.1 implies $\mathsf{IPC}_\Box \vdash ((E^*)^{\boxed{*}})^* \leftrightarrow (E^*)^{\boxed{*}}$. Hence $\mathsf{iK4} \vdash \Box(E^*)^{\boxed{*}} \to \Box((E^*)^{\boxed{*}})^*$, and then by Lemma 4.3 we have $\mathsf{iK4} \vdash \Box(E^{\boxed{*}})^* \to \Box((E^*)^{\boxed{*}})^*$.
- $B = \Box(\Box E \to \Box E^*)$: use necessitation and the previous item to show $\mathsf{iK4} \vdash B^{\boxed{*}}$.

This argument shows that $\mathsf{iGLC_aLe}'_\sigma \vdash A^{\boxed{*}}$. Hence there is some finite set Y of the instances of the axiom schema Le'_σ such that $\mathsf{iGLC_a} \vdash \bigwedge Y \to A^{\boxed{*}}$. Lemma 4.15 implies $\mathsf{iGLC_a} \vdash \bigwedge Y^{\boxed{g}} \to (A^{\boxed{*}})^{\boxed{g}}$. It is enough to show that $\mathsf{iGLC_a} \vdash \bigwedge Y^{\boxed{g}}$. Then we would have $\mathsf{iGLC_a} \vdash (A^{\boxed{*}})^{\boxed{g}}$, as desired. Consider some instance of the axiom schema Le'_σ of the form $B = \Box E \to \Box E^g$, in which $E \in \mathsf{NNIL}$. By definition we have $B^{\boxed{g}} = \Box(E^{\boxed{g}})^g \to \Box((E^g)^{\boxed{g}})^g$. Corollary 4.3 implies $\mathsf{iK4C_a} \vdash B^{\boxed{g}} \leftrightarrow [\Box(E^{\boxed{g}})^g \to (E^{\boxed{g}})^g]$ and hence $\mathsf{iK4C_a} \vdash B^{\boxed{g}}$. $\square$

Theorem 4.5 *The translation* $(.)^{\boxed{h}}$ *has the following properties:*

1. $\mathsf{iH}'_\sigma \vdash A \leftrightarrow A^{\boxed{h}}$.
2. $\mathsf{iH}'_\sigma \vdash A$ *iff* $\mathsf{iGLC_a} \vdash A^{\boxed{h}}$.
3. $(A \to B)^{\boxed{h}} = A^{\boxed{h}} \to B^{\boxed{h}}$.

Proof Item 3 is trivial. Item 2 is a consequence of Lemma 4.19 and first item. So it remains to prove item 1. First by induction on the complexity of A, observe that $\mathsf{iH}'_\sigma \vdash A \leftrightarrow A^{\boxed{*}}$. Also again by induction on the complexity of $A \in \mathsf{NNIL}^\Box$ we have $\mathsf{iH}'_\sigma \vdash A \leftrightarrow A^{\boxed{g}}$. Then by Lemma 4.2 we have $A^{\boxed{*}} \in \mathsf{NNIL}^\Box$ and hence $\mathsf{iH}'_\sigma \vdash A \leftrightarrow (A^{\boxed{*}})^{\boxed{g}}$, as desired. □

Next we show in Theorem 4.6 that iH_σ is an $\mathsf{HNF}^\Box_\sigma$-conservative extension of $\mathsf{iGLC_a}$. The following two technical lemmas will be helpful in the proof of Theorem 4.6.

Lemma 4.20 *For every* $A \in \mathsf{NNIL}^\Box$ *we have* $\mathsf{iK4} \vdash A^{\boxed{*}} \leftrightarrow A$.

Proof Use induction on the complexity of A. We only treat the case $A = \Box B$ here and leave other cases to the reader. Since$A \in \mathsf{NNIL}^\Box$ and $\Box B$ positively occurs in A, we have $B \in \mathsf{NNIL}$. Also by definition $A^{\boxed{*}} := \Box((B^{\boxed{*}})^*)$. Remark 4.2 implies $B^{\boxed{*}} \in \mathsf{NNIL}$. Then Corollary 4.1 implies $\mathsf{IPC}_\Box \vdash (B^{\boxed{*}})^* \leftrightarrow B^{\boxed{*}}$. Then induction hypothesis implies $\mathsf{iK4} \vdash A^{\boxed{*}} \leftrightarrow \Box B$, as desired. □

Lemma 4.21 *For every* $A \in \mathsf{HNF}^\Box_\sigma$ *we have* $\mathsf{iK4} \vdash A^{\boxed{g}} \leftrightarrow A$.

Proof Use induction on the complexity of A. We only treat the case $A = \Box B$ here and leave other cases to the reader. Since $A \in \mathsf{HNF}^\Box_\sigma$, and $\Box B$ positively occurs in A, there is some $B_0 \in \mathsf{NNIL}$ such that $B = B_0^g$. Also by definition we have $A^{\boxed{g}} := \Box((B^{\vec{\boxed{g}}})^g)$. Hence $A^{\boxed{g}} = \Box(((B_0^g)^{\boxed{g}})^g)$ and by Corollary 4.3 we have $\mathsf{iK4C_a} \vdash A^{\boxed{g}} \leftrightarrow \Box((B_0^g)^{\boxed{g}})$. Then induction hypothesis for B implies the desired result. □

Theorem 4.6 iH_σ *is* $\mathsf{HNF}^\Box_\sigma$*-conservative extension of* $\mathsf{iGLC_a}$. *In other words,* iH_σ *and* $\mathsf{iGLC_a}$ *prove the same* $\mathsf{HNF}^\Box_\sigma$*-propositions.*

Proof Let $\mathsf{iH}_\sigma \vdash A$ and $A \in \mathsf{HNF}^\Box_\sigma$. By Lemma 4.19 we have $\mathsf{iGLC_a} \vdash (A^{\boxed{*}})^{\boxed{g}}$. Lemma 4.20 implies $\mathsf{iK4} \vdash A^{\boxed{*}} \leftrightarrow A$ and hence by Lemma 4.15 we have $\mathsf{iK4C_a} \vdash (A^{\boxed{*}})^{\boxed{g}} \leftrightarrow A^{\boxed{g}}$. This implies $\mathsf{iGLC_a} \vdash A^{\boxed{g}}$ and then by Lemma 4.21 we have $\mathsf{iGLC_a} \vdash A$, as desired. □

Our next goal is to strengthen this conservativity result in this way: iH_σ is $\overleftarrow{\mathsf{HNF}^\Box_\sigma}$-conservative over $\mathsf{iGLC_a}$. This is the statement of Corollary 4.6. The following lemmas will be used for this aim.

Lemma 4.22 *For every $A \in \mathcal{L}_\Box$, we have*

1. $A \in \mathsf{NNIL}^{\overrightarrow{\Box}}$ *implies* $\mathsf{iK4} \vdash A^{\overrightarrow{\boxed{g}}} \to A$,
2. $A \in \mathsf{NNIL}^{\overleftarrow{\Box}}$ *implies* $\mathsf{iK4} \vdash A \to A^{\overleftarrow{\boxed{g}}}$,
3. $A \in \mathsf{NNIL}^{\overrightarrow{\Box}}$ *implies* $\mathsf{iK4C_aLe_\sigma} \vdash A \to A^{\overrightarrow{\boxed{g}}}$,
4. $A \in \mathsf{NNIL}^{\overleftarrow{\Box}}$ *implies* $\mathsf{iK4C_aLe_\sigma} \vdash A^{\overleftarrow{\boxed{g}}} \to A$.

Proof Use induction on the complexity of A:

- A is atomic. Then $A^{\overrightarrow{\boxed{g}}} := A^{\overleftarrow{\boxed{g}}} := A$, and trivially we are done.
- $A = \Box B$. Assume that $A \in \mathsf{NNIL}^{\overleftarrow{\Box}}$. Then $(\Box B)^{\overleftarrow{\boxed{g}}} := \Box(B^{\overleftarrow{\boxed{g}}})$, and by induction hypothesis for B, and the necessitation we have items 2 and 4 for A. Next consider $A \in \mathsf{NNIL}^{\overrightarrow{\Box}}$ which implies $B \in \mathsf{NNIL}$ and $B \in \mathsf{NNIL}^{\overrightarrow{\Box}}$. For the proof of item 1, we reason as follows: By induction hypothesis (item 1) we have $\mathsf{iK4} \vdash B^{\overrightarrow{\boxed{g}}} \to B$. Also by Lemma 4.10 we have $\mathsf{iK4} \vdash (B^{\overrightarrow{\boxed{g}}})^g \to B^{\overrightarrow{\boxed{g}}}$, and hence $\mathsf{iK4} \vdash (B^{\overrightarrow{\boxed{g}}})^g \to B$. Then by necessitation we have $\mathsf{iK4} \vdash \Box((B^{\overrightarrow{\boxed{g}}})^g) \to \Box B$. Since by definition $(\Box B)^{\overrightarrow{\boxed{g}}} := \Box((B^{\overrightarrow{\boxed{g}}})^g)$, we have $\mathsf{iK4} \vdash (\Box B)^{\overrightarrow{\boxed{g}}} \to \Box B$, which is the desired result (item 1) for $A = \Box B$. It remains to prove item 3 for $A = \Box B$. By induction hypothesis (item 3) for B, we have $\mathsf{iK4C_aLe_\sigma} \vdash B \to B^{\overrightarrow{\boxed{g}}}$. Hence by necessitation, we have $\mathsf{iK4C_aLe_\sigma} \vdash \Box B \to \Box B^{\overrightarrow{\boxed{g}}}$. By axiom schema $\mathsf{Le_\sigma}$, we have $\mathsf{iK4C_aLe_\sigma} \vdash \Box B^{\overrightarrow{\boxed{g}}} \to \Box((B^{\overrightarrow{\boxed{g}}})^l)$ and hence $\mathsf{iK4C_aLe_\sigma} \vdash \Box B^{\overrightarrow{\boxed{g}}} \to \Box(\boxdot(B^{\overrightarrow{\boxed{g}}})^l)$. Since by Remark 4.5 we have $B^{\overrightarrow{\boxed{g}}} \in \mathsf{NNIL}$, Theorem 4.2 implies $\mathsf{iK4C_aLe_\sigma} \vdash \Box B^{\overrightarrow{\boxed{g}}} \to \Box((B^{\overrightarrow{\boxed{g}}})^g)$. Hence $\mathsf{iK4C_aLe_\sigma} \vdash \Box B \to (\Box B)^{\overrightarrow{\boxed{g}}}$, as desired.
- $A = B \circ C$, and $\circ \in \{\vee, \wedge\}$. Easy by induction hypothesis and left to the reader.
- $A = B \to C$. We only prove item 3. Treatment of other items are similar and we leave them to the reader. Let $A \in \mathsf{NNIL}^{\overrightarrow{\Box}}$. Hence $C \in \mathsf{NNIL}^{\overrightarrow{\Box}}$ and $B \in \mathsf{NNIL}^{\overleftarrow{\Box}}$. By induction hypothesis (item 3) for C, we have $\mathsf{iK4C_aLe_\sigma} \vdash C \to C^{\overrightarrow{\boxed{g}}}$. By item 4 of the induction hypothesis for B, we have $\mathsf{iK4C_aLe_\sigma} \vdash B^{\overleftarrow{\boxed{g}}} \to B$. Hence, $\mathsf{iK4C_aLe_\sigma} \vdash (B \to C) \to (B^{\overleftarrow{\boxed{g}}} \to C^{\overrightarrow{\boxed{g}}})$, and then $\mathsf{iK4C_aLe_\sigma} \vdash A \to A^{\overrightarrow{\boxed{g}}}$, as desired. □

Lemma 4.23 *For every $A \in \mathcal{L}_\Box$, we have*

1. $\mathsf{iK4} \vdash A^{\overrightarrow{\boxed{*}}} \to A$,
2. $\mathsf{iK4} \vdash A \to A^{\overleftarrow{\boxed{*}}}$,
3. $\mathsf{iK4V} \vdash A \to A^{\overrightarrow{\boxed{*}}}$,
4. $\mathsf{iK4V} \vdash A^{\overleftarrow{\boxed{*}}} \to A$.

Proof The proof can be carried out very similar to the proof of Lemma 4.22. One must use the axiom schema $\mathsf{V_\sigma}$ instead of $\mathsf{Le_\sigma}$ and also Theorem 4.1 instead of Lemma 4.10. □

Lemma 4.24 *For every $A \in \mathcal{L}_\Box$,*

- $A \in \mathsf{NNIL}^{\overrightarrow{\Box}}$ *implies* $A^{\overrightarrow{\boxed{g}}}, A^{\overleftarrow{\boxed{g}}} \in \mathsf{NNIL}^{\overrightarrow{\Box}}$,
- $A \in \mathsf{NNIL}^{\overleftarrow{\Box}}$ *implies* $A^{\overrightarrow{\boxed{g}}}, A^{\overleftarrow{\boxed{g}}} \in \mathsf{NNIL}^{\overleftarrow{\Box}}$.

Proof First note that if $A \in \mathsf{NNIL}^{\overrightarrow{\Box}}$ then by Lemma 4.18 we have $A^{\overleftarrow{\boxed{g}}} \in \mathsf{NNIL}^{\overrightarrow{\Box}}$. Also if $A \in \mathsf{NNIL}^{\overleftarrow{\Box}}$ then by Lemma 4.18 we have $A^{\overrightarrow{\boxed{g}}} \in \mathsf{NNIL}^{\overleftarrow{\Box}}$. Prove the remaining statements by induction on the complexity of A. We only treat the boxed case and leave other cases to reader. Let $A = \Box B$ and $A \in \mathsf{NNIL}^{\overrightarrow{\Box}}$. We show $A^{\overrightarrow{\boxed{g}}} \in \mathsf{NNIL}^{\overrightarrow{\Box}}$. By definition we have $A^{\overrightarrow{\boxed{g}}} := \Box((B^{\overrightarrow{\boxed{g}}})^g)$. Consider some positive occurrence of $\Box C$ in $\Box((B^{\overrightarrow{\boxed{g}}})^g)$. We have two cases: □

- $C = (B^{\overrightarrow{\boxed{g}}})^g$. Since $\Box B \in \mathsf{NNIL}^{\overrightarrow{\Box}}$ and $\Box B$ positively occurs in A, we have $B \in \mathsf{NNIL}$. Remark 4.5 implies $B^{\overrightarrow{\boxed{g}}} \in \mathsf{NNIL}^{\overrightarrow{\Box}}$. Then Remark 4.3 implies $C \in \mathsf{NNIL}$.
- $\Box C$ positively occurs in $(B^{\overrightarrow{\boxed{g}}})^g$. Since $B^{\overrightarrow{\boxed{g}}} \in \mathsf{NNIL}$, Lemma 4.8 we have two sub-cases:

  - $\Box C$ positively occurs in $B^{\overrightarrow{\boxed{g}}}$. By induction hypothesis for B we have $C \in \mathsf{NNIL}$.
  - $C = C_0^g \to C_1^g$ and $C_0 \to C_1 \in \mathsf{NNIL}$. Then by Remark 4.3 we have $C \in \mathsf{NNIL}$. □

Lemma 4.25 *Let $A \in \mathsf{HNF}_\sigma^{\overleftarrow{\Box}}$. There is some $A' \in \mathsf{HNF}_\sigma^{\Box}$ such that*

- $\mathsf{iK4C_a} \vdash A' \to A$.
- $\mathsf{iH}_\sigma \vdash A \to A'$.

Proof Let $A' := (A^{\overrightarrow{\boxed{*}}})^{\overrightarrow{\boxed{g}}}$. By Lemma 4.23 we have $\mathsf{iK4} \vdash A^{\overrightarrow{\boxed{*}}} \to A$ and $\mathsf{iK4V} \vdash A \to A^{\overrightarrow{\boxed{*}}}$. Since $A^{\overrightarrow{\boxed{*}}} \in \mathsf{NNIL}^{\overleftarrow{\Box}}$, by Lemma 4.22 we have $\mathsf{iK4C_a} \vdash (A^{\overrightarrow{\boxed{*}}})^{\overrightarrow{\boxed{g}}} \to A^{\overrightarrow{\boxed{*}}}$ and $\mathsf{iK4C_aLe}_\sigma \vdash A^{\overrightarrow{\boxed{*}}} \to (A^{\overrightarrow{\boxed{*}}})^{\overrightarrow{\boxed{g}}}$. Hence we have $\mathsf{iK4C_a} \vdash A' \to A$. and $\mathsf{iH}_\sigma \vdash A \to A'$. It remains only to show that $A' \in \mathsf{HNF}_\sigma^{\Box}$. Since $A \in \mathsf{HNF}_\sigma^{\overleftarrow{\Box}}$, by Lemma 4.11, $A \in \mathsf{NNIL}^{\overleftarrow{\Box}}$. Lemma 4.4 implies $A^{\overrightarrow{\boxed{*}}} \in \mathsf{NNIL}^{\overleftarrow{\Box}}$. By Lemma 4.2 we have $A^{\overrightarrow{\boxed{*}}} \in \mathsf{NNIL}^{\overrightarrow{\Box}}$. Lemma 4.24 implies $(A^{\overrightarrow{\boxed{*}}})^{\overrightarrow{\boxed{g}}} \in \mathsf{NNIL}^{\overrightarrow{\Box}}$ and $(A^{\overrightarrow{\boxed{*}}})^{\overrightarrow{\boxed{g}}} \in \mathsf{NNIL}^{\overleftarrow{\Box}}$. Hence $(A^{\overrightarrow{\boxed{*}}})^{\overrightarrow{\boxed{g}}} \in \mathsf{NNIL}^{\Box}$. Lemma 4.9 implies that $(A^{\overrightarrow{\boxed{*}}})^{\overrightarrow{\boxed{g}}} \in \mathsf{HNF}_\sigma^{\overrightarrow{\Box}}$. On the other hand, since $A \in \mathsf{HNF}_\sigma^{\overleftarrow{\Box}}$, Lemma 4.5 implies $(A^{\overrightarrow{\boxed{*}}})^{\overrightarrow{\boxed{g}}} \in \mathsf{HNF}_\sigma^{\overleftarrow{\Box}}$. Hence by definition we have $(A^{\overrightarrow{\boxed{*}}})^{\overrightarrow{\boxed{g}}} \in \mathsf{HNF}_\sigma^{\Box}$. □

Corollary 4.6 iH_σ *is* $\mathsf{HNF}_\sigma^{\overleftarrow{\Box}}$*-conservative extension of* $\mathsf{iGLC_a}$*, in other words* iH_σ *and* $\mathsf{iGLC_a}$ *prove the same* $\mathsf{HNF}_\sigma^{\overleftarrow{\Box}}$ *propositions.*

Proof Let $\mathsf{iH}_\sigma \vdash A$ for some $A \in \mathsf{HNF}_\sigma^{\overleftarrow{\Box}}$. Let $A' \in \mathsf{HNF}_\sigma^{\Box}$ as provided by Lemma 4.25. Hence $\mathsf{iH}_\sigma \vdash A'$. Theorem 4.6 implies $\mathsf{iGLC_a} \vdash A'$ and hence by Lemma 4.25 we have $\mathsf{iGLC_a} \vdash A$. □

Future works
Let $\underline{\mathsf{S}}$ indicate the axiom schema $\Box A \to A$ and $\underline{\mathsf{P}}$ be the principle of excluded middle $A \vee \neg A$. Mojtahedi (2019) characterizes the truth Σ_1-provability logic of HA as $\mathsf{iH}_\sigma\underline{\mathsf{SP}}$. On the other hand, Visser (1982) characterizes the truth Σ_1-provability logic of PA as $\mathsf{GL\underline{S}C_a}$. We conjecture that $\mathsf{iH}_\sigma\underline{\mathsf{SP}}$ is embeddable in the intuitionistic counterpart of $\mathsf{GL\underline{S}C_a}$, i.e. $\mathsf{iGL\underline{SP}C_a}$.

Also (Ardeshir & Mojtahedi, 2019) characterizes the Σ_1-provability logic of HA^*, the self-completion of HA. Also (Visser, 1982) characterizes the Σ_1-provability logic of PA^*, as iGLCT, in which $\underline{\mathsf{T}} := \Box(A \to B) \to (A \vee (A \to B))$. We conjecture that the Σ_1-provability logic of HA^* is embeddable in the intuitionistic counterpart of iGLCT, i.e. iGLC.

However the provability logic of HA is not yet characterized, we conjecture that it is also embeddable in iGL, the intuitionistic counterpart of the provability logic of PA.

Funding The work is partially funded by FWO grant G0F8421N.

References

Ardeshir, M., & Mojtahedi, M. (2014). The de Jongh property for basic arithmetic. *Archive for Mathematical Logic, 53*(7–8), 881–895.
Ardeshir, M., & Mojtahedi, M. (2018). The Σ_1-provability logic of *HA*. *Annals of Pure and Applied Logic, 169*(10), 997–1043.
Ardeshir, M., & Mojtahedi, M. (2019). The Σ_1-provability logic of HA^*. *Jornal of Symbolic Logic, 84*(3), 118–1135.
Artemov, S., & Beklemishev, L. (2004). Provability logic. In D. Gabbay & F. Guenthner (Eds.), *Handbook of philosophical logic*, 2nd Edn. (Vol. 13, pp. 189–360) Springer.
Beklemishev, L., & Visser, A. (2006). Problems in the logic of provability. In *Mathematical problems from applied logic. I, volume 4 of international series of numerical mathematics (N. Y.)* (pp. 77–136). Springer.
Berarducci, A. (1990). The interpretability logic of Peano arithmetic. *Journal of Symbolic Logic, 55*(3), 1059–1089.
de Jongh, D., & Veltman, F. (1990). Provability logics for relative interpretability. In P. Petkov (Ed.), *Mathematical Logic* (pp. 31–41). Springer.
de Jongh, D. (1970). The maximality of the intuitionistic predicate calculus with respect to heyting's arithmetic. *Journal of Symbolic Logic, 36*, 606.
de Jongh, D., Verbrugge, R., & Visser, A. (2011). Intermediate logics and the de Jongh property. *Archive for Mathematical Logic, 50*, 197–213.
Friedman, H. (1975). The disjunction property implies the numerical existence property. *Proceedings of the National Academy of Sciences of the United States of America,72*(8), 2877–2878.
Gödel, K. (1933). Eine interpretation des intuitionistischen aussagenkalkuls. *Ergebnisse eines mathematischen Kolloquiums, 4*, 39–40. English translation in S. Feferman et al. (Eds.). (1995). Kurt Gödel collected works (Vol. 1, pp. 301–303). Oxford University Press.

Iemhoff, R. (2001). *Provability logic and admissible rules*. Ph.D. thesis, University of Amsterdam.
Iemhoff, R. (2003). Preservativity logic (An analogue of interpretability logic for constructive theories). *Mathematical Logic Quarterly, 49*(3), 1–21.
Japardize, G., & de Jongh, D. (1997). The logic of provability. In S. Buss (Ed.), *Handbook of proof theory* (pp. 475–545). Elsevier.
Leivant, D. (1975). *Absoluteness in intuitionistic logic*. Ph.D. thesis, University of Amsterdam.
Leivant, D. (1979). *Absoluteness of intuitionistic logic, volume 73 of mathematical centre tracts*. Mathematisch Centrum, Amsterdam.
Löb, M. (1955). Solution of a problem of Leon Henkin. *Journal of Symbolic Logic, 20*(2), 115–118.
Mojtahedi, M. (2019). Hard provability logics. https://arxiv.org/abs/1911.04284
Myhill, J. (1973). A note on indicator-functions. *Proceedings of the AMS, 39*, 181–183.
Shavrukov, V. (1988). *The logic of relative interpretability over Peano Arithmetic*. Steklov Mathematical Institute, Moscow: Technical report.
Solovay, R. M. (1976). Provability interpretations of modal logic. *Israel Journal of Mathematics, 25*(3–4), 287–304.
Troelstra, A. S., & van Dalen, D. (1988). *Constructivism in mathematics. Vol. I, volume 121 of Studies in logic and the foundations of mathematics*. North-Holland Publishing Co.
Visser, A. (1990). Interpretability logic. *Mathematical logic* (pp. 175–209). Plenum.
Visser, A. (2002). Substitutions of Σ_1^0 sentences: explorations between intuitionistic propositional logic and intuitionistic arithmetic. *Annals of Pure and Applied Logic, 114*(1–3), 227–271. Commemorative Symposium Dedicated to Anne S. Troelstra (Noordwijkerhout, 1999).
Visser, A., van Benthem, J., de Jongh, D., de Lavalette, R., & G. R. (1995). NNIL, a study in intuitionistic propositional logic. *Modal logic and process algebra (Amsterdam, 1994), volume 53 of CSLI Lecture Notes* (pp. 289–326). CSLI Publ.
Visser, A. (1982). On the completeness principle: A study of provability in Heyting's arithmetic and extensions. *Annals of Mathematical Logic, 22*(3), 263–295.
Visser, A., & Zoethout, J. (2019). Provability logic and the completeness principle. *Annals of Pure and Applied Logic, 170*(6), 718–753.

Chapter 5
An Overview of Verbrugge Semantics, a.k.a. Generalised Veltman Semantics

Joost J. Joosten, Jan Mas Rovira, Luka Mikec, and Mladen Vuković

Abstract This paper gives an overview of what we call *Verbrugge Semantics* which so far in the literature has been called Generalised Veltman semantics. Interpretability logics are endowed with relational semantics à la Kripke: Veltman semantics. For certain applications though, this semantics is not fine-grained enough. Back in 1992, in the research group of de Jongh, the notion of *generalised Veltman semantics* emerged to obtain certain non-derivability results as was first presented by Verbrugge (1992). Henceforth we shall therefore speak of *Verbrugge Semantics*. It has turned out that this semantics has various good properties. In particular, in many cases completeness proofs become simpler and the richer semantics will allow for filtration arguments as opposed to regular Veltman semantics. This paper aims to give an overview of results and applications of Verbrugge semantics up to the current date.

Keywords Modal logic · Interpretability logic · Veltman semantics · Verbrugge semantics · Generalised veltman semantics

5.1 Introduction

This paper deals with interpretability logics and a particular kind of relational semantics for them. Interpretability logics are propositional modal logics with a unary modality $\Box$ for formal provability and a binary modality $\rhd$ used to denote formal interpretability or some related (meta)mathematical notion like, for example, con-

J. J. Joosten
Department of Philosophy, University of Barcelona, Barcelona, Spain

J. M. Rovira
Heliax, Zug, Switzerland

L. Mikec (✉) · M. Vuković
Department of Mathematics, University of Zagreb, Zagreb, Croatia
e-mail: luka.mikec@math.hr; luka.mikec1@gmail.com

N. Bezhanishvili et al. (eds.), *Dick de Jongh on Intuitionistic and Provability Logics*, Outstanding Contributions to Logic 28, https://doi.org/10.1007/978-3-031-47921-2_5

servativity. As such, interpretability logics extend the well-known provability logic **GL**. In particular, Kripke semantics for **GL** can be extended to account for the new binary interpretability operator $\rhd$. Such an extension was first considered by Frank Veltman and now goes by the name of Veltman semantics.

Unary modal logics allow for other abstract semantics like topological or neighbourhood semantics which can be seen as generalisations of the regular Kripke semantics. Although algebraic semantics has not yet been studied for interpretability logics and only recently the first results appeared regarding the topological semantics (Iwata & Kurahashi, 2021), there is some sort of hybrid generalisation which in the literature so far is called *Generalised Veltman Semantics*, and that we shall refer to as Verbrugge Semantics in this paper, VS for short. The notion of VS originated in the research group of Dick de Jongh and was first introduced and studied by Verbrugge (1992). The manuscript has never been published but is included in its original form as appendix to (Mas Rovira, 2020). In VS the $\Box$ modality is dealt with as before and only the $\rhd$ semantics is generalised in a way reminiscent to neighbourhood semantics.

Verbrugge semantics is more fine-grained than Veltman semantics and as such can serve the purpose of distinguishing logical axiom schemes. However, the most important contribution of VS to the field of interpretability logics lies in the fact that it allows for filtration in various systems, contrary to Veltman semantics. This has resulted in a situation where certain logics are known to be complete w.r.t. VS but known to be incomplete with respect to Veltman semantics. In other cases, the only known way of showing decidability of a logic runs via Verbrugge semantics. Thus, VS is proving itself an important tool in the study of interpretability logics. Moreover, completeness proof for VS often turn out to be more uniform and simple than in the case of Veltman semantics as one can appreciate in Sects. 5.5 and 5.6 of this paper.

In the rest of the paper we will give an overview of the uses and occurrences of Verbrugge semantics. Before doing so, we finish this introduction by briefly outlining the general development of interpretability logics and the role of VS therein. The main point of the rest of this introduction is to show that the field of interpretability logics has grown to a mature state having various applications and relations to other fields of logic and we do not claim to give an exhaustive overview. In particular, we hope that Dick de Jongh feels very satisfied of taking notice of the state-of-the-art: to see one of his many intellectual children in early maturity. Thank you for your many contributions Dick, to this beautiful field!

5.1.1 The Beginnings

An interpretation of a theory V in a theory U is roughly a translation j that maps the non-logical symbols in the language of V to formulas in the language of U with exactly the same free variables, so that any theorem φ of V will become provable in U when we apply the translation j to it. Various kinds of interpretations are around in the literature and interpretations are used in practically all branches of mathematics or meta-mathematics though a first methodological treatment is presented by Tarski et al. (1953). We are not too much interested in the variations and uses of interpretations and refer the interested reader to (Visser, 1997) and (Japaridze & de Jongh, 1998). Rather, we shall focus on the logics that describe the structural behaviour of the interpretability notion.

Interpretability logics arose in the eighties/nineties of the last century around Petr Hájek, Dick de Jongh, Franco Montagna, Vítězlav Švejdar, Frank Veltman and Albert Visser. Just like provability logic describes the provably structural behaviour of the formalised provability predicate, the aim was to find a logic that describes exactly all provably structural behaviour of the notion of interpretability.

The first conception of interpretatability as a modality that we know of is given by Hájek (1981) where he considers a modal logic with unary modalities (provability, interpretability in ACA_0) in combination with witness comparison operators. Probably the first published conception of interpretability as a binary modal operator is (Švejdar, 1983).[1] Montagna took this project further in his paper (Montagna, 1987). However, it was not until 1990 when Visser conceived the modal logical framework proposing various principles and axiom schemes (Visser, 1988, 1990) in the format we know today, giving rise to the basic interpretability logic **IL**. In particular, Visser stabilized the modal language containing the unary modality $\Box$ for provability and the binary modality $\rhd$ for interpretability and nothing else. Semantics for these systems was provided by Veltman and de Jongh (1990) who proved completeness for the logics **IL**, and the extensions **IL**P and **IL**M.

Ever since, the field of interpretability logics knows various different logics and has seen a development into quite a mature field interacting with various other branches of logic and mathematics. We shall now mention a few of the most notable developments. Below we shall mention various systems of interpretability logic, the definitions of which shall be postponed to later sections.

5.1.2 Relation to Meta-Mathematics

Various logics akin to **IL** are known to adequately describe well-delimited parts of meta-mathematics lending much importance and applicability to these logics. The first completeness result arose when Visser proved (Visser, 1990; Zambella, 1992) arithmetical completeness for the logic **IL**P for any Σ_1 sound, finitely axiomatisable theory where the superexponential function is provably total.

Any Σ_1-sound theory containing a minimum of arithmetic ($I\Delta_0 + \exp$) will have the same provability logic **GL** (de Jongh et al., 1991). The situation for interpretability is very different. Shavrukov (1988) and independently Berarducci (1990), established that the logic **IL**M generates precisely the set of interpretability principles

[1] We refer in this introduction to the year of publication while often preprints already circulated prior to that.

that are always provable[2] in theories like PA (being Σ_1 sound and having full induction/proving consistency of any of its finite subsystems).

Since different kinds of theories have different kind of interpretability logics, a major question in the field revolves about the logic that generates the collection of principles that is always provable in *any* reasonable arithmetical theory. Here the quantifier "any" is left vague on purpose but can be taken to be any theory containing $I\Delta_0 + \text{exp}$. The target logic is denoted **IL**(All) and even though much progress has been achieved (de Rijke, 1992; Goris & Joosten, 2011, 2020; Joosten, 1998, 2004; Joosten & Visser, 2000) on the question, its exact nature is still a major open problem. Given the plethora of recent principles it is quite conceivable that a natural answer to this question will require an extension of the language as in (Joosten et al., 2020; Joosten & Visser, 2004).

Basically, for any theory that is neither essentially reflexive nor finitely axiomatisable and capable of proving the totality of the superexponentiation function, the corresponding interpretability logic is unknown. Some very modest progress is presented on the interpretability logic of PRA (Bílková et al., 2009; Icard & Joosten, 2012; Joosten, 2005, 2010). Kalsbeek has published some notes (Kalsbeek, 1991) on the interpretability logic of $I\Delta_0 + \text{exp}$ and probably that is about how much is known about interpretability logics for theories that fall out of the two kinds mentioned above.

Interpretability logics are also good for describing (meta-)mathematical phenomena other than or variations of interpretability. Without giving the definitions, we mention that propositional modal logics similar to the logics mentioned above, possibly with modalities different from $\Box$ and $\rhd$, occur when axiomatising phenomena like partial conservativity (Hájek, 1971, 1972; Hájek & Montagna, 1990, 1992; Ignatiev, 1991; Japaridze, 1994; Joosten, 2016; Orey, 1961), cointerpretability, tolerance and cotolerance (Dzhaparidze, 1992, 1993; Japaridze & de Jongh, 1998), Σ_1-interpolability (Ignatiev, 1993), constructive preservativity (Iemhoff et al., 2005), intuitionistic interpretability (Litak, 2014, 2017; Litak & Visser, 2018) and feasible interpretability (Verbrugge, 1993).

5.1.3 Abstract Semantics

Even though interpretability has been studied from a perspective of categories (Visser, 2006) and degrees (Bennet, 1986; Lindström, 1997; Mycielski et al., 1990; Švejdar, 1978; Visser, 2005, 2014) in quite some depth, no abstract algebraic semantics for interpretability logics have been studied or designed yet. Topological semantics has been explored in a preprint (Iwata & Kurahashi, 2021). However, apart from the above-mentioned arithmetical semantics, interpretability logics and its kin come

[2] The way to go from always provable to always true is done as always by adding reflection over the set of theorems. See (Strannegård, 1999) for a slight generalisation.

with natural relational semantics. This relational so-called *Veltman semantics* has seen a considerable development over the past decades.

Completeness proofs and decidability results for various logics abounded since (de Jongh & Veltman, 1990) first results using a kind of semi-canonical model definition to deal with the logics **IL**, **ILM** and **ILP**. Further completeness results yielded ever more complicated proofs culminating in a completeness proof of **ILW** by de Jongh and Veltman (1999).

New model-theoretical techniques were needed to address other systems like the *step-by-step method*[3] (de Jongh & Veltman, 1999; Goris & Joosten, 2008; Joosten, 1998) for logics as $\mathbf{ILM}_0$ and $\mathbf{ILW}^*$ and *assuring labels* for substantial simplifications of general proofs but most notably **ILW** (Bílková et al., 2004; Goris et al., 2020).

Various other logics resisted completeness proofs for Veltman semantics and actually many turned out to be incomplete. Here the technique of generalised Veltman semantics came to the rescue which is the main topic of this paper. Recently, also subsystems of **IL** are being studied where the relational semantics again turned out to be adequate (Kurahashi & Okawa, 2020).

The notion of bisimulation could be applied to both Veltman semantics (Visser, 1990) and Verbrugge semantics (Vrgoč & Vuković, 2009). Furthermore, Čačić and Vrgoč (2011) defined the notion of a game for Veltman models and proved that a winning strategy for the defender in such a game is equivalent to picking out a bisimulation between two models.

In various occasions, results using modal model techniques could be translated back to arithmetic results like in the case of self provers and Σ_1 sentences (de Jongh & Pianigiani, 1998; Goris & Joosten, 2012) and the fixpoint theorem (de Jongh & Veltman, 1990; de Jongh & Visser, 1991).

5.1.4 Decidability and Complexity

For many interpretability logics, decidability follows directly from their modal completeness proofs and the finite model property (FMP). This is the case for **IL**, **ILW**, **ILP**, and **ILM**. For other logics, completeness was obtained with the help of construc-

[3] These step-by-step methods were around and entering main-stream modal-logical literature since the seventies. They were coined as such by Burgess (2002) but before that also went by different names as *Completeness by construction* in a reader from de Jongh and Veltman (1983). The first time the step-by-step method was applied to interpretability logics however, are the completeness proofs for **ILW** by de Jongh and Veltman (1999) and completeness proofs for **ILM** and $\mathbf{ILM}_0$ by Joosten (1998) with later a corrected proof for $\mathbf{ILM}_0$ by Goris and Joosten (2008). Joosten wrote his master thesis under direction of de Jongh and Visser. De Jongh suggested a variation of the step-by-step method so that we would work with infinite maximal consistent sets but only using finitely many of those to build models. In the context of this historic digression it is good to mention Verbrugge's master thesis (Verbrugge, 1988) under direction of de Jongh and Visser where she proves completeness w.r.t. so-called umbelliferous frames using a step-by-step method via the intersection of infinite maximal consistent sets with a finite fragment for a logic Umb which contains modalities for provability, interpretability and witness comparisons as in (Švejdar, 1983).

tions that avoid finite (truncated) maximal consistent sets (Goris & Joosten, 2008; Mikec & Vuković, 2020). In such cases it is not obvious how to make the constructions finite. Here, the method of filtration has proven useful. However, up to today, we only know how to perform filtration for VS and not for regular Veltman semantics. The decidability of the logics $\mathbf{ILM_0}$ (Perkov & Vuković, 2016), $\mathbf{ILW^*}$ (Mikec et al., 2017), $\mathbf{ILP_0}$ and **ILR** (Mikec & Vuković, 2020) has been shown using these methods. In all these cases finite models are obtained as quotient models w.r.t. the largest bisimulation of the given generalised Veltman model. At the time of writing, there is no known example of a complete and undecidable logic.

For a long time nothing non-trivial was known regarding complexity of interpretability logics. The first result concerning complexity was that already the closed fragment of **IL**, unlike the closed fragment of **GL**, is PSPACE-hard (Bou & Joosten, 2011). Recently it was proved that **IL** is PSPACE-complete (Mikec et al., 2018), and so are **ILP** and **ILW** (Mikec, 2022). There is hope to do the same with **ILM** (the proof for which turns out to be significantly harder to complete than expected, see the commentary at the end of (Mikec, 2022)).

5.1.5 Many Classical Results Carry Over to Interpretability Logics

A clear signal of working with the right notion and framework is seen in the fact that various classical results find their analogs in our logics. We already mentioned the Fixpoint Theorem and will in this paragraph mention some others without pretending to give an exhaustive overview.

Areces et al. (2001) proved that arrow interpolation holds for **IL**, i.e. if $\vdash_{\mathbf{IL}} A \to B$ then there is a formula I in the common language of A and B such that $\vdash_{\mathbf{IL}} A \to I$ and $\vdash_{\mathbf{IL}} I \to B$. As corollaries one obtains turnstile interpolation (i.e. if $A \vdash_{\mathbf{IL}} B$ then there is a formula I in the common language such that $A \vdash_{\mathbf{IL}} I$ and $I \vdash_{\mathbf{IL}} B$) and $\rhd$-interpolation (i.e. if $\vdash_{\mathbf{IL}} A \rhd B$ then there is a formula I in the common language such that $\vdash_{\mathbf{IL}} A \rhd I$ and $\vdash_{\mathbf{IL}} I \rhd B$). Areces et al. (2001) also show that all these properties transfer to the system **ILP**.

It is proven that the system **ILW** doesn't have the property of arrow interpolation. Visser (1997) proved that systems between $\mathbf{ILM_0}$ and **ILM** do not have interpolation either, and not even so after enriching the language with a predicate for Σ_1-ness (Goris, 2006). The interpolation property for the system **ILF** is an open problem. For all provability and interpretability logics it is shown by Areces et al. (2001) that the Beth definability property and fixed points property are interderivable. This implies that all extensions of the basic system of provability logic **GL** and all extensions of **IL** have the Beth property.

Perkov and Vuković (2014) proved a version of van Benthem's characterisation theorem (van Benthem, 1983) for interpretability logic. A first-order formula is equivalent to the standard first-order translation of some formula of interpretability logic with respect to Veltman models if and only if it is invariant under bisimulations between Veltman models. To prove this, they used bisimulation games on Veltman models. They provide characteristic formulas which formalise the existence of winning strategies for the defender in finite bisimulation games.

Hájek and Švejdar (1991) determined normal forms for the closed fragment of the system **ILF**, and showed that we can eliminate the modal operator $\rhd$ from closed **IL**-formulas. The normal form for the closed fragment of **IL** is unlikely to have a nice solution given the PSPACE completeness. However, Čačić and Vuković (2012) proved normal forms exist for a wide class of closed **IL** formulas. Čačić and Kovač (2015) quantified asymptotically, in exact numbers, how wide those classes are using results from combinatorics and asymptotic analysis.

5.1.6 Proof Theory

To the best of our knowledge, very few well-behaved proof systems for interpretability logics have been studied. Sasaki (2002) gave a cut-free sequent system for **IL** and prove a cut-elimination theorem for it. Hakoniemi and Joosten (2016) give a treatment of labelled tableaux proof systems and uniformly prove completeness for any logic whose frame condition is given by a Horn formula.

5.2 Logics for Interpretability

In this section we shall lay down the basic definitions.

5.2.1 Modal Interpretability Logics

In interpretability logics, we adopt a reading convention due to Dick de Jongh that will allow us to omit many brackets. As such, we say that the strongest binding 'connectives' are $\neg$, $\Box$ and $\Diamond$ which all bind equally strong. Next come $\wedge$ and $\vee$, followed by $\rhd$ and the weakest connective is $\to$. Thus, for example, $A \rhd B \to A \wedge \Box C \rhd B \wedge \Box C$ will be short for $(A \rhd B) \to \big((A \wedge \Box C) \rhd (B \wedge \Box C)\big)$.

We first define the core logic **IL** which shall be present in any other interpretability logic. As before, we work in a propositional signature where apart from the classical connectives we have a unary modal operator $\Box$ and a binary modal operator $\rhd$.

Definition 5.1 The logic **IL** contains apart from all propositional logical tautologies, all instantiations of the following axiom schemes:

L1 $\Box(A \to B) \to (\Box A \to \Box B)$

L2 $\Box A \to \Box\Box A$
L3 $\Box(\Box A \to A) \to \Box A$
J1 $\Box(A \to B) \to A \rhd B$
J2 $(A \rhd B) \wedge (B \rhd C) \to A \rhd C$
J3 $(A \rhd C) \wedge (B \rhd C) \to A \vee B \rhd C$
J4 $A \rhd B \to (\Diamond A \to \Diamond B)$
J5 $\Diamond A \rhd A$

The rules of the logic are Modus Ponens (from $\vdash A \to B$ and $\vdash A$, conclude $\vdash B$) and Necessitation (from $\vdash A$ conclude $\vdash \Box A$).

Since $\Box$ denotes provability and $\rhd$ denotes interpretability –a base theory together with the left-hand formula interprets the same base theory together with the right-hand formula– we can already see what the J-principles above express. Thus, Principle J1 expresses that the identity translation defines an interpretation. Principle J2 expresses that one can compose interpretations by applying one translation after another. Principle J3 is sometimes referred to as Hájek's principle reflecting so-called Hájek's Theorem[4] about constructing an interpretation by cases. Principle J4 reflects that an interpretation gives rise to relative consistency. Finally, J5 reflects that one can perform the Henkin construction in arithmetic so that consistency provides an inner model from which an interpretation can be distilled.

5.2.2 Arithmetical Semantics

Interpretability logics are related to arithmetics in very much the same way as provability logics are. Thus, we define an arithmetical realisation $*$ as a map that takes propositional variables to sentences in the language of arithmetic.[5] The realisation is extended to act on arbitrary modal formulas by preserving the logical structure thus commuting with the Boolean connectives. The modality $\Box$ is mapped to an arithmetisation of "is provable in the base theory T" for some fixed base theory T and likewise will $\rhd$ be mapped to an arithmetisation of interpretability. The interpretability logic of a theory T is now defined as usual, that is as the set of modal formulas whose realisation is provable in T regardless on the exact nature of the realisation:

Definition 5.2 Let T be a theory in the language of arithmetic that is strong enough to allow for a proper treatment of formalised syntax. We define the *interpretability logic* of T as

$$\mathbf{IL}(T) := \{\varphi \mid \forall * \; T \vdash \varphi^*\}.$$

[4] We thank Vitek Švejdar for pointing this out to us.

[5] We assume that all our theories contain the language of arithmetic in one way or another. We refer the reader to (Japaridze & de Jongh, 1998) for details of this and our Definition 5.2.

Even though the notion of formalised interpretability is Σ^0_3-complete (Shavrukov, 1997), for two classes of theories we have an elegant and decidable characterisation for the corresponding interpretability logic. We say that a theory T is Σ_1-*sound* if all Σ_1-sentences which are provable in T, are true.

By **ILM** we denote the logic that arises by adding Montagna's axiom scheme

$$\mathsf{M} := \quad A \rhd B \to A \wedge \Box C \rhd B \wedge \Box C$$

to **IL**.

Theorem 5.1 (Berarducci, 1990; Shavrukov,[6] 1988) *If T is Σ_1-sound and proves full induction, then* **IL**(T) = **ILM**

The logic **ILP** arises by adding the axiom scheme

$$\mathsf{P} := \quad A \rhd B \to \Box(A \rhd B)$$

to the basic logic **IL**. The logic **ILP** is related to finitely axiomatised theories that can prove the totality of $\texttt{supexp}$, where $\texttt{supexp}(x)$ is defined as $x \mapsto 2^x_x$ with $2^n_0 := n$ and $2^n_{m+1} := 2^{(2^n_m)}$.

Theorem 5.2 (Visser, 1990) *If T is Σ_1-sound, finitely axiomatised and proves the totality of* $\texttt{supexp}$, *then* **IL**(T) = **ILP**.

Since **ILM** $\neq$ **ILP**, it is natural to ask for the core logic that is contained in the interpretability logic of any (strong enough) theory. This results in an additional quantifier in the definition of what is often called *the interpretability logic of all reasonable arithmetical theories*:

Definition 5.3 **IL**(All) $:= \{\varphi \mid \forall T \supseteq \mathrm{I}\Delta_0 + \exp\ \forall * \ T \vdash \varphi^*\}$.

Since it is well known and easy to see that all theorems of **IL** hold in any strong enough arithmetical theory, by the above two theorems we obtain that **IL** $\subseteq$ **IL**(All) $\subseteq$ (**ILP** $\cap$ **ILM**). As to date, a modal characterisation of **IL**(All) is unknown. Most principles in this paper have been considered because of their relation to **IL**(All).

5.2.3 Relational Semantics

We can equip interpretability logics with a natural relational semantics often referred to as Veltman semantics.

---

[6] Shavrukov's arithmetical completeness proof is dated 1988 which is before the official publication of the modal completeness proof of de Jongh and Veltman (1990) and Visser's (1988) simplification thereof using a single accessibility relation S to model the $\rhd$ modality. However, these results were already available in preprint form as *D. H. J. de Jongh, F. J. M. M. Veltman. Provability logics for relative interpretability. ITLI Prepublication Series ML-88-03 (1988)*; and *A. Visser. Preliminary notes on interpretability logic. Logic Group Preprint Series No.29 (1988);* respectively.

Definition 5.4 A *Veltman frame* is a triple $\langle W, R, \{S_w : w \in W\}\rangle$ where W is a non-empty set of *possible worlds*, R a binary relation on W so that R^{-1} is transitive and well-founded. The $\{S_w : w \in W\}$ is a collection of binary relations on $R[w]$ (where $R[w] := \{v \mid wRv\}$). Every S_w is reflexive and transitive and contains the restriction of R to $R[w]$, that is $R \cap R[w] \subseteq S_w$.

A *Veltman model* consists of a Veltman frame together with a valuation $V : \mathtt{Prop} \to \mathcal{P}(W)$ that assigns to each propositional variable $p \in \mathtt{Prop}$ a set of worlds $V(p)$ in W where p is stipulated to be true. This valuation defines a forcing relation $\Vdash \subseteq W \times \mathsf{Form}$ telling us which formulas are true at which particular world:

$$\begin{aligned}
w \Vdash p &:\Leftrightarrow w \in V(p);\\
w \Vdash \bot &\quad \text{for no } w \in W;\\
w \Vdash A \to B &:\Leftrightarrow w \nVdash A \text{ or } w \Vdash B;\\
w \Vdash \Box A &:\Leftrightarrow \forall v\ (wRv \Rightarrow v \Vdash A);\\
w \Vdash A \rhd B &:\Leftrightarrow \forall u\ \Big(wRu \ \&\ u \Vdash A \Rightarrow \exists v\ (uS_w v \ \&\ v \Vdash B)\Big).
\end{aligned}$$

For a Veltman model $\mathfrak{M} = \langle W, R, \{S_w : w \in W\}, V\rangle$, we shall write $\mathfrak{M} \models A$ as short for $(\forall\, w \in W)\ \mathfrak{M}, w \Vdash A$.

De Jongh and Veltman (1990) proved that the logic **IL** is sound and complete with respect to all Veltman models.

Often one is interested in considering all models that can be defined over a frame. Thus, given a frame $\mathfrak{F}$ and a valuation V on $\mathfrak{F}$ we shall denote the corresponding model by $\langle \mathfrak{F}, V\rangle$. A *frame condition* for an axiom scheme A is a formula (A) (first or higher-order) in the language $\{R, \{S_w : w \in W\}\}$ so that $\mathfrak{F} \models (A)$ (as a relational structure) if and only if $\langle \mathfrak{F}, V\rangle \models A$ for every valuation V.

5.3 Verbrugge Semantics

For certain purposes, Veltman semantics is not fine-grained enough. Verbrugge semantics, also known as Generalised Veltman semantics, was originally introduced by Verbrugge (1992) to determine independence between certain interpretability logics as we shall discuss in the next section.

5.3.1 Replacing Worlds by Sets of Worlds

The idea of generalised Veltman semantics is that we will use sets to model the $\rhd$ modality. To be more precise, instead of having the S_w be a relation between worlds, we will use a relation between worlds and *sets of worlds*. Thus, we would have things like $uS_w V$ where V is a set of worlds. The forcing relation would be defined accordingly:

$$w \Vdash A \rhd B \;:\Leftrightarrow\; \forall u\Big(wRu \;\&\; u \Vdash A \Rightarrow \exists V(uS_wV \;\&\; V \Vdash B)\Big),$$

where $V \Vdash B$ is short for $(\forall\, v{\in}V\,)v \Vdash B$.

In doing so, the axiom schema $A \rhd B \to (\Diamond A \to \Diamond B)$ imposes[7] that all of V should be R-above w. The axiom schema $\Box(A \to B) \to A \rhd B$ requires that the S_w relation is what we call *semi-reflexive* in the sense that $uS_w\{u\}$ whenever wRu. The axiom schema $\Diamond A \rhd A$ imposes that whenever $wRuRv$, then $uS_w\{v\}$. Just like in regular Veltman semantics, the axiom schema $(A \rhd C) \wedge (B \rhd C) \to A \vee B \rhd C$ does not impose any requirement on generalised Veltman semantics and is satisfied automatically. It turns out that there is quite some freedom in how to account for the axiom schema $(A \rhd B) \wedge (B \rhd C) \to A \rhd C$. One such choice is the predominant one in the literature and we shall give it here and fix it for the rest of this paper. The definition was already given in the original manuscript by Verbrugge (1992). Variations will be discussed in the next subsection.

Definition 5.5 A *Verbrugge frame* $\mathfrak{F}$ is a structure of the form $\langle W, R, \{S_w : w \in W\}\rangle$, where W is a non-empty set, R is a transitive and conversely well-founded binary relation on W and for all $w \in W$ we have:

(a) $S_w \subseteq R[w] \times (\mathcal{P}(R[w]) \setminus \{\emptyset\})$;
(b) S_w is quasi-reflexive: wRu implies $uS_w\{u\}$;
(c) S_w is quasi-transitive: if uS_wV and vS_wZ_v for all $v \in V$, then $uS_w(\bigcup_{v \in V} Z_v)$;
(d) if $wRuRv$, then $uS_w\{v\}$;
(e) monotonicity: if uS_wV and $V \subseteq Z \subseteq R[w]$, then uS_wZ.

A *Verbrugge model* is a quadruple $\mathfrak{M} = \langle W, R, \{S_w : w \in W\}, V\rangle$, where the first three components form a Verbrugge frame and where V is a valuation mapping propositional variables to subsets of W. The forcing relation $\mathfrak{M}, w \Vdash A$ is defined as before in Definition 5.4 with the sole difference that now

$$w \Vdash A \rhd B \;:\Longleftrightarrow\; \forall u\,\Big(wRu \;\&\; u \Vdash A \Rightarrow \exists V(uS_wV \;\&\; V \Vdash B)\Big).$$

It is easy to see that GVS is adequate for **IL**:

Theorem 5.3 *The logic* **IL** *is sound and complete w.r.t. VS.*

Proof Soundness follows from an easy check on the rules and all the axiom schemas. For completeness we reason as follows. Suppose $\mathbf{IL} \nvdash A$. By using de Jongh and Veltman's theorem from (de Jongh & Veltman, 1990) we get a (regular) Veltman model $\mathfrak{M} = \langle W, R, \{S_w : w \in W\}, V\rangle$ and world $x \in W$ so that $\mathfrak{M}, x \Vdash \neg A$.

We now transform $\mathfrak{M}$ into a Verbrugge model $\mathfrak{M}'$ by only changing the S_w relations so that $\mathfrak{M}' = \langle W, R, \{S'_w : w \in W\}, V\rangle$ and[8]

[7] See Kurahashi and Okawa (2020) for a more detailed discussion.

[8] We would like to emphasise that we use the letter V to denote both the valuation and a subset of W. We are confident no confusion will arise from this.

Table 5.1 Semantic conditions for quasi-transitivity mentioned in the literature

Nr.	Semantic requirement for transitivity	First mentioned in
(1)	$uS_xY \Rightarrow \forall \{Y_y\}_{y\in Y}\Big((\forall\, y \in Y\ yS_xY_y) \Rightarrow \exists Z \subseteq \bigcup_{y\in Y} Y_y\ uS_xZ\Big)$	This paper
(2)	$uS_xY \Rightarrow \forall \{Y_y\}_{y\in Y}\Big((\forall\, y \in Y\ yS_xY_y) \Rightarrow uS_x \bigcup_{y\in Y} Y_y\Big)$	(Verbrugge, 1992)
(3)	$uS_xY \Rightarrow \exists\, y \in Y\, \forall Y'(yS_xY' \Rightarrow \exists\, Y''\subseteq Y'\ uS_xY'')$	This paper
(4)	$uS_xY \Rightarrow \exists\, y \in Y\, \forall Y'(yS_xY' \Rightarrow uS_xY')$	(Joosten, 1998)
(5)	$uS_xY \Rightarrow \forall\, y \in Y\, \forall Y'(yS_xY' \Rightarrow \exists\, Y''\subseteq Y'\ uS_xY'')$	This paper
(6)	$uS_xY \Rightarrow \forall\, y \in Y\, \forall Y'(yS_xY' \Rightarrow uS_xY')$	(Verbrugge, 1992)
(7)	$uS_xY \Rightarrow \forall\, y \in Y\, \forall Y'(yS_xY'\ \&\ y \notin Y' \Rightarrow \exists\, Y''\subseteq Y'\ uS_xY'')$	This paper
(8)	$uS_xY \Rightarrow \forall\, y \in Y\, \forall Y'(yS_xY'\ \&\ y \notin Y' \Rightarrow uS_xY')$	(Goris & Joosten, 2011)

$$uS'_w V \ :\Leftrightarrow\ \exists\, v{\in}V\ uS_w v.$$

It will be clear from the context what the letter V means.

Quasi-transitivity requires a small argument but it is easy to see that this definition of S' yields a Verbrugge model. Furthermore, via an easy induction we can prove that for any formula B and any $w \in W$ we have $\mathfrak{M}, w \Vdash B \Leftrightarrow \mathfrak{M}', w \Vdash B$. In particular $\mathfrak{M}', x \Vdash \neg A$, which completes the proof. □

The above proof tells us that any Veltman model can be transformed into a Verbrugge model preserving truth. Verbrugge has proven that in certain cases, one can also go the other way around and we will discuss this in Sect. 5.3.3.

5.3.2 On Quasi-Transitivity

By requiring the S_x-relations to be quasi-transitive, we see that $(A \rhd B) \wedge (B \rhd C) \rightarrow A \rhd C$ will automatically be true on any Verbrugge model. However, as we mentioned before, there are quite some alternatives to a semantic requirement of the axiom schema $(A \rhd B) \wedge (B \rhd C) \rightarrow A \rhd C$. We will now discuss some of them. In Table 5.1, we should bear in mind that if we have uS_xV for some u, x and V, then this automatically implies that $V \neq \varnothing$.

All of the presented quasi-transitivity requirements are adequate for proving soundness and completeness for **IL** (however, some, in particular (3), (4), (5), and (6), are in a certain sense equivalent to ordinary Veltman semantics, and thus strictly weaker than e.g. (2) in terms of expressiveness). For soundness it is routine to check that every instantiation of J2 holds. For the completeness part it is enough to see that any ordinary Veltman model $\mathfrak{M} = \langle W, R, \{S_w : w \in W\}, V\rangle$ can be transformed into a Verbrugge model $\mathfrak{M}' = \langle W, R, \{S'_w : w \in W\}, V\rangle$ where for all $w \in W$ we have $S'_w := \{\langle x, \{y\}\rangle : \langle x, y\rangle \in S_w\}$ and see[9] that $\mathfrak{M}'$ has the same associated forcing relation as $\mathfrak{M}$. These properties have been verified in the proof assistant Agda and are presented with more detail in (Mas Rovira, 2020).

Agda (Norell, 2007) is a proof assistant based on a constructive type theory with dependent types that allows the paradigm of *propositions as types* (Wadler, 2015) via the Curry-Howard correspondence.

The following theorem tells us why Notion (2) is in a sense the more natural one.

Theorem 5.4 *Let* $\mathfrak{F} = \langle W, R, \{S_w : w \in W\}\rangle$ *be a Verbrugge frame satisfying quasi-transitivity Condition* $(i) \in \{1, \dots, 8\}$. *Let* $\mathfrak{F}' = \langle W, R, \{S'_w : w \in W\}\rangle$ *where for all* $w \in W$ *we define* S'_w *as the monotone closure of* S_w*:*

$$S'_w := \{\langle x, Y'\rangle : \langle x, Y\rangle \in S_w, Y \subseteq Y' \subseteq R[w]\}.$$

Then $\mathfrak{F}'$ *is a Verbrugge frame satisfying quasi-transitivity Condition (2). Furthermore for any formula A and valuation V with* $\mathfrak{M} := \langle \mathfrak{F}, V\rangle$ *and* $\mathfrak{M}' := \langle \mathfrak{F}', V\rangle$ *we have that*

$$\mathfrak{M}, w \Vdash A \textit{ if and only if } \mathfrak{M}', w \Vdash A\ .$$

Proof Details are presented in (Mas Rovira, 2020). □

As we see in Theorem 5.4 taking the monotone closure of each S_w does not change the forcing relation and the resulting frame satisfies quasi-transitivity Condition (2).

Note that taking the monotone closure of each S_w is essentially different than assuming that each S_w is monotone by definition, as then the forcing relation may change. In the following example (Fig. 5.1) we present a Verbrugge model with Condition (8) that showcases such behaviour.

Let $\mathfrak{M}$ be a model based on the frame displayed in Fig. 5.1 such that $V(p) = \{v_0\}$ and $V(q) = \{v_3\}$ (i.e. $\{x : x \Vdash p\} = \{v_0\}, \{x : x \Vdash q\} = \{v_3\}$). We see that $w \Vdash \neg(p \rhd q)$ as p is only true in v_0 and we only have $v_0 S_w \{v_1\}$ with $v_1 \nVdash q$. If we assume that the relation S_w is monotone then we have $v_0 S_w \{v_1, v_2\}$ and by quasi-transitivity (8) we get $v_0 S_w \{v_3\}$. Consequently $w \Vdash \neg(p \rhd q)$ is no longer true.

Fig. 5.1 Example frame: $wRv_0, wRv_1, wRv_2, wRv_3, v_0 S_w\{v_1\}, v_2 S_w\{v_3\}$

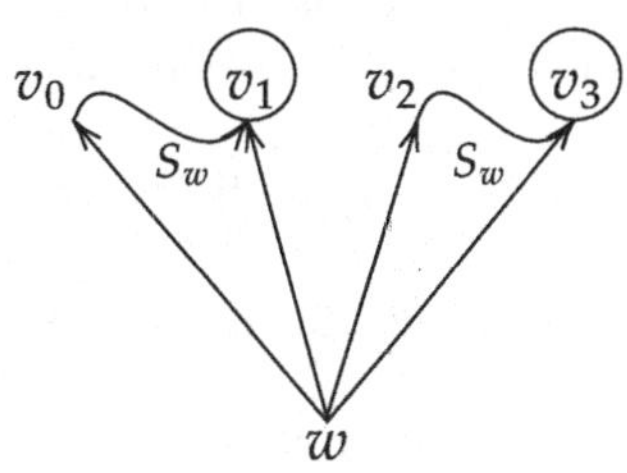

[9] The argument works in virtue that in (1) and (2) we understand that each y has a single Y_y and would break down in case we allowed for various $Y_y^{(\alpha)}$ (yielding yet another semantic variation).

5.3.3 Veltman Semantics Versus Verbrugge Semantics

The completeness proof of VS (Theorem 5.3) tells us that any Veltman model can be transformed into a Verbrugge model preserving truth. Verbrugge has proven that one can also go the other way around for quasi-transitivity Notion 6 from Table 5.1. Below we write $\Vdash$ and $\Vdash'$ instead of V and V' and their respective extensions.

Theorem 5.5 (Verbrugge, 1992) *Let $\langle W, R, \{S_w : w \in W\}, V\rangle$ be a generalised Veltman model with quasi-transitivity Condition $(i) \in \{3, 4, 5, 6\}$ (see Table 5.1). There is a (regular) Veltman model $\langle W', R', \{S'_w : w \in W'\}, V'\rangle$ such that for each $w \in W$ there is $w' \in W'$ such that for any formula B we have that*

$$\mathfrak{M}, w \Vdash B \quad \textit{if and only if} \quad \mathfrak{M}', w' \Vdash' B.$$

Here $\Vdash$ is the forcing relation in $\mathfrak{M}$ based on V and $\Vdash'$ is the forcing relation in $\mathfrak{M}'$ based on V'.

Proof We refer the reader to (Mas Rovira, 2020) for details and mainly present the definition of $\mathfrak{M}'$ here as was given in (Verbrugge, 1992). We will define a regular Veltman model $\mathfrak{M}'$ out of a Verbrugge model $\mathfrak{M}$. The main idea is that we will take many copies of worlds in $\mathfrak{M}$. When we define some $x'S'_w y'$ we should take into account that the single worlds x' and y' from $\mathfrak{M}'$ somehow come from worlds x and y from $\mathfrak{M}$ where these x and y fulfilled many roles as elements of images of the S relation. To capture this richness of the generalised Veltman semantics, we shall choose some representatives from S images. To this end, we first define for every world $x \in W$ a set $\mathsf{SR}(x)$ which contains all sets which are so-called *S-representatives* in a sense that whenever xS_uV, then any S-representatives of x will mention some non-zero number of elements of V. The formal definition reads as follows:

$$\begin{aligned}\mathsf{SR}(x) := \Big\{ A \subset W \times W \mid\ & \forall u \forall V {\subseteq} W \big(xS_uV \Rightarrow \exists\, v {\in} V\ \langle u, v\rangle \in A\big)\ \& \\ & \forall u, v\ \Big(\langle u, v\rangle \in A \Rightarrow \exists\, V {\subseteq} W \big(xS_uV\ \&\ v \in V\big)\Big)\Big\}.\end{aligned}$$

We observe that a world x will typically have many S-representatives. In the new model, we will consider all of them. Thus, we can now define the domain as

$$W' := \{\langle x, A\rangle \mid A {\in} \mathsf{SR}(x) \text{ or } \mathsf{SR}(x) = \varnothing = A\}.$$

To conclude, the relations are defined as

$$\langle x, A\rangle R' \langle y, B\rangle :\Leftrightarrow xRy\ \&\ \forall w, z (wRx\ \&\ \langle w, z\rangle \in B \Rightarrow \langle w, z\rangle \in A);$$

$$\begin{aligned}\langle x, A\rangle S'_{\langle w, C\rangle} \langle y, B\rangle :\Leftrightarrow\ & \langle w, C\rangle R' \langle x, A\rangle\ \&\ \langle w, C\rangle R' \langle y, B\rangle\ \& \\ & \forall v\ \big(\langle w, v\rangle \in B \Rightarrow \langle w, v\rangle \in A\big);\end{aligned}$$

and finally $\langle x, A\rangle \Vdash p :\Leftrightarrow x \Vdash p$. Verbrugge proved that $\mathfrak{M}'$ indeed defines a regular Veltman model and that moreover, for each formula A, for each world x and for each $V \subseteq W$ so that $\langle x, V\rangle \in W'$ we have $x \Vdash A \Leftrightarrow \langle x, V\rangle \Vdash' A$. □

Verbrugge showed the above theorem for generalised Veltman (Verbrugge) models with quasi-transitivity Condition (6). We have slightly improved the result by showing that it also holds for Conditions (3), (4) and (5). The above proof, together with a substantial simplification, has been fully formalised in the proof assistant Agda and is presented in (Mas Rovira, 2020). Vuković (2008) studies[10] how obtaining a Veltman model from a Verbrugge model can be performed for the, by now standard, transitivity Condition 2.

The above observations tell us that when it comes to models, regular Veltman semantics and Verbrugge semantics are equally powerful. With respect to frames the panorama is very different. Before we make this precise, let us first discuss frame conditions for VS.

Let X be a modal axiom schema. We denote by $(\mathsf{X})_{\mathrm{Ver}}$ a formula[11] of first-order or higher-order logic such that for all Verbrugge frames $\mathfrak{F}$ the following holds:

$$\mathfrak{F} \Vdash \mathsf{X} \text{ if and only if } \mathfrak{F} \models (\mathsf{X})_{\mathrm{Ver}}.$$

The formula $(\mathsf{X})_{\mathrm{Ver}}$ is called characteristic property (or frame condition) of the given logic **ILX**. The class of all Verbrugge frames $\mathfrak{F}$ such that $\mathfrak{F} \models (\mathsf{X})_{\mathrm{Ver}}$ is called the characteristic class of generalised frames for **ILX**. If $\mathfrak{F} \models (\mathsf{X})_{\mathrm{Ver}}$ we also say that the frame $\mathfrak{F}$ possesses the property $(\mathsf{X})_{\mathrm{Ver}}$. We say that a Verbrugge model $\mathfrak{M} = \langle W, R, \{S_w : w \in W\}, V\rangle$ is an $\mathbf{IL}_{\mathrm{Ver}}\mathsf{X}$-model,[12] or that model $\mathfrak{M}$ possesses the property $(\mathsf{X})_{\mathrm{Ver}}$, if the frame $\langle W, R, \{S_w : w \in W\}\rangle$ possesses the property $(\mathsf{X})_{\mathrm{Ver}}$.

Vuković (2008) studied in a general setting how to transform a Verbrugge model to an ordinary Veltman model much in the spirit of Theorem 5.5, such that these two models are bisimilar (in some aptly defined sense). Such a program can only yield partial answers w.r.t. frames since, as we shall see in Sect. 5.6, the logic $\mathbf{ILP}_0$ is complete w.r.t. generalised semantics, but incomplete w.r.t. ordinary semantics.

5.4 Verbrugge Semantics for Separating Systems

In this section we briefly mention some results where Verbrugge semantics has been used to prove independence of various systems. However, what makes VS really interesting in our opinion, are its good model-theoretical properties. We will discuss those in later sections.

---

[10] The proof in (Vuković, 2008) contains an omission and in (Mas Rovira, 2020) this is addressed.

[11] In the literature previous to this paper, it was more common to read $(\mathsf{X})_{\mathrm{gen}}$.

[12] Previously, $\mathbf{IL}_{\mathrm{gen}}\mathsf{X}$-model.

5.4.1 Principles and Veltman Models

Visser (1990) studies among others relations between various extensions of the basic interpretability logic **IL**. Among others, he considered the following principles:

$$\begin{aligned} \mathsf{W} &:= A \rhd B \to A \rhd B \wedge \Box\neg A; \\ \mathsf{KW1} &:= A \rhd \Diamond\top \to \top \rhd \neg A; \\ \mathsf{F} &:= A \rhd \Diamond A \to \Box\neg A. \end{aligned}$$

Visser observed[13] that all of W, F and KW1 define proper extensions of **IL** that have the same frame condition w.r.t. Veltman semantics: for each w, the relation $R \circ S_w$ should be conversely well-founded. Further, he noted that **ILW** $\vdash$ KW1 and **ILW** $\vdash$ F (already in (Visser, 1988)) and he posed as an open question if the converse also holds (over **IL**). As a mere curiosity it was mentioned that a slight weakening of F is provable in **IL**. We repeat that here: if we take the contraposition $\Diamond A \to \neg(A \rhd \Diamond A)$ of F and replace the implication by an interpretability modality we obtain an **IL** provable formula[14]: $\Diamond A \rhd \neg(A \rhd \Diamond A)$.

Another family of principles studied in (Visser, 1988) is given by:

$$\begin{aligned} \mathsf{M} &:= A \rhd B \to A \wedge \Box C \rhd B \wedge \Box C; \\ \mathsf{KM1} &:= A \rhd \Diamond B \to \Box(A \to \Diamond B); \\ \mathsf{KM2} &:= A \rhd B \to \big(\Box(B \to \Diamond C) \to \Box(A \to \Diamond C)\big). \end{aligned}$$

It was observed that all of M, KM1 and KM2 have the same frame condition w.r.t. Veltman semantics: $yS_w zRu \;\Rightarrow\; yRu$. Similar to the previous family, it was observed that **ILM** $\vdash$ KM1, KM2 and posed in (Visser, 1988, 1990) as an open question if the converse also holds (over **IL**). For this family it was proven in (Visser, 1988) that KM1 and KM2 are interderivable over **IL**. Moreover, just as W follows from **ILM**, we also have that KW1 follows from **ILKM1**.

Švejdar in 1991 took up the above mentioned questions of Visser's whether certain reversals like **ILKM1** $\vdash$ M hold. Ordinary Veltman models were suitable to distinguish all combinations of the following principles of interpretability (Švejdar, 1991): W, M, KM1, KW1, $\mathsf{KW1}^0$, and F, where

$$\mathsf{KW1}^0 \;:=\; A \wedge B \rhd \Diamond A \to A \rhd (A \wedge \neg B).$$

Unlike most other proofs of independence results which rely on differences in characteristic classes, parts of his proofs are based on exhibiting particular *models* that globally satisfy one of the principles in question. For example, his proof

[13] Even though indeed KW1 is similar in flavour, it turned out that its frame-condition is actually, contrary to what Visser thought and so announced in (Visser, 1990), different from that of W. Švejdar (1991) computed and published the corrected condition: for each wRy there exists $x \in \mathrm{M}(w)$ such that $yS_w x$, where $\mathrm{M}(w) := \{x \in R[w] : \text{there is no } z \in R[w] \text{ such that } x\ (S_w \circ R)\ z\}$.

[14] Principle K10 in (Visser, 1988).

that $\mathbf{IL}\{\mathsf{F}, \mathsf{KW1}\} \nvdash \mathsf{KW1}^0$ exhibits an $\mathbf{IL}\mathsf{F}$-model that globally satisfies $\mathsf{KW1}$, but refutes $\mathsf{KW1}^0$. As a consequence, Švejdar established that the logics $\mathbf{IL}\mathsf{F}$, $\mathbf{IL}\mathsf{KW1}$ and $\mathbf{IL}\mathsf{KM1}$ are incomplete with respect to their class of frames.

5.4.2 Generalised Frame Conditions and Independence

In the previous subsection we saw various principles having the same frame condition. However, their frame conditions for Verbrugge semantics differ and as such this provides a way of telling different logics apart. In this section we will simply present a collection of generalised frame conditions and leave it as an easy exercise that they are all different from each other. Verbrugge (1992) determined $(\mathsf{M})_{\mathsf{Ver}}$, $(\mathsf{KM1})_{\mathsf{Ver}}$, and $(\mathsf{P})_{\mathsf{Ver}}$:

$$\begin{aligned}(\mathsf{M})_{\mathsf{Ver}} &:= uS_wV \Rightarrow (\exists V' \subseteq V)(uS_wV' \ \&\ R[V'] \subseteq R[u]);\\ (\mathsf{KM1})_{\mathsf{Ver}} &:= uS_wV \Rightarrow (\exists v \in V)\forall z(vRz \Rightarrow uRz);\\ (\mathsf{P})_{\mathsf{Ver}} &:= wRw'RuS_wV \Rightarrow (\exists V' \subseteq V)\ uS_{w'}V'.\end{aligned}$$

She proved $\mathbf{IL}\mathsf{KM1} \nvdash \mathsf{M}$, $\mathbf{IL}\mathsf{F} \nvdash \mathsf{W}$, and $\mathbf{IL}\mathsf{F} \nvdash \mathsf{KW1}$ using VS.

There are two more principles that frequently occur in the literature. First, there is

$$\mathsf{M}_0 := A \rhd B \to \Diamond A \wedge \Box C \rhd B \wedge \Box C.$$

And second, there is

$$\mathsf{W}^* := A \rhd B \to B \wedge \Box C \rhd B \wedge \Box C \wedge \Box\neg A.$$

Visser (1991) showed that $\mathbf{IL}\mathsf{W} \nvdash \mathsf{M}_0$ and that $\mathbf{IL}\mathsf{M}_0\mathsf{W} = \mathbf{IL}\mathsf{W}^*$.

Vuković (1996) determined the formula $(\mathsf{M}_0)_{\mathsf{Ver}}$:

$$(\mathsf{M}_0)_{\mathsf{Ver}} := wRuRxS_wV \Rightarrow (\exists V' \subseteq V)(uS_wV' \ \&\ R[V'] \subseteq R[u]).$$

and proved the independence of the principle M_0 with various others principles of interpretability. All connections between principles M, M_0, $\mathsf{KM1}$, $\mathsf{KM2}$, P, W, W^*, $\mathsf{KW1}^0$, $\mathsf{KW1}$ and F were determined in (Vuković, 1999) using GVS. Vuković (1999) provided a comparative modal study of all these principles together using VS. The result of this study can be summarized by the following diagram:

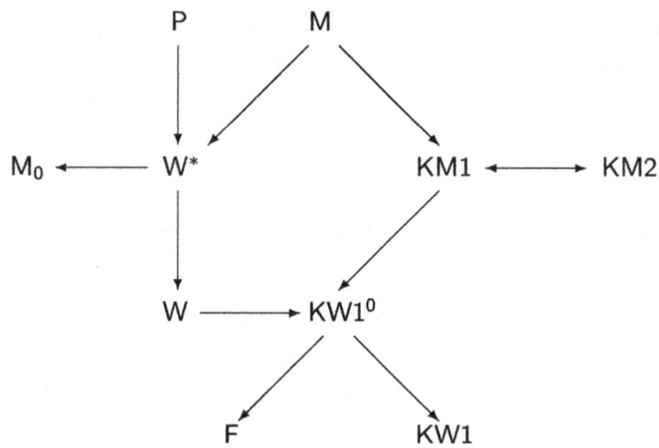

Joosten and Visser presented a new **IL**(All) principle

$$\mathsf{P}_0 \;:=\; A \rhd \Diamond B \to \Box(A \rhd B)$$

in Joosten's master thesis (Joosten, 1998). Using VS but without establishing the frame condition for P_0 Joosten could prove that W, M_0 and P_0 are maximally independent (no two imply the other).

Goris and Joosten considered the principle P_0 in (Goris & Joosten, 2011) and presented a related new principle that has the same frame condition

$$\mathsf{R} \;:=\; A \rhd B \to \neg(A \rhd \neg C) \rhd B \wedge \Box C.$$

They determined formulas $(\mathsf{P}_0)_{\mathsf{Ver}}$ and $(\mathsf{R})_{\mathsf{Ver}}$. Here are slightly reformulated versions from (Mikec & Vuković, 2020):

$$\begin{aligned}(\mathsf{P}_0)_{\mathsf{Ver}} &:= wRxRuS_wV \;\&\; (\forall v \in V)R[v] \cap Z \neq \emptyset \;\Rightarrow\; (\exists Z' \subseteq Z)uS_xZ';\\ (\mathsf{R})_{\mathsf{Ver}} &:= wRxRuS_wV \Rightarrow (\forall C \in \mathcal{C}(x,u))(\exists U \subseteq V)(xS_wU \;\&\; R[U] \subseteq C).\end{aligned}$$

where $\mathcal{C}(x,u) = \{C \subseteq R[x] : (\forall Z)(uS_xZ \Rightarrow Z \cap C \neq \emptyset)\}$ is the family of "choice sets". They proved $\mathbf{ILWP_0M_0} \nvdash \mathsf{R}$ using Verbrugge semantics. Instead of providing a VS frame condition for W, the authors proved a necessary and sufficient VS frame condition for W to fail.

Definition 5.6

$$\begin{aligned}\text{Not-W} \;:=\;& \exists\, w,\; z_0,\; \{Y_i\}_{i\in\omega},\; \{y_i\}_{i\in\omega,\; y_i\in Y_i},\; Z,\; \{z_{i+1}\}_{i\in\omega,\; z_{i+1}\in Z}\\ & [\forall i \in \omega(z_i S_w Y_i \ni y_i R z_{i+i}) \;\&\\ & \forall z \in Z \exists i \in \omega z S_w Y_i \;\&\\ & \forall z \in Z \forall Y\; (z S_w Y \;\&\; Y \subseteq (\cup_{i\in\omega} Y_i) \;\Rightarrow\; \exists z' \in Z \exists y \in Y y R z')]\end{aligned}$$

Lemma 5.1 *For any Verbrugge frame $\mathfrak{F}$ we have that*

$$\mathfrak{F} \models \mathsf{Not\text{-}W} \quad \textit{if and only if} \quad \mathfrak{F} \not\models \mathsf{W}.$$

A positive frame condition for W is presented in Sect. 5.6.4.

5.5 Modal Completeness: Preliminaries

The aim of this and the next section is to explore modal completeness with respect to Verbrugge semantics. We will give the state-of-the-art of completeness results involving VS. Let us first say a few words on the history of modal completeness proofs concerning interpretability logics.

This and the following section are based heavily on the recent paper (Mikec & Vuković, 2020). For this reason we will not cite results; for any definition or result without a reference it is safe to assume it is being quoted from (Mikec & Vuković, 2020).

5.5.1 Overview of Approaches

De Jongh and Veltman proved the completeness of **IL**, **IL**M and **IL**P w.r.t. the corresponding characteristic classes of ordinary (and finite) Veltman frames in (de Jongh & Veltman, 1990). As is usual for extensions of the provability logic **GL**, all completeness proofs suffer from compactness-related issues. One way to go about this is to define a (large enough) adequate set of formulas and let worlds be maximal consistent subsets of such sets (used e.g. in (de Jongh & Veltman, 1990)). With interpretability logics and ordinary Veltman semantics, worlds have not been identified with (only) sets of formulas. It seems that with ordinary Veltman semantics it is sometimes necessary to duplicate worlds (that is, have more than one world correspond to a single maximal consistent set) in order to build models for certain consistent sets (see e.g. (de Jongh & Veltman, 1990)). De Jongh and Veltman (1999) proved completeness of the logic **IL**W w.r.t. its characteristic class of ordinary (and finite) Veltman frames.

Goris and Joosten, inspired by Dick de Jongh, introduced[15] a more robust approach to proving completeness of interpretability logics, the *construction method* or *step-by-step method* (Goris & Joosten, 2008, 2011). In this type of proofs, one builds models step by step, and the final model is retrieved as a union. While closer to the intuition and more informative than the standard proofs, these proofs are hard to produce and verify due to their size. (They might have been shorter if tools from

[15] See our comments in Footnote 3 for some more detailed historical context. Also (de Jongh et al., 2004) provides some comments on construction methods for other modal logics.

(Bílková et al., 2004) and (Goris et al., 2020) have been used from the start.) For the purpose for which this type of proofs was invented (completeness of $\mathbf{ILM_0}$ and $\mathbf{ILW}^*$ w.r.t. the ordinary semantics), this type of proofs is still the only known approach that works.

In (Mikec & Vuković, 2020) a very direct type of proofs of completeness is presented; similar to (de Jongh & Veltman, 1990) in the general approach, but this time with respect to Verbrugge semantics. The so-called *assuring labels* from (Bílková et al., 2004; Goris et al., 2020) were used as a key step. The assuring labels are labels to R-transitions and they collect information about which formulas should receive special attention along those R-transitions. These completeness proofs are the ones that we aim to explore here. An example that illustrates benefits of using the generalised semantics will be given in the section dedicated to $\mathbf{ILM_0}$. The most interesting of these results are completeness of $\mathbf{ILR}$ and $\mathbf{ILP_0}$. The principle R is important because it forms the basis of the best, at the moment, explicit candidate for $\mathbf{IL}(\mathsf{All})$ as discussed in more detail in Sect. 5.8. Results concerning the principle $\mathbf{ILP_0}$ are interesting in a different way; they answer an old question: is there an unravelling technique that transforms generalised $\mathbf{ILX}$-models to ordinary $\mathbf{ILX}$-models, that preserves satisfaction of relevant characteristic properties? The answer is *no*: $\mathbf{ILP_0}$ is complete w.r.t. GVS, but it is known to be incomplete w.r.t. the ordinary semantics (Goris & Joosten, 2011).

5.5.2 *Completeness w.r.t. Generalised Semantics*

In what follows, "formula" will always mean "modal formula". If the ambient logic in some context is $\mathbf{ILX}$, a maximal consistent set w.r.t. $\mathbf{ILX}$ will be called an $\mathbf{ILX}$-MCS. Let us now introduce *assuring labels* from (Bílková et al., 2004) and (Goris et al., 2020). These are any sets of formulas which give us information on the relationship between two $\mathbf{ILX}$-MCS's. Given (an ordered pair of) two $\mathbf{ILX}$-MCS's, there can be zero or more labels describing their relationship.

Definition 5.7 (Bílková et al., 2004, *a slightly modified Definition 3.1*) Let w and u be some $\mathbf{ILX}$-MCS's, and let S be an arbitrary set of formulas. We write $w \prec_S u$ if for any finite $S' \subseteq S$ and any formula A we have both $\neg A \in u$ and $\Box\neg A \in u$ whenever $A \rhd \bigvee_{G \in S'} \neg G \in w$.

Note that the small differences between our Definition 5.7 and Definition 3.1 in (Bílková et al., 2004) do not affect the results of (Bílková et al., 2004) that we use.[16]

Definition 5.8 (Bílková et al., 2004, *p. 4*) Let w be an $\mathbf{ILX}$-MCS, and S an arbitrary set of formulas. Put:

---

[16] The difference is a different strategy of ensuring converse well-foundedness for the relation R. Instead of asking for the existence of some $\Diamond F \in w \setminus u$ whenever wRu, as is usual in the context of provability (and interpretability) logics, we will go for a stronger condition (see Definition 5.9). Since we will later put $R :=\prec$, this choice of ours is reflected already at this point.

$$w_S^{\Box} := \{\Box\neg A : \exists S' \subseteq S,\ S' \text{ finite},\ A \rhd \bigvee_{G\in S'} \neg G \in w\};$$

$$w_S^{\boxdot} := \{\neg A, \Box\neg A : \exists S' \subseteq S,\ S' \text{ finite},\ A \rhd \bigvee_{G\in S'} \neg G \in w\}.$$

Thus, $w \prec_S u$ if and only if $w_S^{\boxdot} \subseteq u$. If $S = \emptyset$ then $w_\emptyset^{\Box} = \{\Box\neg A : A \rhd \bot \in w\}$. Since w is maximal consistent, use of $w_\emptyset^{\Box}$ usually amounts to the same as the use of the set $\{\Box A : \Box A \in w\}$.

We will usually write $w \prec u$ instead of $w \prec_\emptyset u$.

Lemma 5.2 (Bílková et al., 2004, Lemma 3.2) *Let w, u and v be some* **ILX**-*MCS's, and let S and T be some sets of formulas. Then we have:*

(a) if $S \subseteq T$ and $w \prec_T u$, then $w \prec_S u$;
(b) if $w \prec_S u \prec v$, then $w \prec_S v$;
(c) if $w \prec_S u$, then $S \subseteq u$.

We will tacitly use the preceding lemma in most of our proofs. Although not needed in this paper, we mention that in (Goris et al., 2020) it is shown that without loss of generality we may actually assume that labels are full theories.

The following two lemmas can be used to construct (or in our case, find) a MCS with the required properties.

Lemma 5.3 (Bílková et al., 2004, Lemma 3.4) *Let w be an* **ILX**-*MCS, and let $\neg(B \rhd C) \in w$. Then there is an* **ILX**-*MCS u such that $w \prec_{\{\neg C\}} u$ and $B, \Box\neg B \in u$.*

Lemma 5.4 (Bílková et al., 2004, Lemma 3.5) *Let w and u be some* **ILX**-*MCS's such that $B \rhd C \in w$, $w \prec_S u$ and $B \in u$. Then there is an* **ILX**-*MCS v such that $w \prec_S v$ and $C, \Box\neg C \in v$.*

In the rest of this section, we will assume that $\mathcal{D}$ is always a finite set of formulas, closed under taking subformulas and single negations, and $\top \in \mathcal{D}$. The following definition is central to most of the results of this section.

Definition 5.9 Let X be a subset of $\{\mathsf{M}, \mathsf{M}_0, \mathsf{P}, \mathsf{P}_0, \mathsf{R}\}$. We say that $\mathfrak{M} = \langle W, R, \{S_w : w \in W\}, V\rangle$ is *the* **ILX** *-structure for a set of formulas $\mathcal{D}$* if:

- $W := \{w : w \text{ is an } \mathbf{ILX}\text{-MCS and for some } G \in \mathcal{D},\ G \wedge \Box\neg G \in w\}$;
- $wRu :\Leftrightarrow w \prec u$;
- $uS_wV :\Leftrightarrow wRu$ and $V \subseteq R[w]$, and $(\forall S)(w \prec_S u \Rightarrow (\exists v \in V)w \prec_S v)$;
- $w \in V(p) :\Leftrightarrow p \in w$.

We note that the **ILX**-structure for $\mathcal{D}$ is a unique object. In fact, we could work with just one "**ILX**-structure" (that would not depend even on $\mathcal{D}$): the disjoint union of **ILX**-structures for all choices of $\mathcal{D}$. We also observe that the definition entails that when uS_wV, then $V \neq \emptyset$ since $wRu \Rightarrow w \prec_\emptyset u$ so $\exists(v \in V)w \prec_\emptyset v$.

Notice that worlds in the definition above are somewhat more restricted than what is usually found in similar proofs: every world is required to be R-maximal with respect to some formula. That is, for every world $w \in W$ we want to have a formula G_w such that $w \Vdash G_w$ and for any R-successor u of w, $u \nVdash G_w$. This is equivalent to the requirement that for some formula G_w, $w \Vdash G_w \wedge \Box\neg G_w$. Of course, before we prove our truth lemma we can only require that $G_w \wedge \Box\neg G_w \in w$. Because of this we need the following lemma whose proof boils down to an instance of Löb's axiom.

Lemma 5.5 *If* **ILX** $\nvdash \neg A$ *then there is an* **ILX**-*MCS* w *such that* $A \wedge \Box\neg A \in w$.

We are now ready to prove the main lemma of this section, which tells us that the structure defined in Definition 5.9 really is a Verbrugge model. Notice that we do not claim that it is also an $\mathbf{IL}_{\mathrm{Ver}}$X-model; we prove that later.

Lemma 5.6 *Let X be a subset of* {M, M$_0$, P, P$_0$, R}. *The* **ILX**-*structure* $\mathfrak{M}$ *for a set of formulas* $\mathcal{D}$ *is a Verbrugge model. Furthermore, the following truth lemma holds:*

$$\mathfrak{M}, w \Vdash G \text{ if and only if } G \in w,$$

for all $G \in \mathcal{D}$ *and* $w \in W$.

Proof Most of the proof is straightforward. Let us just comment the proof of the truth lemma, more specifically, the following claim in the induction step: $w \Vdash B \rhd C \Rightarrow B \rhd C \in w$. This part is probably the most interesting one, since it explains the motivation behind the definition of S_w in Definition 5.9.

Assume $B \rhd C \notin w$. Lemma 5.3 implies there is u with $w \prec_{\{\neg C\}} u$ and $B, \Box\neg B \in u$ (thus $u \in W$). It is immediate that wRu and the induction hypothesis implies that $u \Vdash B$. Assume uS_wV. We are to show that $V \nVdash C$. Since $w \prec_{\{\neg C\}} u$ and uS_wV, there is $v \in V$ such that $w \prec_{\{\neg C\}} v$. Lemma 5.2 implies $\neg C \in v$. The induction hypothesis implies $v \nVdash C$; thus $V \nVdash C$. □

This lemma is just one step away from a completeness proof:

Theorem 5.6 *Let* $X \subseteq$ {M, M$_0$, P, P$_0$, R}. *Assume that for every set* $\mathcal{D}$ *the* **ILX**-*structure for* $\mathcal{D}$ *possesses the property* (X)$_{\mathrm{Ver}}$. *Then* **ILX** *is complete w.r.t.* $\mathbf{IL}_{\mathrm{Ver}}$X-*models.*

Proof Let A be a formula such that **IL**$X \nvdash \neg A$. Lemma 5.5 implies there is an **ILX**-MCS w such that $A \wedge \Box\neg A \in w$. Let $\mathcal{D}$ have the usual properties, and contain A. Let $\mathfrak{M} = \langle W, R, \{S_w : w \in W\}, V\rangle$ be the **ILX**-structure for $\mathcal{D}$. Since $A \wedge \Box\neg A \in w$ and $A \in \mathcal{D}$, we have $w \in W$. Lemma 5.6 implies $\mathfrak{M}, w \nVdash \neg A$. □

Corollary 5.1 *The logic* **IL** *is complete w.r.t. Verbrugge semantics.*

Note that any method for transforming generalised to ordinary models like presented in Theorem 5.5 or in (Vuković, 2008) now implies completeness of **IL** w.r.t. ordinary Veltman models.

In the next section we comment on the completeness of the following logics w.r.t. VS: **ILM**, **ILM$_0$**, **ILP**, **ILP$_0$**, **ILR**, **ILW** and **ILW***.

5.5.3 A Note on Verbrugge Semantics and Labelling

In all studied extensions of **IL** we have to duplicate maximal consistent sets when building ordinary Veltman models for consistent sets of formulas. More accurately, no one seems to have come up with a natural way of assigning just one purpose to every maximal consistent set of formulas. For example, when building a model where $\{\neg(p \rhd q), \neg(p \rhd r), p \rhd (q \vee r)\}$ is true in some world w, we could try to use the same set/world u visible from w as a witness for the formulas $\neg(p \rhd q)$ and $\neg(p \rhd r)$ in w. For example, this may be the set where the only propositional formula is p, and no formula of the form $\neg(A \rhd B)$ is contained. But, due to $p \rhd (q \vee r)$, in any model where w is there we do require two worlds like u within that model; one of which will have an S_w-successor satisfying q but not r, and the other one an S_w-successor satisfying r but not q.

Verbrugge semantics doesn't share this problem of duplication, at least not in any known case of a complete extension of **IL**. A Verbrugge model for the problem above is simple. Let $w = \{\neg(p \rhd q), \neg(p \rhd r), p \rhd (q \vee r)\}$, $u = \{p\}$, $x = \{q\}$, $y = \{r\}$, and let $wRuS_w\{x, y\}$. Unspecified propositional formulas are assumed to be false, and unspecified $\rhd$-formulas are assumed to be true.

Now, having in mind this Verbrugge model, what can be said about the wRu transition in terms of labels? This might be important if we are building a generalised model step-by-step. Since u has two roles, it would be natural to allow (even with assuringness) two labels: $\{\neg q\}$ and $\{\neg r\}$. And these labels are justified, since indeed $\{x, y\} \nVdash q, r$. Both these labels are expressible without sets (in terms of criticality, for example, the labels would be formulas q, and r, respectively). We recall that the S_w from Definition 5.9 indeed takes multiple labels into account.

However, there is another bit of label-related information that these facts do not express: which labels *do not* hold. Although $\{\neg q\}$ and $\{\neg r\}$ are justified choices, the label $\{\neg q, \neg r\}$ is not a good choice. This label would require $\neg p \notin u$, which is clearly not the case. This is the information the assuringness allows us to express, and criticality does not.[17] Note that such a situation cannot happen in ordinary semantics: if the label $\{\neg q, \neg r\}$ is inappropriate for some wRu transition, that means there is $A \in u$ with $A \rhd q \vee r \in w$. This, since we are now working in ordinary semantics, means there should be an S_w-successor of u satisfying $q \vee r$. So, either this new world satisfies q or r. So, $\{\neg q\}$ or $\{\neg r\}$ had to be inappropriate labels (for wRu) too.

[17] Granted, one might say that the inadequacy of the assuring label $\{\neg q, \neg r\}$ is equivalent to the inadequacy of the critical label $q \vee r$. However, expressing this fact in terms of criticality does not retain structural information of our situation; we see a disjunction where really we are only interested in disjuncts.

5.6 Modal Completeness of Various Systems

In this section we explore completeness proofs for various extensions of **IL**. We also briefly describe a recent preprint where certain subsystems of **IL** are explored.

5.6.1 The Logic ILM

Completeness of the logic **IL**M w.r.t. VS is an easy consequence of the completeness of **IL**M w.r.t. the ordinary semantics, first proved by de Jongh and Veltman (1990). Another proof of the same result was given by Goris and Joosten, using the construction method (Goris & Joosten, 2011; Joosten, 1998).

The frame condition $wRxS_wyRz \Rightarrow xRz$ for M is reflected in the following so-called labelling lemma:

Lemma 5.7 (Bílková et al., 2004, Lemma 3.7) *Let w and u be some* **IL**M*-MCSs, and let S be a set of formulas. If $w \prec_S u$ then $w \prec_{S \cup u^{\Box}_{\emptyset}} u$.*

When we combine this with the main result of the previous section we get a simple, elegant and succinct completeness proof.

Theorem 5.7 *The logic* **IL**M *is complete w.r.t.* $\mathbf{IL}_{\mathsf{Ver}}$M*-models.*

Proof Here we give the whole proof by Mikec and Vuković (2020), to demonstrate the interplay between labelling lemmas and characteristic properties. Proofs for other logics are similar, though usually more complex.

Given Theorem 5.6, it suffices to show that for any set $\mathcal{D}$, the **IL**M-structure for $\mathcal{D}$ possesses the property $(\mathsf{M})_{\mathsf{Ver}}$: $uS_wV \Rightarrow (\exists V' \subseteq V)(uS_wV' \,\&\, R[V'] \subseteq R[u]$. Let $\langle W, R, \{S_w : w \in W\}, V\rangle$ be the **IL**M-structure for $\mathcal{D}$.

Let uS_wV and take $V' = \{v \in V : w \prec_{u^{\Box}_{\emptyset}} v\}$. We claim uS_wV' and $R[V'] \subseteq R[u]$. Suppose $w \prec_S u$. Lemma 5.7 implies $w \prec_{S\cup u^{\Box}_{\emptyset}} u$. Since uS_wV, by Definition 5.9, there is $v \in V$ with $w \prec_{S\cup u^{\Box}_{\emptyset}} v$. So, $v \in V'$. Thus, uS_wV'.

Now let $v \in V'$ and $z \in W$ be such that vRz. Since $v \in V'$, we know $w \prec_{u^{\Box}_{\emptyset}} v$. Then for all $\Box B \in u$ we have $\Box B \in v$. Since vRz, we have $B, \Box B \in z$. So, $u \prec z$ and by Definition 5.9 uRz. □

5.6.2 The Logic $\mathbf{ILM_0}$

Modal completeness of $\mathbf{ILM_0}$ w.r.t. ordinary Veltman semantics was proved by Goris and Joosten (2008). Certain difficulties encountered in this proof were our main motivation for using VS. We will sketch one of these difficulties and show in what way the Verbrugge semantics overcomes it. The frame condition $wRxRyS_wuRz \Rightarrow xRz$ for M_0 is reflected in the following labelling lemma:

Lemma 5.8 (Bílková et al., 2004, Lemma 3.9) *Let w, u and x be* $\mathbf{IL}\mathsf{M}_0$*-MCSs, and S an arbitrary set of formulas. If $w \prec_S u \prec x$ then $w \prec_{S \cup u_\emptyset^\Box} x$.*

To motivate our way of proving completeness (of $\mathbf{IL}\mathsf{M}_0$, but also in general) w.r.t. VS, let us sketch a situation for which there are clear benefits in working with VS. We do this only now because $\mathbf{IL}\mathsf{M}_0$ is sufficiently complex to display (some of) these benefits. Suppose we are building models step-by-step (as in the *construction method* from (Goris & Joosten, 2008), and worlds w, u_1, u_2 and x occur in the configuration displayed in Fig. 5.2. Furthermore, suppose we need to produce an S_w-successor v of x.

With the ordinary semantics, we need to ensure that for our S_w-successor v, for each $\Box B_1 \in u_1$ and $\Box B_2 \in u_2$, we have $\Box B_1, \Box B_2 \in v$. It is not obvious that such a construction is possible. In case of $\mathbf{IL}\mathsf{M}_0$, it was successfully solved by Goris and Joosten (2008) by preserving the invariant that sets of boxed formulas in u_i are linearly ordered. This way, finite (quasi-)models can always be extended by only looking at the last u_i.

With VS, we need to produce a whole set of worlds V, but the requirements from the frame condition $wRuRxS_wV \Rightarrow (\exists V' \subseteq V)(uS_wV' \,\&\, R[V'] \subseteq R[u])$ on each particular world are less demanding. For each u_i, there has to be a corresponding $V_i \subseteq V$ with $\Box B_i$ contained (true) in every world of V_i. Lemma 5.8 gives a recipe for producing such worlds.

Theorem 5.8 *The logic* $\mathbf{IL}\mathsf{M}_0$ *is complete w.r.t.* $\mathbf{IL}_{\mathsf{Ver}}\mathsf{M}_0$*-models.*

Proof Omitted. See (Mikec & Vuković, 2020) for details. □

5.6.3 *The Logics* ILP, ILP$_0$ *and* ILR

The logics $\mathbf{IL}\mathsf{P}$, $\mathbf{IL}\mathsf{P}_0$ and $\mathbf{IL}\mathsf{R}$ can be proven to be complete with respect to their classes of frames in a similar way (see (Mikec & Vuković, 2020) for details).

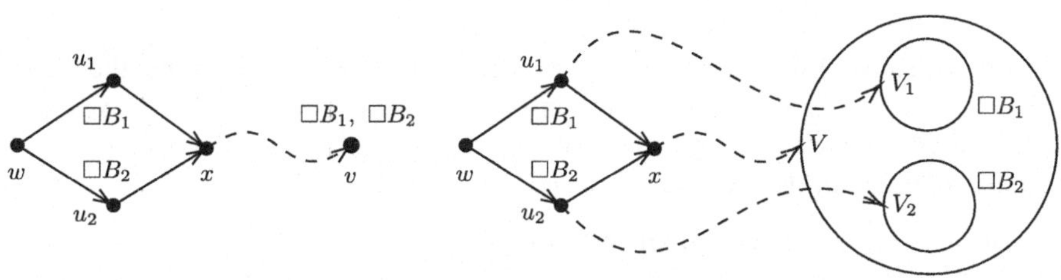

Fig. 5.2 Left: extending an ordinary Veltman model. Right: extending a Verbrugge model. Straight lines represent R-transitions, while curved lines represent S_w-transitions. Full lines represent the starting configuration, and dashed lines represent the transitions that are to be added. This figure is also taken from (Mikec & Vuković, 2020)

We recall that the interpretability logic $\mathbf{ILP_0}$ is incomplete w.r.t. Veltman models (Goris & Joosten, 2011). Since $\mathbf{ILP_0}$ is complete w.r.t. VS, this is the first example of an interpretability logic complete w.r.t. VS, but incomplete w.r.t. ordinary semantics.

5.6.4 The Logics ILW *and* ILW*

To prove that **ILW** is complete, one could try to find a sufficiently strong "labelling lemma" and use Definition 5.9 (**ILX**-structure). One candidate might be the following condition:

$$w \prec_S u \Rightarrow (\exists G \in \mathcal{D}) \Big(w \prec_{S \cup \{\Box\neg G\}} u \;\&\; G \in u\Big),$$

where $\mathcal{D}$ is finite, closed under subformulas and such that each $w \in W$ contains A_w and $\Box\neg A_w$ for some $A_w \in \mathcal{D}$. If there is such a condition, it would greatly simplify proofs of completeness for extensions of **ILW**. Unfortunately, at the moment we do not know if such a condition can be formulated and proved.

Another approach is to use a modified version of Definition 5.9 to work with **ILW** and its extensions. This way we won't require a labelling lemma, but we lose generality in the following sense. To prove the completeness of **ILXW**, for some X, it no longer suffices to simply show that the structure defined in Definition 5.9 has the required characteristic property (where each world is an **ILX**-MCS). Instead, the characteristic property of **ILX** has to be shown to hold on the modified structure. So, to improve compatibility with proofs based on Definition 5.9, we should prove the completeness of **ILW** with a definition as similar to Definition 5.9 as possible. That is what we do in the rest of this section. This approach turns out to be good enough for **ILW*** (which is the same logic as $\mathbf{ILWM_0}$). We didn't succeed in using it to prove the completeness of **ILWR**. However, to the best of our knowledge, **ILWR** might not be complete at all.

We have already mentioned the Not-W frame condition in Definition 5.6 that characterises when W fails on VS. The positive condition $(\mathsf{W})_{\mathsf{Ver}}$ from (Mikec et al., 2017) is given by:

$$(\mathsf{W})_{\mathsf{Ver}} := uS_wV \Rightarrow (\exists V' \subseteq V)\big(uS_wV' \;\&\; R[V'] \cap S_w^{-1}[V] = \emptyset\big).$$

We will use (this formulation of) $(\mathsf{W})_{\mathsf{Ver}}$ in what follows. We note here that the $(\mathsf{W})_{\mathsf{Ver}}$ condition can be formulated in a more informative way. Let $\mathfrak{F} = (W, R, \{S_w : w \in W\})$ be a Verbrugge frame. Whenever there are w, u and V such that (w, u, V) is a counterexample in $\mathfrak{F}$ to $(\mathsf{W})_{\mathsf{Ver}}$, there is $U \subseteq V$ such that:

1. (w, u, U) is a counterexample in $\mathfrak{F}$ to $(\mathsf{W})_{\mathsf{Ver}}$;
2. $R[U] \cap U = \emptyset$;
3. given the set $\overline{U} := \{v \in U : R[v] \cap S_w^{-1}[U] = \emptyset\}$ we have the following:
   a. $U \setminus \overline{U} \neq \emptyset$;
   b. for all $v \in U$ we have $(\forall U' \subseteq U)(\forall z)\big(vRzS_wU' \Rightarrow U' \cap U \setminus \overline{U} \neq \emptyset\big)$.

This new formulation tells us that we can pick a set U and a quasi-partition $\{U_0, \overline{U}\}$ of U such that points in U_0 cannot "return" to U, while the points in $\overline{U}$ can "return" to U, and have an additional property that every such "return" intersects (not only U but also) $\overline{U}$. The proof that such U can always be found will be available in the third author's PhD thesis (Mikec, 2021).

In the proof of completeness of logic **ILW** we will use the following two lemmas. In what follows, **ILWX** denotes an arbitrary extension of **ILW**.

Lemma 5.9 (Bílková et al., 2004, Lemma 3.12) *Let w be an* **ILWX**-*MCS and let B and C be formulas such that $\neg(B \rhd C) \in w$. Then there is an* **ILWX**-*MCS u such that $w \prec_{\{\Box\neg B, \neg C\}} u$ and $B \in u$.*

Lemma 5.10 (Bílková et al., 2004, Lemma 3.13) *Let w and u be some* **ILWX**-*MCS, let B and C be formulas, and let S be a set of formulas such that $B \rhd C \in w$, $w \prec_S u$ and $B \in u$. Then there is an* **ILWX**-*MCS v such that $w \prec_{S \cup \{\Box\neg B\}} v$ and $\{C, \Box\neg C\} \subseteq v$.*

Given a binary relation R on sets of formulas, let $\dot{R}[x] = R[x] \cup \{x\}$. If the set $\dot{R}[x]$ contains maximal consistent sets (which it usually does in this section), then $\bigcup \dot{R}[x]$ is a set of formulas. If satisfaction coincides with formulas contained, then it is useful to think of $\bigcup \dot{R}[x]$ as the set of formulas B such that either B or $\Diamond B$ is satisfied in x (however, one has to be careful with such an interpretation, since we do not claim a truth lemma to hold for all formulas).

Definition 5.10 Let X be W or W*. We say that $\mathfrak{M} = \langle W, R, \{S_w : w \in W\}, V\rangle$ is *the* **ILX** *-structure for a set of formulas $\mathcal{D}$* if:

- $W := \{w : w$ is an **ILX**-MCS and for some $G \in \mathcal{D}$, $G \wedge \Box\neg G \in w\}$;
- $wRu :\Leftrightarrow w \prec u$;
- $uS_wV :\Leftrightarrow wRu$ and, $V \subseteq R[w]$ and, one of the following holds:

(a) $V \cap \dot{R}[u] \neq \emptyset$;
(b) $(\forall S)\big(w \prec_S u \Rightarrow (\exists v \in V)\big(\exists G \in \mathcal{D} \cap \bigcup \dot{R}[u]\big)\ w \prec_{S \cup \{\Box\neg G\}} v\big)$;

- $w \in V(p) :\Leftrightarrow p \in w$.

With this definition, we can now prove a truth lemma.

Lemma 5.11 *Let X be* W *or* W*. *The* **ILX**-*structure $\mathfrak{M}$ for $\mathcal{D}$ is a Verbrugge model. Furthermore, the following holds:*

$$\mathfrak{M}, w \Vdash G \text{ if and only if } G \in w,$$

for each $G \in \mathcal{D}$ and $w \in W$.

Proof The proof of this claim is lengthy (mostly due to the way quasi-transitivity is defined in Definition 5.10). However, the proof is straightforward. See our comment in the proof of Lemma 5.6. For details please refer to (Mikec & Vuković, 2020). □

This lemma brings us one step away from a completeness proof. We first introduce the following notation: Let B be a formula, and w a world in a Verbrugge model. We write $[B]_w$ for $\{u : wRu \text{ and } u \Vdash B\}$.

Theorem 5.9 *The logic* **IL**W *is complete w.r.t.* $\mathbf{IL}_{\mathsf{Ver}}\mathsf{W}$*-models.*

Proof Sketch. In the light of Lemma 5.11, it suffices to show that the **IL**W-structure $\mathfrak{M}$ for $\mathcal{D}$ possesses the property $(\mathsf{W})_{\mathsf{Ver}}$. Recall the characteristic property $(\mathsf{W})_{\mathsf{Ver}}$:

$$uS_wV \Rightarrow (\exists V' \subseteq V)\left(uS_wV' \;\&\; R[V'] \cap S_w^{-1}[V] = \emptyset\right).$$

Suppose for a contradiction that there are w, u and V such that:

$$uS_wV \;\&\; (\forall V' \subseteq V)(uS_wV' \Rightarrow R[V'] \cap S_w^{-1}[V] \neq \emptyset). \tag{5.1}$$

Let $\mathcal{V}$ denote the collection of all such sets V (keeping w and u fixed).

Let $n = 2^{|\mathcal{D}|}$. Fix any enumeration $\mathcal{D}_0, \ldots, \mathcal{D}_{n-1}$ of $\mathcal{P}(\mathcal{D})$ that satisfies $\mathcal{D}_0 = \emptyset$. We define a new relation S_w^i for each $0 \leq i < n$ as follows:

$$yS_w^iU :\Longleftrightarrow yS_wU,\ \mathcal{D}_i \subseteq \bigcup \dot{R}[y],\ U \subseteq \left[\bigvee_{G \in \mathcal{D}_i} \Box\neg G\right]_w.$$

It can be shown that whenever yS_wU, we also have:

$$(\exists U' \subseteq U)(\exists i < n)\ yS_w^iU'. \tag{5.2}$$

Let $m < n$ be maximal such that there are $U \in \mathcal{V}$ and $U' \subseteq U$ with the following properties:

(i) $(\forall x \in U)[(\exists y \in R[x])(\exists Z \subseteq U)(\exists i \leq m)\ yS_w^iZ \Rightarrow x \notin U']$;
(ii) $(\forall x \in W)(xS_wU \Rightarrow xS_wU')$.

Since $\mathcal{D}_0 = \emptyset$, we have $[\bigvee_{G \in \mathcal{D}_0} \Box\neg G]_w = [\bot]_w = \emptyset$. So there are no $Z \subseteq [\bigvee_{G \in \mathcal{D}_0} \Box\neg G]_w$ such that yS_wZ for some $y \in W$. So, if we take $m = 0$ and $U' = U$ for any $U \in \mathcal{V}$, (i) and (ii) are trivially satisfied.

Since n is finite and conditions (i) and (ii) are satisfied for at least one value m, there must be a maximal $m < n$ with the required properties.

Finally, it can be shown that, contrary to the assumption of the maximality of m, also $m + 1$ satisfies properties (i) and (ii). We omit details since the full proof is somewhat cumbersome. For details please refer to (Mikec & Vuković, 2020). □

Goris and Joosten (2008) proved the completeness of **ILW*** (recall that this is equivalent to **ILWM**$_0$) w.r.t. ordinary Veltman semantics. This theorem has its analogue in VS.

Theorem 5.10 *The logic* **ILW*** *is complete w.r.t.* **IL**$_{\mathsf{Ver}}$W**-models.*

Proof From Lemma 5.11, it suffices to prove that the **ILW***-structure for $\mathcal{D}$ possesses the properties $(\mathsf{W})_{\mathsf{Ver}}$ and $(\mathsf{M}_0)_{\mathsf{Ver}}$, for each appropriate $\mathcal{D}$. So, let $\mathfrak{M} = \langle W, R, \{S_w : w \in W\}, V\rangle$ be the **ILW***-structure for $\mathcal{D}$. Theorem 5.9 shows that the model $\mathfrak{M}$ possesses the property $(\mathsf{W})_{\mathsf{Ver}}$. It remains to show that it possesses the property $(\mathsf{M}_0)_{\mathsf{Ver}}$.

The rest of this proof is very similar to the proof of Theorem 5.8. Please refer to (Mikec & Vuković, 2020) for details. □

In (Mikec et al., 2017) it is shown that **ILW*** possesses the finite model property w.r.t. Verbrugge models. To show decidability, (stronger) completeness w.r.t. ordinary Veltman models was used in (Mikec et al., 2017). However, we observe that Theorem 5.10 above suffices for the mere purpose of decidability.

5.6.5 The Logic **ILWR**

In previous subsections we saw that the completeness of **ILR** can be proven using **ILR**-structures, and that the completeness of **ILW** can be proven using **ILW**-structures. These two types of structures are defined differently. However, we saw that **ILW***-structures have the same general form as **ILW**-structures. So, one may hope to prove completeness of **ILWR** with the help of a structure defined similarly to **ILW**-structures.

Unfortunately, it seems that **ILWR**-structures, if by an **ILWR**-structure we mean an **ILW**-structure with the notion of **ILW**-consistency replaced with that of **ILWR**-consistency, does not posses the characteristic property $(\mathsf{R})_{\mathsf{Ver}}$. In (Goris et al., 2020) we call the type of a problem that emerges here "the label iteration problem". In the same paper we demonstrate how to overcome this problem for a simpler logic. With **ILWR** we have some progress, but are not yet sure if we can really solve it.

5.6.6 Logics Below **IL**

Just as this chapter was being prepared, a preprint written by Kurahashi and Okawa (2020) appeared, using VS to prove completeness and decidability of certain subsystems of **IL**. The authors define a new logic, $\mathbf{IL}^-$, similarly to **IL**, but without axiom (schema)s J1, J2, J4 and J5. However, they add new rules to the system:

$$\mathsf{R1}\text{: if } \vdash A \to B \text{ then } \vdash C \rhd A \to C \rhd B;$$
$$\mathsf{R2}\text{: if } \vdash A \to B \text{ then } \vdash B \rhd C \to A \rhd C.$$

These new rules can be seen as approximating J1 and J2. The authors also require $\Box A \leftrightarrow \neg A \rhd \bot$ to hold by definition.

The paper proceeds to study twenty logics between $\mathbf{IL}^-$ and $\mathbf{IL}$. Twelve of these are proven to be complete with respect to a version of ordinary Veltman semantics. The remaining eight are incomplete w.r.t. such semantics, and complete with respect to a version of VS. Their style of proof is similar to (de Jongh & Veltman, 1990) and (Mikec & Vuković, 2020). The authors define their whole models all at once (in a non-iterative construction), and the general structure of the definitions of relations S_w is, roughly, "$u S_w V$ if whenever there is a label Σ between w and u, there should be a corresponding world $v \in V$ with the same label Σ between w and v".

The results from (Kurahashi & Okawa, 2020) on subsystems of **IL** lend support to the conviction that Verbrugge semantics is robust and widely applicable. See also Remark 4 in Section 1.3 of (Litak & Visser, 2020).

5.7 Bisimulations and Filtrations

In this section we introduce and establish basic properties of two notions of bisimulations between Verbrugge models. Next, we use bisimilarity to define equivalence classes when we employ the method of filtrations, which in turn we use to prove finite model property and decidability of various logics. The first attempt at proving results concerning finite model property was made in (Perkov & Vuković, 2016). Just prior to the publishing of this chapter, Perkov and Vuković realised that one of the key proofs in the paper (that modal equivalence up to depth n implies n-bisimilarity) contained a slight oversight, and the notion of a bisimulation had to be weakened in order to correct the proof. The correction is relatively light and does not require any changes in the following proofs of the FMP that used the original definition: **IL**M (Perkov & Vuković, 2016), **IL**M$_0$ (Perkov & Vuković, 2016), **IL**W (Mikec et al., 2017), and **IL**W* (Mikec et al., 2017). For **IL**P$_0$ and **IL**R, the proofs in (Mikec & Vuković, 2020) relied on the parts of the definition that have been weakened, and so these proofs required (mostly straightforward) adaptions.

In the first subsection we showcase the results concerning the notion of a bisimulation that has been studied the most. In the second subsection we describe the weakening of this notion that is required in order to prove the FMP. The details will be published in near future.

5.7.1 *Bisimulations*

Visser (1990) defined the notion of a bisimulation between Veltman models. Vrgoč and Vuković (2009) extended this definition to Verbrugge models.

Definition 5.11 A bisimulation between Verbrugge models $\mathfrak{M} = \langle W, R, \{S_w : w \in W\}, V\rangle$ and $\mathfrak{M}' = \langle W', R', \{S'_{w'} : w' \in W'\}, V'\rangle$ is a non-empty relation $Z \subseteq W \times W'$ such that:

(at) if wZw', then $\mathfrak{M}, w \Vdash p$ if and only if $\mathfrak{M}', w' \Vdash' p$, for all propositional variables p;

(forth) if wZw' and wRu, then there is $u' \in W'$ such that $w'R'u'$, uZu' and for all $V' \subseteq W'$ such that $u'S'_{w'}V'$ there is $V \subseteq W$ such that uS_wV and for all $v \in V$ there is $v' \in V'$ with vZv';

(back) if wZw' and $w'R'u'$, then there is $u \in W$ such that wRu, uZu' and for all $V \subseteq W$ such that uS_wV there is $V' \subseteq W'$ such that $u'S'_{w'}V'$ and for all $v' \in V'$ there is $v \in V$ with vZv'.

We say that $w \in W$ and $w' \in W'$ are *bisimilar* if there is a bisimulation $Z \subseteq W \times W'$ such that wZw'.

The following lemma is proved by Vrgoč and Vuković (2009).

Lemma 5.12 *Let $\mathfrak{M}$, $\mathfrak{M}'$ and $\mathfrak{M}''$ be Verbrugge models.*

(a) If $w \in W$ and $w' \in W'$ are bisimilar, then they are modally equivalent, i.e. they satisfy the same formulas in the language of **IL**.
(b) The identity $\{(w, w) : w \in W\} \subseteq W \times W$ is a bisimulation.
(c) The inverse of a bisimulation between $\mathfrak{M}$ and $\mathfrak{M}'$ is a bisimulation between $\mathfrak{M}'$ and $\mathfrak{M}$.
(d) The composition of bisimulations $Z \subseteq W \times W'$ and $Z' \subseteq W' \times W''$ is a bisimulation between $\mathfrak{M}$ and $\mathfrak{M}''$.
(e) The union of a family of bisimulations between $\mathfrak{M}$ and $\mathfrak{M}'$ is again a bisimulation between $\mathfrak{M}$ and $\mathfrak{M}'$. Thus there exists the largest bisimulation between models $\mathfrak{M}$ and $\mathfrak{M}'$.

The previous lemma shows that this notion of bisimulation has certain desired properties. A property that significantly contributes to whether the notion of bisimulation can be considered well-behaved is the Hennessy–Milner property. We say that a Verbrugge model $\mathfrak{M} = \langle W, R; \{S_w : w \in W\}, V\rangle$ is *image finite* if the set $R[w]$ is finite for all $w \in W$. The following theorem is proved in (Vrgoč & Vuković, 2009).

Theorem 5.11 (Hennessy-Milner[18] property) *Let $\mathfrak{M} = \langle W, R, \{S_w : w \in W\}, V\rangle$ and $\mathfrak{M}' = \langle W', R', \{S'_w : w \in W'\}, V'\rangle$ be two image finite Verbrugge models. If $w \in W$ and $w' \in W'$ are modally equivalent, then there exists a bisimulation Z such that wZw'.*

In (Vrgoč & Vuković, 2009) several other notions of a bisimulation (V-bisimulation, strong and global bisimulation) are considered and the connections between them are explored.

[18] The theorem for unary modal logic was already known to and published by van Benthem (1984).

In (Verbrugge, 1992) and (Vuković, 2008), connections between Veltman semantics and Verbrugge semantics are considered. In (Vuković, 2008) it is shown that for a restricted class of Verbrugge models $\mathfrak{M}$ (the so-called *complete image finite* models) there exists an ordinary Veltman model $\mathfrak{M}'$ that is bisimilar to $\mathfrak{M}$.

In the section concerning completeness, we commented on why in general there cannot exist an ordinary $\mathbf{IL}\mathsf{P}_0$-model that is bisimilar to a given $\mathbf{IL}_{\mathsf{Ver}}\mathsf{P}_0$-model. That example shows that for at least some cases ordinary Veltman semantics is not expressive enough to capture the behaviour of interpretability logics.

In (Perkov & Vuković, 2016) the notion of n-bisimulation is defined similar to various existing notions for other modal logics. This notion will not be used in the corrected proof of the FMP where a weaker notion turns out to be required. We present this version of n-bisimulations for reference, to showcase the changes in the corrected version.

Definition 5.12 An *n-bisimulation* between Verbrugge models $\mathfrak{M} = \langle W, R, \{S_w : w \in W\}, V\rangle$ and $\mathfrak{M}' = \langle W', R', \{S'_{w'} : w' \in W'\}, V'\rangle$ is a decreasing sequence of relations $Z_n \subseteq Z_{n-1} \subseteq \cdots \subseteq Z_1 \subseteq Z_0 \subseteq W \times W'$ such that:

(at) if wZ_0w' then $\mathfrak{M}, w \Vdash p$ if and only if $\mathfrak{M}', w' \Vdash' p$, for all propositional variables p;

(forth) if $0 < i \leqslant n$, wZ_iw' and wRu, then there exists $u' \in R'[w']$ such that $uZ_{i-1}u'$ and for all $V' \in S'_{w'}[u']$ there is $V \in S_w[v]$ such that for all $v \in V$ there is $v' \in V'$ with $vZ_{i-1}v'$;

(back) if $0 < i \leqslant n$, wZ_iw' and $w'R'u'$, then there exists $u \in R[w]$ such that $uZ_{i-1}u'$ and for all $V \in S_w[u]$ there is $V' \in S'_{w'}[u']$ such that for all $v' \in V'$ there is $v \in V$ with $vZ_{i-1}v'$.

We say that $w \in W$ and $w' \in W'$ are *n-bisimilar* if there is an n-bisimulation between $\mathfrak{M}$ and $\mathfrak{M}'$ such that wZ_nw'.

5.7.2 *Filtrations and the Finite Model Property*

The filtration method is often used to prove that a modal logic possesses the finite model property. Perkov and Vuković (2016) applied this technique to VS. Here bisimilarity is used to refine models in order to preserve the structural properties of Verbrugge models. As we already announced, the original proof contained an oversight. This can be remedied by using a weaker definition of a bisimulation compared to Definition 5.11.

Definition 5.13 A w-bisimulation between Verbrugge models $\mathfrak{M} = \langle W, R, \{S_w : w \in W\}, V\rangle$ and $\mathfrak{M}' = \langle W', R', \{S'_{w'} : w' \in W'\}, V'\rangle$ is a non-empty relation $Z \subseteq W \times W'$ such that:

(at) if wZw', then $\mathfrak{M}, w \Vdash p$ if and only if $\mathfrak{M}', w' \Vdash' p$, for all propositional variables p;

(w-forth) if wZw' and wRu, then there is $U' \subseteq W'$ such that for all $u' \in U'$ we have $w'R'u'$, uZu' and for every $V' : U' \to \mathcal{P}(W')$ such that for all $u' \in U'$ we have $u'S'_{w'}V'(u')$, there is $V \subseteq W$ such that uS_wV and for all $v \in V$ there is $v' \in \bigcup_{u' \in U'} V'(u')$ such that vZv';

(w-back) if wZw' and $w'Ru'$, then there is $U \subseteq W$ such that for all $u \in U$ we have wRu, uZu' and for every $V : U \to \mathcal{P}(W)$ such that for all $u \in U$ we have $uS_wV(u)$, there is $V' \subseteq W'$ such that $u'S'_{w'}V'$ and for all $v' \in V'$ there is $v \in \bigcup_{u \in U} V(u)$ such that vZv';

We say that $w \in W$ and $w' \in W'$ are w-*bisimilar* if there is a w-bisimulation $Z \subseteq W \times W'$ such that wZw'.

w-bisimulations satisfy the properties of Lemma 5.12.

In (Perkov & Vuković, 2016) the notion of an n-bisimulation is defined similarly to various existing notions for other modal logics. Then, n-bisimulations are used in the proof of the finite model property of various systems w.r.t. VS. Here we present n-w-bisimulations, since that is the concept required to obtains the FMP in the revised version of the proof.

Definition 5.14 An n-w *-bisimulation* between Verbrugge models $\mathfrak{M} = \langle W, R, \{S_w : w \in W\}, V\rangle$ and $\mathfrak{M}' = \langle W', R', \{S'_{w'} : w' \in W'\}, V'\rangle$ is a decreasing sequence of relations $Z_n \subseteq Z_{n-1} \subseteq \cdots \subseteq Z_1 \subseteq Z_0 \subseteq W \times W'$ such that:

(at) if wZ_0w' then $\mathfrak{M}, w \Vdash p$ if and only if $\mathfrak{M}', w' \Vdash' p$, for all propositional variables p;

(forth) if $0 < i \leqslant n$, wZ_iw' and wRu, then there is $U' \subseteq W'$ such that for all $u' \in U'$ we have $w'R'u'$, $uZ_{i-1}u'$ and for every $V' : U' \to \mathcal{P}(W')$ such that for all $u' \in U'$ we have $u'S'_{w'}V'(u')$, there is $V \subseteq W$ such that uS_wV and for all $v \in V$ there is $v' \in \bigcup_{u' \in U'} V'(u')$ such that $vZ_{i-1}v'$;

(back) if $0 < i \leqslant n$, wZ_iw' and $w'R'u'$, then there is $U \subseteq W$ such that for all $u \in U$ we have $wRu, uZ_{i-1}u'$ and for every $V : U \to \mathcal{P}(W)$ such that for all $u \in U$ we have $uS_wV(u)$, there is $V' \subseteq W'$ such that $u'S'_{w'}V'$ and for all $v' \in V'$ there is $v \in \bigcup_{u \in U} V(u)$ such that $vZ_{i-1}v'$;

We say that $w \in W$ and $w' \in W'$ are n-w-*bisimilar* if there is an n-w-bisimulation $(Z_i)_{0 \leq i \leq n}$ between $\mathfrak{M}$ and $\mathfrak{M}'$ such that wZ_nw'.

We say that w and w' are n-*modally equivalent* and we write $w \equiv_n w'$ if w and w' satisfy exactly the same formulas of the modal depth (that is, the maximal number of nested modalities) up to n. The following lemma is proved in (Perkov & Vuković, 2016). This is the lemma that contained the oversight in the version that was published originally. Here we cite the corrected version, which is based on n-w-bisimulations instead of n-bisimulations.

Lemma 5.13 *Let* $\mathfrak{M} = \langle W, R, \{S_w w \in W\}, V\rangle$ *and* $\mathfrak{M}' = \langle W', R', \{S'_w : w \in W'\}, V'\rangle$ *be Verbrugge models. Let* $w \in W$ *and* $w' \in W'$. *We have:*

1. *if* w *and* w' *are* n-w-*bisimilar, then* w *and* w' *are* n-*modally equivalent;*
2. *if there are only finitely many propositional variables, then the converse also holds: if* w *and* w' *are* n-*modally equivalent, then* w *and* w' *are* n-w-*bisimilar.*

Filtration usually employs partitions whose clusters contain logically equivalent worlds. The equivalence need not be with respect to all formulas; usually a finite set of formulas closed under taking subformulas suffices. In our case, some additional properties of this set of formulas are required. Let A be a formula. If A is not a negation, then $\sim A$ denotes $\neg A$, and otherwise, if A is $\neg B$, then $\sim A$ is B. It is convenient to take $\rhd$ to be the only modality in our language and to define $\Box$ and $\Diamond$ as abbreviations: $\Diamond A$ as $\neg(A \rhd \bot)$ and $\Box A$ as $\sim\Diamond\sim A$, i.e. $\sim A \rhd \bot$. We will give the definition of adequate sets used in (Mikec & Vuković, 2020). It is an extended version of the definition used in (Perkov & Vuković, 2016), where filtrations of Verbrugge models were originally introduced. The extended definition turns out to be important for some logics.

Definition 5.15 Let $\mathcal{D}$ be a finite set of formulas that is closed under taking subformulas and single negations $\sim$, and $\top \in \mathcal{D}$. We say that a set of formulas $\Gamma_\mathcal{D}$ is *an adequate set* (w.r.t. $\mathcal{D}$) if it satisfies the following conditions:[19]

(a) $\Gamma_\mathcal{D}$ is closed under taking subformulas;
(b) if $A \in \Gamma_\mathcal{D}$ then $\sim A \in \Gamma_\mathcal{D}$;
(c) $\bot \rhd \bot \in \Gamma_\mathcal{D}$;
(d) $A \rhd B \in \Gamma_\mathcal{D}$ if A is an antecedent or succedent of some $\rhd$-formula in $\Gamma_\mathcal{D}$, and so is B;
(e) if $A \in \mathcal{D}$ then $\Box\neg A \in \Gamma_\mathcal{D}$.

Since the set of formulas $\mathcal{D}$ is finite, $\Gamma_\mathcal{D}$ is finite too. Now, let $\mathfrak{M} = \langle W, R, \{S_w : w \in W\}, V\rangle$ be a Verbrugge model and let $\Gamma_\mathcal{D}$ be an adequate set of formulas w.r.t. some appropriate set $\mathcal{D}$. For nodes $w, u \in W$, we write $w \equiv_{\Gamma_\mathcal{D}} u$ if for all $A \in \Gamma_\mathcal{D}$ we have $\mathfrak{M}, w \Vdash A$ if and only if $\mathfrak{M}, u \Vdash A$.

Let $\sim\ \subseteq\ \equiv_{\Gamma_\mathcal{D}}$ be an equivalence relation on the set W. Denote the $\sim$-equivalence class of $w \in W$ by $[w]$, and $\widetilde{V} = \{[w] : w \in V\}$ for any $V \subseteq W$.

A *filtration* of a model $\mathfrak{M}$ through $\Gamma_\mathcal{D}, \sim$ is any Verbrugge model $\widetilde{\mathfrak{M}} = \langle \widetilde{W}, \widetilde{R}, \{\widetilde{S}_{[w]} : [w] \in \widetilde{W}\}, V'\rangle$ such that for all $w \in W$ and $A \in \Gamma_\mathcal{D}$ we have $\mathfrak{M}, w \Vdash A$ if and only if $\widetilde{\mathfrak{M}}, [w] \Vdash' A$. Fact (e) of (the adapted version of) Lemma 5.12 implies that the largest w-bisimulation $\sim_\mathfrak{M}$ of model $\mathfrak{M}$ exists. Facts (b), (c) and (d) of the same lemma imply that $\sim_\mathfrak{M}$ is an equivalence relation, while (a) implies $\sim_\mathfrak{M}\ \subseteq\ \equiv_{\Gamma_\mathcal{D}}$.

Lemma 5.14 (Perkov & Vuković, 2016, Lemma 2.3 and Theorem 2.4) *Let $\mathfrak{M} = \langle W, R, \{S_w : w \in W\}, V\rangle$ be a Verbrugge model, $\Gamma_\mathcal{D}$ an adequate set of formulas, and $\sim_\mathfrak{M}$ the largest* w*-bisimulation of model $\mathfrak{M}$. Let us define:*

(a) $\widetilde{R} = \{([w], [u]) : wRu$ and there is $\Box A \in \Gamma_\mathcal{D}$ such that $\mathfrak{M}, w \nVdash \Box A$ and $\mathfrak{M}, u \Vdash \Box A\}$;
(b) $[u]\widetilde{S}_{[w]}\widetilde{V}$ if and only if $[w]\widetilde{R}[u]$, $\widetilde{V} \subseteq \widetilde{R}[[w]]$, and for all $w' \in [w]$ and $u' \in [u]$ such that $w'Ru'$ we have $u'S_{w'}V'$ for some V' such that $\widetilde{V'} \subseteq \widetilde{V}$;
c) for all propositional variables $p \in \Gamma_\mathcal{D}$ put $[w] \in V'(p)$ if and only if $w \in V(p)$, and interpret propositional variables $q \notin \Gamma_\mathcal{D}$ arbitrarily (e.g. put $[w] \notin V(q)$ for all $[w] \in \widetilde{W}$).

---

[19] We note here that most of the complexity in this definition can be eliminated, however it would require changing a number of details in papers that we cite here. Essentially, it suffices to require closure under taking subformulas and single negations. See (Mikec, 2021, Chap. 5) which gives details on how to adapt existing proofs to work with the simpler definition.

Then $\widetilde{\mathfrak{M}} = \langle \widetilde{W}, \widetilde{R}, \{\widetilde{S}_{[w]} : [w] \in \widetilde{W}\}, V' \rangle$ *is a filtration of the model* $\mathfrak{M}$ *through* $\Gamma_{\mathcal{D}}, \sim_{\mathfrak{M}}$.

We can use the construction above to prove the finite model property of many logics. Let us briefly sketch the proof. We start by fixing a formula A satisfied in some model $\mathfrak{M}$ and a finite adequate set $\Gamma_{\mathcal{D}}$ such that $A \in \Gamma_{\mathcal{D}}$. For the purposes of this proof, we assume the language contains only the propositional variables that are contained in A. Let $\widetilde{\mathfrak{M}}$ be a filtration of this model as described above. We first prove that the length of $\widetilde{R}$-chains is bounded by the number of occurrences of boxed formulas in $\Gamma_{\mathcal{D}}$. This implies that w-bisimilarity can be simplified to n-w-bisimilarity, for a sufficiently large n. Lemma 5.13 implies that clusters are n-w-bisimilar if and only if they are modally n-equivalent. Thus there can be only finitely many classes with respect to n-equivalence, implying our model is finite.

Theorem 5.12 (Perkov & Vuković, 2016) *The logic* **IL** *has the finite model property with respect to Verbrugge models.*

For **IL**, **ILM**, **ILP** and **ILW**, the original completeness proofs were proofs of completeness w.r.t. appropriate finite models (de Jongh & Veltman, 1990, 1999). For these logics, the FMP w.r.t. the ordinary semantics and decidability are immediate (and completeness and the FMP w.r.t. VS are easily shown to follow from these results). For more complex logics, not much is known about the FMP w.r.t. the ordinary semantics.

To prove that a specific extension has the FMP, it remains to show that filtration preserves its characteristic property. This approach was successfully used to prove the FMP of $\mathbf{ILM_0}$, **ILW***, $\mathbf{ILP_0}$ and **ILR** w.r.t. VS (Mikec et al., 2017; Mikec & Vuković, 2020; Perkov & Vuković, 2016).

Since we have the finite model property, and "finite" can be taken to mean finite in every sense, we also have decidability. This follows by the standard argument: enumerate all the proofs (which is possible since all the logics **ILX** in question are recursively enumerable) and all the (codes of) finite models simultaneously. Sooner or later, we either find a proof of A, or, because of the completeness and the FMP, a model of $\neg A$.

Corollary 5.2 (Mikec et al., 2017; Mikec & Vuković, 2020; Perkov & Vuković, 2016) *The logics* $\mathbf{ILM_0}$, **ILW***, $\mathbf{ILP_0}$ *and* **ILR** *are decidable.*

5.8 Hierarchies and Frame Conditions

In this final section we will present frame conditions of a new series of principles in **IL**(All) as presented by Goris and Joosten (2020). We shall recall the frame-conditions computed for this series with respect to regular Veltman semantics. Next, we shall present the respective frame conditions for the VS. These novel results are moreover formalised in the proof assistant Agda and are available in the master thesis of Mas Rovira (2020) written under the direction of Joosten and Mikec.

5.8.1 A Broad Series of Principles

Let us present here one of two series from (Goris & Joosten, 2020). By lack of a better name the series was called *the broad series*. To present this series we first define a series of auxiliary formulas. For any $n \geq 1$ we define the schemata U_n as follows.

$$\mathsf{U}_1 := \Diamond\neg(D_1 \rhd \neg C),$$
$$\mathsf{U}_{n+2} := \Diamond((D_{n+1} \rhd D_{n+2}) \wedge \mathsf{U}_{n+1}).$$

Now, for $n \geq 0$ we define the schemata for the broad series R^n as follows.

$$\mathsf{R}^0 := A \rhd B \to \neg(A \rhd \neg C) \rhd B \wedge \Box C,$$
$$\mathsf{R}^{n+1} := A \rhd B \to \mathsf{U}_{n+1} \wedge (D_{n+1} \rhd A) \rhd B \wedge \Box C.$$

As an illustration we shall calculate the first four principles.

$$\mathsf{R}^0 := A \rhd B \to \neg(A \rhd \neg C) \rhd B \wedge \Box C$$
$$\mathsf{R}^1 := A \rhd B \to \Diamond\neg(D_1 \rhd \neg C) \wedge (D_1 \rhd A) \rhd B \wedge \Box C$$
$$\mathsf{R}^2 := A \rhd B \to \Diamond\Big[(D_1 \rhd D_2) \wedge \Diamond\neg(D_1 \rhd \neg C)\Big] \wedge (D_2 \rhd A) \rhd B \wedge \Box C$$
$$\mathsf{R}^3 := A \rhd B \to \Diamond\Big((D_2 \rhd D_3) \wedge \Diamond\Big[(D_1 \rhd D_2) \wedge \Diamond\neg(D_1 \rhd \neg C)\Big]\Big) \wedge (D_3 \rhd A) \rhd B \wedge \Box C$$

It is not hard to determine the frame condition for the first couple of principles in this series and in Fig. 5.3 we have depicted the first three frame-conditions. In this section we shall prove that the correspondence proceeds as expected. Informally, the frame condition for R^n shall be the universal closure of

$$x_{n+1} R x_n \ldots R x_0 R y_0 S_{x_1} y_1 \ldots S_{x_n} y_n S_{x_{n+1}} y_{n+1} R z \Rightarrow y_0 S_{x_0} z. \tag{5.3}$$

We shall first recast the frame condition in a recursive fashion. In writing (5.3) recursively we shall use those variables that will emphasise the relation with (5.3). Of course, free variables can be renamed at the readers liking.

First, we start by introducing a relation $\mathcal{B}_n$ that captures the antecedent of (5.3). Note that this antecedent says that first there is a chain of points x_i related by R, followed by a chain of points y_i related by different S relations. The relation $\mathcal{B}_n$ will be applied to the end-points of both chains where the condition on the intermediate points is imposed by recursion.

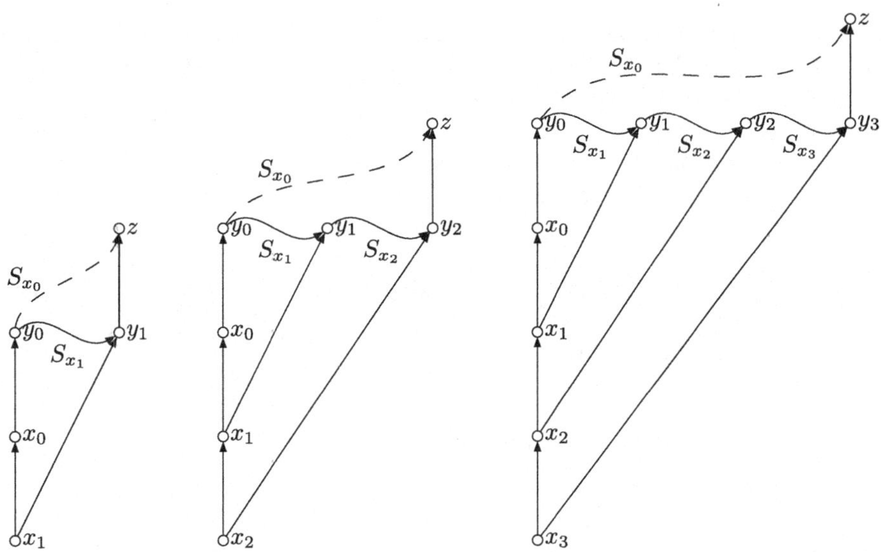

Fig. 5.3 From left to right, this figure depicts the frame conditions (R^0) through (R^2) corresponding to R^0 through R^2. The reading convention is as always: if all the un-dashed relations are present as in the picture, then also the dashed relation should be there

$$\mathcal{B}_0(x_1, x_0, y_0, y_1) := x_1 R x_1 R y_0 S_{x_1} y_1,$$

$$\mathcal{B}_{n+1}(x_{n+2}, x_0, y_0, y_{n+2}) := \exists x_{n+1}, y_{n+1} \big(x_{n+2} R x_{n+1} \;\&\; \mathcal{B}_n(x_{n+1}, x_0, y_0, y_{n+1}) \;\&\; y_{n+1} S_{x_{n+2}} y_{n+2}\big).$$

For every $n \geq 0$ we can now define the first order frame condition (R^n) as follows.

$$(\mathsf{R}^n) := \forall x_{n+1}, x_0, y_0, y_{n+1} \ \big(\mathcal{B}_n(x_{n+1}, x_0, y_0, y_{n+1}) \Rightarrow \forall z\, (y_{n+1} R z \Rightarrow y_0 S_{x_0} z)\big).$$

Sometimes we shall write $x_{n+1}\mathcal{B}_n[x_0, y_0]\, y_{n+1}$ conceiving the quaternary relation $\mathcal{B}_n$ as a binary relation indexed by the pair x_0, y_0.

Theorem 5.13 *We have* $\mathfrak{F} \models (\mathsf{R}^n)$ *if and only if* $\mathfrak{F} \models \mathsf{R}^n$, *for each Veltman frame* $\mathfrak{F}$ *and* $n \in \mathbb{N}$.

5.8.2 *Frame Conditions for VS*

In this section we present Verbrugge-frame conditions for the above presented series R^i. We observe that the mere definition of what it means to be a frame condition of an axiom scheme can be stated in second order logic where an arbitrary valuation corresponds to an arbitrary subset. Since VS is second order in nature, it may raise a question what actually constitutes a natural frame condition for a principle other than just writing down the definition. This question is discussed in (Mas Rovira, 2020). The $(\mathsf{R}^n)_{gen}$ condition reads as follows:

$$\begin{aligned}
&\forall w, x_0, \ldots, x_{n-1}, y, z, \mathbb{A}, \mathbb{B}, \mathbb{C}, \mathbb{D}_0, \ldots, \mathbb{D}_{n-1}.\\
&wRx_{n-1}R\ldots Rx_0RyRz,\\
&(\forall u.wRu, u \in \mathbb{A} \Rightarrow \exists V.uS_wV \subseteq \mathbb{B}),\\
&(\forall u.x_{n-1}Ru \in \mathbb{D}_{n-1} \Rightarrow \exists V.uS_{x_{n-1}}V \subseteq \mathbb{A}),\\
&(\forall i \in \{1, \ldots, n-2\}\forall u.x_iRu \in \mathbb{D}_i \Rightarrow \exists V.uS_{x_i}V \subseteq \mathbb{D}_{i+1}),\\
&(\forall V.zS_yV \Rightarrow V \cap \mathbb{C} \neq 0),\\
&z \in \mathbb{D}_0\\
\Rightarrow\ &\exists V \subseteq \mathbb{B}.x_{n-1}S_wV, \{w : \exists v \in V.vRw\} \subseteq \mathbb{C}
\end{aligned}$$

Lemma 5.15 *Let $\mathfrak{M}$ be a Verbrugge model, let x be a world and let $n \in \mathbb{N}$. For any $i \leq n$ we have that if $\mathfrak{M}, x \Vdash U_i$ then there exist some worlds $y, z, x_0, \ldots, x_i$ such that:*

1. $x_i = x$;
2. $x_iR\ldots Rx_0RyRz$;
3. *for all* $j \leq i$ *we have that* $\mathfrak{M}, x_j \Vdash U_j$;
4. *for all* $j < i$ *we have that* $\mathfrak{M}, x_j \Vdash D_j \rhd D_{j+1}$;
5. *for all* V *we have that if* zS_yV *then* $V \cap \{w : \mathfrak{M}, w \Vdash C\} \neq \emptyset$;
6. $\mathfrak{M}, z \Vdash D_0$.

Proof By induction on i. For $i = 0$ we have that $x \Vdash \Diamond\neg(D_0 \rhd \neg C)$. It follows that there exists some y such that $xRy \Vdash \neg(D_0 \rhd \neg C)$ and therefore there exists some z such that $yRz \Vdash D_0$ and for any V, if zS_yV, then $V \cap \{w : \mathfrak{M}, w \Vdash C\} \neq \emptyset$. It is clear that all claims are met.

For $i + 1$ we have that $x \Vdash \Diamond(D_i \rhd D_{i+1} \wedge U_i)$. It follows that there exists some x_i such that $x_i \Vdash D_i \rhd D_{i+1} \wedge U_i$. By the inductive hypothesis there exist $y, z, x_0, \ldots, x_i$ such that satisfy claims 1 ... 6. We set $x_{i+1} := x$. It is trivial to observe that by using the inductive hypothesis all conditions are met for $i + 1$. □

Theorem 5.14 *For any Verbrugge frame $\mathfrak{F}$, we have that $\mathfrak{F}$ satisfies the $(\mathsf{R}^n)_{gen}$ condition if and only if any model based on $\mathfrak{F}$ forces every instantiation of the R^n principle. In symbols:*

$$\mathfrak{F} \models (\mathsf{R}^n)_{gen} \quad \textit{if and only if} \quad \mathfrak{F} \models \mathsf{R}^n$$

Proof If $n = 0$ we refer to Sect. 5.4.1 or (Mas Rovira, 2020) for full details.

For $n + 1$ proceed as follows. Suppose first that we have a Verbrugge model and a world w such that $w \Vdash A \rhd B$. Then assume also that $wRx \Vdash ((D_n \rhd A) \wedge U_n)$. By

Lemma 5.15 it follows that there exist $y, z, x_0, \ldots, x_n$ satisfying $1, \ldots, 6$. For a formula F, define $[\![F]\!] := \{x \in \mathfrak{M} : x \Vdash F\}$. Then let $\mathbb{A} := [\![A]\!]$, $\mathbb{B} := [\![B]\!]$, $\mathbb{C} := [\![C]\!]$ and for $i \leq n$ let $\mathbb{D}_i := [\![D_i]\!]$.

It is routine to check that the left part of the $(\mathsf{R}^{n+1})_{gen}$ holds and thus we get that there exists some $V \subseteq \mathbb{B}$ such that $x_n S_w V$ and $R[V] \subseteq \mathbb{C}$. Since $V \subseteq \mathbb{B}$ we have that $V \Vdash B$ and since $R[V] \subseteq \mathbb{C}$ we have $V \Vdash \Box C$. Finally, since $x_n = x$ we conclude $x S_w V \Vdash B \wedge \Box C$.

Let us now prove the opposite direction. Fix a Verbrugge frame $\mathfrak{F}$ and let $a, b, c, d_0, \ldots, d_n$ be propositional variables and assume $\mathfrak{F} \Vdash \mathsf{R}^{n+1}$. Assume that the left part of the implication of $(\mathsf{R}^{n+1})_{gen}$ holds. Now consider a model based on $\mathfrak{F}$ that satisfies the following:

$$[\![a]\!] = \mathbb{A}, \ [\![b]\!] = \mathbb{B}, \ [\![c]\!] = \mathbb{C}, \ [\![d_i]\!] = \mathbb{D}_i, \ \text{ for all } i \in \{0 \ldots n\}$$

Now one can routinely check that $w \Vdash A \rhd B$ and $x \Vdash ((D_n \rhd A) \wedge U_n)$, hence there exists U such that $x S_w U$ and $U \Vdash B \wedge \Box C$. From that we derive that $U \subseteq \mathbb{B}$ and $R[U] \subseteq \mathbb{C}$.

This proof has been formalised in Agda. Full details can be found in (Mas Rovira, 2020). □

Acknowledgements We thank Dick de Jongh, Volodya Shavrukov, Vítěslav Švejdar, Rineke Verbrugge and Albert Visser for fruitful discussions and for comments on the origins of the field. Many thanks also to Taishi Kurahashi and Rineke Verbrugge for a careful cover-to-cover reading of the document pointing out various inaccuracies and room for improvement. An anonymous referee helped improving the presentation of the text considerably with many excellent comments and suggestions. Joosten received support from grants RTC-2017-6740-7, FFI2015-70707P and 2017 SGR 270. Mikec was supported by Croatian Science Foundation (HRZZ) under the projects UIP–05–2017–9219 and IP–01–2018–7459. Vuković was supported by Croatian Science Foundation (HRZZ) under the project IP–01–2018–7459.

References

Areces, C., Hoogland, E., & de Jongh, D. H. J. (2001). Interpolation, definability and fixed-points in interpretability logics. In M. Zakharyaschev, K. Segerberg, M. de Rijke & H. Wansing (Eds.), *Advances in modal logic*, (Vol. 2, pp. 35–58). CSLI.

Bennet, C. (1986). *On some orderings of extensions of arithmetic*. Department of Philosophy, University of Göteborg.

van Benthem, J. (1983). *Modal logic and classical logic*. Bibliopolis.

van Benthem, J. (1984). Correspondence theory. In F. Guenthner & D. Gabbay (Eds.), *Handbook of philosophical logic* (pp. 167–247). Springer.

Berarducci, A. (1990). The interpretability logic of peano arithmetic. *The Journal of Symbolic Logic, 55*(3), 1059–1089.

Bou, F., & Joosten, J. J. (2011). The closed fragment of IL is PSPACE hard. *Electronical Notes Theoretical Computer Science, 278*, 47–54.

Burgess, J. P. (2002). Basic tense logic. In D. Gabbay & F. Guenthner (Eds.), *Handbook of philosophical logic, Second Edition: Volume VII: Extensions of classical logic* (pp. 1–42). Kluwer Academic Publishers.

Bílková, M., Goris, E., & Joosten, J. J. (2004). Smart labels. In L. Afanasiev & M. Marx (Eds.), *Liber Amicorum for Dick de Jongh. Intitute for Logic, Language and Computation*. Electronically published. ISBN: 90 5776 1289.

Bílková, M., de Jongh, D. H. J., & Joosten, J. J. (2009). Interpretability in PRA. *Annals of Pure and Applied Logic, 161*(2), 128–138.

Čačić, V., & Kovač, V. (2015). On the share of closed IL formulas which are also in GL. *Archive for Mathematical Logic, 54*, 741–767.

Čačić, V., & Vrgoč, D. (2011). A note on bisimulation and modal equivalence in provability logic and interpretability logic. *Studia Logica, 101*, 31–44.

Čačić, V., & Vuković, M. (2012). A note on normal forms for closed fragment of system IL. *Mathematical Communications, 17*, 195–204.

de Jongh, D. H. J., Jumelet, M., & Montagna, F. (1991). On the proof of Solovay's theorem. *Studia Logica, 50*(1), 51–69.

de Jongh, D. H. J., & Pianigiani, D. (1998). Solution of a problem of David Guaspari. Studies in fuzziness and soft computing. In E. Orłowska (Ed.), *Logic at work* (pp. 246–254). Physica-Verlag.

de Jongh, D. H. J., & Veltman, F. J. M. M. (1983). *Intensional logic*. University of Amsterdam, unpublished course text.

de Jongh, D. H. J., & Veltman, F. J. M. M. (1990). Provability logics for relative interpretability. In P. P. Petkov (Ed.), *Mathematical logic, proceedings of the heyting 1988 summer school in Varna* (pp. 31–42). Plenum Press.

de Jongh, D. H. J., & Veltman, F. J. M. M. (1999). Modal completeness of ILW. In J. Gerbrandy, M. Marx, M. Rijke, & Y. Venema (Eds.), *Essays dedicated to Johan van Benthem on the occasion of his 50th birthday*. Amsterdam University Press.

de Jongh, D. H. J., Veltman, F. J. M. M., & Verbrugge, R. (2004). Completeness by construction for tense logics of linear time. In J. van Benthem, A. Troelstra, F. J. M. M. Veltman, & A. Visser (Eds.), *Liber Amicorum for Dick de Jongh*. Institute of Logic, Language and Computation.

de Jongh, D. H. J., & Visser, A. (1991). Explicit fixed points in interpretability logic. *Studia Logica: An International Journal for Symbolic Logic, 50*(1), 39–49.

de Rijke, M. (1992). Unary interpretability logic. *Notre Dame Journal of Formal Logic, 33*, 249–272.

Dzhaparidze (Japaridze), G. (1992). The logic of linear tolerance. *Studia Logica, 51*, 249–277.

Dzhaparidze (Japaridze), G. (1993). A generalized notion of weak interpretability and the corresponding logic. *Annals of Pure and Applied Logic, 61*, 113–160.

Goris, E. (2006). Interpolation and the interpretability logic of PA. *Notre Dame Journal of Formal Logic, 47*(2), 179–195.

Goris, E., Bílková, M., Joosten, J. J., Mikec, L. (2020). Assuring and critical labels for relations between maximal consistent sets for interpretability logics. https://arxiv.org/2003.04623

Goris, E., & Joosten, J. J. (2008). Modal matters for interpretability logics. *Logic Journal of the IGPL, 16*(4), 371–412.

Goris, E., & Joosten, J. J. (2011). A new principle in the interpretability logic of all reasonable arithmetical theories. *Logic Journal of the IGPL, 19*(1), 14–17.

Goris, E., & Joosten, J. J. (2012). Self provers and Σ_1 sentences. *Logic Journal of the IGPL, 20*(1), 1–21.

Goris, E., & Joosten, J. J. (2020). Two new series of principles in the interpretability logic of all reasonable arithmetical theories. *The Journal of Symbolic Logic, 85*(1), 1–25.

Hájek, P. (1971). On interpretability in set theories I. *Commentationes Mathematicae Universitatis Carolinae, 12*, 73–79.

Hájek, P. (1972). On interpretability in set theories II. *Commentationes Mathematicae Universitatis Carolinae, 13*, 445–455.

Hájek, P. (1981). Interpretability in theories containing arithmetic II. *Commentationes Mathematicae Universitatis Carolinae, 22*, 667–688.

Hájek, P., & Montagna, F. (1990). The logic of Π_1-conservativity. *Archiv für Mathematische Logik und Grundlagenforschung, 30*, 113–123.
Hájek, P., & Montagna, F. (1992). The logic of Π_1-conservativity continued. *Archiv für Mathematische Logik und Grundlagenforschung, 32*, 57–63.
Hájek, P., & Švejdar, V. (1991). A note on the normal form of closed formulas of interpretability logic. *Studia Logica, 50*(1), 25–28.
Hakoniemi, T. A., & Joosten, J. J. (2016). Labelled tableaux for interpretability logics. A Tribute to Albert Visser. In J. van Eijck, R. Iemhoff & J. J. Joosten (Eds.), *Liber Amicorum Alberti* (pp. 141–154). College Publications.
Icard, T., & Joosten, J. J. (2012). Provabilty and interpretability logics with re-stricted substitutions. *Notre Dame Journal of Formal Logic, 53*(2), 133–154.
Iemhoff, R., de Jongh, D. H. J., & Zhou, C. (2005). Properties of intuitionistic provability and preservativity logics. *Logic Journal of the IGPL, 13*(6), 615–636.
Ignatiev, K. N. (1991). Partial conservativity and modal logics. Technical Report X-91-04, ILLC, University of Amsterdam.
Ignatiev, K. N. (1993). The provability logic of Σ_1-interpolability. *Annals of Pure and Applied Logic, 64*, 1–25.
Iwata, S., & Kurahashi, T. (2021). Topological semantics of conservativity and interpretability logics.
Japaridze, G., & de Jongh, D. H. J. (1998). The logic of provability. In S. Buss (Ed.), *Handbook of proof theory* (pp. 475–546). North-Holland Publishing Co.
Japaridze (Dzhaparidze), G. (1994). A simple proof of arithmetical completeness for Π_1-conservativity logic. *Notre Dame Journal of Formal Logic, 35*, 346–354.
Joosten, J.J. (1998). Towards the interpretability logic of all reasonable arithmetical theories. Master's thesis, University of Amsterdam.
Joosten, J. J. (2004). *Interpretability Formalized*. Ph.D. thesis, Utrecht University.
Joosten, J. J. (2005). The closed cragment of the interpretability logic of PRA with a constant for $I\Sigma_1$. *Notre Dame Journal of Formal Logic, 46*(2), 127–146.
Joosten, J. J. (2010). Consistency statements and iterations of computable functions in $I\Sigma_1$ and PRA. *Archive for Mathematical Logic, 49*(7, 8), 773–798.
Joosten, J. J. (2016). On formalizations of the Orey-Hájek characterization for interpretability. In P. Cegielski, A. Enayat, & R. Kossak (Eds.), *Studies in weak arithmetics* (pp. 57–90). CSLI Publications.
Joosten, J. J., Mikec, L., & Visser, A. (2020). Feferman axiomatisations, definable cuts and principles of interpretability. Forthcoming.
Joosten, J. J., & Visser, A. (2000). The interpretability logic of all reasonable arithmetical theories. The new conjecture. *Erkenntnis, 53*(1–2), 3–26.
Joosten, J. J., & Visser, A. (2004). How to derive principles of interpretability logic, A toolkit. In J. van Benthem, F. J. M. M. Veltman, A. Troelstra & A. Visser (Eds.), *Liber Amicorum for Dick de Jongh*. Intitute for Logic, Language and Computation. Electronically published. ISBN: 90 5776 1289.
Kalsbeek, M. B. (1991). Towards the interpretability logic of $I\Delta_0 + \mathsf{EXP}$. Logic Group Preprint Series 61, Faculty of Humanities, Philosophy, Utrecht University, https://lgps.sites.uu.nl
Kurahashi, T., & Okawa, Y. (2020). Modal completeness of sublogics of the interpretability logic **IL**. https://arxiv.org/2004.03813
Lindström, P. (1997). *Aspects of incompleteness* (Vol. 10). Springer.
Litak, T. (2014). Constructive modalities with provability smack. In G. Bezhanishvili (Ed.), *Leo Esakia on duality in modal and intuitionistic logics, volume 4 of outstanding contributions to logic* (pp. 179–208). Springer.
Litak, T. (2017). Constructive modalities with provability smack (author's cut). Unabridged and extended version of a chapter in the Esakia volume of Outstanding Contributions to Logic.
Litak, T., & Visser, A. (2018). Lewis meets Brouwer: constructive strict implication. *Indagationes Mathematicae, 29*, 36–90. A special issue "L.E.J. Brouwer, fifty years later".

Litak, T., & Visser, A. (2020). Lewisian fixed points I: Two incomparable constructions. This volume.

Mas Rovira, J. (2020). Frame conditions for interpretability logics using generalised veltman semantics and the agda proof assistant. Master's thesis, Master of Pure and Applied Logic, University of Barcelona. Available on http://diposit.ub.edu/dspace/handle/2445/133559

Mikec, L. (2021). *On logics and semantics for interpretability*. PhD thesis. University of Barcelona and University of Zagreb.

Mikec, L. (2022). Complexity of the interpretability logics ILW and ILP. *Logic Journal of the IGPL*, Accepted, volume and issue number not assigned yet.

Mikec, L., Pakhomov, F., & Vuković, M. (2018). Complexity of the interpretability logic IL. *Logic Journal of the IGPL, 27*(1), 1–7.

Mikec, L., Perkov, T., & Vuković, M. (2017). Decidability of interpretability logics IL M_0 and IL W*. *Logic Journal of the IGPL, 25*(5), 758–772.

Mikec, L., & Vuković, M. (2020). Interpretability logics and generalised Veltman semantics. *The Journal of Symbolic Logic, 85*(2), 749–772.

Montagna, F. (1987). Provability in finite subtheories of PA and relative interpretability: A modal investigation. *The Journal of Symbolic Logic, 52*(2), 494–511.

Mycielski, J., Pudlák, P., & Stern, A. S. (1990). *A lattice of chapters of mathematics (interpretations between theorems) Memoirs of the American Mathematical Society* (Vol. 426). Providence, Rhode Island: AMS.

Norell, U. (2007). *Towards a practical programming language based on dependent type theory*. Ph.D. thesis, Department of Computer Science and Engineering, Chalmers University of Technology, SE-412 96 Göteborg, Sweden.

Orey, S. (1961). Relative interpretations. *Zeitschrift für mathematische Logik und Grundlagen der Mathematik, 7*, 146–153.

Perkov, T., & Vuković, M. (2014). A bisimulation characterization for interpretability logic. *Logic Journal of the IGPL, 22*, 872–879.

Perkov, T., & Vuković, M. (2016). Filtrations of generalized Veltman models. *Mathematical Logic Quarterly, 62*(4–5), 412–419.

Sasaki, K. (2002). A cut-free sequent system for the smallest interpretability logic. *Studia Logica, 70*(3), 353–372.

Shavrukov, V. Y. (1988). *The logic of relative interpretability over Peano arithmetic*. Steklov Mathematical Institute, Preprint. In Russian.

Shavrukov, V. Y. (1997). Interpreting reflexive theories in finitely many axioms. *Fundamenta Mathematicae, 152*, 99–116.

Strannegård, C. (1999). Interpretability over peano arithmetic. *The Journal of Symbolic Logic, 64*(4), 1407–1425.

Švejdar, V. (1978). Degrees of interpretability. *Commentationes Mathematicae Universitatis Carolinae, 19*, 789–813.

Švejdar, V. (1983). Modal analysis of generalized Rosser sentences. *The Journal of Symbolic Logic, 48*, 986–999.

Švejdar, V. (1991). Some independence results in interpretability logic. *Studia Logica, 50*, 29–38.

Tarski, A., Mostowski, A., & Robinson, R. (1953). *Undecidable Theories*. North-Holland.

Verbrugge, L. C. (1992). Verzamelingen-Veltman frames en modellen (Set-Veltman Frames and Models). Unpublished Manuscript, Amsterdam, Original Scan included in [58].

Verbrugge, L. C. (1988). Does Solovay's completeness theorem extend to bounded arithmetic? Master's thesis, University of Amsterdam.

Verbrugge, R. (1993). Feasible interpretability. In P. Clote & J. Krajíček (Eds.), *Arithmetic, proof theory and computational complexity* (pp. 387–428). Oxford University Press.

Visser, A. (1988). Preliminary notes on Interpretability Logic. Technical Report LGPS 29, Department of Philosophy, Utrecht University.

Visser, A. (1990). Interpretability logic. In P. P. Petkov (Ed.), *Proceedings of the Heyting 1988 summer school in Varna. Mathematical logic* (pp. 175–209). Plenum Press.

Visser, A. (1991). The formalization of interpretability. *Studia Logica, 50*(1), 81–106.
Visser, A. (1997). An overview of Interpretability Logic. In M. Kracht, M. de Rijke, & H. Wansing (Eds.), *Advances in modal logic '96* (pp. 307–359). CSLI Publications.
Visser, A. (2005). Faith and Falsity: A study of Faithful Interpretations and false Σ^0_1-sentences. *Annals of Pure and Applied Logic, 131*(1–3), 103–131.
Visser, A. (2006). Categories of theories and interpretations. In A. Enayat, I. Kalantari & M. Moniri (Eds.), *Logic in Tehran. Proceedings of the workshop and conference on Logic, Algebra and Arithmetic, held October 18–22, 2003, volume 26 of Lecture Notes in Logic* (pp. 284–341). ASL, A.K. Peters, Ltd.
Visser, A. (2014). Interpretability degrees of finitely axiomatized sequential theories. *Archive for Mathematical Logic, 53*, 23–42.
Vrgoč, D., & Vuković, M. (2009). Bisimulations and bisimulation quotients of generalized Veltman models. *Logic Journal of the IGPL, 18*(6), 870–880.
Vuković, M. (1996). Some correspondence of principles in interpretability logic. *Glasnik Matematički, 31*(51), 193–200.
Vuković, M. (1999). The principles of interpretability. *Notre Dame Journal of Formal Logic, 40*(2), 227–235.
Vuković, M. (2008). Bisimulations between generalized Veltman models and Veltman models. *Mathematical Logic Quarterly, 54*(4), 368–373.
Wadler, P. (2015). Propositions as types. *Communications of the ACM, 58*(12), 75–84.
Zambella, D. (1992). On the proofs of arithmetical completeness of interpretability logic. *Notre Dame Journal of Formal Logic, 35*, 542–551.

Chapter 6
Deciding Dependence in Logic and Algebra

George Metcalfe and Naomi Tokuda

Abstract We introduce a universal algebraic generalization of de Jongh's notion of dependence for formulas of intuitionistic propositional logic, relating it to a notion of dependence defined by Marczewski for elements of an algebraic structure. Following ideas of de Jongh and Chagrova, we show how constructive proofs of (weak forms of) uniform interpolation can be used to decide dependence for varieties of abelian ℓ-groups, MV-algebras, semigroups, and modal algebras. We also consider minimal provability results for dependence, obtaining in particular a complete description and decidability of dependence for the variety of lattices.

Keywords Dependence · Uniform interpolation · Coherence · Lattices

6.1 Introduction

In de Jongh & Chagrova (1995) de Jongh and Chagrova introduced an intriguing notion of dependence of formulas for intuitionistic propositional logic IPC as an analogue of the usual notion of dependence of vectors in linear algebra.[1] Formulas $\varphi_1, \ldots, \varphi_n$ are said to be IPC-*dependent* if there exists a formula $\psi(p_1, \ldots, p_n)$ such that $\vdash_{\mathsf{IPC}} \psi(\varphi_1, \ldots, \varphi_n)$, but $\nvdash_{\mathsf{IPC}} \psi(p_1, \ldots, p_n)$; otherwise, $\varphi_1, \ldots, \varphi_n$ are

[1]This notion was previously defined in de Jongh (1982) using different terminology and appears also in Lemmon's textbook *Beginning Logic* (Lemmon, 1965) in the context of comparing two formulas of classical propositional logic. For a detailed history and comparison of this and other notions of dependence in logic, we refer to Humberstone (2020). Details of the seemingly unrelated but fascinating field of *dependence logic* may be found in Väänänen (2007); Yang & Väänänen (2016).

This research was supported by Swiss National Science Foundation grant 200021_165850.

G. Metcalfe (✉) · N. Tokuda
Mathematical Institute, University of Bern, Sidlerstrasse 5, 3012 Bern, Switzerland
e-mail: george.metcalfe@unibe.ch

N. Tokuda
e-mail: naomi.tokuda@math.unibe.ch

N. Bezhanishvili et al. (eds.), *Dick de Jongh on Intuitionistic and Provability Logics*, Outstanding Contributions to Logic 28, https://doi.org/10.1007/978-3-031-47921-2_6

said to be IPC-*independent*. In de Jongh (1982) it is shown that a single formula φ is IPC-dependent if, and only if, either $\vdash_{\mathsf{IPC}} \neg\neg\varphi \to \varphi$ or $\vdash_{\mathsf{IPC}} \neg\neg\varphi$, and de Jongh & Chagrova (1995) provides a "reasonably simple" sequence of formulas $(\psi_i(p_1, p_2))_{i \in \mathbb{N}}$ such that any two formulas φ_1, φ_2 are dependent if, and only if, $\vdash_{\mathsf{IPC}} \psi_i(\varphi_1, \varphi_2)$ for some $i \in \mathbb{N}$. Notably, it is also proved in de Jongh & Chagrova (1995) that checking the IPC-dependence of any finite number of formulas is decidable, making ingenious use of Pitts' constructive proof of uniform interpolation for IPC (Pitts, 1992).

In this paper, we extend the ideas of de Jongh & Chagrova (1995) to a general universal algebraic setting (recalled in Sect. 6.2), defining the $\mathcal{V}$-dependence of terms $t_1, \ldots, t_n$ for an arbitrary variety (equational class) $\mathcal{V}$. In Sect. 6.3, we show that $t_1, \ldots, t_n$ are $\mathcal{V}$-independent if, and only if, a certain homomorphism between finitely generated $\mathcal{V}$-free algebras is an embedding, or, equivalently, a certain finitely generated subalgebra of a $\mathcal{V}$-free algebra is $\mathcal{V}$-free over n generators. It then follows that $\mathcal{V}$-dependence is a special case of Marczewski's notion of dependence for elements of an algebraic structure (Marczewski, 1958, 1959, 1961), and a converse is obtained by relativizing $\mathcal{V}$-dependence to a given set of equations. Marczewski-dependence has been studied extensively in universal algebra (see Grätzer, 2008, Chap. 5), leading in particular to the introduction of v^*-algebras (Narkiewicz, 1961/1962), later rediscovered by semigroup theorists in the guise of independence algebras (Gould, 1995).

In Sect. 6.4, we turn our attention to the problem of deciding $\mathcal{V}$-dependence. Generalizing the proof strategy of de Jongh & Chagrova (1995), we show that $\mathcal{V}$-dependence is decidable (even relative to a finite set of equations) if there is a constructive proof that $\mathcal{V}$ is *coherent*, i.e., that every finitely generated subalgebra of a finitely presented algebra in $\mathcal{V}$ is finitely presented. The property of coherence originated in sheaf theory and has been studied widely in algebra (see, e.g., Soublin, 1970; Schmidt, 1983; Gould, 1992), and from a more general model-theoretic perspective in Wheeler (1976, 1978). In Kowalski and Metcalfe (2019) it was proved that coherence and deductive interpolation are jointly equivalent to the property of right uniform deductive interpolation considered in van Gool et al. (2017).

Constructive proofs of coherence are implicit in uniform interpolation proofs for, e.g., IPC (Pitts, 1992), the modal logics GL and S4Grz (Bílkova, 2007), and the varieties of abelian ℓ-groups and MV-algebras (Metcalfe et al., 2014; van Gool et al., 2017). However, coherence is a quite rare property for non-locally finite varieties, at least those corresponding to modal, substructural, and other non-classical logics (Kowalski & Metcalfe, 2019, 2018). We therefore also consider a weaker condition for deciding $\mathcal{V}$-dependence that requires only a constructive proof that finitely generated $\mathcal{V}$-free algebras are coherent, and is satisfied, for example, by the non-coherent variety of groups. Even when such a constructive proof is not available, however, there may exist other methods for deciding $\mathcal{V}$-dependence. Notably, the dependence problem for the variety of semigroups corresponds precisely to checking whether a finite set of words is a code, solved by the famous Sardinas-Patterson algorithm (Sardinas & Patterson, 1953), and the dependence problem for the variety of modal algebras can be decided using bisimulation-based methods described in Lutz and Wolter (2011) for calculating uniform interpolants (when they exist) for the description logic ALC.

In Sect. 6.5, following again ideas of de Jongh & Chagrova (1995), we consider minimal sets of equations for checking $\mathcal{V}$-dependence for some variety $\mathcal{V}$. In particular, we provide finite minimal sets of equations for dependence in both the locally finite variety of distributive lattices and the non-locally finite variety of lattices, obtaining a first proof that the dependence problem for the variety of lattices is decidable. Finally, in Sect. 6.6, we conclude the paper with a short list of open problems.

6.2 Equational Consequence and Free Algebras

Let us begin by recalling some elementary material on universal algebra, referring to Burris and Sankappanavar (1981) for further details and references. We assume that $\mathcal{L}$ is an algebraic language and that an *$\mathcal{L}$-algebra* $\mathbf{A}$ is a (first-order) structure for this language with universe A and fundamental operations $f^{\mathbf{A}}$ for each function symbol f of $\mathcal{L}$. For any set of variables $\overline{x}$, we denote by $\mathrm{Tm}(\overline{x})$ the set of *$\mathcal{L}$-terms over* $\overline{x}$, and by $\mathrm{Eq}(\overline{x}) := \mathrm{Tm}(\overline{x}) \times \mathrm{Tm}(\overline{x})$, the set of *$\mathcal{L}$-equations over* $\overline{x}$. For $\mathrm{Tm}(\overline{x}) \neq \emptyset$ (i.e., when $\overline{x} \neq \emptyset$ or $\mathcal{L}$ contains a constant), we denote by $\mathbf{Tm}(\overline{x})$ the *$\mathcal{L}$-term algebra over* $\overline{x}$. We also write $t(\overline{x})$, $\varepsilon(\overline{x})$, or $\Sigma(\overline{x})$ to mean that the variables occurring in an $\mathcal{L}$-term t, an $\mathcal{L}$-equation ε, or a set of $\mathcal{L}$-equations Σ, are included in $\overline{x}$, and assume that $\overline{x}$, $\overline{y}$, etc. are disjoint sets, writing $\overline{x}, \overline{y}$ to denote their disjoint union. Given an $\mathcal{L}$-term $t(x_1, \ldots, x_n)$ and an $\mathcal{L}$-algebra $\mathbf{A}$, we denote by $t^{\mathbf{A}}$ the induced *term-function* from A^n to A.

For any homomorphism $h \colon \mathbf{A} \to \mathbf{B}$ between $\mathcal{L}$-algebras $\mathbf{A}$ and $\mathbf{B}$, its *kernel* $\ker(h) := \{\langle a, b\rangle \in A \times A \mid h(a) = h(b)\}$ forms a *congruence* of $\mathbf{A}$: that is, an equivalence relation that is preserved by the fundamental operations of $\mathbf{A}$. Moreover, the converse is also true; every congruence of $\mathbf{A}$ is the kernel of some homomorphism with domain $\mathbf{A}$. For any $\mathcal{L}$-algebra $\mathbf{A}$, the set $\mathrm{Con}(\mathbf{A})$ of congruences of $\mathbf{A}$ forms a complete lattice ordered by inclusion $\subseteq$ with greatest element $\nabla_A := A \times A$ and least element $\Delta_A := \{\langle a, a\rangle \mid a \in A\}$. Given $S \subseteq A \times A$, the congruence $\mathrm{Cg}_{\mathbf{A}}(S)$ of $\mathbf{A}$ *generated by* S is the smallest congruence of $\mathbf{A}$ containing S. A congruence Θ of $\mathbf{A}$ is *finitely generated* if $\Theta = \mathrm{Cg}_{\mathbf{A}}(S)$ for some finite $S \subseteq A \times A$.

Let $\mathbb{H}$, $\mathbb{I}$, $\mathbb{S}$, and $\mathbb{P}$ denote the class operators of taking homomorphic images, isomorphic images, subalgebras, and products, respectively. A class of $\mathcal{L}$-algebras $\mathcal{K}$ is called a *variety* if it is closed under $\mathbb{H}$, $\mathbb{S}$, and $\mathbb{P}$. By theorems of Birkhoff and Tarski, respectively, $\mathcal{K}$ is a variety if, and only if, it is an equational class, and $\mathbb{HSP}(\mathcal{K})$ is the variety generated by $\mathcal{K}$.

Example 6.1 A *Heyting algebra* is an algebra $\langle H, \wedge, \vee, \to, 0, 1\rangle$ such that $\langle H, \wedge, \vee, 0, 1\rangle$ is a bounded distributive lattice (with $a \leq b :\Leftrightarrow a \wedge b = a$) and $\to$ is the residual of $\wedge$; that is, $a \leq b \to c$ if, and only if, $a \wedge b \leq c$ for all $a, b, c \in H$.

Heyting algebras form a variety $\mathcal{HA}$ that provides the algebraic semantics for intuitionistic propositional logic.

Equational consequence for a class $\mathcal{K}$ of $\mathcal{L}$-algebras may be defined as follows. For a set of $\mathcal{L}$-equations $\Sigma \cup \{\varepsilon\}$ containing exactly the variables in a set $\overline{x}$,

$$\Sigma \models_{\mathcal{K}} \varepsilon :\Longleftrightarrow \text{for every } \mathbf{A} \in \mathcal{K} \text{ and homomorphism } e \colon \mathbf{Tm}(\overline{x}) \to \mathbf{A},$$
$$\Sigma \subseteq \ker(e) \implies \varepsilon \in \ker(e).$$

For a set of $\mathcal{L}$-equations $\Sigma \cup \Delta$, we write $\Sigma \models_{\mathcal{K}} \Delta$ if $\Sigma \models_{\mathcal{K}} \varepsilon$ for all $\varepsilon \in \Delta$.

In the case where $\mathcal{K}$ is a variety, we may reformulate equational consequence in terms of congruences of $\mathcal{K}$-free algebras. Let us first recall the construction of $\mathcal{K}$-free algebras for an arbitrary class $\mathcal{K}$ of $\mathcal{L}$-algebras over a set $\overline{x}$, assuming that either $\mathcal{L}$ contains a constant or $\overline{x}$ is non-empty. Let $\Theta_{\mathcal{K}}(\overline{x})$ be the smallest congruence of $\mathbf{Tm}(\overline{x})$ such that the quotient by this congruence embeds into a member of $\mathcal{K}$, i.e.,

$$\Theta_{\mathcal{K}}(\overline{x}) := \bigcap\{\Theta \in \mathrm{Con}(\mathbf{Tm}(\overline{x})) \mid \mathbf{Tm}(\overline{x})/\Theta \in \mathbb{IS}(\mathcal{K})\}.$$

The *$\mathcal{K}$-free algebra over $\overline{x}$* may then be defined as

$$\mathbf{F}_{\mathcal{K}}(\overline{x}) := \mathbf{Tm}(\overline{x})/\Theta_{\mathcal{K}}(\overline{x}).$$

It follows that for any $s, t \in \mathrm{Tm}(\overline{x})$,

$$\langle s, t\rangle \in \Theta_{\mathcal{K}}(\bar{x}) \iff \models_{\mathcal{K}} s \approx t \iff \models_{\mathbf{F}_{\mathcal{K}}(\overline{x})} s \approx t.$$

Where appropriate, we deliberately confuse $\mathcal{L}$-terms, $\mathcal{L}$-equations, and sets of $\mathcal{L}$-equations with the corresponding elements, pairs of elements, and sets of pairs of elements from $\mathbf{F}_{\mathcal{K}}(\overline{x})$. Hence for $s, t \in \mathrm{Tm}(\overline{x})$,

$$\models_{\mathcal{K}} s \approx t \iff s = t \text{ in } \mathbf{F}_{\mathcal{K}}(\overline{x}).$$

Also, when the class of algebras $\mathcal{K}$ is clear from the context, we drop the subscript and write simply $\mathbf{F}(\overline{x})$.

The following lemma expresses the crucial connection between equational consequence in a variety and congruences on the free algebras of that variety.

Lemma 6.1 (cf. Metcalfe et al., 2014, Lemma 2) *For any variety $\mathcal{V}$ and $\Sigma \cup \Delta \subseteq \mathrm{Eq}(\overline{x})$,*

$$\Sigma \models_{\mathcal{V}} \Delta \iff \mathrm{Cg}_{\mathbf{F}(\overline{x})}(\Delta) \subseteq \mathrm{Cg}_{\mathbf{F}(\overline{x})}(\Sigma).$$

In what follows, we will omit mention of the language $\mathcal{L}$, assuming throughout that a class of algebras $\mathcal{K}$ is a class of $\mathcal{L}$-algebras, and that terms, equations, and sets of equations are defined over this language. We will also adopt the useful notation $[n]$ to denote the set $\{1, \ldots, n\}$ for $n \in \mathbb{N}$.

6.3 An Algebraic Theory of Dependence

We now have the tools available to formulate and study the de Jongh notion of dependence in a more general algebraic setting. Let $\mathcal{V}$ be any variety. We call terms $t_1, \ldots, t_n \in \mathrm{Tm}(\overline{x})$ *$\mathcal{V}$-dependent* if for some equation $\varepsilon(y_1, \ldots, y_n)$,

$$\models_{\mathcal{V}} \varepsilon(t_1, \ldots, t_n) \quad \text{and} \quad \not\models_{\mathcal{V}} \varepsilon;$$

otherwise, we call $t_1, \ldots, t_n$ *$\mathcal{V}$-independent.*

Example 6.2 If $\mathcal{V}$ is the variety of vector spaces over some fixed field $\mathbf{K}$, this notion of independence coincides with the usual notion in linear algebra. Just observe that terms $t_1, \ldots, t_n \in \mathrm{Tm}(\overline{x})$ in the language with the usual group operations and scalar multiplication for each $\lambda \in K$ are $\mathcal{V}$-independent if, and only if, (without loss of generality) for any equation $\varepsilon(y_1, \ldots, y_n)$ of the form $\lambda_1 y_1 + \cdots + \lambda_n y_n \approx 0$ with $\lambda_1, \ldots, \lambda_n \in K$,

$$\models_{\mathcal{V}} \lambda_1 t_1 + \cdots + \lambda_n t_n \approx 0 \implies \models_{\mathcal{V}} \lambda_1 y_1 + \cdots + \lambda_n y_n \approx 0,$$

and that $\models_{\mathcal{V}} \lambda_1 y_1 + \cdots + \lambda_n y_n \approx 0$ if, and only if, $\lambda_1 = \cdots = \lambda_n = 0$.

Example 6.3 Let $\mathcal{L}at$ be the variety of lattices and $\mathcal{DL}at$ the variety of distributive lattices, and consider the lattice terms

$$t_1 := x_1 \wedge (x_2 \vee x_3) \quad \text{and} \quad t_2 := x_2 \vee (x_1 \wedge x_3).$$

Defining $\varepsilon(y_1, y_2) := y_1 \leq y_2$ (where $s \leq t$ denotes $s \wedge t \approx s$), we have $\models_{\mathcal{DL}at} \varepsilon(t_1, t_2)$ and $\not\models_{\mathcal{DL}at} \varepsilon$, so t_1 and t_2 are $\mathcal{DL}at$-dependent. However, no such equation exists in the case of lattices, so t_1 and t_2 are $\mathcal{L}at$-independent.

The next result provides equivalent characterizations of $\mathcal{V}$-independence.

Proposition 6.1 *Let $\mathcal{V}$ be a variety. For any terms $t_1, \ldots, t_n \in \mathrm{Tm}(\overline{x})$ and variables $\overline{y} = \{y_1, \ldots, y_n\}$, the following are equivalent:*

(1) $t_1, \ldots, t_n$ *are $\mathcal{V}$-independent.*
(2) *The homomorphism $h \colon \mathbf{F}(\overline{y}) \to \mathbf{F}(\overline{x})$ defined by mapping y_i to t_i for each $i \in [n]$ is injective.*
(3) $\mathrm{Cg}_{\mathbf{F}(\overline{x},\overline{y})}(\langle y_1, t_1\rangle, \ldots, \langle y_n, t_n\rangle) \cap F(\overline{y})^2 = \Delta_{F(\overline{y})}$.

Proof (1)$\Rightarrow$(2) Suppose that $t_1, \ldots, t_n$ are $\mathcal{V}$-independent and that $h(u) = h(v)$ in $\mathbf{F}(\overline{x})$ for some $u, v \in \mathrm{Tm}(\overline{y})$. That is, $u(t_1, \ldots, t_n) = v(t_1, \ldots, t_n)$ in $\mathbf{F}(\overline{x})$, and hence $\models_{\mathcal{V}} u(t_1, \ldots, t_n) \approx v(t_1, \ldots, t_n)$. But $t_1, \ldots, t_n$ are $\mathcal{V}$-independent, so $\models_{\mathcal{V}} u \approx v$ and $u = v$ in $\mathbf{F}(\overline{y})$. Hence h is injective.

(2)$\Rightarrow$(3) Suppose that h is injective. Let $f \colon \mathbf{F}(\overline{y}) \to \mathbf{F}(\overline{x}, \overline{y})$ be the natural inclusion homomorphism and let $g \colon \mathbf{F}(\overline{x}, \overline{y}) \to \mathbf{F}(\overline{x})$ be the homomorphism mapping $x \in$

$\overline{x}$ to x and y_i to t_i for each $i \in [n]$. Clearly, $h = g \circ f$ and $\mathrm{Cg}_{\mathbf{F}(\overline{x},\overline{y})}(\langle y_1, t_1\rangle, \ldots, \langle y_n, t_n\rangle) \subseteq \ker(g)$, and hence, since h is injective,

$$\mathrm{Cg}_{\mathbf{F}(\overline{x},\overline{y})}(\langle y_1, t_1\rangle, \ldots, \langle y_n, t_n\rangle) \cap F(\overline{y})^2 \subseteq \ker(h) = \Delta_{F(\overline{y})}.$$

(3)$\Rightarrow$(1) Suppose that $\mathrm{Cg}_{\mathbf{F}(\overline{x},\overline{y})}(\langle y_1, t_1\rangle, \ldots, \langle y_n, t_n\rangle) \cap F(\overline{y})^2 = \Delta_{F(\overline{y})}$ and consider $\varepsilon \in \mathrm{Eq}(\overline{y})$ with $\models_{\mathcal{V}} \varepsilon(t_1, \ldots, t_n)$. Then $\{y_1 \approx t_1, \ldots, y_n \approx t_n\} \models_{\mathcal{V}} \varepsilon$ and an application of Lemma 6.1 yields

$$\varepsilon \in \mathrm{Cg}_{\mathbf{F}(\overline{x},\overline{y})}(\langle y_1, t_1\rangle, \ldots, \langle y_n, t_n\rangle) \cap F(\overline{y})^2 = \Delta_{F(\overline{y})}.$$

Hence, by Lemma 6.1, also $\models_{\mathcal{V}} \varepsilon$. So $t_1, \ldots, t_n$ are $\mathcal{V}$-independent. □

Remark 6.1 Condition (3) of Proposition 6.1 can also be understood as a property of equational consequence. Given terms $t_1, \ldots, t_n \in \mathrm{Tm}(\overline{x})$ and variables $\overline{y} = \{y_1, \ldots, y_n\}$, define

$$\Gamma := \{y_1 \approx t_1, \ldots, y_n \approx t_n\} \quad \text{and} \quad \Pi := \{\varepsilon \in \mathrm{Eq}(\overline{y}) \mid \Gamma \models_{\mathcal{V}} \varepsilon\}.$$

Then for any $\varepsilon \in \mathrm{Eq}(\overline{y})$,

$$\models_{\mathcal{V}} \varepsilon(t_1, \ldots, t_n) \iff \Gamma \models_{\mathcal{V}} \varepsilon \iff \Pi \models_{\mathcal{V}} \varepsilon,$$

and, corresponding to condition (3),

$$t_1, \ldots, t_n \text{ are } \mathcal{V}\text{-independent} \iff \models_{\mathcal{V}} \Pi.$$

Let us now consider the relationship between this notion of dependence and the general algebraic notion introduced in Marczewski (1958). We say that elements $a_1, \ldots, a_n$ of an algebra $\mathbf{A}$ are *Marczewski-dependent* in $\mathbf{A}$ if there exist terms $u(y_1, \ldots, y_n)$, $v(y_1, \ldots, y_n)$ satisfying

$$u^{\mathbf{A}}(a_1, \ldots, a_n) = v^{\mathbf{A}}(a_1, \ldots, a_n) \quad \text{and} \quad u^{\mathbf{A}} \neq v^{\mathbf{A}};$$

otherwise, we call $a_1, \ldots, a_n$ *Marczewski-independent* in $\mathbf{A}$.

Remark 6.2 Equivalently, $a_1, \ldots, a_n \in A$ are Marczewski-independent in $\mathbf{A}$ if, and only if, they are distinct and generate a subalgebra of $\mathbf{A}$ that is $\mathbb{HSP}(\mathbf{A})$-free over the set of generators $\{a_1, \ldots, a_n\}$ (Marczewski, 1959).

It is not hard to see that $\mathcal{V}$-dependence for a variety $\mathcal{V}$ corresponds to the Marczewski-dependence of elements of finitely generated $\mathcal{V}$-free algebras.

Proposition 6.2 *Let $\mathcal{V}$ be a variety. Terms $t_1, \ldots, t_n \in \mathrm{Tm}(\overline{x})$ are $\mathcal{V}$-dependent if, and only if, $t_1, \ldots, t_n$ are Marczewski-dependent in $\mathbf{F}(\overline{x})$.*

Proof It suffices to observe that $t_1, \dots, t_n \in \mathrm{Tm}(\overline{x})$ are Marczewski-dependent in $\mathbf{F}(\overline{x})$ if, and only if, there exist terms $u(y_1, \dots, y_n)$, $v(y_1, \dots, y_n)$ satisfying

$$u^{\mathbf{F}(\overline{x})}(t_1, \dots, t_n) = v^{\mathbf{F}(\overline{x})}(t_1, \dots, t_n) \quad \text{and} \quad u^{\mathbf{F}(\overline{x})} \neq v^{\mathbf{F}(\overline{x})},$$

or, equivalently, there exist terms $u(y_1, \dots, y_n)$, $v(y_1, \dots, y_n)$ satisfying

$$\models_{\mathcal{V}} u(t_1, \dots, t_n) \approx v(t_1, \dots, t_n) \quad \text{and} \quad \models_{\mathcal{V}} u \not\approx v,$$

which holds if, and only if, $t_1, \dots, t_n$ are $\mathcal{V}$-dependent. □

Let us now consider a more general version of dependence defined for a variety $\mathcal{V}$ relative to a fixed set of equations. We call $t_1, \dots, t_n \in \mathrm{Tm}(\overline{x})$ *$\mathcal{V}$-dependent over* $\Sigma \subseteq \mathrm{Eq}(\overline{x})$ if for some equation $\varepsilon(y_1, \dots, y_n)$,

$$\Sigma \models_{\mathcal{V}} \varepsilon(t_1, \dots, t_n) \quad \text{and} \quad \not\models_{\mathcal{V}} \varepsilon;$$

otherwise, we call $t_1, \dots, t_n$ *$\mathcal{V}$-independent over* Σ.

To obtain a reformulation of this property analogous to Proposition 6.1, we consider for $\Sigma \subseteq \mathrm{Eq}(\overline{x})$, the quotient algebra $\mathbf{F}(\overline{x})/\mathrm{Cg}_{\mathbf{F}(\overline{x})}(\Sigma)$, denoting its elements by $[t]_\Sigma$ for each $t \in \mathrm{Tm}(\overline{x})$.

Proposition 6.3 *Let $\mathcal{V}$ be a variety. For any $\Sigma \subseteq \mathrm{Eq}(\overline{x})$, $t_1, \dots, t_n \in \mathrm{Tm}(\overline{x})$, and $\overline{y} = \{y_1, \dots, y_n\}$, the following are equivalent:*

(1) $t_1, \dots, t_n$ are $\mathcal{V}$-independent over Σ.
(2) The homomorphism $h\colon \mathbf{F}(\overline{y}) \to \mathbf{F}(\overline{x})/\mathrm{Cg}_{\mathbf{F}(\overline{x})}(\Sigma)$ mapping y_i to $[t_i]_\Sigma$ for each $i \in [n]$ is injective.
(3) $\mathrm{Cg}_{\mathbf{F}(\overline{x},\overline{y})}(\Sigma \cup \{\langle y_1, t_1\rangle, \dots, \langle y_n, t_n\rangle\}) \cap F(\overline{y})^2 = \Delta_{F(\overline{y})}$.

Proof (1)$\Rightarrow$(2) Suppose that $t_1, \dots, t_n$ are $\mathcal{V}$-independent over Σ and $h(u) = h(v)$ for some $u, v \in \mathrm{Tm}(\overline{y})$. It follows that $[u(t_1, \dots, t_n)]_\Sigma = [v(t_1, \dots, t_n)]_\Sigma$ in $\mathbf{F}(\overline{x})/\mathrm{Cg}_{\mathbf{F}(\overline{x})}(\Sigma)$ and, by Lemma 6.1, that $\Sigma \models_{\mathcal{V}} u(t_1, \dots, t_n) \approx v(t_1, \dots, t_n)$. Since, by assumption, $t_1, \dots, t_n$ are $\mathcal{V}$-independent over Σ, we have $\models_{\mathcal{V}} u \approx v$. So $u = v$ in $\mathbf{F}(\overline{y})$. That is, h is injective.

(2)$\Rightarrow$(3) Suppose that h is injective. Let $f : \mathbf{F}(\overline{y}) \to \mathbf{F}(\overline{x}, \overline{y})$ be the natural inclusion homomorphism and let $g\colon \mathbf{F}(\overline{x}, \overline{y}) \to \mathbf{F}(\overline{x})/\mathrm{Cg}_{\mathbf{F}(\overline{x})}(\Sigma)$ be the homomorphism mapping $x \in \overline{x}$ to $[x]_\Sigma$ and y_i to $[t_i]_\Sigma$ for each $i \in [n]$. Clearly, $h = g \circ f$ and $\mathrm{Cg}_{\mathbf{F}(\overline{x},\overline{y})}(\Sigma \cup \{\langle y_1, t_1\rangle, \dots, \langle y_n, t_n\rangle\}) \subseteq \ker(g)$, and hence, since h is injective,

$$\mathrm{Cg}_{\mathbf{F}(\overline{x},\overline{y})}(\Sigma \cup \{\langle y_1, t_1\rangle, \dots, \langle y_n, t_n\rangle\}) \cap F(\overline{y})^2 \subseteq \ker(h) = \Delta_{F(\overline{y})}.$$

(3)$\Rightarrow$(1) Suppose that $\mathrm{Cg}_{\mathbf{F}(\overline{x},\overline{y})}(\Sigma \cup \{\langle y_1, t_1\rangle, \dots, \langle y_n, t_n\rangle\}) \cap F(\overline{y})^2 = \Delta_{F(\overline{y})}$ and $\Sigma \models_{\mathcal{V}} \varepsilon(t_1, \dots, t_n)$ for some $\varepsilon \in \mathrm{Eq}(\overline{y})$. Then $\Sigma \cup \{y_1 \approx t_1, \dots, y_n \approx t_n\} \models_{\mathcal{V}} \varepsilon$ and, by Lemma 6.1,

$$\varepsilon \in \mathrm{Cg}_{\mathbf{F}(\overline{x},\overline{y})}(\Sigma \cup \{\langle y_1, t_1\rangle, \dots, \langle y_n, t_n\rangle\}) \cap F(\overline{y})^2 = \Delta_{F(\overline{y})}.$$

Hence, by Lemma 6.1, also $\models_{\mathcal{V}} \varepsilon$. So $t_1, \dots, t_n$ are $\mathcal{V}$-independent over Σ. □

Remark 6.3 $\mathcal{V}$-dependence over Σ can again be understood as a property of equational consequence. For $t_1, \dots, t_n \in \mathrm{Tm}(\overline{x})$ and $\overline{y} = \{y_1, \dots, y_n\}$, let

$$\Gamma := \Sigma \cup \{y_1 \approx t_1, \dots, y_n \approx t_n\} \quad \text{and} \quad \Pi := \{\varepsilon \in \mathrm{Eq}(\overline{y}) \mid \Gamma \models_{\mathcal{V}} \varepsilon\}.$$

Then for any $\varepsilon \in \mathrm{Eq}(\overline{y})$,

$$\Sigma \models_{\mathcal{V}} \varepsilon(t_1, \dots, t_n) \iff \Gamma \models_{\mathcal{V}} \varepsilon \iff \Pi \models_{\mathcal{V}} \varepsilon,$$

and, corresponding to condition (3),

$$t_1, \dots, t_n \text{ are } \mathcal{V}\text{-independent over } \Sigma \iff \models_{\mathcal{V}} \Pi.$$

A natural question to ask at this point is whether this more general notion is related to Marczewski-dependence. Below we show that this is indeed the case, although the relationship is unlikely to be of any practical value.

Recall that the *positive diagram* $\mathrm{Diag}^+(\mathbf{A})$ of an algebra $\mathbf{A}$ can be identified with the set of equations $s(a_1, \dots, a_n) \approx t(a_1, \dots, a_n) \in \mathrm{Eq}(A)$ such that $s^{\mathbf{A}}(a_1, \dots, a_n) = t^{\mathbf{A}}(a_1, \dots, a_n)$.

Proposition 6.4 *Let* $\mathbf{A}$ *be any algebra. Then* $a_1, \dots, a_n \in A$ *are Marczewski-dependent in* $\mathbf{A}$ *if, and only if,* $a_1, \dots, a_n \in \mathrm{Tm}(A)$ *are* $\mathbb{HSP}(\mathbf{A})$*-dependent over* $\mathrm{Diag}^+(\mathbf{A})$.

Proof It suffices to observe that for any set of variables $\overline{y} = \{y_1, \dots, y_n\}$, equation $u \approx v \in \mathrm{Eq}(\overline{y})$, and $a_1, \dots, a_n \in A$,

(i) $\mathrm{Diag}^+(\mathbf{A}) \models_{\mathbb{HSP}(\mathbf{A})} u(a_1, \dots, a_n) \approx v(a_1, \dots, a_n)$

$$\iff u(a_1, \dots, a_n) \approx v(a_1, \dots, a_n) \in \mathrm{Diag}^+(\mathbf{A})$$
$$\iff u^{\mathbf{A}}(a_1, \dots, a_n) = v^{\mathbf{A}}(a_1, \dots, a_n)$$

(ii) $\models_{\mathbb{HSP}(\mathbf{A})} u \approx v \iff \mathbf{A} \models u \approx v \iff u^{\mathbf{A}} = v^{\mathbf{A}}$. □

6.4 Deciding Dependence

In de Jongh & Chagrova (1995) de Jongh and Chagrova established the decidability of IPC-dependence for finitely many formulas (equivalently, the $\mathcal{HA}$-dependence of finitely many terms) using Pitts' constructive proof of uniform interpolation for IPC (Pitts, 1992). More concretely, they proved that formulas $\varphi_1, \dots, \varphi_n$ are

IPC-independent if, and only if, the right uniform interpolant of $(p_1 \leftrightarrow \varphi_1) \wedge \cdots \wedge (p_n \leftrightarrow \varphi_n)$ with respect to a new set of variables $\{p_1, \ldots, p_n\}$ is a theorem of IPC. In this section, we generalize their approach to an arbitrary variety $\mathcal{V}$ and finite set of equations Σ, showing that to check the $\mathcal{V}$-dependence of terms $t_1, \ldots, t_n$ over Σ, a constructive proof of the weaker property of *coherence* in $\mathcal{V}$ — or, when $\Sigma = \emptyset$, just coherence for finitely generated $\mathcal{V}$-free algebras — suffices.

A finitely presented algebra $\mathbf{A} \in \mathcal{V}$ is said to be *coherent* if every finitely generated subalgebra of $\mathbf{A}$ is finitely presented, and a variety $\mathcal{V}$ is called coherent if all of its finitely presented members are coherent. The following result from Kowalski and Metcalfe (2019) relates this notion to finitely generated congruences of finitely generated $\mathcal{V}$-free algebras and equational consequence.

Theorem 6.1 (Kowalski and Metcalfe, 2019, Theorem 2.3) *Let $\mathcal{V}$ be a variety. The following are equivalent:*

(1) $\mathcal{V}$ is coherent.
(2) For any finite sets $\overline{x}, \overline{y}$ and finitely generated congruence Θ of $\mathbf{F}(\overline{x}, \overline{y})$, the congruence $\Theta \cap F(\overline{y})^2$ of $\mathbf{F}(\overline{y})$ is finitely generated.
(3) For any finite sets $\overline{x}, \overline{y}$ and finite set of equations $\Gamma(\overline{x}, \overline{y})$, there exists a finite set of equations $\Pi(\overline{y})$ such that for any equation $\varepsilon(\overline{y})$,

$$\Gamma \models_{\mathcal{V}} \varepsilon \iff \Pi \models_{\mathcal{V}} \varepsilon.$$

Remark 6.4 Condition (3) is closely related to the property of *right uniform deductive interpolation*, which is obtained by replacing "any equation $\varepsilon(\overline{y})$" with "any equation $\varepsilon(\overline{y}, \overline{z})$". Indeed, a variety has right uniform deductive interpolation if, and only if, it is coherent and has deductive interpolation. Let us note also that deductive interpolation (obtained from right uniform deductive interpolation by dropping the requirement that $\Pi(\overline{y})$ be finite) is equivalent to the amalgamation property in the presence of the congruence extension property. We refer to Metcalfe et al. (2014), van Gool et al. (2017), and Kowalski and Metcalfe (2019) for further details and discussion of these relationships.

Let us call the problem of finding for any finite sets $\overline{x}, \overline{y}$ and finite set of equations $\Gamma(\overline{x}, \overline{y})$, a finite set of equations $\Pi(\overline{y})$ satisfying the equivalence in condition (3) of Theorem 6.1, the *coherence problem* for $\mathcal{V}$. Note that this is equivalent to the problem of finding a finite presentation for a finitely generated subalgebra of a finitely presented algebra of $\mathcal{V}$.

Proposition 6.5 *Let $\mathcal{V}$ be a variety and let Σ be any finite set of equations. If the coherence problem for $\mathcal{V}$ and the equational theory of $\mathcal{V}$ are both decidable, then $\mathcal{V}$-dependence over Σ is decidable.*

Proof Given $t_1, \ldots, t_n \in \mathrm{Tm}(\overline{x})$, define $\Gamma := \Sigma \cup \{y_1 \approx t_1, \ldots, y_n \approx t_n\}$. It follows from Remark 6.3 that $t_1, \ldots, t_n$ are $\mathcal{V}$-independent over Σ if, and only if, for any equation $\varepsilon(\overline{y})$,

$$\Gamma \models_{\mathcal{V}} \varepsilon \iff \models_{\mathcal{V}} \varepsilon.$$

However, by assumption, a finite set of equations $\Pi(\overline{y})$ can be constructed such that for any equation $\varepsilon(\overline{y})$,

$$\Gamma \models_{\mathcal{V}} \varepsilon \iff \Pi \models_{\mathcal{V}} \varepsilon.$$

It follows that $t_1, \ldots, t_n \in \mathrm{Tm}(\overline{x})$ are $\mathcal{V}$-independent over Σ if, and only if, $\models_{\mathcal{V}} \Pi$, which, by assumption, is decidable. □

The coherence problem is clearly decidable for any locally finite variety. Also, as mentioned above, Pitts' constructive proof of uniform interpolation for IPC provides an algorithm that decides the coherence problem for the variety $\mathcal{HA}$ of Heyting algebras (Pitts, 1992) and hence also an algorithm for deciding $\mathcal{HA}$-dependence over any finite set of equations.

Pitts-style proofs of uniform interpolation have been obtained for various intermediate, modal, and substructural logics (see in particular Visser, 1996; Shavrukov, 1993; Bílkova, 2007; Alizadeh et al., 2014), but typically establish an implication-based uniform interpolation property that does not imply coherence. However, for the modal logics GL and S4Grz, the proof-theoretic proofs of uniform interpolation by Bílkova (2007) originally proved semantically by Shavrukov (1993) and Visser (1996), respectively provide constructive proofs of coherence for the associated varieties and hence also decidability of dependence over finite sets of equations.

Example 6.4 (Abelian ℓ-groups) An *abelian ℓ-group* is an algebraic structure $\langle L, \wedge, \vee, +, -, 0\rangle$ such that $\langle L, \wedge, \vee\rangle$ is a lattice with order $a \le b :\Leftrightarrow a \wedge b = a$, $\langle L, +, -, 0\rangle$ is an abelian group, and $a \le b$ implies $a + c \le b + c$ for all $a, b, c \in L$. They form a variety $\mathcal{LA}$ that is generated as a quasivariety by $\mathbf{R} = \langle \mathbb{R}, \wedge, \vee, +, -, 0\rangle$ (cf. Anderson and Feil, 1988, Lemma 6.2) and has a decidable equational theory. In Mundici (2017) it is proved that checking whether the subalgebra generated by n elements of a finitely generated free abelian ℓ-group is isomorphic to the n-generated free abelian ℓ-group is decidable, and hence, although this is not explicitly stated, that $\mathcal{LA}$-dependence is decidable. However, a stronger version of this result, the decidability of $\mathcal{LA}$-dependence over a finite set of equations, follows already from a (quite easy) constructive proof of coherence given implicitly in Metcalfe et al. (2014). Note first that it suffices to show that for any finite set $\overline{y}$ and finite set of equations $\Gamma(x, \overline{y})$, there exists a finite set of equations $\Pi(\overline{y})$ such that for any equation $\varepsilon(\overline{y})$,

$$\Gamma \models_{\mathbf{R}} \varepsilon \iff \Pi \models_{\mathbf{R}} \varepsilon.$$

Moreover, we may assume (with a little work, omitted here) that Γ consists of inequations $0 \le s_i + nx$ $(i \in I)$, $0 \le t_j - nx$ $(j \in J)$, and $0 \le u_k$ $(k \in K)$ for some $n \ge 1$, finite sets I, J, K, and terms $s_i(\overline{y}), t_j(\overline{y}), u_k(\overline{y})$. The desired set $\Pi(\overline{y})$ is then $\{0 \le s_i + t_j \mid i \in I; j \in J\} \cup \{0 \le u_k \mid k \in K\}$.

Example 6.5 (MV-algebras) The variety $\mathcal{MV}$ of *MV-algebras* consists of algebraic structures $\langle M, \oplus, \neg, 0\rangle$ satisfying the equations

(M1) $x \oplus (y \oplus z) \approx (x \oplus y) \oplus z$ (M4) $\neg\neg x \approx x$
(M2) $x \oplus y \approx y \oplus x$ (M5) $x \oplus \neg 0 \approx \neg 0$
(M3) $x \oplus 0 \approx x$ (M6) $\neg(\neg x \oplus y) \oplus y \approx \neg(\neg y \oplus x) \oplus x.$

It is generated as a quasivariety by $[\mathbf{0}, \mathbf{1}] = \langle [0, 1], \oplus, \neg, 0\rangle$, where $a \oplus b = \min(1, a + b)$ and $\neg a = 1 - a$, with defined operations $1 := \neg 0$, $a \odot b := \neg(\neg a \oplus \neg b)$, $a \vee b := \neg(\neg a \oplus b) \oplus b$, and $a \wedge b := \neg(\neg a \vee \neg b)$ (Di Nola, 1991). Let $\mathbf{R}^u$ be the unital abelian ℓ-group consisting of $\mathbf{R}$ with an additional constant 1. It follows from McNaughton's representation theorem (or see Metcalfe et al., 2014, Sect. 6 for a direct proof) that (i) the interpretation of any term s in $[\mathbf{0}, \mathbf{1}]$ is equivalent on $[0, 1]$ to the interpretation of some term $(t \wedge 0) \vee 1$ in $\mathbf{R}^u$, and, conversely, (ii) the interpretation of any term $(t \wedge 0) \vee 1$ in $\mathbf{R}^u$ is equivalent on $[0, 1]$ to the interpretation of some term s in $[\mathbf{0}, \mathbf{1}]$. Coherence for $\mathcal{MV}$ may then be established constructively as in the case of abelian ℓ-groups described in the previous example. Since $\mathcal{MV}$ also has a decidable equational theory, $\mathcal{MV}$-dependence over a finite set of equations is decidable.

As shown in Kowalski and Metcalfe (2019, 2018), coherence for a non-locally finite variety is a rather exceptional property. In particular, the varieties of lattices, semigroups, and groups, as well as broad families of varieties of modal algebras and residuated lattices are not coherent. However, for checking $\mathcal{V}$-dependence (i.e., over the empty set of equations), it is not necessary to have an algorithm that decides the full coherence problem; it suffices to consider the coherence of finitely generated $\mathcal{V}$-free algebras.

Lemma 6.2 *Let $\mathcal{V}$ be a variety. The following are equivalent for any finite set $\overline{x}$:*

(1) $\mathbf{F}(\overline{x})$ *is coherent.*
(2) For any $t_1, \dots, t_n \in \mathrm{Tm}(\overline{x})$ *and* $\overline{y} = \{y_1, \dots, y_n\}$*, the congruence* $\mathrm{Cg}_{\mathbf{F}(\overline{x},\overline{y})}(\langle y_1, t_1\rangle, \dots, \langle y_n, t_n\rangle) \cap F(\overline{y})^2$ *is finitely generated.*
(3) For any $t_1, \dots, t_n \in \mathrm{Tm}(\overline{x})$ *and* $\overline{y} = \{y_1, \dots, y_n\}$*, there exists a finite set of equations* $\Pi(\overline{y})$ *such that for any equation* $\varepsilon(\overline{y})$*,*

$$\{y_1 \approx t_1, \dots, y_n \approx t_n\} \models_{\mathcal{V}} \varepsilon \iff \Pi \models_{\mathcal{V}} \varepsilon.$$

Proof For the equivalence of (1) and (2), consider any $t_1, \dots, t_n \in \mathrm{Tm}(\overline{x})$ and $\overline{y} = \{y_1, \dots, y_n\}$. Let $h\colon \mathbf{F}(\overline{y}) \to \mathbf{F}(\overline{x})$ be the homomorphism mapping y_i to t_i for each $i \in [n]$. Then $\ker(h) = \mathrm{Cg}_{\mathbf{F}(\overline{x},\overline{y})}(\langle y_1, t_1\rangle, \dots, \langle y_n, t_n\rangle) \cap F(\overline{y})^2$ and, by the homomorphism theorem (see Burris and Sankappanavar, 1981), the quotient $\mathbf{F}(\overline{y})/\ker(h)$ is isomorphic to the subalgebra $\mathbf{A}$ of $\mathbf{F}(\overline{x})$ generated by $\{t_1, \dots, t_n\}$. We now recall the following:

Fact 6.1 (Kowalski and Metcalfe, 2019, Lemma 2.2) If $\mathbf{B} \in \mathcal{V}$ is finitely presented and isomorphic to a quotient $\mathbf{F}(\overline{z})/\Theta$ for some finite set $\overline{z}$, then Θ is finitely generated.

Hence, if $\mathbf{F}(\overline{x})$ is coherent, $\mathbf{A}$ is finitely presented and, by the fact stated above, $\mathrm{Cg}_{\mathbf{F}(\overline{x},\overline{y})}(\langle y_1, t_1\rangle, \ldots, \langle y_n, t_n\rangle) \cap F(\overline{y})^2$ is finitely generated. Conversely, assuming (2), if $\mathbf{A}$ is a subalgebra of $\mathbf{F}(\overline{x})$ generated by $\{t_1, \ldots, t_n\}$, then the congruence $\mathrm{Cg}_{\mathbf{F}(\overline{x},\overline{y})}(\langle y_1, t_1\rangle, \ldots, \langle y_n, t_n\rangle) \cap F(\overline{y})^2$ is finitely generated and $\mathbf{A}$ is finitely presented, so $\mathbf{F}(\overline{x})$ is coherent. The equivalence of (2) and (3) follows directly from Lemma 6.1. □

Let us therefore call the problem of finding for any $t_1, \ldots, t_n \in \mathrm{Tm}(\overline{x})$ and $\overline{y} = \{y_1, \ldots, y_n\}$, a finite set of equations $\Pi(\overline{y})$ satisfying the equivalence in condition (3), the *free coherence problem* for $\mathcal{V}$. (Note that this is equivalent to finding a finite presentation for a finitely generated subalgebra of a finitely generated $\mathcal{V}$-free algebra.) It follows directly from the proof of Proposition 6.5 that if the free coherence problem for $\mathcal{V}$ and the equational theory of $\mathcal{V}$ are decidable, then $\mathcal{V}$-dependence is decidable.

Example 6.6 (Groups) The variety $\mathcal{G}rp$ of *groups* is not coherent; e.g., the wreath product $\mathbb{Z}\,\mathrm{wr}\,\mathbb{Z}$ is a finitely generated subgroup of a finitely presented group that is not finitely presented (Cleary & Pacific, 2006). However, by the Nielsen-Schreier theorem, every finitely generated subgroup of a finitely generated free group is again a finitely generated free group, so finitely generated free groups are coherent. Moreover, Nielsen's proof in Nielsen (1981) also determines the rank of a finitely generated subgroup of a free group, so the free coherence problem for $\mathcal{G}rp$ is decidable. Hence also $\mathcal{G}rp$-dependence is decidable.

It may not be the case that all finitely generated free algebras of a variety $\mathcal{V}$ are coherent. However, assuming that the equational theory of $\mathcal{V}$ is decidable, $\mathcal{V}$-dependence is decidable (again considering the proof of Proposition 6.5) if there exists an algorithm to check for $t_1, \ldots, t_n \in \mathrm{Tm}(\overline{x})$ and $\overline{y} = \{y_1, \ldots, y_n\}$ whether for any equation $\varepsilon(\overline{y})$,

$$\{y_1 \approx t_1, \ldots, y_n \approx t_n\} \models_{\mathcal{V}} \varepsilon \iff \models_{\mathcal{V}} \varepsilon.$$

Example 6.7 (Semigroups) The variety $\mathcal{SG}$ of *semigroups* is not coherent. Indeed, even the n-generated free semigroups are not coherent for $n \geq 3$; e.g., the subsemigroup of the free semigroup $\mathbf{F}(x, y, z)$ generated by yx, yx^2, x^3, xz, and x^2z is not finitely presented (Campbell et al., 1996). However, the $\mathcal{SG}$-dependence problem corresponds precisely to the problem of checking whether a finite subset X of a finitely generated free semigroup $\mathbf{F}(\overline{x})$ is a *code*, solved by the famous Sardinas-Patterson algorithm (Sardinas & Patterson, 1953). For finite subsets $Y, Z \subseteq F(\overline{x})$, let $Y^{-1}Z := \{t \in F(\overline{x}) \mid st \in Z;\ s \in Y\}$. The algorithm starts with $U_0 := X$ and continues iteratively with $U_{n+1} := X^{-1}U_n \cup U_n^{-1}X$ for each $n \in \mathbb{N}$. It can be proved that X is *not* a code if, and only if, $X \cap U_n \neq \emptyset$ for some $n \geq 1$, and, since there can only be finitely many different U_n's, the algorithm is terminating.

Example 6.8 (Modal algebras) *Modal algebras*—Boolean algebras with an additional unary operation $\Box$ satisfying $\Box(x \wedge y) \approx \Box x \wedge \Box y$ and $\Box 1 \approx 1$—form a variety $\mathcal{MA}$ that provides algebraic semantics for the modal logic K. This variety is not coherent (Kowalski & Metcalfe, 2019), and the coherence of finitely generated free modal algebras seems to be an open problem. Nevertheless, $\mathcal{MA}$-dependence can be decided using a bisimulation-based method given in Lutz and Wolter (2011) for calculating existing right uniform deductive interpolants for the description logic ALC. It is well known that ALC restricted to a single role may be viewed as a syntactic variant of the logic K. Hence to decide $\mathcal{MA}$-dependence, it suffices to observe that modal formulas $\varphi_1, \ldots, \varphi_n$ (corresponding to terms) are $\mathcal{MA}$-independent if, and only if, the formula $(p_1 \leftrightarrow \varphi_1) \wedge \cdots \wedge (p_n \leftrightarrow \varphi_n)$ has a right uniform deductive interpolant with respect to the new variables $p_1, \ldots, p_n$ that is a theorem of K. This latter claim can be checked using the algorithm described in Lutz and Wolter (2011).

6.5 Dependence and Minimal Provability

In de Jongh (1982) it is shown that a single formula φ is IPC-dependent if, and only if, either $\vdash_{\mathsf{IPC}} \neg\neg\varphi \to \varphi$ or $\vdash_{\mathsf{IPC}} \neg\neg\varphi$, and in de Jongh & Chagrova (1995) a family $(\psi_i(p_1, p_2))_{i \in \mathbb{N}}$ of formulas is given such that two formulas φ_1, φ_2 are IPC-dependent if, and only if, $\vdash_{\mathsf{IPC}} \psi_i(\varphi_1, \varphi_2)$ for some $i \in \mathbb{N}$. In both cases, no proper subset of the given set of formulas suffices for checking IPC-dependence. In this section, we provide a general framework for describing such "minimal provability" results, obtaining a complete description and decidability of dependence for the variety of lattices.

Let $\mathcal{V}$ be a variety. Given any sets of equations Γ, Δ with variables in a set $\overline{y}$, we write $\Gamma \mathrel{|\!\!\sim}_{\mathcal{V}} \Delta$ to denote that for any substitution (i.e., homomorphism) $\sigma\colon \mathbf{Tm}(\overline{y}) \to \mathbf{Tm}(\omega)$ extended to $\mathrm{Eq}(\overline{y})$ by $\sigma(s \approx t) = \sigma(s) \approx \sigma(t)$,

$$\models_{\mathcal{V}} \sigma[\Gamma] \implies \models_{\mathcal{V}} \sigma(\delta) \text{ for some } \delta \in \Delta.$$

It is not hard to check (see, e.g., Iemhoff, 2016) that $\mathrel{|\!\!\sim}_{\mathcal{V}}$ is a (finitary) multiple-conclusion consequence relation over $\mathrm{Eq}(\overline{y})$; that is, for any equation ε and finite sets of equations $\Gamma, \Gamma', \Delta, \Delta'$ with variables in $\overline{y}$,

(i) $\{\varepsilon\} \mathrel{|\!\!\sim}_{\mathcal{V}} \{\varepsilon\}$
(ii) if $\Gamma \mathrel{|\!\!\sim}_{\mathcal{V}} \Delta$, then $\Gamma \cup \Gamma' \mathrel{|\!\!\sim}_{\mathcal{V}} \Delta \cup \Delta'$
(iii) if $\Gamma \cup \{\varepsilon\} \mathrel{|\!\!\sim}_{\mathcal{V}} \Delta$ and $\Gamma' \mathrel{|\!\!\sim}_{\mathcal{V}} \{\varepsilon\} \cup \Delta'$, then $\Gamma \cup \Gamma' \mathrel{|\!\!\sim}_{\mathcal{V}} \Delta \cup \Delta'$
(iv) if $\Gamma \mathrel{|\!\!\sim}_{\mathcal{V}} \Delta$, then $\sigma[\Gamma] \mathrel{|\!\!\sim}_{\mathcal{V}} \sigma[\Delta]$ for any substitution $\sigma\colon \mathbf{Tm}(\overline{y}) \to \mathbf{Tm}(\overline{y})$.

It is also easy to see that for any set of equations $\Gamma \cup \{\varepsilon\}$ with variables in $\overline{y}$,

$$\Gamma \models_{\mathcal{V}} \varepsilon \implies \Gamma \mathrel{|\!\!\sim}_{\mathcal{V}} \{\varepsilon\}.$$

Remark 6.5 The relation $\mathrel{|\!\sim}_{\mathcal{V}}$ describes the *admissibility* of universal formulas (or multiple-conclusion rules) in the variety $\mathcal{V}$. Indeed, for finite sets of equations Γ, Δ, it is the case that $\Gamma \mathrel{|\!\sim}_{\mathcal{V}} \Delta$ is equivalent to the validity of the implication from the conjunction of the equations in Γ to the disjunction of the equations in Δ in the free algebra $\mathbf{F}(\omega)$ (see, e.g., Cabrer & Metcalfe, 2015).

Let us call $\Delta \subseteq \mathrm{Eq}(\overline{y})$ $\mathcal{V}$*-refuting* for a set $\overline{y}$ if for any equation $\varepsilon(\overline{y})$,

$$\not\models_{\mathcal{V}} \varepsilon \iff \{\varepsilon\} \mathrel{|\!\sim}_{\mathcal{V}} \Delta,$$

and *minimal* if, additionally, no proper subset of Δ is $\mathcal{V}$-refuting for $\overline{y}$.

Lemma 6.3 *Let $\mathcal{V}$ be a variety and let $\Delta(\overline{y})$ be a $\mathcal{V}$-refuting set of equations for $\overline{y} = \{y_1, \ldots, y_n\}$. Then $t_1, \ldots, t_n \in \mathrm{Tm}(\overline{x})$ are $\mathcal{V}$-dependent if, and only if, $\models_{\mathcal{V}} \delta(t_1, \ldots, t_n)$ for some $\delta \in \Delta$.*

Proof Suppose first that $t_1, \ldots, t_n$ are $\mathcal{V}$-dependent. Then $\models_{\mathcal{V}} \varepsilon(t_1, \ldots, t_n)$ and $\not\models_{\mathcal{V}} \varepsilon$ for some equation $\varepsilon(\overline{y})$. Since Δ is a $\mathcal{V}$-refuting set of equations for $\overline{y}$, also $\{\varepsilon\} \mathrel{|\!\sim}_{\mathcal{V}} \Delta$. Hence $\models_{\mathcal{V}} \delta(t_1, \ldots, t_n)$ for some $\delta \in \Delta$. For the converse, suppose that $\models_{\mathcal{V}} \delta(t_1, \ldots, t_n)$ for some $\delta \in \Delta$. Clearly $\{\delta\} \mathrel{|\!\sim}_{\mathcal{V}} \Delta$ and hence $\not\models_{\mathcal{V}} \delta$, so $t_1, \ldots, t_n$ are $\mathcal{V}$-dependent. □

If $\mathcal{V}$ has a decidable equational theory and a finite $\mathcal{V}$-refuting set of equations can be found for $\overline{y} = \{y_1, \ldots, y_n\}$ for each $n \in \mathbb{N}$, then $\mathcal{V}$-dependence is clearly decidable. In the case where $\mathcal{V}$ is locally finite, a finite $\mathcal{V}$-refuting set of equations Δ_n can be obtained for each $n \in \mathbb{N}$ and $\overline{y} = \{y_1, \ldots, y_n\}$ by considering all pairs of distinct elements s, t from the finite free algebra $\mathbf{F}(\overline{y})$. A minimal $\mathcal{V}$-refuting set for $\overline{y}$ can then be obtained by iteratively removing any $\delta \in \Delta_n$ such that $\{\delta\} \mathrel{|\!\sim}_{\mathcal{V}} \Delta_n \setminus \{\delta\}$.

Example 6.9 Let us illustrate this idea with the simple case of the (locally finite) variety $\mathcal{DL}at$ of distributive lattices. It is straightforward to show that for each $n \in \mathbb{N}$,

$$\Delta_n := \{\bigwedge_{i \in I} y_i \leq \bigvee_{j \in [n] \setminus I} y_j \mid \emptyset \neq I \subsetneq [n]\}$$

is a minimal $\mathcal{DL}at$-refuting set of equations for $\overline{y} = \{y_1, \ldots, y_n\}$. We first observe that, using distributivity, the set of equations of the form $s \leq t$, where s is a join of meets of variables, and t is a meet of joins of variables, is $\mathcal{DL}at$-refuting for $\overline{y}$. We then obtain the minimal $\mathcal{DL}at$-refuting set Δ_n using the fact that for $i \in \{1, 2\}$,

$$\{s_1 \vee s_2 \leq t\} \mathrel{|\!\sim}_{\mathcal{DL}at} \{s_i \leq t\} \quad \text{and} \quad \{s \leq t_1 \wedge t_2\} \mathrel{|\!\sim}_{\mathcal{DL}at} \{s \leq t_i\}.$$

It follows that $t_1, \ldots, t_n \in \mathrm{Tm}(\overline{x})$ are $\mathcal{DL}at$-dependent if, and only if, for some $\emptyset \neq I \subsetneq [n]$,

$$\models_{\mathcal{DL}at} \bigwedge_{i \in I} t_i \leq \bigvee_{j \in [n] \setminus I} t_j.$$

We therefore obtain further confirmation that dependence in the variety of distributive lattices is decidable.

We devote the rest of this section to the more interesting case of the (non-locally finite) variety $\mathcal{L}at$ of all lattices.[2] The equational theory of $\mathcal{L}at$ is decidable. However, it is not coherent. For example, consider the congruence Θ of $\mathbf{F}(x, y, z, u, w)$ generated by $\{y \leq x, x \leq z, x \leq u \vee (w \wedge (u \vee x))\}$. It can be shown that the congruence $\Psi := \Theta \cap F(y, z, u, w)^2$ of $\mathbf{F}(y, z, u, w)$ is not finitely generated, yielding a finitely generated sublattice of $\mathbf{F}(x, y, z, u, w)/\Theta$, isomorphic to $\mathbf{F}(y, z, u, w)/\Psi$, that is not finitely presented (Kowalski & Metcalfe, 2019).[3] On the other hand, since finitely generated sublattices of free lattices are projective (Kostinsky, 1972) and finitely generated projective algebras are finitely presented (see Freese and Nation, 2016), every finitely generated free lattice is coherent. To the best of our knowledge, however, there is no known algorithm that constructs a finite presentation for a finitely generated sublattice of a finitely generated free lattice.

We prove that $\mathcal{L}at$-dependence is decidable by providing a finite minimal $\mathcal{L}at$-refuting set of equations for $\overline{y} = \{y_1, \ldots, y_n\}$ for each $n \in \mathbb{N}$. The following admissibility properties for $\mathcal{L}at$ will be useful:

(i) $\{x_1 \leq y, x_2 \leq y\} \mathrel{|\!\sim}_{\mathcal{L}at} \{x_1 \vee x_2 \leq y\}$
(ii) $\{x_1 \vee x_2 \leq y\} \mathrel{|\!\sim}_{\mathcal{L}at} \{x_1 \leq y\}$ and $\{x_1 \vee x_2 \leq y\} \mathrel{|\!\sim}_{\mathcal{L}at} \{x_2 \leq y\}$
(iii) $\{x \leq y_1\} \mathrel{|\!\sim}_{\mathcal{L}at} \{x \leq y_1 \vee y_2\}$ and $\{x \leq y_2\} \mathrel{|\!\sim}_{\mathcal{L}at} \{x \leq y_1 \vee y_2\}$
(iv) $\{x \leq y_1, x \leq y_2\} \mathrel{|\!\sim}_{\mathcal{L}at} \{x \leq y_1 \wedge y_2\}$
(v) $\{x \leq y_1 \wedge y_2\} \mathrel{|\!\sim}_{\mathcal{L}at} \{x \leq y_1\}$ and $\{x \leq y_1 \wedge y_2\} \mathrel{|\!\sim}_{\mathcal{L}at} \{x \leq y_2\}$
(vi) $\{x_1 \leq y\} \mathrel{|\!\sim}_{\mathcal{L}at} \{x_1 \wedge x_2 \leq y\}$ and $\{x_2 \leq y\} \mathrel{|\!\sim}_{\mathcal{L}at} \{x_1 \wedge x_2 \leq y\}$.

We will also need the fact that each generator y of a free lattice is both join- and meet-irreducible, which can be expressed as follows:

(vii) $\models_{\mathcal{V}} y \leq t_1 \vee t_2 \implies \models_{\mathcal{V}} y \leq t_1$ or $\models_{\mathcal{V}} y \leq t_2$
(viii) $\models_{\mathcal{V}} s_1 \wedge s_2 \leq y \implies \models_{\mathcal{V}} s_1 \leq y$ or $\models_{\mathcal{V}} s_2 \leq y$.

Finally, we will make crucial use of the following property of admissibility in lattices, known as *Whitman's condition* (Whitman, 1941):

$$\{x_1 \wedge x_2 \leq y_1 \vee y_2\} \mathrel{|\!\sim}_{\mathcal{L}at} \{x_1 \leq y_1 \vee y_2,\ x_2 \leq y_1 \vee y_2,\ x_1 \wedge x_2 \leq y_1,\ x_1 \wedge x_2 \leq y_2\}.$$

All these properties can be established directly as properties of free lattices (see, e.g., Freese and Nation, 2016) or follow as easy consequences of the completeness of a simple analytic Gentzen system for lattices (see, e.g., Metcalfe, 2011).

Theorem 6.2 *For each $n \in \mathbb{N}$, the following is a minimal $\mathcal{L}at$-refuting set of equations for $\overline{y} = \{y_1, \ldots, y_n\}$:*

---

[2] Some general properties of Marczewski-dependence in lattices are explored in Szász (1963), Marczewski (1963), but decidability issues are not considered in these papers.

[3] An earlier example given by R. McKenzie of a finitely generated sublattice of a finitely presented lattice that is not finitely presented can be found in Freese and Nation (2016).

$$\Delta_n := \{y_i \leq \bigvee_{j\in[n]\setminus\{i\}} y_j \mid i \in [n]\} \cup \{\bigwedge_{j\in[n]\setminus\{i\}} y_j \leq y_i \mid i \in [n]\}.$$

Proof Note first that clearly $\not\models_{\mathcal{L}at} \delta$ for each $\delta \in \Delta_n$. Also $\{\delta\} \not\vdash_{\mathcal{V}} \Delta_n \setminus \{\delta\}$ for each $\delta \in \Delta_n$. E.g., if δ is $y_1 \leq y_2 \vee \cdots \vee y_n$, let σ be the substitution mapping y_1 to $(y_2 \wedge z) \vee \cdots \vee (y_n \wedge z)$ and y_i to y_i for $i \in \{2, \ldots, n\}$. Then $\models_{\mathcal{L}at} \sigma(\delta)$, but $\not\models_{\mathcal{L}at} \sigma(\varepsilon)$ for each $\varepsilon \in \Delta_n$.

Hence it suffices now without loss of generality (since an equation $s \approx t$ can always be replaced by inequations $s \leq t$ and $t \leq s$) to prove that for any inequation $\varepsilon(\overline{y})$,

$$\not\models_{\mathcal{V}} \varepsilon \implies \{\varepsilon\} \vdash_{\mathcal{V}} \Delta_n,$$

proceeding by induction on the number of symbols in ε. For the base case, observe that if ε is an inequation with one variable on the left and a join of variables on the right, or one variable on the right and a meet of variables on the left, the claim follows directly from the definition of Δ_n. For the induction step, we consider the following cases:

(a) Suppose that ε is $s_1 \vee s_2 \leq t$ and $\not\models_{\mathcal{V}} \varepsilon$. Then, by property (i), $\not\models_{\mathcal{V}} s_1 \leq t$ or $\not\models_{\mathcal{V}} s_2 \leq t$, and, by the induction hypothesis, $\{s_1 \leq t\} \vdash_{\mathcal{V}} \Delta_n$ or $\{s_2 \leq t\} \vdash_{\mathcal{V}} \Delta_n$. By property (ii), $\{\varepsilon\} \vdash_{\mathcal{V}} \{s_1 \leq t\}$ and $\{\varepsilon\} \vdash_{\mathcal{V}} \{s_2 \leq t\}$, so also $\{\varepsilon\} \vdash_{\mathcal{V}} \Delta_n$.

(b) Suppose that ε is $s \leq t_1 \wedge t_2$ and $\not\models_{\mathcal{V}} \varepsilon$. Then, by property (iv), $\not\models_{\mathcal{V}} s \leq t_1$ or $\not\models_{\mathcal{V}} s \leq t_2$, and, by the induction hypothesis, $\{s \leq t_1\} \vdash_{\mathcal{V}} \Delta_n$ or $\{s \leq t_2\} \vdash_{\mathcal{V}} \Delta_n$. By property (v), $\{\varepsilon\} \vdash_{\mathcal{V}} s \leq t_1$ and $\{\varepsilon\} \vdash_{\mathcal{V}} s \leq t_2$, so also $\{\varepsilon\} \vdash_{\mathcal{V}} \Delta_n$.

(c) Suppose that ε is $s_1 \wedge s_2 \leq t_1 \vee t_2$ and $\not\models_{\mathcal{V}} \varepsilon$. Then, by properties (iii) and (vi), $\not\models_{\mathcal{V}} s_1 \leq t_1 \vee t_2$, $\not\models_{\mathcal{V}} s_2 \leq t_1 \vee t_2$, $\not\models_{\mathcal{V}} s_1 \wedge s_2 \leq t_1$, and $\not\models_{\mathcal{V}} s_1 \wedge s_2 \leq t_2$. So, by the induction hypothesis, $\{s_1 \leq t_1 \vee t_2\} \vdash_{\mathcal{V}} \Delta_n$, $\{s_2 \leq t_1 \vee t_2\} \vdash_{\mathcal{V}} \Delta_n$, $\{s_1 \wedge s_2 \leq t_1\} \vdash_{\mathcal{V}} \Delta_n$, and $\{s_1 \wedge s_2 \leq t_2\} \vdash_{\mathcal{V}} \Delta_n$. But also $\{\varepsilon\} \vdash_{\mathcal{V}} \{s_1 \leq t_1 \vee t_2, s_2 \leq t_1 \vee t_2, s_1 \wedge s_2 \leq t_1, s_1 \wedge s_2 \leq t_2\}$, by Whitman's condition, so $\{\varepsilon\} \vdash_{\mathcal{V}} \Delta_n$.

(d) Suppose that ε is (up to a permutation of meets) $(s_1 \vee s_2) \wedge s_3 \leq y_i$ and $\not\models_{\mathcal{V}} \varepsilon$. Then, by properties (vi) and (i), $\not\models_{\mathcal{V}} s_3 \leq y_i$ and either $\not\models_{\mathcal{V}} s_1 \leq y_i$ or $\not\models_{\mathcal{V}} s_2 \leq y_i$. Hence, by property (viii), either $\not\models_{\mathcal{V}} s_1 \wedge s_3 \leq y_i$ or $\not\models_{\mathcal{V}} s_2 \wedge s_3 \leq y_i$, and, by the induction hypothesis, either $\{s_1 \wedge s_3 \leq y_i\} \vdash_{\mathcal{V}} \Delta_n$ or $\{s_2 \wedge s_3 \leq y_i\} \vdash_{\mathcal{V}} \Delta_n$. By the monotonicity of the lattice operations, $\{(s_1 \vee s_2) \wedge s_3 \leq y_i\} \models_{\mathcal{L}at} s_j \wedge s_3 \leq y_i$ for $j \in \{1, 2\}$, so also $\{\varepsilon\} \vdash_{\mathcal{V}} \{s_j \wedge s_3 \leq y_i\}$ for $j \in \{1, 2\}$. Hence $\{\varepsilon\} \vdash_{\mathcal{V}} \Delta_n$.

(e) Suppose that ε is (up to a permutation of joins) $y_i \leq (t_1 \wedge t_2) \vee t_3$ and $\not\models_{\mathcal{V}} \varepsilon$. Then, by properties (iii) and (v), $\not\models_{\mathcal{V}} y_i \leq t_3$ and either $\not\models_{\mathcal{V}} y_i \leq t_1$ or $\not\models_{\mathcal{V}} y_i \leq t_2$. Hence $\not\models_{\mathcal{V}} y_i \leq t_1 \vee t_3$ or $\not\models_{\mathcal{V}} y_i \leq t_2 \vee t_3$, and, by the induction hypothesis, either $\{y_i \leq t_1 \vee t_3\} \vdash_{\mathcal{V}} \Delta_n$ or $\{y_i \leq t_2 \vee t_3\} \vdash_{\mathcal{V}} \Delta_n$. By the monotonicity of the lattice operations, $\{y_i \leq (t_1 \wedge t_2) \vee t_3\} \models_{\mathcal{L}at} y_i \leq t_j \vee t_3$ for $j \in \{1, 2\}$, so also $\{\varepsilon\} \vdash_{\mathcal{V}} \{y_i \leq t_j \vee t_3\}$ for $j \in \{1, 2\}$. Hence $\{\varepsilon\} \vdash_{\mathcal{V}} \Delta_n$. □

Corollary 6.1 *The terms $t_1, \ldots, t_n \in \mathrm{Tm}(\overline{x})$ are $\mathcal{L}at$-dependent if, and only if, for some $i \in \{1, \ldots, n\}$,*

$$\models_{\mathcal{L}at} t_i \leq \bigvee_{j \in [n] \setminus \{i\}} t_j \quad or \quad \models_{\mathcal{L}at} \bigwedge_{j \in [n] \setminus \{i\}} t_j \leq t_i.$$

Hence dependence in the variety of lattices is decidable.

6.6 Open Problems

We conclude with a short list of open problems:

(1) Proofs of uniform interpolation can be quite intricate. In particular, Pitts' constructive proof of this property for IPC involves a complicated definition of left and right interpolants that is checked by induction on derivations in a terminating sequent calculus. It therefore makes sense to seek a simpler proof that IPC-dependence is decidable. Such a proof might perhaps use the fact that subalgebras of finitely generated free Heyting algebras are projective (Ghilardi, 1999) and hence that finitely generated subalgebras of finitely generated free Heyting algebras are finitely presented. The challenge would then be to provide a simple algorithm for producing finite presentations of these subalgebras. Similarly, we expect that there should be a more direct proof of decidability for dependence in the variety of modal algebras that does not rely on the bisimulation-based method of Lutz and Wolter (2011) for calculating uniform interpolants.

(2) It follows from Theorem 6.2 and Whitman's condition that the minimal $\mathcal{DL}at$-refuting sets of equations described in Example 6.9 also serve as (non-minimal) $\mathcal{L}at$-refuting sets of equations. This raises the question of whether it is the case for *any* variety of lattices $\mathcal{V}$ that $t_1, \ldots, t_n \in \mathrm{Tm}(\overline{x})$ are $\mathcal{V}$-dependent if, and only if, for some $\emptyset \neq I \subsetneq [n]$,

$$\models_{\mathcal{V}} \bigwedge_{i \in I} t_i \leq \bigvee_{j \in [n] \setminus I} t_j.$$

Note that it is easily checked (since the free lattice on two generators is finite) that this is true for $n = 2$; that is, in any variety $\mathcal{V}$ of lattices, t_1, t_2 are $\mathcal{V}$-dependent if, and only if, $\models_{\mathcal{V}} t_1 \leq t_2$ or $\models_{\mathcal{V}} t_2 \leq t_1$.

(3) For all the varieties considered in this paper, the dependence problem is decidable. It would therefore be interesting to know of an example (if one exists) of a variety with a decidable equational theory for which dependence is undecidable.

(4) The decidability of dependence is an open problem for most non-locally finite varieties associated to non-classical logics. In particular, varieties of modal algebras for modal logics such as T, K4, S4, and KD and varieties of pointed residuated lattices for substructural logics such as FL (the full Lambek calculus), MTL (monoidal t-norm logic), R (relevant logic), and MALL (multiplicative additive linear logic), all fail to be coherent (Kowalski & Metcalfe, 2019, 2018) and it is not known if their finitely generated free algebras are coherent.

References

Alizadeh, M., Derakhshan, F., & Ono, H. (2014). Uniform interpolation in substructural logics. *The Review of Symbolic Logic, 7*(3), 455–483.

Anderson, M. E., & Feil, T. H. (1988). *Lattice-ordered groups: An introduction*. Springer.

Bílkova, M. (2007). Uniform interpolation and propositional quantifiers in modal logics. *Studia Logica, 85*, 1–31.

Burris, S., & Sankappanavar, H. P. (1981). *A course in universal algebra, Graduate texts in mathematics* (Vol. 78). Springer.

Cabrer, L. M., & Metcalfe, G. (2015). Admissibility via natural dualities. *Journal of Pure and Applied Algebra, 219*(9), 4229–4253.

Campbell, C. M., Robertson, E. F., Ruškuc, N. (1996). On subsemigroups of finitely presented semigroups. *Journal of Algebra, 180*, 1–21.

Cleary, S. (2006). Distortion of wreath products in some finitely presented groups. *Pacific Journal of Mathematics, 228*, 53–61

de Jongh, D., Chagrova, L. A. (1995). The decidability of dependency in intuitionistic propositional logic. *Journal of Symbolic Logic, 60*(2), 498–504

de Jongh, D. (1982). Formulas of one propositional variable in intuitionistic arithmetic. In *Studies in logic and the foundations of mathematics* (Vol. 110, pp. 51–64). The L.E.J. Brouwer Centenary Symposium, Elsevier.

Di Nola, A. (1991). Representation and reticulation by quotients of MV-algebras. *Ricerche di Matematica, 40*, 291–297

Freese, R., Nation, J. B. (2016). Free and finitely presented lattices. In G. Grätzer & F. Wehrung (Eds.), Lattice theory: Special topics and applications (Vol. 2). Springer.

Ghilardi, S. (1999). Unification in intuitionistic logic. *Journal of Symbolic Logic, 2*(64), 859–880.

van Gool, S., Metcalfe, G., & Tsinakis, C. (2017). Uniform interpolation and compact congruences. *Annals of Pure and Applied Logic, 168*, 1927–1948.

Gould, V. (1992). Coherent monoids. *Journal of the Australian Mathematical Society, 53*(2), 166–182.

Gould, V. (1995). Independence algebras. *Algebra Universalis, 33*, 294–318.

Grätzer, G. (2008). *Universal algebra* (Vol. 2). Springer.

Humberstone, L. (2020). Explicating logical independence. *Journal of Philosophical Logic, 49*, 135–218.

Iemhoff, R. (2016). Consequence relations and admissible rules. *Journal of Philosophical Logic, 45*(3), 327–348.

Kostinsky, A. (1972). Projective lattices and bounded homomorphisms. *Pacific Journal of Mathematics, 40*(1), 111–119.

Kowalski, T., Metcalfe, G. (2018). Coherence in modal logic. In *Proceedings of AiML'18* (pp. 236–251). College Publications.

Kowalski, T., Metcalfe, G. (2019). Uniform interpolation and coherence. *Annals of Pure and Applied Logic, 170*(7), 825–841.

Lemmon, E. J. (1965). *Beginning logic*. Thomas Nelson and Sons.
Lutz, C., Wolter, F. (2011). Foundations for uniform interpolation and forgetting in expressive description logics. *Proceedings of IJCAI* (pp. 989–995).
Marczewski, E. (1958). A general scheme of the notions of independence in mathematics. *Bulletin of the Polish Academy of Sciences Mathematics, 6*, 731–736.
Marczewski, E. (1959). Independence in some abstract algebras. *Bulletin of the Polish Academy of Sciences Mathematics, 7*, 611–616
Marczewski, E. (1961). Independence and homomorphisms in abstract algebras. *Fundamenta Mathematicae, 50*(1), 45–61.
Marczewski, E. (1963). Concerning the independence in lattices. *Colloquium Mathematicum, 10*(10), 21–23
Metcalfe, G. (2011). Proof theory of mathematical fuzzy logic. *Handbook of mathematical fuzzy logic* (Vol. I, pp. 209–282). King's College Publications.
Metcalfe, G., Montagna, F., Tsinakis, C. (2014). Amalgamation and interpolation in ordered algebras. *Journal of Algebra, 402*, 21–82
Mundici, D. (2017). Fans, decision problems and generators of free abelian ℓ-groups. *Forum of Mathematics, 29*, 1429–1439.
Narkiewicz, W. (1961/1962). Independence in a certain class of abstract algebras. *Fundamenta Mathematicae, 50*, 333–340.
Nielsen, J. (1981) 77–94, Om regning med ikke-kommutative faktorer og dens anvendelse i gruppeteorien, Math. Tidsskrift B (1921). English translation: Math. Scientist 6, 73–85
Pitts, A. M. (1992). On an interpretation of second-order quantification in first-order intuitionistic propositional logic. *Journal of Symbolic Logic, 57*, 33–52.
Sardinas, A. A., Patterson, G. W. (1953). A necessary and sufficient condition for the unique decomposition of coded messages. *Convention Record of the I.R.E., 8*, 104–108.
Schmidt, P. H. (1983). Algebraically complete lattices. *Algebra Universalis, 17*, 135–142.
Shavrukov, V. Y. (1993). Subalgebras of diagonizable algebras theories containing arithmetic. *Dissertationes Mathematicae (Rozprawy Mat.), 323*.
Soublin, J.-P. (1970). Anneaux et modules coherents. *Journal of Algebra, 15*, 455–472.
Szász, G. (1963). Marczewski independence in lattices and semilattices. *Colloquium Mathematicum, 10*(1), 15–20.
Väänänen, J. (2007). *Dependence logic: A new approach to independence friendly logic*. London Mathematical Society Student Texts. Cambridge University Press.
Visser, A. (1996). Uniform interpolation and layered bisimulation. In P. Hájek (Ed.), *Gödel '96: Logical foundations on mathematics, computer science and physics—Kurt Gödel's legacy*. Springer.
Wheeler, W. H. (1976). Model-companions and definability in existentially complete structures. *Israel Journal of Mathematics, 25*(3), 305–330.
Wheeler, W. H. (1978). A characterization of companionable, universal theories. *Journal of Symbolic Logic, 43*(3), 402–429.
Whitman, P. (1941). Free lattices. *Annals of Mathematics, 42*, 325–329.
Yang, F., & Väänänen, J. (2016). Propositional logics of dependence. *Annals of Pure and Applied Logic, 167*(7), 557–589.

Chapter 7
About the Unification Types of Modal Logics

Philippe Balbiani and Çiğdem Gencer

Abstract The unification problem in a modal logic is to determine whether given formulas have instances belonging to this modal logic. In this paper, we survey some known results about the unification types of modal logics.

Keywords Modal logics · Unification types

7.1 Introduction

The unification problem in a modal logic **L** is to determine, given a formula φ, whether there exists a substitution σ such that $\sigma(\varphi)$ is in **L**. In that case, φ is unifiable and σ is a unifier of φ. An important question is the following: is it computable to determine whether a given formula is unifiable? When the answer is "yes", how complex is it? Suppose **L** is consistent. Obviously, for all Boolean formulas φ, φ is unifiable in **L** if and only if φ is consistent in Boolean Logic. As a result, when **L** is consistent, the unification problem in **L** is **NP**-hard. Can we say more about its computability?

In some popular modal logics, to solve the unification problem is not a mere formality. While attacking it, little, if anything, from the standard tools in modal logics (canonical models, filtrations, etc) is helpful. For instance, the computability of the unification problem remains open in **K**. The truth is that contrary to the case of Boolean Logic, there exists **K**-consistent formulas which are not unifiable in **K**.[1] To

[1] As is well-known, in Boolean Logic, a formula is unifiable if and only if it is consistent. It immediately follows that in Boolean Logic, the unification problem is **NP**-complete.

P. Balbiani (✉) · Ç. Gencer
Toulouse Institute of Computer Science Research, CNRS — Toulouse University, Toulouse, France
e-mail: Philippe.Balbiani@irit.fr

Ç. Gencer
Faculty of Arts and Sciences, Istanbul Aydın University, Istanbul, Turkey

N. Bezhanishvili et al. (eds.), *Dick de Jongh on Intuitionistic and Provability Logics*, Outstanding Contributions to Logic 28, https://doi.org/10.1007/978-3-031-47921-2_7

see why, take the formula $\Diamond x \wedge \Diamond \neg x$. Obviously, $\Diamond x \wedge \Diamond \neg x$ is **K**-consistent, seeing that $\mathbf{Alt}_1 \not\subseteq \mathbf{K}$. Nevertheless, $\Diamond x \wedge \Diamond \neg x$ is not **K**-unifiable, seeing that $\mathbf{KD} \not\subseteq \mathbf{K}$.[2]

Luckily, in some other popular modal logics, to solve the unification problem is relatively easy. For instance, if either $\mathbf{KD} \subseteq \mathbf{L}$, or $\mathbf{K5} \subseteq \mathbf{L}$, or $\mathbf{KB} \subseteq \mathbf{L}$ then the unification problem in **L** is in **NP**. This can be proved by reducing the unification problem in **L** to the problem of determining whether given a formula $\varphi(x_1, \ldots, x_n)$, there are variable-free formulas $\psi_1, \ldots, \psi_n$ such that $\varphi(\psi_1, \ldots, \psi_n)$ is in **L**.[3] Moreover, Balbiani and Tinchev (2016) have proved that the $\mathbf{Alt}_1$-unification problem is in **PSPACE** by reducing it to the problem of determining whether a graph contains an Hamiltonian path. In other respect, by means of the so-called "universal model", Gencer and de Jongh (2009) and Rybakov et al. (1999) have designed concrete decision procedures for solving the unification problem in some transitive modal logics like **GL** and **K**4.3 for which they have shown that the unification problem is in **NP**. In connection with the computability of the unification problem, see also Rybakov (1997) for a general introduction to the admissibility problem and its computability in intermediate logics and in transitive modal logics.

However, the computability of the unification problem is not the only significant question in this area. A set of unifiers of a given **L**-unifiable formula is complete if every unifier of the formula can be obtained as an instance of a unifier of that set. When a unifiable formula has minimal complete sets of unifiers, the formula is either of type 1, or of type ω, or of type ∞, depending on the cardinality of its minimal complete sets of unifiers in **L**. Otherwise, the formula is of type 0. These types being ordered by $1 < \omega < \infty < 0$, the unification type of **L** is the greatest type among the types of its unifiable formulas. The significance of the unification types of modal logics stems from the fact that if a modal logic is either unitary, or finitary, minimal complete sets of unifiers for **L**-unifiable formulas can be effectively computed and membership in **L** can be decided then the problem of checking the admissibility of inference rules can be decided.

---

[2] Denoting for all formulas φ, as formally defined in Sect. 7.2.2, $\mathbf{L} \oplus \varphi$ the least modal logic containing **L** and φ, the point is that $(*)$ for all formulas φ, $\mathbf{L} \oplus \Diamond\varphi \wedge \Diamond\neg\varphi$ is inconsistent, as firstly remarked by Porte (1981) within the context of **S5** and secondly generalized by Fagin et al. (1992) within the context of all modal logics. Interestingly, as proved by French & Humberstone (2015), $(**)$ for all formulas φ, either $\mathbf{L} \oplus \Diamond\varphi$ is inconsistent, or $\mathbf{L} \oplus \Diamond\neg\varphi$ is inconsistent. It is worth asking what becomes of $(*)$ and $(**)$ in the case of a modal logic which language contains two modalities instead of one. In the case of a tense modal logic with the modalities $\Box$ and $\Box^{-1}$, $(**)$ becomes false. This can be proved by considering the formula $\varphi = \Box^{-1}\Diamond^{-1}\top$. It is not known whether the same happens to $(*)$.

[3] Indeed, it is a well-known fact that if $\mathbf{KD} \subseteq \mathbf{L}$ then every variable-free formula is either **L**-equivalent to $\bot$, or **L**-equivalent to $\top$ and if either $\mathbf{K5} \subseteq \mathbf{L}$, or $\mathbf{KB} \subseteq \mathbf{L}$ then every variable-free formula is either **L**-equivalent to $\bot$, or **L**-equivalent to $\top$, or **L**-equivalent to $\Box\bot$, or **L**-equivalent to $\Diamond\top$. Moreover, the simple formula equivalent in the considered modal logic to the given variable-free formula can be easily computed in polynomial time. Hence, if either $\mathbf{KD} \subseteq \mathbf{L}$, or $\mathbf{K5} \subseteq \mathbf{L}$, or $\mathbf{KB} \subseteq \mathbf{L}$ then the following decision problem is in **P**: given a variable-free $\varphi \in \mathbf{FOR}$, determine whether φ is in **L**. For example, suppose $\mathbf{KD} \subseteq \mathbf{L}$. Hence, every variable-free formula is either **L**-equivalent to $\bot$, or **L**-equivalent to $\top$. Thus, the variable-free formulas $\psi_1, \ldots, \psi_n$ uniformly replacing the variables $x_1, \ldots, x_n$ in φ can be restricted to $\bot$ and $\top$. This gives rise to a nondeterministic algorithm able to solve the **L**-unification problem in polynomial time.

Taking a look at the literature about unification types in modal logics (Baader & Ghilardi, 2011; Dzik, 2007), one will quickly note that much remains to be done. It is known that $\mathbf{Alt}_1$ is nullary (Balbiani & Tinchev, 2016), **S5** and **S4.3** are unitary (Dzik, 2003, 2007; Dzik & Wojtylak, 2012), transitive modal logics like **K4** and **S4** are finitary (Ghilardi, 2000; Iemhoff, 2016), $\mathbf{K4.2}^+$ is unitary (Ghilardi & Sacchetti, 2004; Jeřábek, 2013), **K** is nullary (Jeřábek, 2015) and **K4D1** is unitary (Kost, 2018). The types of simple modal logics like **KD**, **KT**, **KB**, **KDB**, **KTB**, $\mathbf{DAlt}_1$ and $\mathbf{DAlt}_n$ when $n \geq 2$ are nevertheless unknown. Even, the types of the least modal logics containing $\Box^d\bot$ when $d \geq 2$ are unknown.[4] As well, the types of most of the modal logics containing **K5** are unknown.[5]

In this paper, we survey some known results about the unification types of modal logics.

7.2 Modal Logics

7.2.1 Formulas

Let **VAR** be a countably infinite set of *variables* (with typical members denoted x, y, etc.). The set **FOR** of all *formulas* (with typical members denoted φ, ψ, etc.) is inductively defined by

- $\varphi := x \mid \bot \mid \neg\varphi \mid (\varphi \vee \varphi) \mid \Box\varphi$.

We adopt the standard rules for omission of the parentheses. The Boolean connectives $\top$, $\wedge$, $\rightarrow$ and $\leftrightarrow$ are defined as usual. For all formulas φ, we use $\Diamond\varphi$ as a shorthand for $\neg\Box\neg\varphi$. For all $d \geq 0$, the modal connective $\Box^d$ is inductively defined by

- $\Box^0\varphi := \varphi$,
- $\Box^{d+1}\varphi := \Box\Box^d\varphi$.

For all $d \geq 0$, the modal connective $\Box^{<d}$ is inductively defined by

- $\Box^{<0}\varphi := \top$,
- $\Box^{<d+1}\varphi := \varphi \wedge \Box\Box^{<d}\varphi$.

For all $\varphi \in \mathbf{FOR}$, the *degree* of φ is the maximal number $\mathtt{deg}(\varphi)$ of nested $\Box$ in φ. In particular, we shall say that a formula φ is *Boolean* if $\mathtt{deg}(\varphi) = 0$. For all $\varphi \in \mathbf{FOR}$, let $\mathtt{var}(\varphi)$ be the finite set of all variables occurring in φ. In particular, we shall say that a formula φ is *variable-free* if $\mathtt{var}(\varphi) = \emptyset$. For all $d \geq 0$, let $\mathbf{FOR}_d$ be the set of all formulas φ such that $\mathtt{deg}(\varphi) \leq d$. For all finite $X \subseteq \mathbf{VAR}$, let $\mathbf{FOR}_X$ be the set of all formulas φ such that $\mathtt{var}(\varphi) \subseteq X$. We adopt the standard definitions for *subformulas* and *instances*.

[4] We are indebted to Silvio Ghilardi for his suggestion to consider the question of the unification type of the least modal logic containing $\Box^d\bot$ when $d \geq 2$. See Balbiani et al. (2022, 2023).

[5] See Alizadeh et al. (2023).

7.2.2 Modal Logics

We shall say that $\mathbf{L} \subseteq \mathbf{FOR}$ is a *modal logic* if the following conditions hold[6]:

- $\mathbf{L}$ contains all tautologies,
- $\mathbf{L}$ contains the formula $\Box(x \to y) \to (\Box x \to \Box y)$,
- $\mathbf{L}$ is closed for *modus ponens*: if $\varphi \to \psi \in \mathbf{L}$ and $\varphi \in \mathbf{L}$ then $\psi \in \mathbf{L}$,
- $\mathbf{L}$ is closed for *generalization*: if $\varphi \in \mathbf{L}$ then $\Box\varphi \in \mathbf{L}$,
- $\mathbf{L}$ is closed for *uniform substitution*: if $\varphi \in \mathbf{L}$ and ψ is an instance of φ then $\psi \in \mathbf{L}$.

We write $\mathbf{K}$ for the least modal logic. We shall say that a modal logic $\mathbf{L}$ is *consistent* if $\mathbf{L} \neq \mathbf{FOR}$. From now on in this paper,

let L be an arbitrary consistent modal logic.

For all $\varphi \in \mathbf{FOR}$, we write $\mathbf{L} \oplus \varphi$ for the least modal logic containing $\mathbf{L}$ and φ. For all $S \subseteq \mathbf{FOR}$, we write $\mathbf{L} \oplus S$ for the least modal logic containing $\mathbf{L}$ and S. In order of appearance from now on, the following extensions of $\mathbf{K}$ are considered in this paper:

- $\mathbf{S4} = \mathbf{K} \oplus \{\Box x \to x, \Box x \to \Box\Box x\}$,
- $\mathbf{S5} = \mathbf{S4} \oplus \Diamond x \to \Box\Diamond x$,
- $\mathbf{Triv} = \mathbf{K} \oplus \Box x \leftrightarrow x$,
- $\mathbf{Ver} = \mathbf{K} \oplus \Box x$,
- $\mathbf{S4.2} = \mathbf{S4} \oplus \Diamond\Box x \to \Box\Diamond x$,
- $\mathbf{S4.3} = \mathbf{S4} \oplus \Box(\Box x \to y) \vee \Box(\Box y \to x)$,
- $\mathbf{K4} = \mathbf{K} \oplus \Box x \to \Box\Box x$,
- $\mathbf{K4.3} = \mathbf{K4} \oplus \Box(\Box x \wedge x \to y) \vee \Box(\Box y \wedge y \to x)$,
- $\mathbf{K4D1} = \mathbf{K4} \oplus \Box(\Box x \to y) \vee \Box(\Box y \to x)$,
- $\mathbf{K4.2^+} = \mathbf{K4} \oplus \Diamond(\Box x \wedge x) \vee (\Box x \wedge x) \to \Box(\Diamond x \vee x) \wedge (\Diamond x \vee x)$,
- $\mathbf{K45} = \mathbf{K4} \oplus \Diamond x \to \Box\Diamond x$,
- $\mathbf{K5} = \mathbf{K} \oplus \Diamond x \to \Box\Diamond x$,
- $\mathbf{KD45} = \mathbf{K45} \oplus \Diamond\top$,
- $\mathbf{Alt}_n = \mathbf{K} \oplus \bigwedge\{\Diamond x_i : 0 \leq i \leq n\} \to \bigvee\{\Diamond(x_i \wedge x_j) : 0 \leq i < j \leq n\}$ when $n \geq 1$,
- $\mathbf{KD} = \mathbf{K} \oplus \Diamond\top$,
- $\mathbf{DAlt}_n = \mathbf{Alt}_n \oplus \Diamond\top$ when $n \geq 1$,
- $\mathbf{KT} = \mathbf{K} \oplus \Box x \to x$,
- $\mathbf{KB} = \mathbf{K} \oplus x \to \Box\Diamond x$,
- $\mathbf{KDB} = \mathbf{KD} \oplus x \to \Box\Diamond x$,
- $\mathbf{KTB} = \mathbf{KT} \oplus x \to \Box\Diamond x$.

We shall say that $\varphi \in \mathbf{FOR}$ is $\mathbf{L}$-*consistent* if $\neg\varphi \notin \mathbf{L}$. We shall say that $\varphi \in \mathbf{FOR}$ is $\mathbf{L}$-*derivable* from $\Gamma \subseteq \mathbf{FOR}$ (in symbols $\Gamma \vdash_{\mathbf{L}} \varphi$) if there exists $n \geq 1$ and there exists $\varphi_1, \ldots, \varphi_n \in \mathbf{FOR}$ such that $\varphi_n = \varphi$ and for all $k \in 1, \ldots, n$, at least one of the following conditions holds:

---

[6] The modal logics considered here are exactly the *normal* modal logics considered in standard textbooks such as Blackburn and de Rijke (2001), Chagrov (1997), Kracht (1999).

- $\varphi_k \in \mathbf{L}$,
- $\varphi_k \in \Gamma$,
- there exists $i, j \in 1, \ldots, n$ such that $i, j < k$ and $\varphi_i = \varphi_j \to \varphi_k$,
- there exists $i \in 1, \ldots, n$ such that $i < k$ and $\varphi_k = \Box\varphi_i$.

Let $\equiv_{\mathbf{L}}$ be the equivalence relation on **FOR** defined by

- $\varphi \equiv_{\mathbf{L}} \psi$ if and only if $\varphi \leftrightarrow \psi \in \mathbf{L}$,

φ and ψ ranging over **FOR**. We shall say that **L** is *locally tabular* if for all finite $X \subseteq \mathbf{VAR}$, $\equiv_{\mathbf{L}}$ possesses finitely many equivalence classes on $\mathbf{FOR}_X$.

7.2.3 Substitutions

A *substitution* is a triple (X, Y, σ) where $X, Y \subseteq \mathbf{VAR}$ are finite and σ is a homomorphism from $\mathbf{FOR}_X$ to $\mathbf{FOR}_Y$.[7] To apply a substitution (X, Y, σ) to $\varphi \in \mathbf{FOR}_X$ amounts to replace each occurrence in φ of a variable $x \in X$ by the corresponding formula $\sigma(x) \in \mathbf{FOR}_Y$. The sets X and Y are respectively the *domain* and the *codomain* of the substitution (X, Y, σ). Let **SUB** be the set of all substitutions. We shall say that $(X, Y, \sigma) \in \mathbf{SUB}$ is *variable-free* if $Y = \emptyset$. Substitutions being homomorphisms, it is possible to compose two substitutions if the codomain of the first is equal to the domain of the second. The *composition* of $(X, Y, \sigma), (Y, Z, \tau) \in \mathbf{SUB}$ (in symbols $(X, Y, \sigma) \circ (Y, Z, \tau)$) is the substitution $(X, Z, \upsilon) \in \mathbf{SUB}$ such that for all $x \in X$, $\upsilon(x) = \tau(\sigma(x))$. Substitutions being completely defined by the restrictions to their domains, it is possible to compare two substitutions by means of these restrictions if their domains are equal. Let $\simeq_{\mathbf{L}}$ be the equivalence relation on **SUB** defined by

- $(X, Y, \sigma) \simeq_{\mathbf{L}} (X, Z, \tau)$ if and only if for all $x \in X$, $\sigma(x) \leftrightarrow \tau(x) \in \mathbf{L}$,

(X, Y, σ) and (X, Z, τ) ranging over **SUB**. Obviously, if $(X, Y, \sigma) \simeq_{\mathbf{L}} (X, Z, \tau)$ then for all $\varphi \in \mathbf{FOR}_X$, $\sigma(\varphi) \leftrightarrow \tau(\varphi) \in \mathbf{L}$. Let $\preccurlyeq_{\mathbf{L}}$ be the quasi-order on **SUB** defined by

- $(X, Y, \sigma) \preccurlyeq_{\mathbf{L}} (X, Z, \tau)$ if and only if there exists $(Z, T, \upsilon) \in \mathbf{SUB}$ such that for all $x \in X$, $\sigma(x) \leftrightarrow \upsilon(\tau(x)) \in \mathbf{L}$,

(X, Y, σ) and (X, Z, τ) ranging over **SUB**. Obviously, if $(X, Y, \sigma) \preccurlyeq_{\mathbf{L}} (X, Z, \tau)$ then there exists $(Z, T, \upsilon) \in \mathbf{SUB}$ such that for all $\varphi \in \mathbf{FOR}_X$, $\sigma(\varphi) \leftrightarrow \upsilon(\tau(\varphi)) \in \mathbf{L}$. When the domain and the codomain can be guessed from the context of the

---

[7] Substitutions could also be defined as functions from **VAR** to **FOR** almost everywhere equal to the identity function (Baader & Snyder, 2001, Sect. 2). In this paper, we follow the definition of substitutions used, for example, by Babenyshev and Rybakov (2011), Dzik (2007) and Ghilardi (1999, 2000).

discussion, the substitution (X, Y, σ) will be simply indicated by σ. However, when we write that two substitutions are equal, this will imply in any case that their domains are equal and their codomains are equal.

7.3 Unifiability

7.3.1 *Unifiers*

An **L**-*unifier* of $\varphi \in \mathbf{FOR}$ is a substitution $(\mathtt{var}(\varphi), X, \sigma)$ such that $\sigma(\varphi) \in \mathbf{L}$. We write $\Sigma_{\mathbf{L}}(\varphi)$ to mean the set of all **L**-unifiers of $\varphi \in \mathbf{FOR}$. We shall say that $\varphi \in \mathbf{FOR}$ is **L**-*unifiable* if $\Sigma_{\mathbf{L}}(\varphi) \neq \emptyset$. Notice that since **L** is closed for uniform substitution, for all **L**-unifiable $\varphi \in \mathbf{FOR}$, $\Sigma_{\mathbf{L}}(\varphi)$ contains variable-free substitutions. In other respect, for all consistent modal logics $\mathbf{L}'$, if $\mathbf{L} \subseteq \mathbf{L}'$ then for all $\varphi \in \mathbf{FOR}$, $\Sigma_{\mathbf{L}}(\varphi) \subseteq \Sigma_{\mathbf{L}'}(\varphi)$.

Example The substitutions $(\{x\}, \emptyset, \sigma_{\bot})$ and $(\{x\}, \emptyset, \sigma_{\top})$ defined by $\sigma_{\bot}(x) = \bot$ and $\sigma_{\top}(x) = \top$ are **S4**-unifiers of $\Box x \vee \Box\neg x$.

Example The substitution $(\{x\}, \{x\}, \epsilon)$ defined by $\epsilon(x) = \Box x$ is an **S5**-unifier of $\Box x \vee \Box\neg x$.

We shall say that an **L**-unifier σ of $\varphi \in \mathbf{FOR}$ is a *most general* **L**-*unifier* of φ if for all **L**-unifiers τ of φ, $\tau \preccurlyeq_{\mathbf{L}} \sigma$.

Example $\Box x \vee \Box\neg x$ has no most general **S4**-unifier. Why? Remind the **S4**-unifiers $\sigma_{\bot}$ and $\sigma_{\top}$ of $\Box x \vee \Box\neg x$ defined above. Let σ be an arbitrary **S4**-unifier of $\Box x \vee \Box\neg x$. Hence, $\Box\sigma(x) \vee \Box\neg\sigma(x) \in \mathbf{S4}$. Since **S4** has the modal disjunction property (Chagrov, 1997, Chap. 3), either $\sigma(x) \in \mathbf{S4}$, or $\neg\sigma(x) \in \mathbf{S4}$. In the former case, $\sigma_{\bot} \not\preccurlyeq_{\mathbf{S4}} \sigma$. In the latter case, $\sigma_{\top} \not\preccurlyeq_{\mathbf{S4}} \sigma$. For this reason, $\Box x \vee \Box\neg x$ has no most general **S4**-unifier.

Example The substitution ϵ defined above is a most general **S5**-unifier of $\Box x \vee \Box\neg x$. Why? Let τ be an arbitrary **S5**-unifier of $\Box x \vee \Box\neg x$. Hence, $\Box\tau(x) \vee \Box\neg\tau(x) \in \mathbf{S5}$. Thus, $\tau(x) \leftrightarrow \Box\tau(x) \in \mathbf{S5}$. Consequently, $\tau \preccurlyeq_{\mathbf{S5}} \epsilon$. Since τ was arbitrary, ϵ is a most general **S5**-unifier of $\Box x \vee \Box\neg x$.

7.3.2 *Bases*

We shall say that a set Σ of **L**-unifiers of an **L**-unifiable $\varphi \in \mathbf{FOR}$ is *complete* if for all **L**-unifiers σ of φ, there exists $\tau \in \Sigma$ such that $\sigma \preccurlyeq_{\mathbf{L}} \tau$. Notice that for all **L**-unifiable $\varphi \in \mathbf{FOR}$, $\Sigma_{\mathbf{L}}(\varphi)$ is a complete set of **L**-unifiers of φ.

Example Remind the **S4**-unifiers $\sigma_\bot$ and $\sigma_\top$ of $\Box x \vee \Box\neg x$ defined above. They constitute a complete set of **S4**-unifiers of $\Box x \vee \Box\neg x$. Why? Let σ be an arbitrary **S4**-unifier of $\Box x \vee \Box\neg x$. Hence, $\Box\sigma(x) \vee \Box\neg\sigma(x) \in$ **S4**. Thus, as we know, either $\sigma(x) \in$ **S4**, or $\neg\sigma(x) \in$ **S4**. In the former case, $\sigma_\top \preccurlyeq_{\mathbf{S4}} \sigma$. In the latter case, $\sigma_\bot \preccurlyeq_{\mathbf{S4}} \sigma$. Since σ was arbitrary, $\sigma_\bot$ and $\sigma_\top$ constitute a complete set of **S4**-unifiers of $\Box x \vee \Box\neg x$.

Example As a most general **S5**-unifier of $\Box x \vee \Box\neg x$, the substitution ϵ defined above constitutes a complete set of **S5**-unifiers of $\Box x \vee \Box\neg x$.

We shall say that a complete set Σ of **L**-unifiers of an **L**-unifiable $\varphi \in$ **FOR** is a *basis* for φ if for all $\sigma, \tau \in \Sigma$, if $\sigma \preccurlyeq_{\mathbf{L}} \tau$ then $\sigma = \tau$.

Example Remind the **S4**-unifiers $\sigma_\bot$ and $\sigma_\top$ of $\Box x \vee \Box\neg x$ defined above. They constitute a basis for $\Box x \vee \Box\neg x$. Why? Just because neither $\sigma_\bot \preccurlyeq_{\mathbf{S4}} \sigma_\top$, nor $\sigma_\top \preccurlyeq_{\mathbf{S4}} \sigma_\bot$.

Example As a most general **S5**-unifier of $\Box x \vee \Box\neg x$, the substitution ϵ defined above constitutes a basis for $\Box x \vee \Box\neg x$.

Proposition 7.1 *Let Σ be a complete set of* **L** *unifiers of an* **L**-*unifiable $\varphi \in$* **FOR**. *The following conditions are equivalent:*

1. *Σ is a basis for φ,*
2. *Σ is a minimal complete set of* **L**-*unifiers of φ, i.e. for all $\Delta \subseteq \Sigma$, if Δ is a complete set of* **L**-*unifiers of φ then $\Delta = \Sigma$.*

Proof This is a standard result.

Suppose Σ is a basis for φ. For the sake of the contradiction, suppose there exists $\Delta \subseteq \Sigma$ such that Δ is a complete set of **L**-unifiers of φ and $\Delta \neq \Sigma$. Hence, there exists $\sigma \in \Sigma$ such that $\sigma \notin \Delta$. Since Δ is a complete set of **L**-unifiers of φ, there exists $\tau \in \Delta$ such that $\sigma \preccurlyeq_{\mathbf{L}} \tau$. Since $\Delta \subseteq \Sigma$, $\tau \in \Sigma$. Since Σ is a basis for φ and $\sigma \preccurlyeq_{\mathbf{L}} \tau$, $\sigma = \tau$. Thus, $\tau \notin \Delta$: a contradiction.

Reciprocally, suppose for all $\Delta \subseteq \Sigma$, if Δ is a complete set of **L**-unifiers of φ then $\Delta = \Sigma$. For the sake of the contradiction, suppose Σ is not a basis for φ. Consequently, there exists $\sigma, \tau \in \Sigma$ such that $\sigma \preccurlyeq_{\mathbf{L}} \tau$ and $\sigma \neq \tau$. Let $\Lambda = \Sigma \setminus \{\sigma\}$. Obviously, $\Lambda \subseteq \Sigma$ and $\Lambda \neq \Sigma$. Moreover, since Σ is a complete set of **L**-unifiers of φ, $\sigma \preccurlyeq_{\mathbf{L}} \tau$ and $\sigma \neq \tau$, Λ is a complete set of **L**-unifiers of φ. Since $\Lambda \subseteq \Sigma$ and for all $\Delta \subseteq \Sigma$, if Δ is a complete set of **L**-unifiers of φ then $\Delta = \Sigma$, $\Lambda = \Sigma$: a contradiction. $\dashv$

Proposition 7.2 *For all* **L**-*unifiable $\varphi \in$* **FOR** *and for all bases Σ, Δ for φ, Σ and Δ have the same cardinality.*

Proof This is a standard result.

Suppose $\varphi \in$ **FOR** is **L**-unifiable and Σ, Δ are bases for φ. Hence, by Proposition 7.1, Σ and Δ are minimal complete sets of **L**-unifiers of φ. By the completeness of Σ and Δ, one can readily define functions $f : \Sigma \longrightarrow \Delta$ and

$g : \Delta \longrightarrow \Sigma$ such that $\sigma \preccurlyeq_{\mathbf{L}} f(\sigma)$ for each $\sigma \in \Sigma$ and $\delta \preccurlyeq_{\mathbf{L}} g(\delta)$ for each $\delta \in \Delta$. For the sake of the contradiction, suppose Σ and Δ have not the same cardinality. Thus, either f is not injective, or g is not injective. Without loss of generality, suppose f is not injective. Let $\sigma, \tau \in \Sigma$ be such that $f(\sigma) = f(\tau)$ and $\sigma \neq \tau$. Thus, $\sigma \preccurlyeq_{\mathbf{L}} f(\sigma) \preccurlyeq_{\mathbf{L}} g(f(\sigma))$ and $\tau \preccurlyeq_{\mathbf{L}} f(\tau) \preccurlyeq_{\mathbf{L}} g(f(\tau))$. Since Σ is a basis for φ, $\sigma = g(f(\sigma))$ and $\tau = g(f(\tau))$. Since $f(\sigma) = f(\tau)$, $\sigma = \tau$: a contradiction. $\dashv$

7.3.3 Types of Formulas

Owing to Proposition 7.2, an important question is the following: when $\varphi \in \mathbf{FOR}$ is **L**-unifiable, is there a basis for φ? When the answer is "yes", how large is this basis? For all **L**-unifiable $\varphi \in \mathbf{FOR}$, we shall say that:

- φ is **L**-*unitary—or of type* 1—if there exists a basis for φ with cardinality 1,
- φ is **L**-*finitary—or of type* ω—if there exists a basis for φ with finite cardinality ≥ 2,
- φ is **L**-*infinitary—or of type* ∞—if there exists a basis for φ with infinite cardinality,
- φ is **L**-*nullary—or of type* 0—if there exists no basis for φ.

Obviously, the types 1, ω, ∞ and 0 constitute a set of jointly exhaustive and pairwise distinct situations for each **L**-unifiable $\varphi \in \mathbf{FOR}$.

Example Remind the **S4**-unifiers $\sigma_\bot$ and $\sigma_\top$ of $\Box x \vee \Box\neg x$ defined above. Since they constitute a basis for $\Box x \vee \Box\neg x$ in **S4**, $\Box x \vee \Box\neg x$ is of type ω in **S4**.

Example Remind the **S5**-unifier ϵ of $\Box x \vee \Box\neg x$ defined above. Since it constitutes a basis for $\Box x \vee \Box\neg x$ in **S5**, $\Box x \vee \Box\neg x$ is of type 1 in **S5**.

Proposition 7.3 *Let $\varphi \in$ **FOR** be **L**-unifiable. If φ is Boolean then φ is **L**-unitary.*

Proof This is a standard result.

Suppose φ is Boolean. Since φ is **L**-unifiable, there exists a variable-free substitution $(\mathtt{var}(\varphi), \emptyset, \sigma)$ such that $\sigma(\varphi) \in \mathbf{L}$. Since **L** is consistent, it follows by Makinson (1971) that either **L** is contained in **Triv**, or **L** is contained in **Ver**. Let $(\mathtt{var}(\varphi), \emptyset, \sigma')$ and $(\mathtt{var}(\varphi), \emptyset, \sigma'')$ be the variable-free substitutions such that

- for all $x \in \mathtt{var}(\varphi)$, $\sigma'(x)$ is the formula obtained from $\sigma(x)$ after having iteratively replaced each subformula of the form $\Box\psi$ by ψ,
- for all $x \in \mathtt{var}(\varphi)$, $\sigma''(x)$ is the formula obtained from $\sigma(x)$ after having iteratively replaced each subformula of the form $\Box\psi$ by $\top$.

Notice that for all $x \in \mathtt{var}(\varphi)$, $\mathtt{deg}(\sigma'(x)) = 0$ and $\mathtt{deg}(\sigma''(x)) = 0$. Since φ is Boolean, $\mathtt{deg}(\sigma'(\varphi)) = 0$ and $\mathtt{deg}(\sigma''(\varphi)) = 0$. Moreover, the substitutions σ' and σ'' being "$\Box$-free", they can be considered as substitutions for the language of **BL**.

Obviously, $\sigma \simeq_{\mathbf{Triv}} \sigma'$ and $\sigma \simeq_{\mathbf{Ver}} \sigma''$. Since $\sigma(\varphi) \in \mathbf{L}$, $\sigma'(\varphi) \in \mathbf{Triv}$ when $\mathbf{L}$ is contained in **Triv** and $\sigma''(\varphi) \in \mathbf{Ver}$ when $\mathbf{L}$ is contained in **Ver**. Hence, φ is **BL**-unifiable, **Triv** and **Ver** being conservative extensions of **BL**. As is well-known (Martin & Nipkow, 1989), there exists a **BL**-unifier ϵ of φ such that for all $x \in \mathtt{var}(\varphi)$, either $\epsilon(x) = \varphi \rightarrow x$, or $\epsilon(x) = \varphi \wedge x$. Thus, $\epsilon(\varphi) \in \mathbf{BL}$ and therefore $\epsilon(\varphi) \in \mathbf{L}$. Moreover, for all **L**-unifiers τ of φ and for all $x \in \mathtt{var}(\varphi)$, $\tau(x) \leftrightarrow \tau(\epsilon(x)) \in \mathbf{L}$. Consequently, for all **L**-unifiers τ of φ, $\tau \preccurlyeq_{\mathbf{L}} \epsilon$. As a result, ϵ is a most general **L**-unifier of φ. Hence, φ is **L**-unitary. $\dashv$

7.3.4 Types of Logics

The types 1, ω, ∞ and 0 being ordered by $1<\omega<\infty<0$, the unification type of **L** is the greatest type among the types of its unifiable formulas. More precisely, we shall say that:

- **L** is *unitary – or of type* 1 – if every **L**-unifiable formula is of type 1,
- **L** is *finitary – or of type* ω – if every **L**-unifiable formula is either of type 1, or of type ω and there exists an **L**-unifiable formula of type ω,
- **L** is *infinitary – or of type* ∞ – if every **L**-unifiable formula is either of type 1, or of type ω, or of type ∞ and there exists an **L**-unifiable formula of type ∞,
- **L** is *nullary – or of type* 0 – if there exists an **L**-unifiable formula of type 0.

The significance of the unification types of modal logics in the research on the problem of checking the unifiability of formulas stems from the fact that if a modal logic is either unitary, or finitary, bases for **L**-unifiable formulas can be effectively computed and membership in **L** can be decided then the problem of checking the admissibility of inference rules can be decided.[8] Owing to Proposition 7.3, it is natural to ask the question whether there exists $d \geq 0$ such that the type of **L** is the greatest type among the types of all **L**-unifiable formulas of degree at most d. In Sect. 7.3.5, following a line of reasoning that can be found in several papers devoted to the problem of checking the unifiability of formulas, we will explain why $d = 1$ is the answer.

7.3.5 Reduced Formulas

We explain why the type of **L** is the greatest type among the types of all **L**-unifiable formulas of degree at most 1. Let $\varphi \in \mathbf{FOR}$. Let $l \geq 1$ and $(\varphi_1, \ldots, \varphi_l)$ be an enumer-

---

[8] Indeed, if **L** is either unitary, or finitary then an arbitrary inference rule $\frac{\varphi_1,\ldots,\varphi_m}{\psi}$ is **L**-admissible if and only if either $\varphi_1 \wedge \ldots \wedge \varphi_m$ is not **L**-unifiable, or $\varphi_1 \wedge \ldots \wedge \varphi_m$ is **L**-unifiable and there exists a basis Σ for $\varphi_1 \wedge \ldots \wedge \varphi_m$ such that for all $\sigma \in \Sigma$, $\sigma(\psi) \in \mathbf{L}$. Of course, this can be effectively checked only if bases for **L**-unifiable formulas can be effectively computed and membership in **L** can be decided.

ation of the set of all subformulas of φ in which each subformula occurs exactly once. Without loss of generality, assume for all $i, j \in \{1, \ldots, l\}$, if φ_i is a strict subformula of φ_j then $i<j$. Hence, $\varphi = \varphi_l$. We associate the new variables $y_1, \ldots, y_l$ to the enumeration $(\varphi_1, \ldots, \varphi_l)$, i.e. for all $i \in \{1, \ldots, l\}$, $y_i \notin \mathtt{var}(\varphi)$. For all $i \in \{1, \ldots, l\}$, let φ_i° be the formula defined by case as follows:

case $\varphi_i = x$: $\varphi_i^\circ := y_i \leftrightarrow x$,
case $\varphi_i = \bot$: $\varphi_i^\circ := \neg y_i$,
case $\varphi_i = \neg\varphi_j$: $\varphi_i^\circ := y_i \leftrightarrow \neg y_j$,
case $\varphi_i = \varphi_j \vee \varphi_k$: $\varphi_i^\circ := y_i \leftrightarrow y_j \vee y_k$,
case $\varphi_i = \Box\varphi_j$: $\varphi_i^\circ := y_i \leftrightarrow \Box y_j$.

Let $\varphi^\circ = \varphi_1^\circ \wedge \ldots \wedge \varphi_l^\circ \wedge y_l$. Notice that the size of φ° is linear in the size of φ. Moreover, $\mathtt{deg}(\varphi^\circ) \leq 1$ and $\mathtt{var}(\varphi^\circ) = \mathtt{var}(\varphi) \cup \{y_1, \ldots, y_l\}$.

Example With respect to the enumeration $(x, \neg x, \Box x, \Box\neg x, \Box x \vee \Box\neg x)$ of all subformulas of $\Box x \vee \Box\neg x$ and the associated new variables y_1, y_2, y_3, y_4, y_5, $(\Box x \vee \Box\neg x)^\circ$ is the conjunction of the following formulas:

- $y_1 \leftrightarrow x$,
- $y_2 \leftrightarrow \neg y_1$,
- $y_3 \leftrightarrow \Box y_1$,
- $y_4 \leftrightarrow \Box y_2$,
- $y_5 \leftrightarrow y_3 \vee y_4$,
- y_5.

For all substitutions $(\mathtt{var}(\varphi), X, \sigma)$, let $\sigma^\uparrow$ be the substitution $(\mathtt{var}(\varphi^\circ), X, \tau)$ such that

- for all $x \in \mathtt{var}(\varphi)$, $\tau(x) = \sigma(x)$,
- for all $i \in \{1, \ldots, l\}$, $\tau(y_i) = \sigma(\varphi_i)$.

For all substitutions $(\mathtt{var}(\varphi^\circ), X, \sigma)$, let $\sigma^\downarrow$ be the substitution $(\mathtt{var}(\varphi), X, \tau)$ such that

- for all $x \in \mathtt{var}(\varphi)$, $\tau(x) = \sigma(x)$.

Example With respect to the enumeration $(x, \neg x, \Box x, \Box\neg x, \Box x \vee \Box\neg x)$ of all subformulas of $\Box x \vee \Box\neg x$ and the associated new variables y_1, y_2, y_3, y_4, y_5, considering the **S4**-unifiers $\sigma_\bot$ and $\sigma_\top$ of $\Box x \vee \Box\neg x$ defined above, the substitutions $\sigma_\bot^\uparrow$ and $\sigma_\top^\uparrow$ defined as follows are **S4**-unifiers of $(\Box x \vee \Box\neg x)^\circ$:

- $\sigma_\bot^\uparrow(x) = \bot$ and $\sigma_\top^\uparrow(x) = \top$,
- $\sigma_\bot^\uparrow(y_1) = \bot$ and $\sigma_\top^\uparrow(y_1) = \top$,
- $\sigma_\bot^\uparrow(y_2) = \neg\bot$ and $\sigma_\top^\uparrow(y_2) = \neg\top$,
- $\sigma_\bot^\uparrow(y_3) = \Box\bot$ and $\sigma_\top^\uparrow(y_3) = \Box\top$,
- $\sigma_\bot^\uparrow(y_4) = \Box\neg\bot$ and $\sigma_\top^\uparrow(y_4) = \Box\neg\top$,

- $\sigma_{\bot}^{\uparrow}(y_5) = \Box\bot \vee \Box\neg\bot$ and $\sigma_{\top}^{\uparrow}(y_5) = \Box\top \vee \Box\neg\top$.

Proposition 7.4 *For all* $\varphi \in$ **FOR**,

- *if* φ *is* **L**-*unifiable then* φ° *is* **L**-*unifiable too, seeing that for all* **L**-*unifiers* σ *of* φ, $\sigma^{\uparrow}$ *is an* **L**-*unifier of* φ°,
- *if* φ° *is* **L**-*unifiable then* φ *is* **L**-*unifiable too, seeing that for all* **L**-*unifiers* σ *of* φ°, $\sigma^{\downarrow}$ *is an* **L**-*unifier of* φ.

Proof Let $\varphi \in$ **FOR**.

Suppose φ is **L**-unifiable. Let σ be an **L**-unifier of φ. Hence, $\sigma(\varphi) \in$ **L**. Since $\varphi = \varphi_l$ and $\sigma^{\uparrow}(y_l) = \sigma(\varphi_l)$, $\sigma^{\uparrow}(y_l) \in$ **L**. As the reader may easily check, for all $i \in \{1, \ldots, l\}$, $\sigma^{\uparrow}(\varphi_i^{\circ}) \in$ **L**. Since $\sigma^{\uparrow}(y_l) \in$ **L**, $\sigma^{\uparrow}(\varphi^{\circ}) \in$ **L**. Thus, $\sigma^{\uparrow}$ is an **L**-unifier of φ°.

Suppose φ° is **L**-unifiable. Let σ be an **L**-unifier of φ°. Consequently, $\sigma(\varphi^{\circ}) \in$ **L**. Hence, $\sigma(y_l) \in$ **L**. As the reader may easily check, for all $i \in \{1, \ldots, l\}$, $\sigma^{\downarrow}(\varphi_i) \leftrightarrow \sigma(y_i) \in$ **L**. Thus, $\sigma^{\downarrow}(\varphi_l) \leftrightarrow \sigma(y_l) \in$ **L**. Since $\varphi = \varphi_l$ and $\sigma(y_l) \in$ **L**, $\sigma^{\downarrow}(\varphi) \in$ **L**. Consequently, $\sigma^{\downarrow}$ is an **L**-unifier of φ. ⊣

Proposition 7.5 *For all* $\varphi \in$ **FOR**,

- *for all complete sets* Σ *of* **L**-*unifiers of* φ, $\{\sigma^{\uparrow} : \sigma \in \Sigma\}$ *is a complete set of* **L**-*unifiers of* φ°,
- *for all complete sets* Σ *of* **L**-*unifiers of* φ°, $\{\sigma^{\downarrow} : \sigma \in \Sigma\}$ *is a complete set of* **L**-*unifiers of* φ.

Proof Let $\varphi \in$ **FOR**.

Let Σ be a complete set of **L**-unifiers of φ. Let $(\mathtt{var}(\varphi) \cup \{y_1, \ldots, y_l\}, X, \tau)$ be an arbitrary **L**-unifier of φ°. Hence, by Proposition 7.4, $(\mathtt{var}(\varphi), X, \tau^{\downarrow})$ is an **L**-unifier of φ. Moreover, as the reader may easily check, for all $i \in \{1, \ldots, l\}$, $\tau^{\downarrow}(\varphi_i) \leftrightarrow \tau(y_i) \in$ **L**. Since Σ is a complete set of **L**-unifiers of φ, let $(\mathtt{var}(\varphi), Y, \sigma) \in \Sigma$ be such that $\tau^{\downarrow} \preccurlyeq_{\mathbf{L}} \sigma$. Let $(Y, Z, \upsilon) \in$ **SUB** be such that for all $x \in \mathtt{var}(\varphi)$, $\tau^{\downarrow}(x) \leftrightarrow \upsilon(\sigma(x)) \in$ **L**. Since for all $x \in \mathtt{var}(\varphi)$, $\tau^{\downarrow}(x) = \tau(x)$ and $\sigma(x) = \sigma^{\uparrow}(x)$, for all $x \in \mathtt{var}(\varphi)$, $\tau(x) \leftrightarrow \upsilon(\sigma^{\uparrow}(x)) \in$ **L**. Since for all $x \in \mathtt{var}(\varphi)$, $\tau^{\downarrow}(x) \leftrightarrow \upsilon(\sigma(x)) \in$ **L**, for all $i \in \{1, \ldots, l\}$, $\tau^{\downarrow}(\varphi_i) \leftrightarrow \upsilon(\sigma(\varphi_i)) \in$ **L**. Since for all $i \in \{1, \ldots, l\}$, $\tau^{\downarrow}(\varphi_i) \leftrightarrow \tau(y_i) \in$ **L** and $\sigma(\varphi_i) = \sigma^{\uparrow}(y_i)$, for all $i \in \{1, \ldots, l\}$, $\tau(y_i) \leftrightarrow \upsilon(\sigma^{\uparrow}(y_i)) \in$ **L**. Since for all $x \in \mathtt{var}(\varphi)$, $\tau(x) \leftrightarrow \upsilon(\sigma^{\uparrow}(x)) \in$ **L**, $\tau \preccurlyeq_{\mathbf{L}} \sigma^{\uparrow}$. Since τ was arbitrary, $\{\sigma^{\uparrow} : \sigma \in \Sigma\}$ is a complete set of **L**-unifiers of φ°.

Let Σ be a complete set of **L**-unifiers of φ°. Let $(\mathtt{var}(\varphi), X, \tau)$ be an arbitrary **L**-unifier of φ. Thus, by Proposition 7.4, $(\mathtt{var}(\varphi) \cup \{y_1, \ldots, y_l\}, X, \tau^{\uparrow})$ is an **L**-unifier of φ°. Since Σ is a complete set of **L**-unifiers of φ°, let $(\mathtt{var}(\varphi) \cup \{y_1, \ldots, y_l\}, Y, \sigma) \in \Sigma$ be such that $\tau^{\uparrow} \preccurlyeq_{\mathbf{L}} \sigma$. Let $(Y, Z, \upsilon) \in$ **SUB** be such that for all $x \in \mathtt{var}(\varphi)$, $\tau^{\uparrow}(x) \leftrightarrow \upsilon(\sigma(x)) \in$ **L**. Since for all $x \in \mathtt{var}(\varphi)$, $\tau^{\uparrow}(x) = \tau(x)$ and $\sigma(x) = \sigma^{\downarrow}(x)$, for all $x \in \mathtt{var}(\varphi)$, $\tau(x) \leftrightarrow \upsilon(\sigma^{\downarrow}(x)) \in$ **L**. Consequently, $\tau \preccurlyeq_{\mathbf{L}} \sigma^{\downarrow}$. Since τ was arbitrary, $\{\sigma^{\downarrow} : \sigma \in \Sigma\}$ is a complete set of **L**-unifiers of φ. ⊣

Proposition 7.6 *For all $\varphi \in$ **FOR**,*

- *for all bases Σ of* **L**-*unifiers of φ, $\{\sigma^{\uparrow} : \sigma \in \Sigma\}$ is a basis of* **L**-*unifiers of φ° with the same cardinality,*
- *for all bases Σ of* **L**-*unifiers of φ°, $\{\sigma^{\downarrow} : \sigma \in \Sigma\}$ is a basis of* **L**-*unifiers of φ with the same cardinality.*

Proof Let $\varphi \in \mathbf{FOR}$.

Let Σ be a basis of $\mathbf{L}$-unifiers of φ. Hence, by Proposition 7.1, Σ is a minimal complete set of $\mathbf{L}$-unifiers of φ. Thus, by Proposition 7.5, $\{\sigma^{\uparrow} : \sigma \in \Sigma\}$ is a complete set of $\mathbf{L}$-unifiers of φ°. For the sake of the contradiction, suppose $\{\sigma^{\uparrow} : \sigma \in \Sigma\}$ is not a minimal complete set of $\mathbf{L}$-unifiers of φ°. Let $\Delta \subseteq \{\sigma^{\uparrow} : \sigma \in \Sigma\}$ be such that Δ is a complete set of $\mathbf{L}$-unifiers of φ° and $\Delta \neq \{\sigma^{\uparrow} : \sigma \in \Sigma\}$. Let $(\mathtt{var}(\varphi), X, \sigma) \in \Sigma$ be such that $\sigma^{\uparrow} \notin \Delta$. As we already know, $\sigma^{\uparrow}$ is an $\mathbf{L}$-unifier of φ°. Since Δ is a complete set of $\mathbf{L}$-unifiers of φ°, let $(\mathtt{var}(\varphi), Y, \tau) \in \Sigma$ be such that $\tau^{\uparrow} \in \Delta$ and $\sigma^{\uparrow} \preccurlyeq_{\mathbf{L}} \tau^{\uparrow}$. Let $(Y, Z, \upsilon) \in \mathbf{SUB}$ be such that for all $x \in \mathtt{var}(\varphi)$, $\sigma^{\uparrow}(x) \leftrightarrow \upsilon(\tau^{\uparrow}(x)) \in \mathbf{L}$. Since for all $x \in \mathtt{var}(\varphi)$, $\sigma^{\uparrow}(x) = \sigma(x)$ and $\tau^{\uparrow}(x) = \tau(x)$, for all $x \in \mathtt{var}(\varphi)$, $\sigma(x) \leftrightarrow \upsilon(\tau(x)) \in \mathbf{L}$. Consequently, $\sigma \preccurlyeq_{\mathbf{L}} \tau$. Since Σ is a basis of $\mathbf{L}$-unifiers of φ, $\sigma = \tau$. Hence, $\sigma^{\uparrow} = \tau^{\uparrow}$. Since $\sigma^{\uparrow} \notin \Delta$, $\tau^{\uparrow} \notin \Delta$: a contradiction. Thus, $\{\sigma^{\uparrow} : \sigma \in \Sigma\}$ is a minimal complete set of $\mathbf{L}$-unifiers of φ°. Consequently, by Proposition 7.1, $\{\sigma^{\uparrow} : \sigma \in \Sigma\}$ is a basis of $\mathbf{L}$-unifiers of φ°. Let $f : \Sigma \longrightarrow \{\sigma^{\uparrow} : \sigma \in \Sigma\}$ be the function such that $f(\sigma) = \sigma^{\uparrow}$ for each $\sigma \in \Sigma$. We claim that f is injective. Indeed, suppose $\sigma, \tau \in \Sigma$ are such that $\sigma \neq \tau$ and $f(\sigma) = f(\tau)$. Hence, $\sigma^{\uparrow} = \tau^{\uparrow}$. Since $\sigma \neq \tau$, let $x \in \mathtt{var}(\varphi)$ be such that $\sigma(x) \neq \tau(x)$. Since $\sigma(x) = \sigma^{\uparrow}(x)$ and $\tau(x) = \tau^{\uparrow}(x)$, $\sigma^{\uparrow}(x) \neq \tau^{\uparrow}(x)$. Thus, $\sigma^{\uparrow} \neq \tau^{\uparrow}$: a contradiction. In the end, f is injective and Σ and $\{\sigma^{\uparrow} : \sigma \in \Sigma\}$ have the same cardinality.

Let Σ be a basis of $\mathbf{L}$-unifiers of φ°. Hence, by Proposition 7.1, Σ is a minimal complete set of $\mathbf{L}$-unifiers of φ°. Thus, by Proposition 7.5, $\{\sigma^{\downarrow} : \sigma \in \Sigma\}$ is a complete set of $\mathbf{L}$-unifiers of φ. For the sake of the contradiction, suppose $\{\sigma^{\downarrow} : \sigma \in \Sigma\}$ is not a minimal complete set of $\mathbf{L}$-unifiers of φ. Let $\Delta \subseteq \{\sigma^{\downarrow} : \sigma \in \Sigma\}$ be such that Δ is a complete set of $\mathbf{L}$-unifiers of φ and $\Delta \neq \{\sigma^{\downarrow} : \sigma \in \Sigma\}$. Let $(\mathtt{var}(\varphi) \cup \{y_1, \ldots, y_l\}, X, \sigma) \in \Sigma$ be such that $\sigma^{\downarrow} \notin \Delta$. As we already know, $\sigma^{\downarrow}$ is an $\mathbf{L}$-unifier of φ. Since Δ is a complete set of $\mathbf{L}$-unifiers of φ, let $(\mathtt{var}(\varphi) \cup \{y_1, \ldots, y_l\}, Y, \tau) \in \Sigma$ be such that $\tau^{\downarrow} \in \Delta$ and $\sigma^{\downarrow} \preccurlyeq_{\mathbf{L}} \tau^{\downarrow}$. Let $(Y, Z, \upsilon) \in \mathbf{SUB}$ be such that for all $x \in \mathtt{var}(\varphi)$, $\sigma^{\downarrow}(x) \leftrightarrow \upsilon(\tau^{\downarrow}(x)) \in \mathbf{L}$. Consequently, for all $i \in \{1, \ldots, l\}$, $\sigma^{\downarrow}(\varphi_i) \leftrightarrow \upsilon(\tau^{\downarrow}(\varphi_i)) \in \mathbf{L}$. Since for all $x \in \mathtt{var}(\varphi)$, $\sigma^{\downarrow}(x) \leftrightarrow \upsilon(\tau^{\downarrow}(x)) \in \mathbf{L}$ and for all $x \in \mathtt{var}(\varphi)$, $\tau^{\downarrow}(x) = \tau(x)$ and $\sigma^{\downarrow}(x) = \sigma(x)$, for all $x \in \mathtt{var}(\varphi)$, $\sigma(x) \leftrightarrow \upsilon(\tau(x)) \in \mathbf{L}$. As the reader may easily check, for all $i \in \{1, \ldots, l\}$, $\sigma^{\downarrow}(\varphi_i) \leftrightarrow \sigma(y_i) \in \mathbf{L}$ and $\tau^{\downarrow}(\varphi_i) \leftrightarrow \tau(y_i) \in \mathbf{L}$. Since for all $i \in \{1, \ldots, l\}$, $\sigma^{\downarrow}(\varphi_i) \leftrightarrow \upsilon(\tau^{\downarrow}(\varphi_i)) \in \mathbf{L}$, for all $i \in \{1, \ldots, l\}$, $\sigma(y_i) \leftrightarrow \upsilon(\tau(y_i)) \in \mathbf{L}$. Since for all $x \in \mathtt{var}(\varphi)$, $\sigma(x) \leftrightarrow \upsilon(\tau(x)) \in \mathbf{L}$, $\sigma \preccurlyeq_{\mathbf{L}} \tau$. Since Σ is a basis of $\mathbf{L}$-unifiers of φ°, $\sigma = \tau$. Hence, $\sigma^{\downarrow} = \tau^{\downarrow}$. Since $\sigma^{\downarrow} \notin \Delta$, $\tau^{\downarrow} \notin \Delta$: a contradiction. Thus, $\{\sigma^{\downarrow} : \sigma \in \Sigma\}$ is a minimal complete set of $\mathbf{L}$-unifiers of φ. Consequently, by Proposition 7.1, $\{\sigma^{\downarrow} : \sigma \in \Sigma\}$ is a basis of $\mathbf{L}$-unifiers of φ. Let $g : \Sigma \longrightarrow \{\sigma^{\downarrow} : \sigma \in \Sigma\}$ be the function such that $g(\sigma) = \sigma^{\downarrow}$ for each $\sigma \in \Sigma$. We claim that g is injective. Indeed, suppose $\sigma, \tau \in \Sigma$

are such that $\sigma \neq \tau$ and $g(\sigma) = g(\tau)$. Hence, $\sigma^{\downarrow} = \tau^{\downarrow}$. Thus, for all $x \in \mathtt{var}(\varphi)$, $\sigma^{\downarrow}(x) = \tau^{\downarrow}(x)$. Since for all $x \in \mathtt{var}(\varphi)$, $\sigma^{\downarrow}(x) = \sigma(x)$ and $\tau^{\downarrow}(x) = \tau(x)$, for all $x \in \mathtt{var}(\varphi)$, $\sigma(x) = \tau(x)$. Moreover, for all $i \in \{1, \ldots, l\}$, $\sigma^{\downarrow}(\varphi_i) = \tau^{\downarrow}(\varphi_i)$. Since for all $i \in \{1, \ldots, l\}$, $\sigma^{\downarrow}(\varphi_i) \leftrightarrow \sigma(y_i) \in \mathbf{L}$ and $\tau^{\downarrow}(\varphi_i) \leftrightarrow \tau(y_i) \in \mathbf{L}$, for all $i \in \{1, \ldots, l\}$, $\sigma(y_i) \leftrightarrow \tau(y_i) \in \mathbf{L}$. Since for all $x \in \mathtt{var}(\varphi)$, $\sigma(x) = \tau(x)$, $\sigma \simeq_{\mathbf{L}} \tau$. Consequently, $\sigma \preccurlyeq_{\mathbf{L}} \tau$. Since Σ is a basis of $\mathbf{L}$-unifiers of φ°, $\sigma = \tau$: a contradiction. In the end, g is injective and Σ and $\{\sigma^{\downarrow} : \sigma \in \Sigma\}$ have the same cardinality. ⊣

As a result, the unification type of **L** is the greatest type among the types of its unifiable formulas of degree at most 1.

7.4 Projective, Transparent, Directed and Reasonable Unification

In Sects. 7.5 and 7.6, we will provide examples for each of the types $1, \omega$ and 0.[9] In the meantime, in order to understand the relationships between these types, let us introduce the concepts of projective unification (Ghilardi, 1997, 1999, 2000), transparent unification (Dzik, 2006, 2011; Wroński, 1995, 2008), directed unification (Ghilardi & Sacchetti, 2004; Jeřábek, 2013) and reasonable unification (Balbiani et al., 2022, 2023).

7.4.1 Projective Unification

For all $\varphi \in \mathbf{FOR}$, a substitution $(\mathtt{var}(\varphi), \mathtt{var}(\varphi), \sigma)$ is **L**-*projective* for φ if for all $x \in \mathtt{var}(\varphi)$, $\varphi \vdash_{\mathbf{L}} x \leftrightarrow \sigma(x)$.[10] For all consistent modal logics $\mathbf{L}'$, if $\mathbf{L} \subseteq \mathbf{L}'$ then for all $\varphi \in \mathbf{FOR}$ and for all substitutions $(\mathtt{var}(\varphi), \mathtt{var}(\varphi), \sigma)$, if σ is **L**-projective for φ then σ is $\mathbf{L}'$-projective for φ. The predicate "is **L**-projective for" has properties that deserve to be noticed.[11]

Proposition 7.7 *Let $\varphi \in \mathbf{FOR}$ and σ be an **L**-projective substitution for φ. For all $\psi \in \mathbf{FOR}_{\mathtt{var}(\varphi)}$, $\varphi \vdash_{\mathbf{L}} \psi \leftrightarrow \sigma(\psi)$.*

---

[9] No modal logic is known to be infinitary (Dzik, 2007, Chap. 5).

[10] The term "projective" comes from the algebraic and categorical approach to unification introduced and developed by Ghilardi (1997, 1999, 2000).

[11] Concerning the proof of Proposition 7.7, it can be done by induction on ψ. As for the proof of Proposition 7.8, let τ be an **L**-projective substitution for φ. Since σ is an **L**-projective substitution for φ, for all $x \in \mathtt{var}(\varphi)$, $\varphi \vdash_{\mathbf{L}} x \leftrightarrow \sigma(x)$. Since τ is an **L**-projective substitution for φ, by Proposition 7.7, for all $x \in \mathtt{var}(\varphi)$, $\varphi \vdash_{\mathbf{L}} \sigma(x) \leftrightarrow \tau(\sigma(x))$. Since for all $x \in \mathtt{var}(\varphi)$, $\varphi \vdash_{L} x \leftrightarrow \sigma(x) \in \mathbf{L}$, for all $x \in \mathtt{var}(\varphi)$, $\varphi \vdash_{\mathbf{L}} x \leftrightarrow \tau(\sigma(x))$. Hence, $\sigma \circ \tau$ is **L**-projective for φ. About the proof of Proposition 7.9, let τ be an **L**-unifier of φ. Hence, $\tau(\varphi) \in \mathbf{L}$. Since σ is an **L**-projective substitution for φ, for all $x \in \mathtt{var}(\varphi)$, $\varphi \vdash_{\mathbf{L}} x \leftrightarrow \sigma(x)$. Since **L** is closed for uniform substitution, it follows that for all $x \in \mathtt{var}(\varphi)$, $\tau(\varphi) \vdash_{\mathbf{L}} \tau(x) \leftrightarrow \tau(\sigma(x))$. Since $\tau(\varphi) \in \mathbf{L}$, for all $x \in \mathtt{var}(\varphi)$, $\tau(x) \leftrightarrow \tau(\sigma(x)) \in \mathbf{L}$. Thus, $\tau \preccurlyeq_{\mathbf{L}} \sigma$.

Proposition 7.8 *Let $\varphi \in$ **FOR** and σ be an **L**-projective substitution for φ. For all **L**-projective substitutions τ for φ, $\sigma \circ \tau$ is **L**-projective for φ.*

Proposition 7.9 *Let $\varphi \in$ **FOR** and σ be an **L**-projective substitution for φ. For all **L**-unifiers τ of φ, $\tau \preccurlyeq_{\mathbf{L}} \sigma$.*

For all **L**-unifiable $\varphi \in$ **FOR**, we shall say that φ is **L**-*projective* if there exists an **L**-projective **L**-unifier of φ.

Example If $\Box x \lor \Box\neg x$ were **S4**-projective then there would exist an **S4**-unifier σ of $\Box x \lor \Box\neg x$ such that $\Box x \lor \Box\neg x \rightarrow (x \leftrightarrow \sigma(x)) \in$ **S4**. Hence, $\Box\sigma(x) \lor \Box\neg\sigma(x) \in$ **S4**. As we already know, this would imply that either $\sigma(x) \in$ **S4**, or $\neg\sigma(x) \in$ **S4**. In the former case, together with the fact that $\Box x \lor \Box\neg x \rightarrow (x \leftrightarrow \sigma(x)) \in$ **S4**, this would imply that $\Box\neg x \rightarrow x \in$ **S4**: a contradiction. In the latter case, together with the fact that $\Box x \lor \Box\neg x \rightarrow (x \leftrightarrow \sigma(x)) \in$ **S4**, this would imply that $\Box x \rightarrow \neg x \in$ **S4**: a contradiction. For this reason, although $\Box x \lor \Box\neg x$ is **S4**-unifiable, it is not **S4**-projective.

Example Remind the **S5**-unifier ϵ of $\Box x \lor \Box\neg x$ defined above. Obviously, $\Box x \lor \Box\neg x \rightarrow (x \leftrightarrow \epsilon(x)) \in$ **S5**. For this reason, $\Box x \lor \Box\neg x$ is **S5**-projective.

The significance of projectivity stems from the following result.[12]

Proposition 7.10 *Let $\varphi \in$ **FOR** be **L**-unifiable. If φ is **L**-projective then φ is of type* 1.

We shall say that **L** has *projective unification* if for all **L**-unifiable $\varphi \in$ **FOR**, φ is **L**-projective. From this definition, it follows that

Proposition 7.11 *If* **L** *has projective unification then* **L** *is of type* 1.

The predicates "has projective unification" and "is of type 1" are not equivalent, seeing that **S**4.2 is of type 1 (Ghilardi & Sacchetti, 2004) whereas the **S**4.2-unifiable formula $\Box(\Box x \rightarrow \Box y) \rightarrow \Box x \lor \Box z$ is not **S**4.2-projective (Dzik, 2007, Chap. 5). When **L** contains **S4**, a simple condition is equivalent to the predicate "has projective unification".

Proposition 7.12 *If* **L** *contains* **S**4 *then the following conditions are equivalent:*

1. **L** *has projective unification,*
2. **L** *contains* **S**4.3.

The proof of Proposition 7.12 has been discovered by Dzik and Wojtylak (2012). It consists in 3 steps. Suppose **L** contains **S4**. Firstly, using standard properties of **S**4.3, Dzik and Wojtylak have proved that if the unifiable formula $\Box(\Box x \rightarrow \Box y) \lor \Box(\Box y \rightarrow \Box x)$ is **L**-projective then **L** contains **S**4.3.[13] Secondly, using several kinds of conjunctive normal forms of formulas in **S**4.3, Dzik and Wojtylak have proved that

[12] Proposition 7.10 is an immediate consequence of Proposition 7.9.

[13] The easy proof of this first step shows that $1 \Rightarrow 2$.

S4.3 has projective unification.[14] Thirdly, considering that **L** and **S4.3** contain the same variable-free formulas, Dzik and Wojtylak have proved that if **L** contains **S4.3** then **L** has projective unification.[15] As a result of Proposition 7.12, **S4.3** has projective unification and by Proposition 7.11, we obtain that **S4.3** is unitary. Surprisingly, in Proposition 7.12, if we replace **S4** by **K4** and **S4.3** by **K4.3** then the result stated there does not hold any more. The truth is that the **K4.3**-unifiable formula $\Box x$ is not **K4.3**-projective.[16] Nevertheless, when **L** contains **K4**, a simple condition is equivalent to the predicate "has projective unification".

Proposition 7.13 *If* **L** *contains* **K4** *then the following conditions are equivalent:*

1. **L** *has projective unification,*
2. **L** *contains* **K4D1**.

The proof of Proposition 7.13 has been discovered by Kost (2018). It bears a resemblance to the proof of Proposition 7.12. As a result of Proposition 7.13, **K4D1** has projective unification (Kost, 2018) and by Proposition 7.11, we obtain that **K4D1** is unitary.

Projective substitutions have been introduced by Ghilardi (1999, 2000) in Intuitionistic Logic and in **K4**. In Intuitionistic Logic (and also in some other intermediate logics), Ghilardi's proof that every unifiable formula is either unitary, or finitary can be broken up into 2 steps. Firstly, Ghilardi has proved that a given formula φ is **IL**-projective if and only if a suitable composition of simple projective substitutions obtained from φ is a unifier of φ.[17] Secondly, Ghilardi has proved that every unifier in Intuitionistic Logic of a unifiable formula φ is also a unifier of a projective formula ψ such that $\psi \to \varphi \in$ **IL**, $\texttt{var}(\psi) \subseteq \texttt{var}(\varphi)$ and the $\to$-complexity of ψ is not greater than the $\to$-complexity of φ.[18] In **K4** (and also in some other transitive modal logics), Ghilardi's proof that every unifiable formula is either unitary, or finitary can be broken up into similar steps. Notice that both proofs use in their first step a semantic property called the extension property.[19] In Iemhoff (2016), Iemhoff has

---

[14] Necessitating not less than 16 pages in Dzik and Wojtylak (2012), the proof of this second step is a real *tour de force*.

[15] Indeed, the proof of this step is easy. Suppose **L** contains **S4.3** and take an **L**-unifiable formula φ. Let σ be a variable-free **L**-unifier of φ. Since **L** and **S4.3** contain the same variable-free formulas, σ is a variable-free **S4.3**-unifier of φ. Since **S4.3** is projective, φ has a projective unifier ϵ in **S4.3**. Since **L** contains **S4.3**, ϵ is a projective unifier of φ in **L** too.

[16] Indeed, if $\Box x$ were **K4.3**-projective then there would exist a **K4.3**-projective **K4.3**-unifier σ of $\Box x$. Hence, $\Box\sigma(x) \in$ **K4.3**. Moreover, $\Box x \to (x \leftrightarrow \sigma(x)) \in$ **K4.3**. Thus, $\Box\Box x \to (\Box x \leftrightarrow \Box\sigma(x)) \in$ **K4.3**. Since $\Box\sigma(x) \in$ **K4.3**, $\Box\Box x \to \Box x \in$ **K4.3**: a contradiction.

[17] This first step shows that projectivity in Intuitionistic Logic of a given formula can be effectively checked.

[18] As is well-known, there are finitely many non-**IL**-equivalent formulas ψ such that $\texttt{var}(\psi) \subseteq \texttt{var}(\varphi)$ and the $\to$-complexity of ψ is not greater than the $\to$-complexity of φ. As a result, this second step shows that every **IL**-unifiable formula is either unitary, or finitary.

[19] See Ghilardi (1999, p. 866) and Ghilardi (2000, p. 189). Roughly speaking, a formula φ has the *extension property* if any finite disjoint union of finite rooted Kripke models validating φ can be extended to a finite rooted Kripke model validating φ by adding a new root. Indeed, the extension

proved that in **S4** (and also in some other transitive modal logics), every unifiable formula is either unitary, or finitary by using a proof-theoretic analysis of unification in transitive modal logics.

7.4.2 Transparent Unification

For all **L**-unifiable $\varphi \in \mathbf{FOR}$, a substitution $(\mathtt{var}(\varphi), \mathtt{var}(\varphi), \sigma)$ is **L**-*transparent* for φ if for all **L**-unifiers τ of φ and for all $x \in \mathtt{var}(\varphi)$, $\tau(x) \leftrightarrow \tau(\sigma(x)) \in \mathbf{L}$.[20] The predicate "is **L**-transparent for" has properties that deserve to be noticed.[21]

Proposition 7.14 *Let $\varphi \in \mathbf{FOR}$ and σ be an* **L***-transparent substitution for φ. For all* **L***-unifiers τ of φ and for all $\psi \in \mathbf{FOR}_{\mathtt{var}(\varphi)}$, $\tau(\psi) \leftrightarrow \tau(\sigma(\psi)) \in \mathbf{L}$.*

Proposition 7.15 *Let $\varphi \in \mathbf{FOR}$ and σ be an* **L***-transparent substitution for φ. For all* **L***-transparent substitutions τ for φ, $\sigma \circ \tau$ is* **L***-transparent for φ.*

Proposition 7.16 *Let $\varphi \in \mathbf{FOR}$ and σ be an* **L***-transparent substitution for φ. For all* **L***-unifiers τ of φ, $\tau \preccurlyeq_{\mathbf{L}} \sigma$.*

For all **L**-unifiable $\varphi \in \mathbf{FOR}$, we shall say that φ is **L**-*transparent* if there exists an **L**-transparent **L**-unifier of φ.

Example If $\Box x \vee \Box\neg x$ were **S4**-transparent then there would exist an **S4**-unifier σ of $\Box x \vee \Box\neg x$ such that for all **S4**-unifiers τ of $\Box x \vee \Box\neg x$, $\tau(x) \leftrightarrow \tau(\sigma(x)) \in$ **S4**. Hence, $\Box\sigma(x) \vee \Box\neg\sigma(x) \in$ **S4**. As we already know, this would imply that either $\sigma(x) \in$ **S4**, or $\neg\sigma(x) \in$ **S4**. In the former case, together with the fact that for all **S4**-unifiers τ of $\Box x \vee \Box\neg x$, $\tau(x) \leftrightarrow \tau(\sigma(x)) \in$ **S4**, using the **S4**-unifier $\sigma_\bot$ of $\Box x \vee \Box\neg x$ defined above, this would imply that $\bot \leftrightarrow \top \in$ **S4**: a contradiction. In the latter case, together with the fact that for all **S4**-unifiers τ of $\Box x \vee \Box\neg x$,

---

property has been introduced in Intuitionistic Logic by de Jongh and Visser (1996) in their study of the embeddings of Heyting algebras. Among the important results in Ghilardi (1999, 2000) are the proofs that in some intermediate logics and in some transitive modal logics, every extendible formula is projective. See also Bezhanishvili and De Jongh (2012) where is proposed an alternative proof in Intuitionistic Logic that every extendible formula containing at most 2 variables is projective.

[20] Transparent substitutions have been introduced by Wroński (1995, 2008) and used by Dzik (2006, 2011). The term "transparent" comes from the fact that the membership in **L** of the equivalence $\tau(x) \leftrightarrow \tau(\sigma(x))$ for each $x \in \mathtt{var}(\varphi)$ suggests that τ "cannot see" σ.

[21] Concerning the proof of Proposition 7.14, it can be done by induction on ψ. As for the proof of Proposition 7.15, let τ be an **L**-transparent substitution for φ. Since σ is an **L**-transparent substitution for φ, for all **L**-unifiers υ of φ and for all $x \in \mathtt{var}(\varphi)$, $\upsilon(x) \leftrightarrow \upsilon(\sigma(x)) \in \mathbf{L}$. Since τ is an **L**-transparent substitution for φ, by Proposition 7.14, for all **L**-unifiers υ of φ and for all $x \in \mathtt{var}(\varphi)$, $\upsilon(\sigma(x)) \leftrightarrow \upsilon(\tau(\sigma(x))) \in \mathbf{L}$. Since for all **L**-unifiers υ of φ and for all $x \in \mathtt{var}(\varphi)$, $\upsilon(x) \leftrightarrow \upsilon(\sigma(x)) \in \mathbf{L}$, for all **L**-unifiers υ of φ and for all $x \in \mathtt{var}(\varphi)$, $\upsilon(x) \leftrightarrow \upsilon(\tau(\sigma(x))) \in \mathbf{L}$. Hence, $\sigma \circ \tau$ is **L**-transparent for φ. About the proof of Proposition 7.16, let τ be an **L**-unifier of φ. Since σ is an **L**-transparent substitution for φ, for all $x \in \mathtt{var}(\varphi)$, $\tau(x) \leftrightarrow \tau(\sigma(x)) \in \mathbf{L}$. Hence, $\tau \preccurlyeq_{\mathbf{L}} \sigma$.

$\tau(x) \leftrightarrow \tau(\sigma(x)) \in$ **S4**, using the **S4**-unifier $\sigma_\top$ of $\Box x \vee \Box\neg x$ defined above, this would imply that $\top \leftrightarrow \bot \in$ **S4**: a contradiction. For this reason, although $\Box x \vee \Box\neg x$ is **S4**-unifiable, it is not **S4**-transparent.

Example Remind the **S5**-unifier ϵ of $\Box x \vee \Box\neg x$ defined above. Obviously, for all **S5**-unifiers τ of $\Box x \vee \Box\neg x$, $\tau(x) \leftrightarrow \tau(\epsilon(x)) \in$ **S5**. For this reason, $\Box x \vee \Box\neg x$ is **S5**-transparent.

The significance of transparency stems from the following result.[22]

Proposition 7.17 *Let* $\varphi \in$ **FOR** *be* **L**-*unifiable. If* φ *is* **L**-*transparent then* φ *is of type* 1.

We shall say that **L** has *transparent unification* if for all **L**-unifiable $\varphi \in$ **FOR**, φ is **L**-transparent. From this definition, it follows that

Proposition 7.18 *If* **L** *has transparent unification then* **L** *is of type* 1.

Obviously, for all **L**-unifiable $\varphi \in$ **FOR**, if a substitution is **L**-projective for φ then it is **L**-transparent for φ.[23] No transparent substitution is known to be non-projective (Dzik & Wojtylak, 2012).

Transparent substitutions have been introduced by Wroński (1995, 2008) while he was showing that the intermediate logics having transparent unification are exactly those intermediate logics containing the linear logic of Dummett and Gödel. They have been also used by Dzik (2006, 2011) within the context of modal logics.

7.4.3 Directed Unification

For all **L**-unifiable $\varphi \in$ **FOR**, we shall say that φ is **L**-*directed* if for all **L**-unifiers σ, τ of φ, there exists an **L**-unifier υ of φ such that $\sigma \preccurlyeq_{\mathbf{L}} \upsilon$ and $\tau \preccurlyeq_{\mathbf{L}} \upsilon$.[24]

Example If $\Box x \vee \Box\neg x$ were **S4**-directed then there would exist an **S4**-unifier υ of $\Box x \vee \Box\neg x$ such that $\sigma_\bot \preccurlyeq_{\mathbf{S4}} \upsilon$ and $\sigma_\top \preccurlyeq_{\mathbf{S4}} \upsilon$. Hence, $\Box\upsilon(x) \vee \Box\neg\upsilon(x) \in$ **S4**. As we already know, this would imply that either $\upsilon(x) \in$ **S4**, or $\neg\upsilon(x) \in$ **S4**. In the former case, together with the fact that $\sigma_\bot \preccurlyeq_{\mathbf{S4}} \upsilon$, this would imply that $\bot \leftrightarrow \top \in$ **S4**: a contradiction. In the latter case, together with the fact that $\sigma_\top \preccurlyeq_{\mathbf{S4}} \upsilon$, this would imply that $\top \leftrightarrow \bot \in$ **S4**: a contradiction. For this reason, although $\Box x \vee \Box\neg x$ is **S4**-unifiable, it is not **S4**-directed.

---

[22] Proposition 7.17 is an immediate consequence of Proposition 7.16.

[23] Indeed, let the substitution σ be **L**-projective for φ. Hence, for all $x \in \mathtt{var}(\varphi)$, $\varphi \vdash_{\mathbf{L}} x \leftrightarrow \sigma(x)$. Let τ be an **L**-unifier of φ and $y \in \mathtt{var}(\varphi)$. We have to prove that $\tau(y) \leftrightarrow \tau(\sigma(y)) \in$ **L**. Since τ is an **L**-unifier of φ, $\tau(\varphi) \in$ **L**. Since for all $x \in \mathtt{var}(\varphi)$, $\varphi \vdash_{\mathbf{L}} x \leftrightarrow \sigma(x)$ and $y \in \mathtt{var}(\varphi)$, $\varphi \vdash_{\mathbf{L}} y \leftrightarrow \sigma(y)$. Since **L** is closed for uniform substitution, $\tau(\varphi) \vdash_{\mathbf{L}} \tau(y) \leftrightarrow \tau(\sigma(y))$. Since $\tau(\varphi) \in$ **L**, $\tau(y) \leftrightarrow \tau(\sigma(y)) \in$ **L**.

[24] Directed formulas have been introduced by Jeřábek (2013) whereas Ghilardi and Sacchetti (2004) are talking about *filtering formulas*. The term "directed" comes from "directed preorders" in mathematics.

Example Consider arbitrary **S5**-unifiers $(\{x\}, X, \sigma)$ and $(\{x\}, Y, \tau)$ of the **S5**-unifiable formula $\Box x \vee \Box\neg x$ defined above. Hence, $\Box\sigma(x) \vee \Box\neg\sigma(x) \in \mathbf{S5}$ and $\Box\tau(x) \vee \Box\neg\tau(x) \in \mathbf{S5}$. Let $(\{x\}, X \cup Y \cup \{z\}, \upsilon)$ be the substitution defined by $\upsilon(x) = (\Box z \wedge \sigma(x)) \vee (\Diamond\neg z \wedge \tau(x))$ where z is a new variable, i.e. neither $z = x$, nor $z \in X \cup Y$. Obviously, $\sigma \preccurlyeq_{\mathbf{S5}} \upsilon$ and $\tau \preccurlyeq_{\mathbf{S5}} \upsilon$. Moreover, by induction on $\psi \in \mathbf{FOR}_{\{x\}}$, it is easy to show that $\Box z \rightarrow (\upsilon(\psi) \leftrightarrow \sigma(\psi)) \in \mathbf{S5}$ and $\Diamond\neg z \rightarrow (\upsilon(\psi) \leftrightarrow \tau(\psi)) \in \mathbf{S5}$. Since $\Box\sigma(x) \vee \Box\neg\sigma(x) \in \mathbf{S5}$ and $\Box\tau(x) \vee \Box\neg\tau(x) \in \mathbf{S5}$, $\Box\upsilon(x) \vee \Box\neg\upsilon(x) \in \mathbf{S5}$. Thus, υ is an **S5**-unifier of $\Box x \vee \Box\neg x$. Since σ and τ were arbitrary, $\Box x \vee \Box\neg x$ is **S5**-directed.

The significance of directed unification stems from the following result.

Proposition 7.19 *Let $\varphi \in$ **FOR** be **L**-unifiable. If φ is **L**-directed then φ is either of type* 1, *or of type* 0.

Proof Suppose φ is **L**-directed. For the sake of the contradiction, suppose φ is neither of type 1, nor of type 0. Hence, φ is either of type ω, or of type ∞. Thus, there exists a basis Σ for φ either with finite cardinality ≥ 2, or with infinite cardinality. Consequently, there exists $\sigma, \tau \in \Sigma$ such that $\sigma \neq \tau$. Since φ is **L**-directed, there exists an **L**-unifier υ of φ such that $\sigma \preccurlyeq_{\mathbf{L}} \upsilon$ and $\tau \preccurlyeq_{\mathbf{L}} \upsilon$. Since Σ is a basis for φ, there exists $\lambda \in \Sigma$ such that $\upsilon \preccurlyeq_{\mathbf{L}} \lambda$. Since $\sigma \preccurlyeq_{\mathbf{L}} \upsilon$ and $\tau \preccurlyeq_{\mathbf{L}} \upsilon$, we obtain $\sigma \preccurlyeq_{\mathbf{L}} \lambda$ and $\tau \preccurlyeq_{\mathbf{L}} \lambda$. Since Σ is a basis, $\sigma = \lambda$ and $\tau = \lambda$. Hence, $\sigma = \tau$: a contradiction. $\dashv$

We shall say that **L** has *directed unification* if for all **L**-unifiable $\varphi \in$ **FOR**, φ is **L**-directed. From this definition, it follows that

Proposition 7.20 *If* **L** *has directed unification then* **L** *is either of type* 1, *or of type* 0.

Obviously, if **L** is of type 1 then **L** has directed unification. Nevertheless, the predicates "has directed unification" and "is either of type 1, or of type 0" are not equivalent, seeing that **K** is of type 0 (Jeřábek, 2015) whereas the **K**-unifiable formula $\Box x \vee \Box\neg x$ is not **K**-directed[25]. When **L** contains **K**4, a simple condition is equivalent to the predicate "has directed unification".

Proposition 7.21 *If* **L** *contains* **K**4 *then the following conditions are equivalent:*

1. **L** *has directed unification,*
2. **L** *contains* **K**4.2^{+}.

[25] Indeed, if $\Box x \vee \Box\neg x$ were **K**-directed then there would exist a **K**-unifier υ of $\Box x \vee \Box\neg x$ such that $\sigma_{\bot} \preccurlyeq_{\mathbf{K}} \upsilon$ and $\sigma_{\top} \preccurlyeq_{\mathbf{K}} \upsilon$, $\sigma_{\bot}$ and $\sigma_{\top}$ being the substitutions defined above—these substitutions are obvious **K**-unifiers of $\Box x \vee \Box\neg x$. Hence, $\Box\upsilon(x) \vee \Box\neg\upsilon(x) \in \mathbf{K}$. Since **K** has the modal disjunction property (Chagrov, 1997, Chap. 3), this would imply that either $\upsilon(x) \in \mathbf{K}$, or $\neg\upsilon(x) \in \mathbf{K}$. In the former case, together with the fact that $\sigma_{\bot} \preccurlyeq_{\mathbf{K}} \upsilon$, this would imply that $\bot \leftrightarrow \top \in \mathbf{K}$: a contradiction. In the latter case, together with the fact that $\sigma_{\top} \preccurlyeq_{\mathbf{K}} \upsilon$, this would imply that $\top \leftrightarrow \bot \in \mathbf{K}$: a contradiction. For this reason, although $\Box x \vee \Box\neg x$ is **K**-unifiable, it is not **K**-directed.

The proof of Proposition 7.21 has been firstly discovered by Ghilardi and Sacchetti (2004). It is based on a sophisticated argument involving universal algebra and category theory. Later, using standard properties of **K**4, Jeřábek (2013) has proved that the unifiable formula $(x \wedge \Box x) \vee (\neg x \wedge \Box \neg x)$ is **L**-directed if and only if **L** contains **K**4.2$^+$. As a result of Proposition 7.21, **K**4.2$^+$ has directed unification (Ghilardi & Sacchetti, 2004; Jeřábek, 2013) and by Proposition 7.20, we obtain that **K**4.2$^+$ is either unitary, or finitary. In actual fact, as proved by Ghilardi and Sacchetti (2004), unification is unitary in consistent extensions of **K**4.2$^+$ having the finite model property and the 2-glueing property.[26]

Example Obviously, **K**45 contains **K**4.2$^+$. Hence, by Proposition 7.21, if **L** contains **K**45 then **L** has directed unification.

Example Obviously, **K**5 does not contain **K**4. Hence, Proposition 7.21 cannot be used to conclude, as proved by Alizadeh et al. (2023), that if **L** contains **K**5 then **L** has directed unification. Indeed, suppose **L** contains **K**5. Let $\varphi \in \mathbf{FOR}$ be **L**-unifiable and $(\mathtt{var}(\varphi), X, \sigma)$, $(\mathtt{var}(\varphi), Y, \tau)$ be arbitrary **L**-unifiers of φ. Thus, $\sigma(\varphi) \in \mathbf{L}$ and $\tau(\varphi) \in \mathbf{L}$. We have to prove that there exists an **L**-unifier υ of φ such that $\sigma \preccurlyeq_{\mathbf{L}} \upsilon$ and $\tau \preccurlyeq_{\mathbf{L}} \upsilon$. Let $(\mathtt{var}(\varphi), X \cup Y \cup \{z\}, \upsilon)$ be the substitution defined for all $x \subset \mathtt{var}(\varphi)$, by $\upsilon(x) = ((\Box\Box z \wedge (z \vee \Diamond\top)) \wedge \sigma(x)) \vee ((\Diamond\Diamond\neg z \vee (\neg z \wedge \Box\bot)) \wedge \tau(x))$ where z is a new variable, i.e. neither $z \in \mathtt{var}(\varphi)$, nor $z \in X \cup Y$. Obviously, $\sigma \preccurlyeq_{\mathbf{L}} \upsilon$ and $\tau \preccurlyeq_{\mathbf{L}} \upsilon$. Moreover, by induction on $\psi \in \mathbf{FOR}_{\mathtt{var}(\varphi)}$, it is easy to show that $(\Box\Box z \wedge (z \vee \Diamond\top)) \rightarrow (\upsilon(\psi) \leftrightarrow \sigma(\psi)) \in \mathbf{L}$ and $(\Diamond\Diamond\neg z \vee (\neg z \wedge \Box\bot)) \rightarrow (\upsilon(\psi) \leftrightarrow \tau(\psi)) \in \mathbf{L}$. Since $\sigma(\varphi) \in \mathbf{L}$ and $\tau(\varphi) \in \mathbf{L}$, $\upsilon(\varphi) \in \mathbf{L}$. Consequently, υ is an **L**-unifier of φ. Since σ and τ were arbitrary, φ is **L**-directed. Since φ was arbitrary, **L** has directed unification.

7.4.4 Reasonable Unification

For all **L**-unifiable $\varphi \in \mathbf{FOR}$, we shall say that φ is **L**-*reasonable* if there exists $k \geq 0$ such that for all **L**-unifiers σ of φ, there exists an **L**-unifier $(\mathtt{var}(\varphi), Y, \tau)$ of φ such that $\sigma \preccurlyeq_{\mathbf{L}} \tau$ and $\parallel Y \parallel \leq k$.[27]

Example Remind the **S4**-unifiers $\sigma_\bot$ and $\sigma_\top$ of $\Box x \vee \Box\neg x$ defined above. Since they constitute a basis for $\Box x \vee \Box\neg x$ in **S4**, for all **L**-unifiers σ of φ, either $\sigma \preccurlyeq_{\mathbf{L}} \sigma_\top$, or $\sigma \preccurlyeq_{\mathbf{L}} \sigma_\bot$. Hence, $\Box x \vee \Box\neg x$ is **S4**-reasonable.

---

[26] See Ghilardi and Sacchetti, 2004, p. 897. Roughly speaking, an extension of **K**4.2$^+$ has the 2-glueing property if the class of its finite frames is closed under the operation of taking disjoint union, adding a new root and identifying final clusters.

[27] Reasonable formulas have been introduced by Balbiani et al. (2022, 2023) within the context of the modal logics $\mathbf{Alt}_1 \oplus \Box^d\bot$ and $\mathbf{K} \oplus \Box^d\bot$ for each $d \geq 2$. The term "reasonable" comes from the fact that **L**-reasonable formulas do not require **L**-unifiers with arbitrary large codomains.

Example Remind the **S5**-unifier ϵ of $\Box x \vee \Box\neg x$ defined above. Since it constitutes a basis for $\Box x \vee \Box\neg x$ in **S5**, for all **L**-unifiers σ of φ, $\sigma \preccurlyeq_{\mathbf{L}} \epsilon$. Hence, $\Box x \vee \Box\neg x$ is **S5**-reasonable.

The significance of reasonable unification stems from the following result.

Proposition 7.22 *Let* $\varphi \in$ **FOR** *be* **L**-*unifiable. If* **L** *is locally tabular and* φ *is* **L**-*reasonable then* φ *is either of type* 1, *or of type* ω.

Proof Suppose **L** is locally tabular and φ is **L**-reasonable. Hence, there exists $k \geq 0$ such that for all **L**-unifiers σ of φ, there exists an **L**-unifier $(\mathtt{var}(\varphi), Y, \tau)$ of φ such that $\sigma \preccurlyeq_{\mathbf{L}} \tau$ and $\parallel Y \parallel \leq k$. Let Σ be the set of all **L**-unifiers $(\mathtt{var}(\varphi), Z, \upsilon)$ of φ such that $\parallel Z \parallel \leq k$. Since for all **L**-unifiers σ of φ, there exists an **L**-unifier $(\mathtt{var}(\varphi), Y, \tau)$ of φ such that $\sigma \preccurlyeq_{\mathbf{L}} \tau$ and $\parallel Y \parallel \leq k$, Σ is complete. Let Σ' be the set of substitutions obtained from Σ by keeping only one representative of each equivalence class modulo $\simeq_{\mathbf{L}}$. Since Σ is complete, therefore Σ' is complete. Moreover, since **L** is locally tabular, therefore Σ' is finite. Hence, φ is either of type 1, or of type ω. $\dashv$

We shall say that **L** has *reasonable unification* if for all **L**-unifiable $\varphi \in$ **FOR**, φ is **L**-reasonable. From this definition, it follows that

Proposition 7.23 *If* **L** *is locally tabular and has reasonable unification then* **L** *is either of type* 1, *or of type* ω.

Obviously, if **L** is either of type 1, or of type ω then **L** has reasonable unification. Nevertheless, the predicates "is locally tabular and has reasonable unification" and "is either of type 1, or of type ω" are not equivalent, seeing that **K4** is of type ω (Ghilardi, 2000) whereas **K4** is not locally tabular (Chagrov, 1997, Theorem 12.21).

7.5 Some Unitary and Finitary Modal Logics

7.5.1 Preamble

By Propositions 7.10, 7.17, 7.19 and 7.22, the following **L**-unifiable formulas are either of type 1, or of type ω: **L**-projective **L**-unifiable formulas, **L**-transparent **L**-unifiable formulas and—when **L** is locally tabular—**L**-reasonable **L**-unifiable formulas. Let us notice that for all **L**-unifiable $\varphi \in$ **FOR**, if φ is Boolean then φ is **L**-projective. This follows from the proof of Proposition 7.3, noticing that the substitution ϵ considered there is **L**-projective. Hence, by Proposition 7.10, for all **L**-unifiable $\varphi \in$ **FOR**, if φ is Boolean then φ is of type 1. Thus, rephrasing the last sentence in Sect. 7.3, the unification type of **L** is the greatest type among the types of its unifiable formulas of degree 1. Let us see in which case this unification type is either 1, or ω.

7.5.2 *Extensions of* S5

Suppose **L** contains **S5**. Let $\varphi \in \mathbf{FOR}$ be **L**-unifiable and σ be a variable-free **L**-unifier of φ. Hence, $\sigma(\varphi) \in \mathbf{L}$. Let $(\mathtt{var}(\varphi), \mathtt{var}(\varphi), \epsilon)$ be the substitution such that

- for all $x \in \mathtt{var}(\varphi)$, $\epsilon(x) = (\Box\varphi \wedge x) \vee (\Diamond\neg\varphi \wedge \sigma(x))$.

Since **L** contains **S5** and σ is variable-free, for all $x \in \mathtt{var}(\varphi)$, either $\sigma(x) \equiv_{\mathbf{L}} \top$, or $\sigma(x) \equiv_{\mathbf{L}} \bot$. Hence, for all $x \in \mathtt{var}(\varphi)$, either $\epsilon(x) \equiv_{\mathbf{L}} \Box\varphi \to x$ (when $\sigma(x) \equiv_{\mathbf{L}} \top$), or $\epsilon(x) \equiv_{\mathbf{L}} \Box\varphi \wedge x$ (when $\sigma(x) \equiv_{\mathbf{L}} \bot$). For this reason, ϵ is a variant of the so-called *Löwenheim substitution* used in **BL** (Baader & Ghilardi, 2011; Dzik, 2003). The following fact can be easily proved:

- for all $x \in \mathtt{var}(\varphi)$, $\varphi \vdash_{\mathbf{L}} x \leftrightarrow \epsilon(x)$.

Thus, ϵ is **L**-projective for φ. Moreover, by induction on $\psi \in \mathbf{FOR}_{\mathtt{var}(\varphi)}$, it is easy to show that $\Box\varphi \to (\psi \leftrightarrow \epsilon(\psi)) \in \mathbf{L}$ and $\Diamond\neg\varphi \to (\sigma(\psi) \leftrightarrow \epsilon(\psi)) \in \mathbf{L}$. Since $\sigma(\varphi) \in \mathbf{L}$, $\epsilon(\varphi) \in \mathbf{L}$. All in all, we have proved the classical result saying that if **L** contains **S5** then for all **L**-unifiable $\varphi \in \mathbf{FOR}$, φ is **L**-projective. Consequently, as is well-known, if **L** contains **S5** then **L** has projective unification and by Proposition 7.10, **L** is of type 1.

The consistent modal logics containing **S5** are completely known. Apart from **S5** itself, there are those modal logics of the form $\mathbf{S5} \oplus \mathbf{Alt}_n$ for some $n \geq 1$. This is a rather restricted family. We have seen in Sect. 7.4, an argument showing that the consistent extensions of **K5** are directed. Hence, by Proposition 7.20, they are either of type 1, or of type 0. The consistent extensions of **K5** possess numerous interesting properties: they have an **NP**-complete consistency problem, they are locally tabular, etc. (Halpern & Rêgo, 2007; Nagle, 1981; Nagle & Thomason, 1985). Hence, it would rather be a surprise if one of them is nullary. The problem is that the argument showing that the substitution ϵ defined above is a projective unifier of φ cannot be repeated for extensions of **K5** that do not contain **S5**[28]; in actual fact, it cannot be repeated even for **KD45**. Nevertheless, the truth is that, like **S5** itself, multifarious consistent non-transitive extensions of **K5** have projective unification (Alizadeh et al., 2023). In Sect. 7.5.3, we will show that as is well-known, each consistent extension of **K45** has projective unification.

7.5.3 *Extensions of* K45

We show that each consistent extension of **K45** has projective unification. Our first proof shows that this result is an immediate consequence of Proposition 7.13.

[28] That part of the argument showing by induction on $\psi \in \mathbf{FOR}_{\mathtt{var}(\varphi)}$ that $\Box\varphi \to (\psi \leftrightarrow \epsilon(\psi)) \in \mathbf{L}$ and $\Diamond\neg\varphi \to (\sigma(\psi) \leftrightarrow \epsilon(\psi)) \in \mathbf{L}$ uses the fact that $\Box\varphi \to \Box\Box\varphi \in \mathbf{L}$ and $\Diamond\neg\varphi \to \Box\Diamond\neg\varphi \in \mathbf{L}$ whereas that part of the argument showing that $\epsilon(\varphi) \in \mathbf{L}$ uses the fact that $\Box\varphi \to \varphi \in \mathbf{L}$.

Our second proof is an alternative independent argument based on the properties of transitive Euclidean Kripke frames.[29]

Proposition 7.24 *If* **L** *contains* **K**45 *then* **L** *has projective unification.*

Proof 1: Suppose **L** contains **K**45. Hence, **L** contains **K**4. Moreover, as is well-known, **L** contains **K4D**1. Thus, by Proposition 7.13, **L** has projective unification. ⊣

Proof 2: Let $\varphi \in \mathbf{FOR}$ be **L**-unifiable and σ be a variable-free **L**-unifier of φ. Hence, $\sigma(\varphi) \in \mathbf{L}$. Let $(\mathtt{var}(\varphi), \emptyset, \tau)$ be an arbitrary variable-free substitution. Let $(\mathtt{var}(\varphi), \mathtt{var}(\varphi), \epsilon_\tau)$ be the substitution such that

- for all $x \in \mathtt{var}(\varphi)$, $\epsilon_\tau(x) = ((\varphi \wedge \Box\varphi) \wedge x) \vee ((\neg\varphi \vee \Diamond\neg\varphi) \wedge \tau(x))$.

The following fact can be easily proved:

- for all $x \in \mathtt{var}(\varphi)$, $\varphi \vdash_{\mathbf{L}} x \leftrightarrow \epsilon_\tau(x)$.

Thus, ϵ_τ is **L**-projective for φ. Moreover, by induction on $\psi \in \mathbf{FOR}_{\mathtt{var}(\varphi)}$, it is easy to show that $\varphi \wedge \Box\varphi \vdash_{\mathbf{L}} \psi \leftrightarrow \epsilon_\tau(\psi)$ and $\neg\varphi \vee \Diamond\neg\varphi \vdash_{\mathbf{L}} \tau(\psi) \leftrightarrow \epsilon_\tau(\psi)$. Consequently, if $\tau(\varphi) \in \mathbf{L}$ then $\Box\varphi \to (\varphi \to \epsilon_\tau(\varphi)) \in \mathbf{L}$ and $\Diamond\neg\varphi \to \epsilon_\tau(\varphi) \in \mathbf{L}$. In that case, $\varphi \to \epsilon_\tau(\varphi) \in \mathbf{L}$ and $\Box\varphi \vee \epsilon_\tau(\varphi) \in \mathbf{L}$— which implies that $\Box\epsilon_\tau(\varphi) \in \mathbf{L}$. Hence, for all variable-free substitutions $(\mathtt{var}(\varphi), \emptyset, \tau)$, if $\tau(\varphi) \in \mathbf{L}$ then $\Box\epsilon_\tau(\varphi) \in \mathbf{L}$. Since $\sigma(\varphi) \in \mathbf{L}$, $\Box\epsilon_\sigma(\varphi) \in \mathbf{L}$. Let $l \geq 1$ and $((\mathtt{var}(\varphi), \emptyset, \tau_1), \ldots, (\mathtt{var}(\varphi), \emptyset, \tau_l))$ be an enumeration of the set of all variable-free substitutions $(\mathtt{var}(\varphi), \emptyset, \tau)$ such that for all $x \in \mathtt{var}(\varphi)$, either $\tau(x) = \top$, or $\tau(x) = \bot$. Let $\epsilon = \epsilon_\sigma \circ \epsilon_{\tau_l} \circ \ldots \circ \epsilon_{\tau_1} \circ \epsilon_\sigma$. Since $\Box\epsilon_\sigma(\varphi) \in \mathbf{L}$, $\Box\epsilon(\varphi) \in \mathbf{L}$. Moreover, since $\epsilon_\sigma, \epsilon_{\tau_1}, \ldots, \epsilon_{\tau_l}$ are **L**-projective for φ, by Proposition 7.8, ϵ is **L**-projective for φ. Indeed, ϵ is also an **L**-unifier of φ. Why? For the sake of the contradiction, suppose $\epsilon(\varphi) \notin \mathbf{L}$. Since **L** contains **K**45, **L** is complete with respect to a class of transitive Euclidean Kripke frames (Halpern & Rêgo, 2007; Nagle, 1981; Nagle & Thomason, 1985). Thus, there exists a transitive Euclidean Kripke frame $\mathcal{F} = (W, R)$ such that $\mathcal{F} \models \mathbf{L}$ and $\mathcal{F} \not\models \epsilon(\varphi)$. Let V be a valuation on $\mathcal{F}$ and $s \in W$ be such that $\mathcal{F}, V, s \not\models \epsilon(\varphi)$. Since $\Box\epsilon(\varphi) \in \mathbf{L}$ and $\mathcal{F} \models \mathbf{L}$, not sRs. Consequently, we have to consider the following 2 cases.

Case "$R(s) = \emptyset$": Since $\varphi \wedge \Box\varphi \vdash_{\mathbf{L}} \varphi \leftrightarrow \epsilon(\varphi)$, $\mathcal{F} \models \mathbf{L}$, $R(s) = \emptyset$ and $\mathcal{F}, V, s \not\models \epsilon(\varphi)$, we obtain $\mathcal{F}, V, s \not\models \varphi$. Since $\neg\varphi \vee \Diamond\neg\varphi \vdash_{\mathbf{L}} \sigma(\varphi) \leftrightarrow \epsilon_\sigma(\varphi)$, $\sigma(\varphi) \in \mathbf{L}$, $\mathcal{F} \models \mathbf{L}$, $R(s) = \emptyset$ and $\mathcal{F}, V, s \not\models \varphi$, we obtain $\mathcal{F}, V, s \models \epsilon_\sigma(\varphi)$. Let V^{ϵ_σ} be a valuation on $\mathcal{F}$ such that for all $x \in \mathtt{var}(\varphi)$, $V^{\epsilon_\sigma}(x) = \{t \in W : \mathcal{F}, V, t \models \epsilon_\sigma(x)\}$. By induction on $\psi \in \mathbf{FOR}_{\mathtt{var}(\varphi)}$, it is easy to show the following property:

- for all $t \in W$, $\mathcal{F}, V, t \models \epsilon_\sigma(\psi)$ if and only if $\mathcal{F}, V^{\epsilon_\sigma}, t \models \psi$.[30]

---

[29] It has been adapted by Alizadeh et al. (2023) to the case of multifarious consistent non-transitive extensions of **K**5.

[30] This property—that we will use again in the 2nd case of our argument—can be also stated as follows.

Claim: Let $(\mathtt{var}(\varphi), \mathtt{var}(\varphi), \mu)$ be a substitution and V^μ be a valuation on $\mathcal{F}$ such that for all $x \in \mathtt{var}(\varphi)$, $V^\mu(x) = \{t \in W : \mathcal{F}, V, t \models \mu(x)\}$. For all $\psi \in \mathbf{FOR}_{\mathtt{var}(\varphi)}$ and for all $t \in W$, $\mathcal{F}, V, t \models \mu(\psi)$ if and only if $\mathcal{F}, V^\mu, t \models \psi$.

Since $\mathcal{F}, V, s \models \epsilon_\sigma(\varphi)$, $\mathcal{F}, V^{\epsilon_\sigma}, s \models \varphi$. Since $\varphi \wedge \Box\varphi \vdash_{\mathbf{L}} \varphi \leftrightarrow \epsilon_{\tau_1}(\ldots \epsilon_{\tau_l}(\epsilon_\sigma(\varphi))\ldots)$, $\mathcal{F} \models \mathbf{L}$, $R(s) = \emptyset$ and $\mathcal{F}, V^{\epsilon_\sigma}, s \models \varphi$, we obtain $\mathcal{F}, V^{\epsilon_\sigma}, s \models \epsilon_{\tau_1}(\ldots \epsilon_{\tau_l}(\epsilon_\sigma(\varphi))\ldots)$.

Hence, $\mathcal{F}, V, s \models \epsilon_\sigma(\epsilon_{\tau_1}(\ldots \epsilon_{\tau_l}(\epsilon_\sigma(\varphi))\ldots))$, i.e. $\mathcal{F}, V, s \models \epsilon(\varphi)$: a contradiction.

Case "$R(s) \neq \emptyset$": For all $i \leq l$, let $\epsilon_i = \epsilon_{\tau_i} \circ \ldots \circ \epsilon_{\tau_1} \circ \epsilon_\sigma$. Thus, for all $i \leq l$, if $i = 0$ then $\epsilon_i = \epsilon_\sigma$ else $\epsilon_i = \epsilon_{\tau_i} \circ \epsilon_{i-1}$. Since $\Box\epsilon_\sigma(\varphi) \in \mathbf{L}$ and $\mathcal{F} \models \mathbf{L}$, $\mathcal{F}, V, s \models \Box\epsilon_\sigma(\varphi)$. Consequently, $\mathcal{F}, V^{\epsilon_\sigma}, s \models \Box\varphi$. Since $R(s) \neq \emptyset$, let $t \in W$ be such that sRt. Since $\mathcal{F}, V^{\epsilon_\sigma}, s \models \Box\varphi$, $\mathcal{F}, V^{\epsilon_\sigma}, t \models \varphi \wedge \Box\varphi$. Since not sRs, let V' be a valuation on $\mathcal{F}$ such that for all $x \in \mathtt{var}(\varphi)$,

- $V'(x) \cap R(s) = V^{\epsilon_\sigma}(x) \cap R(s)$,
- $s \in V'(x)$ if and only if $t \in V^{\epsilon_\sigma}(x)$.

By induction on $\psi \in \mathbf{FOR}_{\mathtt{var}(\varphi)}$, it is easy to show that $\mathcal{F}, V', s \models \psi$ if and only if $\mathcal{F}, V^{\epsilon_\sigma}, t \models \psi$. Since $\mathcal{F}, V^{\epsilon_\sigma}, t \models \varphi \wedge \Box\varphi$, $\mathcal{F}, V', s \models \varphi \wedge \Box\varphi$. Let $j \in \{1, \ldots, l\}$ be such that for all $x \in \mathtt{var}(\varphi)$,

- if $\mathcal{F}, V', s \models x$ then $\tau_j(x) = \top$,
- if $\mathcal{F}, V', s \not\models x$ then $\tau_j(x) = \bot$.

Since $\mathcal{F}, V, s \not\models \epsilon(\varphi)$, $\mathcal{F}, V, s \not\models \epsilon_{j-1}(\epsilon_{\tau_j}(\ldots \epsilon_{\tau_l}(\epsilon_\sigma(\varphi))\ldots))$. Hence, $\mathcal{F}, V^{\epsilon_{j-1}}, s \not\models \epsilon_{\tau_j}(\ldots \epsilon_{\tau_l}(\epsilon_\sigma(\varphi))\ldots)$. Since $\varphi \wedge \Box\varphi \vdash_{\mathbf{L}} \varphi \leftrightarrow \epsilon_{\tau_j}(\ldots \epsilon_{\tau_l}(\epsilon_\sigma(\varphi))\ldots)$, $\mathcal{F} \models \mathbf{L}$ and $\mathcal{F}, V^{\epsilon_{j-1}}, s \not\models \epsilon_{\tau_j}(\ldots \epsilon_{\tau_l}(\epsilon_\sigma(\varphi))\ldots)$, $\mathcal{F}, V^{\epsilon_{j-1}}, s \not\models \varphi \wedge \Box\varphi$. By induction on $i \in \{1, \ldots, l\}$, it is easy to show that for all $\psi \in \mathbf{FOR}_{\mathtt{var}(\varphi)}$ and for all $u \in R(s)$, $\mathcal{F}, V^{\epsilon_i}, u \models \psi$ if and only if $\mathcal{F}, V^{\epsilon_\sigma}, u \models \psi$. By induction on $\psi \in \mathbf{FOR}_{\mathtt{var}(\varphi)}$, it is easy to show that $\mathcal{F}, V^{\epsilon_j}, s \models \psi$ if and only if $\mathcal{F}, V', s \models \psi$. Since $\mathcal{F}, V', s \models \varphi \wedge \Box\varphi$, $\mathcal{F}, V^{\epsilon_j}, s \models \varphi \wedge \Box\varphi$. Since $\varphi \wedge \Box\varphi \vdash_{\mathbf{L}} \varphi \leftrightarrow \epsilon_{\tau_{j+1}}(\ldots \epsilon_{\tau_l}(\epsilon_\sigma(\varphi))\ldots)$, $\mathcal{F} \models \mathbf{L}$ and $\mathcal{F}, V^{\epsilon_j}, s \models \varphi \wedge \Box\varphi$, $\mathcal{F}, V^{\epsilon_j}, s \models \epsilon_{\tau_{j+1}}(\ldots \epsilon_{\tau_l}(\epsilon_\sigma(\varphi))\ldots)$. Thus, $\mathcal{F}, V, s \models \epsilon_j(\epsilon_{\tau_{j+1}}(\ldots \epsilon_{\tau_l}(\epsilon_\sigma(\varphi))\ldots))$, i.e. $V, s \models \epsilon(\varphi)$: a contradiction.

All in all, we have proved that if **L** contains **K**45 then for all **L**-unifiable $\varphi \in \mathbf{FOR}$, φ is **L**-projective. Consequently, **L** has projective unification. $\dashv$

7.5.4 Some Non-transitive Modal Logics

Most of the unitary modal logics and the finitary modal logics considered in Sects. 7.5.2 and 7.5.3 as well as the above-mentioned unitary modal logics **S**4.3 and **K4D**1 contain **K**4. Hence, it is natural to ask whether there exists either non-transitive unitary modal logics, or non-transitive finitary modal logics. With the exception of the non-transitive extensions of **K**5 mentioned in Sect. 7.5.2,[31] most of the non-transitive

Its simple proof can be done by induction on $\psi \in \mathbf{FOR}_{\mathtt{var}(\varphi)}$. Such property is standard within the context of the problem of determining the unification types of modal logics (Babenyshev & Rybakov, 2011; Dzik, 2007; Ghilardi, 1999, 2000).

[31] Notice that although **K**5 does not contain **K**4, it contains all formulas of the form $\Box(\Box\varphi \rightarrow \Box\Box\varphi)$.

modal logics considered in this paper are nullary (see Sect. 7.6). Nevertheless, there exists non-nullary non-transitive modal logics, namely for all $d \geq 3$, the modal logics $\mathbf{Alt}_1 \oplus \Box^d\bot$ and $\mathbf{K} \oplus \Box^d\bot$. The arguments showing that for all $d \geq 3$, $\mathbf{Alt}_1 \oplus \Box^d\bot$ is unitary and $\mathbf{K} \oplus \Box^d\bot$ is finitary have been described by Balbiani et al. (2022, 2023). They mainly consist in firstly showing that for all $d \geq 3$, $\mathbf{Alt}_1 \oplus \Box^d\bot$ has directed and reasonable unification, secondly showing that for all $d \geq 3$, $\mathbf{K} \oplus \Box^d\bot$ has reasonable unification and thirdly using Propositions 7.20 and 7.23.[32]

In the arguments showing that for all $d \geq 3$, $\mathbf{Alt}_1 \oplus \Box^d\bot$ is unitary and $\mathbf{K} \oplus \Box^d\bot$ is finitary, the difficulties lie in the proofs that for all $d \geq 3$, $\mathbf{Alt}_1 \oplus \Box^d\bot$ and $\mathbf{K} \oplus \Box^d\bot$ have reasonable unification. What are the properties peculiar to these modal logics giving them the quality of having reasonable unification? By analyzing the proofs described in Balbiani et al. (2022, 2023), one will quickly understand that the following properties are essential:

- for all $d \geq 3$, for all finite $X \subseteq \mathbf{VAR}$ and for all $\varphi \in \mathbf{FOR}_X$, φ is consistent in $\mathbf{Alt}_1 \oplus \Box^d\bot$ if and only if there exists a finite chain of length at most d which nodes are labelled by subsets of X and at the root of which φ is satisfied,
- for all $d \geq 3$, for all finite $X \subseteq \mathbf{VAR}$ and for all $\varphi \in \mathbf{FOR}_X$, φ is consistent in $\mathbf{K} \oplus \Box^d\bot$ if and only if there exists a finite tree of depth at most $d - 1$ which nodes are labelled by subsets of X and at the root of which φ is satisfied.

Moreover, the finite chains as well as the finite trees mentioned in the above properties can be precisely characterized by modal formulas. But the decisive factor in the proofs described in Balbiani et al. (2022, 2023), is the fact that for all $d \geq 3$, $\mathbf{Alt}_1 \oplus \Box^d\bot$ and $\mathbf{K} \oplus \Box^d\bot$ are locally tabular. For this reason, we believe that in general, if $\mathbf{L}$ is locally tabular then $\mathbf{L}$ is either unitary, or finitary.

7.6 Some Nullary Modal Logics

In Sect. 7.5, we have seen that there are a great many unitary modal logics and that there is just as much finitary modal logics. When these modal logics have a decidable membership problem and allow bases for unifiable formulas to be effectively computed, their admissibility problem is decidable. What about the other modal logics, i.e. those not mentioned in Sect. 7.5? What is the type of $\mathbf{K}$? What is the type of $\mathbf{Alt}_1$? What are the types of $\mathbf{KD}$ and $\mathbf{DAlt}_1$? We will see that with very few exceptions, it is not often that a modal logic can be proved to be nullary.

The proof that $\mathbf{K}$ is of type 0 has been obtained by Jeřábek (2015) who has considered the $\mathbf{K}$-unifiable formula $\Box x \vee \neg x$. Its most obvious unifiers in $\mathbf{K}$ are the substitutions $\sigma_\bot$ and $\sigma_\top$ defined above. Indeed, lots of substitutions are $\mathbf{K}$-unifiers

[32] Of course, $\mathbf{Alt}_1 \oplus \Box^2\bot$ and $\mathbf{K} \oplus \Box^2\bot$ contain $\mathbf{K4}$. For all $d \geq 3$, although neither $\mathbf{Alt}_1 \oplus \Box^d\bot$, nor $\mathbf{K} \oplus \Box^d\bot$ contain $\mathbf{K4}$, they contain all formulas of the form $\Box^{d-2}(\Box\varphi \to \Box\Box\varphi)$. In any case, as shown in Balbiani et al. (2022, 2023), the proofs that $\mathbf{Alt}_1 \oplus \Box^2\bot$ is unitary and $\mathbf{K} \oplus \Box^2\bot$ is finitary are similar to the proofs that for all $d \geq 3$, $\mathbf{Alt}_1 \oplus \Box^d\bot$ is unitary and $\mathbf{K} \oplus \Box^d\bot$ is finitary.

of $\Box x \vee \neg x$, seeing that for all $k \geq 0$, the substitution $(\{x\}, \{x\}, \sigma_k)$ defined by $\sigma_k(x) = \Box^{<k} x \wedge \Box^k \bot$ is a **K**-unifier of $\Box x \vee \neg x$. After having noticed that $\sigma_\bot \simeq_{\mathbf{K}} \sigma_0$, it is natural to ask whether the set $\{\sigma_\top\} \cup \{\sigma_k : k \geq 0\}$ of **K**-unifiers of $\Box x \vee \neg x$ is complete. Indeed, it is. And the proof of this completeness is an immediate consequence of the following simple facts:

- **K** satisfies an infinitary version of the *inference rule of margins* investigated by Williamson (1994), i.e. for all $\varphi \in \mathbf{FOR}$, if $\Box\varphi \vee \neg\varphi \in \mathbf{K}$ then either $\varphi \in \mathbf{K}$, or for all $k \geq 0$, if $k \geq \mathtt{deg}(\varphi)$ then $\Box^k \bot \vee \neg\varphi \in \mathbf{K}$,
- for all substitutions $(\{x\}, X, \sigma)$, if $\sigma(x) \in \mathbf{K}$ then $\sigma \simeq_{\mathbf{K}} \sigma_\top$,
- for all **K**-unifiers σ of $\Box x \vee \neg x$ and for all $k \geq 0$, if $\Box^k \bot \vee \neg\sigma(x) \in \mathbf{K}$ then $\sigma \preccurlyeq_{\mathbf{K}} \sigma_k$.

But the set $\{\sigma_\top\} \cup \{\sigma_k : k \geq 0\}$ can also be used to prove that there exists no basis for $\Box x \vee \neg x$ in **K**. How? For the sake of the contradiction, suppose there exists a basis Σ for $\Box x \vee \neg x$ in **K**. Since σ_0 is a **K**-unifier of $\Box x \vee \neg x$, there exists $\sigma \in \Sigma$ such that $\sigma_0 \preccurlyeq_{\mathbf{K}} \sigma$. Hence, $\Box\sigma(x) \vee \neg\sigma(x)$ in **K**. Thus, either $\sigma(x) \in \mathbf{K}$, or for all $k \geq 0$, if $k \geq \mathtt{deg}(\sigma(x))$ then $\Box^k \bot \vee \neg\sigma(x) \in \mathbf{K}$. In the former case, $\sigma \simeq_{\mathbf{K}} \sigma_\top$. Since $\sigma_0 \preccurlyeq_{\mathbf{K}} \sigma$, $\sigma_0 \preccurlyeq_{\mathbf{K}} \sigma_\top$. Consequently, $\bot \leftrightarrow \top \in \mathbf{K}$: a contradiction. In the latter case, let $k = \mathtt{deg}(\sigma(x))$. Hence, $\Box^k \bot \vee \neg\sigma(x) \in \mathbf{K}$ and $\sigma \preccurlyeq_{\mathbf{K}} \sigma_k$. Since σ_{k+1} is a **K**-unifier of $\Box x \vee \neg x$, there exists $\tau \in \Sigma$ such that $\sigma_{k+1} \preccurlyeq_{\mathbf{K}} \tau$. By an easy modal reasoning, one can prove that $\sigma_k \preccurlyeq_{\mathbf{K}} \sigma_{k+1}$. Since $\sigma \preccurlyeq_{\mathbf{K}} \sigma_k$ and $\sigma_{k+1} \preccurlyeq_{\mathbf{K}} \tau$, we obtain $\sigma \preccurlyeq_{\mathbf{K}} \tau$. Since Σ is a basis for $\Box x \vee \neg x$ in **K**, $\sigma = \tau$. Since $\sigma \preccurlyeq_{\mathbf{K}} \sigma_k$ and $\sigma_{k+1} \preccurlyeq_{\mathbf{K}} \tau$, we obtain $\sigma_{k+1} \preccurlyeq_{\mathbf{K}} \sigma_k$. Thus, there exists a substitution $(\{x\}, T, \upsilon)$ such that $\Box^{<k+1} x \wedge \Box^{k+1} \bot \leftrightarrow \Box^{<k} \upsilon(x) \wedge \Box^k \bot \in \mathbf{K}$. Consequently, $\Box^{k+1} \bot \to \Box^k \bot \in$ **K**: a contradiction.

Owing to this argument, it is natural to ask the question whether it can be repeated for other modal logics. It is quite easy to prove that it can be repeated for $\mathbf{Alt}_1$ (Balbiani & Tinchev, 2016). In actual fact, it can be repeated for $\mathbf{Alt}_n$ when $n \geq 2$ as well. As a result, $\mathbf{Alt}_1$ is of type 0 and for all $n \geq 2$, $\mathbf{Alt}_n$ is of type 0 as well. Alas! Apart from the modal logics mentioned above, no other modal logic is known to be nullary. What about the modal logics **KD** and $\mathbf{DAlt}_1$? What about the modal logics $\mathbf{DAlt}_n$ when $n \geq 2$? These modal logics seem to be closely related to the modal logics **K**, $\mathbf{Alt}_1$ and $\mathbf{Alt}_n$ when $n \geq 2$. Nevertheless, they are not known to be nullary.[33] And the same can be said for **KT**, **KB**, **KDB** and **KTB**. The thing is that the property peculiar to the modal logics **K**, $\mathbf{Alt}_1$ and $\mathbf{Alt}_n$ when $n \geq 2$ giving them the quality of being nullary is the fact that for all $k \geq 0$, $\Box^{k+1} \bot \to \Box^k \bot$ does not belong to them. As soon as a modal logic contains either the formula $\Diamond\top$, or the formula $x \to \Box\Diamond x$, it ceases to satisfy this property: **KD**, $\mathbf{DAlt}_1$ and $\mathbf{DAlt}_n$ when $n \geq 2$ are of that kind, but also **KT**, **KB**, **KDB** and **KTB**.

If one puts the question of the unification types of **KD**, $\mathbf{DAlt}_1$ and $\mathbf{DAlt}_n$ when $n \geq 2$, the answer will most probably be that like **K**, $\mathbf{Alt}_1$ and $\mathbf{Alt}_n$ when $n \geq 2$, these

[33] It is only known that these modal logics are not unitary. Indeed, for example in $\mathbf{DAlt}_1$, the substitutions $\sigma_\bot$ and $\sigma_\top$ defined above constitute a basis for the unifiable formula $\Box x \vee \neg x$, seeing that for all $\varphi \in \mathbf{FOR}$, if $\Box\varphi \vee \neg\varphi \in \mathbf{DAlt}_1$ then either $\varphi \in \mathbf{DAlt}_1$, or $\neg\varphi \in \mathbf{DAlt}_1$. Hence, $\Box x \vee \neg x$ is finitary in $\mathbf{DAlt}_1$.

modal logics are of type 0. Several signs seem to indicate that these modal logics are finitary. Let us see some of these signs in the case of $\mathbf{DAlt}_1$. As we already know,

- the unification type of $\mathbf{DAlt}_1$ is the greatest type among the types of its unifiable formulas of degree 1,
- $\Box x \vee \neg x$ is finitary in $\mathbf{DAlt}_1$.

In actual fact, it can be easily proved that the unification type of $\mathbf{DAlt}_1$ is the greatest type among the types of its unifiable formulas of the form

- $\bigvee\{x_i \wedge \Box x_j : 1 \leq i, j \leq n \text{ and } i \rhd j\} \wedge \bigwedge\{\neg(x_i \wedge x_j) : 1 \leq i, j \leq n \text{ and } i \neq j\}$,

n ranging over the set of all positive integers and $\rhd$ ranging over the set of all binary relations on $\{1, \ldots, n\}$. Well, we can never be sure what tomorrow will bring and we leave the question of the unification types of **KD**, $\mathbf{DAlt}_1$ and $\mathbf{DAlt}_n$ when $n \geq 2$ and also the unification types of **KT**, **KB**, **KDB** and **KTB** as an open problem.[34]

Special Acknowledgement

Special acknowledgement is heartily granted to Dick de Jongh. His research on intermediate logics and modal logics, the reference papers that he has written on these subjects and especially his papers about universal models are exerting a profound influence on our research and a great deal of it is directly motivated and influenced by his ideas and achievements.

Acknowledgements We make a point of thanking Mojtaba Mojtahedi (Ghent University, Ghent, Belgium), Maryam Rostamigiv (Toulouse Institute of Computer Science Research, Toulouse, France) and Tinko Tinchev (Sofia University St. Kliment Ohridski, Sofia, Bulgaria) for many stimulating discussions about modal logics and the unification problem. We also make a point of thanking our anonymous referee for her feedback: her useful suggestions have been essential for improving the correctness and the readability of a preliminary version of this paper.

References

Alizadeh, M., Ardeshir, M., Balbiani, P., & Mojtahedi, M. (2023). Unification types in Euclidean modal logics. *Logic Journal of the IGPL, 31*, 422–440.

Baader, F., & Ghilardi, S. (2011). Unification in modal and description logics. *Logic Journal of the IGPL, 19*, 705–730.

[34] Notice that some of these modal logics have been proved to be nullary within the context of unification with parameters (Balbiani, 2019; Balbiani & Gencer, 2017, 2020; Rostamigiv, 2020). The difference between elementary unification—as considered in this paper—and unification with parameters is that in elementary unification, all variables are likely to be replaced by formulas when one applies a substitution whereas in unification with parameters, some variables—called parameters—remain unchanged.

Baader, F., & Snyder, W. (2001). Unification theory. In *Handbook of automated reasoning* (pp. 439–526). Elsevier.

Babenyshev, S., & Rybakov, V. (2011). Unification in linear temporal logic **LTL**. *Annals of Pure and Applied Logic, 162*, 991–1000.

Balbiani, P. (2019). Remarks about the unification type of several non-symmetric non-transitive modal logics. *Logic Journal of the IGPL, 27*, 639–658.

Balbiani, P., & Gencer, Ç. (2017). **KD** is nullary. *Journal of Applied Non-classical Logics, 27*, 196–205.

Balbiani, P., & Gencer, Ç. (2020). About the unification type of modal logics between **KB** and **KTB**. *Studia Logica, 108*, 941–966.

Balbiani, P., Gencer, Ç., Rostamigiv, M., & Tinchev, T. (2022). About the unification type of $\mathbf{K} + \Box\Box\bot$. *Annals of Mathematics and Artificial Intelligence, 90*, 481–497.

Balbiani, P., Gencer, Ç., Rostamigiv, M., & Tinchev, T. (2023). Remarks about the unification types of some locally tabular normal modal logics. *Logic Journal of the IGPL, 31*, 115–139.

Balbiani, P., & Tinchev, T. (2016). Unification in modal logic $\mathbf{Alt}_1$. In *Advances in modal logic* (pp. 117–134). College Publications.

Balbiani, P., & Tinchev, T. (2018). Elementary unification in modal logic **KD**45. *Journal of Applied Logics, 5*, 301–317.

Bezhanishvili, N., & De Jongh, D. (2012). Extendible formulas in two variables in intuitionistic logic. *Studia Logica, 100*, 61–89.

Blackburn, P., de Rijke, M., & Venema, Y. (2001). *Modal logic*. Cambridge University Press.

Chagrov, A., & Zakharyaschev, M. (1997). *Modal logic*. Oxford University Press.

Dzik, W. (2003). Unitary unification of **S5** modal logics and its extensions. *Bulletin of the Section of Logic, 32*, 19–26.

Dzik, W. (2006). Transparent unifiers in modal logics with self-conjugate operators. *Bulletin of the Section of Logic, 35*, 73–83.

Dzik, W. (2007). *Unification types in logic,* Wydawnicto Uniwersytetu Slaskiego.

Dzik, W. (2011). Remarks on projective unifiers. *Bulletin of the Section of Logic, 40*, 37–46.

Dzik, W., & Wojtylak, P. (2012). Projective unification in modal logic. *Logic Journal of the IGPL, 20*, 121–153.

Fagin, R., Halpern, J., & Vardi, M. (1992). What is an inference rule? *Journal of Symbolic Logic, 57*, 1018–1045.

French, R., & Humberstone, L. (2015). An observation concerning Porte's rule in modal logics. *Bulletin of the Section of Logic, 44*, 25–31.

Gencer, Ç., & de Jongh, D. (2009). Unifiability in extensions of **K**4. *Logic Journal of the IGPL, 17*, 159–172.

Ghilardi, S. (1997). Unification through projectivity. *Journal of Logic and Computation, 7*, 733–752.

Ghilardi, S. (1999). Unification in intuitionistic logic. *Journal of Symbolic Logic, 64*, 859–880.

Ghilardi, S. (2000). Best solving modal equations. *Annals of Pure and Applied Logic, 102*, 183–198.

Ghilardi, S., & Sacchetti, L. (2004). Filtering unification and most general unifiers in modal logic. *Journal of Symbolic Logic, 69*, 879–906.

Halpern, J., & Rêgo, L. (2007). Characterizing the **NP**-**PSPACE** gap in the satisfiability problem for modal logic. *Journal of Logic and Computation, 17*, 795–806.

Iemhoff, R. (2016). A syntactic approach to unification in transitive reflexive modal logics. *Notre Dame Journal of Formal Logic, 57*, 233–247.

Jeřábek, E. (2013) Logics with directed unification. In *Algebra and coalgebra meet proof theory*.

Jeřábek, E. (2015) Blending margins: The modal logic **K** has nullary unification type. *Journal of Logic and Computation,* **25**, 1231–1240.

de Jongh, D., & Visser, A. (1996). Embeddings of Heyting algebras. In *Logic: From foundations to applications* (187–213). Clarendon Press.

de Jongh, D., & Yang, F. (2011). Jankov's theorems for intermediate logics in the setting of universal models. In *Logic, language, and computation* (pp. 53–76). Springer.

Kost, S. (2018). Projective unification in transitive modal logics. *Logic Journal of the IGPL, 26*, 548–566.
Kracht, M. (1999). *Tools and techniques in modal logic*. Elsevier.
Makinson, D. (1971). Some embedding theorems for modal logic. *Notre Dame Journal of Formal Logic, 12*, 252–254.
Martin, U., & Nipkow, T. (1989). Boolean unification—the story so far. *Journal of Symbolic Computation, 7*, 275–293.
Nagle, M. (1981). The decidability of normal **K5** logics. *Journal of Symbolic Logic, 46*, 319–328.
Nagle, M., & Thomason, S. (1985). The extensions of the modal logic **K5**. *Journal of Symbolic Logic, 50*, 102–109.
Porte, J. (1981). The deducibilities of **S5**. *Journal of Philosophical Logic, 10*, 409–422.
Rostamigiv, M. (2020). *About the Type of Modal Logics for the Unification Problem,* Doctoral thesis of the University of Toulouse 3.
Rybakov, V. (1997). *Admissibility of logical inference rules*. Elsevier.
Rybakov, V., Terziler, M., & Gencer, Ç. (1999). An essay on unification and inference rules for modal logics. *Bulletin of the Section of Logic, 28*, 145–157.
Williamson, T. (1994). Non-genuine MacIntosh logics. *Journal of Philosophical Logic, 23*, 87–101.
Wroński, A. (1995). Transparent unification problem. *Reports on Mathematical Logic, 29*, 105–107.
Wroński, A. (2008). Transparent verifiers in intermediate logics. In: *54th conference in history of mathematics*. Cracow

Chapter 8
Proof Theory for Lax Logic

Rosalie Iemhoff

Abstract In this paper some proof theory for propositional Lax Logic is developed. A cut free terminating sequent calculus is introduced for the logic, and based on that calculus it is shown that the logic has uniform interpolation. Furthermore, a separate, simple proof of interpolation is provided that also uses the sequent calculus. From the literature it is known that Lax Logic has interpolation, but all known proofs use models rather than proof systems.

Keywords Intuitionistic modal logic · Lax Logic · Sequent calculus · Uniform interpolation

8.1 Introduction

This paper develops proof theory for an intuitionistic modal logic, Propositional Lax Logic, that models several phenomena in logic and computer science. When I was a PhD student, Dick taught me about intuitionistic and other intermediate logics and the numerous methods and techniques that exist to study them. Dick's love for this topic is something I have inherited from him. Many other topics to love and study would follow, but I am still grateful for Dick introducing me early on to this particular one.

Propositional Lax Logic (PLL) is an intuitionistic modal logic with a single unary modal operator $\bigcirc$ and is axiomatized by the principles of intuitionistic propositional logic IPC plus the following three modal principles:

Support by the Netherlands Organisation for Scientific Research under grant 639.073.807 as well as partial support by the MOSAIC project (EU H2020-MSCA-RISE-2020 Project 101007627) is gratefully acknowledged.

R. Iemhoff (✉)
Utrecht University, Utrecht, The Netherlands
e-mail: r.iemhoff@uu.nl

N. Bezhanishvili et al. (eds.), *Dick de Jongh on Intuitionistic and Provability Logics*, Outstanding Contributions to Logic 28, https://doi.org/10.1007/978-3-031-47921-2_8

$$\varphi \to \bigcirc\varphi \ (\bigcirc R) \quad \bigcirc\bigcirc\varphi \to \bigcirc\varphi \ (\bigcirc M) \quad \bigcirc\varphi \wedge \bigcirc\psi \to \bigcirc(\varphi \wedge \psi) \ (\bigcirc S).$$

In type theory, the modal operator describes a certain type constructor, and there exists a Curry-Howard like correspondence between Lax Logic and the computational lambda calculus (Benton et al., 1993). In algebraic logic the operator is a so-called nucleus, and in recent work in that area, embeddings of superintuitionistic logics into extensions of PLL have been used to provide new characterizations of subframe superintuitionistic logics (Bezhanishvili et al., 2019), while in Melzer (2020) the method of canonical formulas for Lax Logic has been developed.

The origins of the modality can be traced back to Curry (1957) but the logic received its name *Lax Logic* in the PhD-thesis of Mendler (1993), where the logic was introduced in the setting of hardware verification. In that context Lax Logic is used to model reasoning about statements of the form φ *holds under some constraint*, denoted as $\bigcirc\varphi$. Further investigation led to Fairtlough and Mendler (1994, 1997), in which, among other things, a sequent calculus for the logic is provided, a fact that is relevant for this paper.

In this paper we introduce a terminating sequent calculus G4iLL for PLL which is weakly analytic (in a way explained in Sect. 8.2.1) and use it to prove that the logic has uniform interpolation. Results on uniform interpolation for intuitionistic modal logics are rare, in particular for those that are not normal. In Iemhoff (2019b) it is shown that the logics $\mathsf{iK}_\Box$ and $\mathsf{iKD}_\Box$ have uniform interpolation, which is generalized in Jalali and Tabatabai (2018b) to the K and KD versions of substructural logics weaker than IPC, but whether other well-known intuitionistic modal logics, such as $\mathsf{iK4}_\Box$, $\mathsf{iS4}_\Box$ and $\mathsf{iGL}_\Box$, have uniform interpolation has, as far as we know, not been established, and the same holds for most other intuitionistic modal logics. This is in sharp contrast with the situation for classical modal and intermediate logic, about which much more is known. For example, for K4, S4 and GL, which can be seen as classical counterparts of the logics mentioned above, it is known that the last one does have uniform interpolation (Shavrukov, 1993) while the first two do not (Bílková, 2006; Ghilardi & Zawadowski, 1995). We hope that this paper's proof-theoretic results for Lax Logic may lead to analogous results for other interesting members of the class of intuitionistic modal logics.

Besides its use in the proof of uniform interpolation, the terminating sequent calculus that we develop is also interesting in connection with a paper of Howe (2001) in which a decision method for PLL is provided that is based on an annotated sequent calculus for the logic. In that paper it is shown that the calculus is sound and complete for Lax Logic and that backwards proof search in the calculus is terminating. Our sequent calculus G4iLL is also terminating, but consists of standard sequents instead of annotated ones.

Finally, this work is part of the larger project of *Universal Proof Theory*, which aim is to develop general methods to answer universal questions about proof systems. Questions such as: When does a theory has a certain type of proof system? What is the connection between the logical properties of a theory and its proof systems? How can it be established that a theory has no proof systems of a certain type? One possible way to obtain results in this direction is by applying a method that uses the fact that certain well-known logical properties are rare among the class of logics considered (Iemhoff, 2019b; Jalali & Tabatabai, 2018b). In intermediate and (intuitionistic) modal logics, uniform interpolation is such a property. If a logic does not have uniform interpolation, then the method identifies a class of proof systems that are excluded for the logic. If, on the other hand, a logic has uniform interpolation, then that particular method does not apply. Thus we have found that the positive result that Lax Logic has uniform interpolation turns out to be a negative result in terms of the project just described.

8.2 Preliminaries

The language $\mathcal{L}$ for propositional lax logic PLL contains a *constant* $\bot$, *propositional variables* or *atoms* $p, q, r, \ldots$, the *modal operator* $\bigcirc$ and the *connectives* $\wedge, \vee, \rightarrow$, where $\neg\varphi$ is defined as $(\varphi \rightarrow \bot)$. $\bot$ is by definition not an atom. A *circle formula* is a formula of the form $\bigcirc\varphi$. The set of formulas in $\mathcal{L}$ is denoted by $\mathcal{F}$.

We denote finite multisets of formulas by $\Gamma, \Pi, \Delta, \Sigma$. $\Gamma \cup \Pi$ is the multiset that contains only formulas φ that belong to Γ or Π and the number of occurrences of φ in $\Gamma \cup \Pi$ is the sum of the occurrences of φ in Γ and in Π. Γ, Π is short for $\Gamma \cup \Pi$. Multiset Γ, φ^n, Π denotes the union of Γ, Π and the multiset that consists of n copies of φ. We let $(\Gamma \Rightarrow \Delta) \subseteq (\Gamma' \Rightarrow \Delta')$ denote that $\Gamma \subseteq \Gamma'$ and $\Delta \subseteq \Delta'$. Furthermore (a for antecedent, s for succedent):

$$(\Gamma \Rightarrow \Delta)^a \equiv_{df} \Gamma \quad (\Gamma \Rightarrow \Delta)^s \equiv_{df} \Delta \quad \bigcirc\Gamma \equiv_{df} \{\bigcirc\varphi \mid \varphi \in \Gamma\}.$$

When working with sequents with a superscript, such as S^i, then S^{ia} is short for $(S^i)^a$, and similarly for S^{is}. We only consider (single-conclusion) sequents, which are expressions $(\Gamma \Rightarrow \Delta)$, where Γ and Δ are finite multisets of formulas and Δ contains at most one formula. They are interpreted as $I(\Gamma \Rightarrow \Delta) \equiv_{df} (\bigwedge\Gamma \rightarrow \bigvee\Delta)$, where $\bigvee\varnothing$ is interpreted as $\bot$. Expression $(\Gamma \Rightarrow \Delta)\cdot(\Pi \Rightarrow \Sigma)$ denotes $(\Gamma, \Pi \Rightarrow \Sigma, \Delta)$. When sequents are used in the setting of formulas, we often write S for $I(S)$, such as in $\vdash \bigvee_i S_i$, which thus denotes $\vdash \bigvee_i I(S_i)$.

The *degree* of a formula φ is inductively defined as $d(\bot) = 0$, $d(p) = 1$, $d(\bigcirc\varphi) = d(\varphi) + 1$, and $d(\varphi \circ \psi) = d(\varphi) + d(\psi) + 1$ for $\circ \in \{\wedge, \vee, \rightarrow\}$. In the setting of $\mathsf{G4iLL}$ we need an order on sequents based on the *weight function* $w(\cdot)$ on formulas from Dyckhoff (1992), which is inductively defined as: the weight of an atom and the constant $\bot$ is 1, $w(\bigcirc\varphi) = w(\varphi) + 1$, and $w(\varphi \circ \psi) = w(\varphi) + w(\psi) + i$, where $i = 1$ in case $\circ \in \{\vee, \rightarrow\}$ and $i = 2$ otherwise. We use the following ordering on sequents: $S_0 \ll S_1$ if and only if $S_0^a \cup S_0^s \ll S_1^a \cup S_1^s$, where $\ll$ is the order on multisets determined by weight as in Dershowitz and Manna (1979) (where they in fact define $\gg$): for multisets Γ, Δ we have $\Delta \ll \Gamma$ if Δ is the result of replacing one or more formulas in Γ by zero or more formulas of lower weight.

$$\Gamma, p \Rightarrow p \quad Ax \quad (p \text{ an atom}) \qquad \Gamma, \bot \Rightarrow \Delta \quad L\bot$$

$$\frac{\Gamma \Rightarrow \varphi \quad \Gamma \Rightarrow \psi}{\Gamma \Rightarrow \varphi \wedge \psi} R\wedge \qquad \frac{\Gamma, \varphi, \psi \Rightarrow \Delta}{\Gamma, \varphi \wedge \psi \Rightarrow \Delta} L\wedge$$

$$\frac{\Gamma \Rightarrow \varphi_i}{\Gamma \Rightarrow \varphi_0 \vee \varphi_1} R\vee \ (i = 0, 1) \qquad \frac{\Gamma, \varphi \Rightarrow \Delta \quad \Gamma, \psi \Rightarrow \Delta}{\Gamma, \varphi \vee \psi \Rightarrow \Delta} L\vee$$

$$\frac{\Gamma, \varphi \Rightarrow \psi}{\Gamma \Rightarrow \varphi \to \psi} R{\to} \qquad \frac{\Gamma, \varphi \to \psi \Rightarrow \varphi \quad \Gamma, \psi \Rightarrow \Delta}{\Gamma, \varphi \to \psi \Rightarrow \Delta} L{\to}$$

Fig. 8.1 The Gentzen calculus G3ip

8.2.1 The Calculi G3iLL *and* G4iLL

One of the standard calculi without structural rules for IPC is G3ip (Fig. 8.1), which is the propositional part of the calculus G3i from Troelstra and Schwichtenberg (1996). The calculus G4ip (Fig. 8.2) for IPC from Dyckhoff (1992) is similar to G3ip except for the left implication rule, which is replaced by rules that make G4ip terminating, a feature that G3ip does not possess (the definition of *terminating* is given in Sect. 8.2.2). The calculus G3iLL is G3ip extended by the two modal rules $R\bigcirc$ and $L\bigcirc$ given in Fig. 8.3, which are the two modal rules of the calculus GPLL in Fairtlough and Mendler (1997). The calculus G4iLL is G4ip extended by the four rules given in Fig. 8.3. G3iLL is an analytic calculus, meaning that it has the subformula property. G4iLL does not have the subformula property, but the words of Dyckhoff (1992) (last paragraph Sect. 1) about G4ip apply here as well: "Nevertheless, it has it in an obvious weak sense: one can work out what formulae are able to appear in a proof of a particular end-sequent."

8.2.2 Properties of Calculi

A calculus is *terminating* with respect to an order $\prec$ on sequents if the following hold: the calculus is finite (has finitely many axioms and rules); for all sequents S there are at most finitely many instances of a rule in the calculus with conclusion S; in every instance of a rule in the calculus the premisses come before the conclusion in the order $\prec$.

Fact 8.1 Calculus G4iLL is terminating in order $\ll$ and calculus G3iLL is not.

In the section on uniform interpolation we use that G4iLL is a reductive calculus. Here an order on sequents $\prec$ is *reductive* if the following hold: it is well-founded; all proper subsequents of a sequent come before that sequent; whenever all formulas in S occur boxed in S', then $S \prec S'$; $(\Gamma, \varphi \Rightarrow \Delta) \prec (\Gamma, q \to \varphi \Rightarrow \Delta)$ for all multisets

$$\Gamma, p \Rightarrow p \quad Ax \quad (p \text{ an atom}) \qquad\qquad \Gamma, \bot \Rightarrow \Delta \quad L\bot$$

$$\frac{\Gamma \Rightarrow \varphi \quad \Gamma \Rightarrow \psi}{\Gamma \Rightarrow \varphi \wedge \psi}\ R\wedge \qquad\qquad \frac{\Gamma, \varphi, \psi \Rightarrow \Delta}{\Gamma, \varphi \wedge \psi \Rightarrow \Delta}\ L\wedge$$

$$\frac{\Gamma \Rightarrow \varphi_i}{\Gamma \Rightarrow \varphi_0 \vee \varphi_1}\ R\vee\ (i = 0, 1) \qquad\qquad \frac{\Gamma, \varphi \Rightarrow \Delta \quad \Gamma, \psi \Rightarrow \Delta}{\Gamma, \varphi \vee \psi \Rightarrow \Delta}\ L\vee$$

$$\frac{\Gamma, \varphi \Rightarrow \psi}{\Gamma \Rightarrow \varphi \to \psi}\ R\!\to \qquad\qquad \frac{\Gamma, p, \varphi \Rightarrow \Delta}{\Gamma, p, p \to \varphi \Rightarrow \Delta}\ Lp\!\to\ (p \text{ an atom})$$

$$\frac{\Gamma, \varphi \to (\psi \to \gamma) \Rightarrow \Delta}{\Gamma, \varphi \wedge \psi \to \gamma \Rightarrow \Delta}\ L\wedge\!\to \qquad\qquad \frac{\Gamma, \varphi \to \gamma, \psi \to \gamma \Rightarrow \Delta}{\Gamma, \varphi \vee \psi \to \gamma \Rightarrow \Delta}\ L\vee\!\to$$

$$\frac{\Gamma, \psi \to \gamma \Rightarrow \varphi \to \psi \quad \gamma, \Gamma \Rightarrow \Delta}{\Gamma, (\varphi \to \psi) \to \gamma \Rightarrow \Delta}\ L\!\to\!\to$$

Fig. 8.2 The Gentzen calculus G4ip

$$\frac{\Gamma \Rightarrow \varphi}{\Gamma \Rightarrow \bigcirc\varphi}\ R\bigcirc \qquad\qquad \frac{\Gamma, \psi \Rightarrow \bigcirc\varphi}{\Gamma, \bigcirc\psi \Rightarrow \bigcirc\varphi}\ L\bigcirc$$

$$\frac{\Gamma \Rightarrow \varphi \quad \Gamma, \psi \Rightarrow \Delta}{\Gamma, \bigcirc\varphi \to \psi \Rightarrow \Delta}\ R\bigcirc^{\to} \qquad\qquad \frac{\Gamma, \chi \Rightarrow \bigcirc\varphi \quad \Gamma, \bigcirc\chi, \psi \Rightarrow \Delta}{\Gamma, \bigcirc\chi, \bigcirc\varphi \to \psi \Rightarrow \Delta}\ L\bigcirc^{\to}$$

Fig. 8.3 G3iLL is G3ip plus $R\bigcirc$ and $L\bigcirc$, G4iLL is G4ip plus all four rules

Γ, Δ, formulas φ and atoms q. Clearly, $\ll$ is reductive. A calculus is *reductive* if it is terminating with respect to an order that is reductive.

Fact 8.2 Calculus G4iLL is reductive with respect to order $\ll$.

A typical example of a rule that in general cannot belong to a reductive calculus is the cut rule, as in most common orders on sequents the premisses of that rule do not come before its conclusion.

In Iemhoff (2019b) (Sect. 7.1), the notion of a *balanced calculus* G is defined as: G is reductive, Cut and Weakening are admissible in it, and G satisfies two conditions that G4iLL trivially satisfies. Therefore, using Fact 8.2 and Corollary 8.1 below, we can conclude the following.

Fact 8.3 Calculus G4iLL is a balanced calculus with respect to order $\ll$.

A rule is *nonflat* if its conclusion contains at least one connective or modal operator. A set of rules or a calculus is nonflat if all of its rules are. Clearly, calculi G3iLL and G4iLL are nonflat.

8.2.3 Structural Rules

In a calculus *weakening is admissible* if the following two rules are admissible:

$$\frac{\Gamma \Rightarrow \Delta}{\Gamma, \varphi \Rightarrow \Delta} \qquad \frac{\Gamma \Rightarrow}{\Gamma \Rightarrow \varphi}$$

In a calculus *contraction is admissible* if the following rule is admissible:

$$\frac{\Gamma, \varphi, \varphi \Rightarrow \Delta}{\Gamma, \varphi \Rightarrow \Delta}$$

A calculus has *cut-elimination* or *cut is admissible* in it if the following Cut Rule is admissible:

$$\frac{\Gamma_1 \Rightarrow \varphi \quad \Gamma_2, \varphi \Rightarrow \Delta}{\Gamma_1, \Gamma_2 \Rightarrow \Delta}\ \mathit{Cut}$$

In a calculus the *structural rules are admissible* if weakening, contraction and cut are admissible in it.

Lemma 8.1 *Weakening and contraction as well as the following rule are admissible in* G3iLL.

$$\frac{\Gamma \Rightarrow \bot}{\Gamma \Rightarrow \Delta}$$

Proof The proof is standard, analogous to the proof for G3ip in (Troelstra & Schwichtenberg 1996) and therefore omitted. ⊣

8.3 Cut-Elimination

In this section we prove that the Cut Rule is admissible in G3iLL. The proof is by induction on the degree and a subinduction on the level of cuts. The degree of an application of the Cut Rule is the degree of the cut formula. The *level* of a cut in a derivation is the sum of the heights of its two premisses, where the *height* of a derivation is the length of its longest branch, where branches consisting of one node are considered to have height 1. In the rules of G3iLL, *principal formulas* are defined as usual for the rules in G3ip (Troelstra & Schwichtenberg, 1996) and for the modal rules $R\bigcirc$ and $L\bigcirc$ as given in Fig. 8.3, the principal formulas are $\bigcirc\varphi$ and $\bigcirc\psi$, respectively.

Theorem 8.1 (Cut-elimination) *The Cut Rule is admissible in* G3iLL.

Proof Let $\vdash$ denote $\vdash_{\mathsf{G3iLL}}$. Following the corrected version (Troelstra, 1998) of the cut elimination proof for G3ip in Troelstra and Schwichtenberg (1996), we successively eliminate cuts from the proof, always considering those cuts that have no cuts above them, the *topmost* cuts. For this it suffices to show that for cut free proofs $\mathcal{D}_1$ and $\mathcal{D}_2$, the following proof $\mathcal{D}$ of sequent $S = (\Gamma_1, \Gamma_2 \Rightarrow \Delta)$ can be transformed into a cut free proof $\mathcal{D}'$ of the same endsequent:

$$\dfrac{\begin{array}{cc}\mathcal{D}_1 & \mathcal{D}_2 \\ \Gamma_1 \Rightarrow \varphi & \varphi, \Gamma_2 \Rightarrow \Delta\end{array}}{\Gamma_1, \Gamma_2 \Rightarrow \Delta}\ \mathit{Cut}$$

This is proved by induction on the degree of the cut formula, with a subinduction to the level of the cut. We use the fact that weakening and contraction are admissible in G3iLL implicitly at various places.

There are three possibilities:

1. at least one of the premises is an axiom;
2. both premises are not axioms and the cut formula is not principal in at least one of the premises;
3. the cut formula is principal in both premises, which are not axioms.

1. As in Troelstra and Schwichtenberg (1996), straightforward, by checking all possible cases: If $\mathcal{D}_1$ is axiom $L\bot$ or $\mathcal{D}_2$ is axiom $L\bot$ and $\bot$ is not the cut formula, then we let $\mathcal{D}'$ be the instance S of that axiom. If $\bot$ is the cut formula, then $(\Gamma_1 \Rightarrow \bot)$ has a cut free derivation. Thus so has $(\Gamma_1, \Gamma_2 \Rightarrow \Delta)$ by Lemma 8.1, where we use the additional fact that in this proof no new cuts are introduced with respect to $\mathcal{D}_1$. Therefore we can let $\mathcal{D}'$ be this proof.

Assume both premises are not instances of $L\bot$. If $\mathcal{D}_1$ is an instance of Ax, then φ is an atom. If $\mathcal{D}_2$ also is an instance of Ax, then S is an instance of Ax, so we let $\mathcal{D}'$ be that instance. If $\mathcal{D}_2$ is not an axiom, then φ cannot be principal in its last inference, because it is an atom. We obtain $\mathcal{D}'$ by *cutting at a lower level* in $\mathcal{D}_2$, by which mean the following. Suppose the last inference of $\mathcal{D}_2$ is $R\rightarrow$:

$$\dfrac{\begin{array}{cc}\mathcal{D}_1 & \dfrac{\begin{array}{c}\mathcal{D}_2' \\ \Gamma_2, \varphi, \psi_1 \Rightarrow \psi_2\end{array}}{\Gamma_2, \varphi \Rightarrow \psi_1 \rightarrow \psi_2}\ R\rightarrow \\ \Gamma_1 \Rightarrow \varphi & \end{array}}{\Gamma_1, \Gamma_2 \Rightarrow \psi_1 \rightarrow \psi_2}\ \mathit{Cut}$$

Consider the proof $\mathcal{D}_3$:

$$\dfrac{\begin{array}{cc}\mathcal{D}_1 & \mathcal{D}_2' \\ \Gamma_1 \Rightarrow \varphi & \Gamma_2, \varphi, \psi_1 \Rightarrow \psi_2\end{array}}{\Gamma_1, \Gamma_2, \psi_1 \Rightarrow \psi_2}\ \mathit{Cut}$$

Since the cut in $\mathcal{D}_3$ is of the same degree but of lower level, there is a cut free proof $\mathcal{D}_3'$ of $(\Gamma_1, \Gamma_2, \psi_1 \Rightarrow \psi_2)$. Hence the following derivation is the desired cut free proof $\mathcal{D}'$:

$$\frac{\begin{array}{c}\mathcal{D}_3'\\ \Gamma_1, \Gamma_2, \psi_1 \Rightarrow \psi_2\end{array}}{\Gamma_1, \Gamma_2 \Rightarrow \psi_1 \to \psi_2}$$

The cases where the last inference of $\mathcal{D}_2$ is another rule than $R\to$ are treated in a similar way.

The case that $\mathcal{D}_1$ is not an axiom and $\mathcal{D}_2$ is an instance of *Ax* remains. Here we also cut at a lower level, the proof is analogous to the case just treated.

2. First, the case that φ is not principal in the last inference of $\mathcal{D}_1$. Thus the last inference in $\mathcal{D}_1$ is one of the nonmodal rules $\mathcal{R}$ of G3iLL or $L\bigcirc$. We treat the latter, where the last part of the proof looks as follows and $\varphi = \bigcirc\chi$:

$$\frac{\dfrac{\begin{array}{c}\mathcal{D}_1'\\ \Pi, \psi \Rightarrow \bigcirc\chi\end{array}}{\Pi, \bigcirc\psi \Rightarrow \bigcirc\chi}\, L\bigcirc \qquad \begin{array}{c}\mathcal{D}_2\\ \Gamma_2, \bigcirc\chi \Rightarrow \Delta\end{array}}{\Pi, \bigcirc\psi, \Gamma_2 \Rightarrow \Delta}\, Cut$$

where $\Gamma_1 = \Pi, \bigcirc\psi$. In case Δ consists of a circle formula, we can cut at a lower level in $\mathcal{D}_1$:

$$\frac{\dfrac{\begin{array}{c}\mathcal{D}_1'\\ \Pi, \psi \Rightarrow \bigcirc\chi\end{array} \quad \begin{array}{c}\mathcal{D}_2\\ \Gamma_2, \bigcirc\chi \Rightarrow \Delta\end{array}}{\Pi, \psi, \Gamma_2 \Rightarrow \Delta}\, Cut}{\Pi, \bigcirc\psi, \Gamma_2 \Rightarrow \Delta}\, L\bigcirc$$

Thus we obtain a proof of $(\Gamma_1, \Gamma_2 \Rightarrow \Delta)$ with a cut of the same degree as the cut in $\mathcal{D}$ but of lower level, and the induction hypothesis can be applied. In case Δ is not a circle formula, first note that $\bigcirc\chi$ cannot be principal in the last inference of $\mathcal{D}_2$ as that would imply that Δ is a circle formula, which by assumption it is not. Thus $\bigcirc\chi$ is not principal in the last inference of $\mathcal{D}_2$ and we therefore can cut at a lower level in $\mathcal{D}_2$.

In the case that φ is not principal in the last inference of $\mathcal{D}_2$, we only treat the following two cases, where $\varphi = \bigcirc\chi$:

$$\frac{\Gamma_1 \Rightarrow \bigcirc\chi \quad \dfrac{\Gamma_2, \bigcirc\chi \Rightarrow \psi}{\Gamma_2, \bigcirc\chi \Rightarrow \bigcirc\psi}\, R\bigcirc}{\Gamma_1, \Gamma_2 \Rightarrow \bigcirc\psi} \qquad \frac{\Gamma_1 \Rightarrow \bigcirc\chi \quad \dfrac{\Pi, \bigcirc\chi, \psi' \Rightarrow \Delta}{\Pi, \bigcirc\chi, \bigcirc\psi' \Rightarrow \Delta}\, L\bigcirc}{\Gamma_1, \Gamma_2 \Rightarrow \Delta}$$

where on the right side Δ is a circle formula and $\Gamma_2 = \Pi, \bigcirc\psi'$. Clearly, in both cases one can cut at a lower level in $\mathcal{D}_2$ (cf. case 1.).

3. The cut formula is principal in both premises, which are not axioms. We distinguish by cases according to the form of the cut formula and only treat the circle formulas. Thus the proof has the following form:

$$\dfrac{\dfrac{\begin{matrix}\mathcal{D}_1'\\ \Gamma_1 \Rightarrow \chi\end{matrix}}{\Gamma_1 \Rightarrow \bigcirc\chi} \quad \dfrac{\begin{matrix}\mathcal{D}_2'\\ \Gamma_2, \chi \Rightarrow \bigcirc\psi\end{matrix}}{\Gamma_2, \bigcirc\chi \Rightarrow \bigcirc\psi}}{\Gamma_1, \Gamma_2 \Rightarrow \bigcirc\psi}$$

This can be replaced by a proof in which the only cut has a lower degree:

$$\dfrac{\begin{matrix}\mathcal{D}_1'\\ \Gamma_1 \Rightarrow \chi\end{matrix} \quad \begin{matrix}\mathcal{D}_2'\\ \Gamma_2, \chi \Rightarrow \bigcirc\psi\end{matrix}}{\Gamma_1, \Gamma_2 \Rightarrow \bigcirc\psi}$$

⊣

8.4 Equivalence of G3iLL and G4iLL

In the short note (Iemhoff, 2020), a theorem has been proven that implies that G3iLL and G4iLL are equivalent. To formulate the theorem, we first need some terminology. In that paper, we define a *right modal rule* as a rule of the form (the S_i are sequents)

$$\dfrac{S_1 \quad \dots \quad S_n}{\Gamma \Rightarrow \bigcirc\varphi}\ \mathcal{R}$$

Because under this definition the rule $L\bigcirc$ is a right modal rule, which is an unfortunate clash in terminology, we here use the name *succedent modal rule* for such rules instead. With such a succedent modal rule we associate the following implication rule

$$\dfrac{S_1 \quad \dots \quad S_n \quad \Gamma, \psi \Rightarrow \Delta}{\Gamma, \bigcirc\varphi \to \psi \Rightarrow \Delta}\ \mathcal{R}^{\to}$$

If X is a set of rules, then G3iX denotes the calculus G3ip extended by the rules in X and G4iX is the calculus G4ip extended by the rules in X *plus* the rules $\mathcal{R}^{\to}$ for those $\mathcal{R} \in X$ that are succedent modal rules. Note that our notation G3iLL and G4iLL is consistent with this definition, that is, G4iLL is the result of adding the two modal rules $R\bigcirc$ and $L\bigcirc$ of G3iLL to G4ip as well as the rules $R\bigcirc^{\to}$ and $L\bigcirc^{\to}$, which are added because both $R\bigcirc$ and $L\bigcirc$ have been added.

In Iemhoff (2020), the following is shown.

Theorem 8.2 (Equivalence) *If $\mathcal{R}$ is a set of nonflat rules, the structural rules are admissible in* G3ip + $\mathcal{R}$, *and* G4ip + $\mathcal{R}$ *is terminating and closed under weakening, then* G3ip + $\mathcal{R}$ *and* G4iX + $\mathcal{R}$ *are equivalent (derive exactly the same sequents) and the structural rules are admissible in* G4iX.

Corollary 8.1 G3iLL *and* G4iLL *are equivalent and the structural rules are admissible in both of them. In particular,* G4iLL *is a terminating cut free sequent calculus for Lax Logic in which the Cut Rule is admissible.*

8.5 Interpolation

That PLL has interpolation follows from the fact that it has uniform interpolation, as shown in the next section. Because the proof of the latter is complicated and the proof of the former is not, we present in this section a simple proof of interpolation for PLL. Also the algorithm to compute interpolants that can be derived from it is much simpler than the one obtained from the proof of uniform interpolation. The method we present here is a straightforward adaptation of the *Maehara method*, a method to prove interpolation based on sequent calculi first presented in Maehara (1960), in Japanese, and later in English in Ono (1998).

We introduce some terminology well-known from the literature on interpolation. Given multisets Γ, Δ and formula φ, let $V(\varphi)$ denote all atoms that occur in φ, let $V(\Gamma) = \bigcup\{V(\varphi) \mid \varphi \in \Gamma\}$, let $V(\Gamma, \Delta)$ be short for $V(\Gamma \cup \Delta)$, and $V(\Gamma, \varphi)$ and $V(\Gamma, \varphi, \psi)$ for $V(\Gamma, \{\varphi\})$ and $V(\Gamma, \{\varphi, \psi\})$, respectively. To be able to refer to the particular partition of the antecedent of a sequent of which a formula χ is an interpolant, we say that χ is an *interpolant* of $\Gamma; \Gamma' \Rightarrow \Delta$ if $\Gamma \Rightarrow \chi$ and $\Gamma', \chi \Rightarrow \Delta$ are derivable in G3iLL and $V(\chi) \subseteq V(\Gamma) \cap V(\Gamma', \Delta)$.

Theorem 8.3 G3iLL *has sequent interpolation: if* $\Gamma, \Gamma' \Rightarrow \Delta$ *is derivable in* G3iLL, *then there is a formula* χ *such that* $\Gamma \Rightarrow \chi$ *and* $\Gamma', \chi \Rightarrow \Delta$ *are derivable in* G3iLL *and all atoms in* χ *occur both in* Γ *and in* $\Gamma' \cup \Delta$.

Proof We prove the theorem by constructing an algorithm that given a derivation $\mathcal{D}$ of $\Gamma, \Gamma' \Rightarrow \Delta$ in G3iLL produces for any partition $\Gamma; \Gamma' \Rightarrow \Delta$ an interpolant. We use induction to the height of $\mathcal{D}$. The case that $\mathcal{D}$ is an instance of an axiom is straightforward, and so are the cases where the last inference of $\mathcal{D}$ is one of the propositional rules. We treat the case that it is an implication rule as an example, followed by the case that it is one of the modal rules.

If the last inference is $R{\rightarrow}$, then Δ consists of $\varphi \rightarrow \psi$. If the partition is $\Gamma; \Gamma' \Rightarrow \varphi \rightarrow \psi$, then consider the partition of the premise $\Gamma; \Gamma', \varphi \Rightarrow \psi$. By the induction hypothesis there is an interpolant χ such that both $\Gamma \Rightarrow \chi$ and $\Gamma', \chi, \varphi \Rightarrow \psi$ are derivable and $V(\chi) \subseteq V(\Gamma) \cap V(\Gamma', \varphi, \psi)$. This implies that $\Gamma', \chi \Rightarrow \varphi \rightarrow \psi$ is derivable and $V(\chi) \subseteq V(\Gamma', \varphi \rightarrow \psi)$. Thus proving that χ is an interpolant for $\Gamma; \Gamma' \Rightarrow \varphi \rightarrow \psi$.

If the last inference is $L{\rightarrow}$ with principal formula $\varphi \rightarrow \psi$, then there are two possible partitions of the last sequent: $\Gamma, \varphi \rightarrow \psi; \Gamma' \Rightarrow \Delta$ and $\Gamma; \varphi \rightarrow \psi, \Gamma' \Rightarrow \Delta$. In the first case, by the induction hypothesis there are interpolants χ_1 and χ_2 of the partitions of the premises $\Gamma'; \Gamma, \varphi \rightarrow \psi \Rightarrow \varphi$ (note the changed order in the partition) and $\Gamma, \psi; \Gamma' \Rightarrow \Delta$. Thus $V(\chi_1) \subseteq V(\Gamma, \varphi \rightarrow \psi) \cap V(\Gamma')$, $V(\chi_2) \subseteq V(\Gamma, \psi) \cap V(\Gamma', \Delta)$, and the following are derivable:

$$\Gamma' \Rightarrow \chi_1 \quad \Gamma, \varphi \to \psi, \chi_1 \Rightarrow \varphi \qquad \Gamma, \psi \Rightarrow \chi_2 \quad \Gamma', \chi_2 \Rightarrow \Delta.$$

This implies that $\Gamma', \chi_1 \to \chi_2 \Rightarrow \Delta$ and $\Gamma, \varphi \to \psi, \chi_1 \Rightarrow \chi_2$ are derivable. Thus so is $\Gamma, \varphi \to \psi \Rightarrow \chi_1 \to \chi_2$. Since $V(\chi_1 \to \chi_2) \subseteq V(\Gamma, \varphi \to \psi) \cap V(\Gamma', \Delta)$, this proves that $\chi_1 \to \chi_2$ is an interpolant of $\Gamma, \varphi \to \psi; \Gamma' \Rightarrow \Delta$.

In the second case, $\Gamma; \varphi \to \psi, \Gamma' \Rightarrow \Delta$, the induction hypothesis implies that there are interpolants χ_1 and χ_2 of the partitions of the premises $\Gamma; \varphi \to \psi, \Gamma' \Rightarrow \varphi$ and $\Gamma; \psi, \Gamma' \Rightarrow \Delta$. Thus $V(\chi_1) \subseteq V(\Gamma) \cap V(\Gamma', \varphi \to \psi)$, $V(\chi_2) \subseteq V(\Gamma) \cap V(\Gamma', \psi, \Delta)$, and the following are derivable:

$$\Gamma \Rightarrow \chi_1 \quad \Gamma', \varphi \to \psi, \chi_1 \Rightarrow \varphi \qquad \Gamma \Rightarrow \chi_2 \quad \Gamma', \psi, \chi_2 \Rightarrow \Delta.$$

It is not hard to see that $\chi = (\chi_1 \wedge \chi_2)$ is the desired interpolant.

If the last inference is $R\bigcirc$, then for the interpolant of the conclusion one can take the interpolant of the premise.

If the last inference is $L\bigcirc$, then Δ consists of a circle formula $\bigcirc\psi$ and the principal formula is $\bigcirc\varphi$. For partition $\Gamma; \Gamma', \bigcirc\varphi \Rightarrow \bigcirc\psi$ of the conclusion, let χ be the interpolant for the corresponding partition of the premise, namely such that the following are derivable:

$$\Gamma \Rightarrow \chi \quad \Gamma', \varphi, \chi \Rightarrow \bigcirc\psi.$$

Clearly, χ is an interpolant for the conclusion as well. Similarly, for partition $\Gamma, \bigcirc\varphi; \Gamma' \Rightarrow \bigcirc\psi$ of the conclusion, let χ be the interpolant for the corresponding partition of the premise, namely such that the following are derivable:

$$\Gamma, \varphi \Rightarrow \chi \quad \Gamma', \chi \Rightarrow \bigcirc\psi.$$

If χ is a circle formula, then it is an interpolant for the conclusion. Otherwise we can take $\bigcirc\chi$ instead. ⊣

Corollary 8.2 *Lax Logic has Craig interpolation. In particular, there is an algorithm that given a derivation of $\varphi \Rightarrow \psi$ in* G3iLL *produces an interpolant for $\varphi \to \psi$.*

8.6 Uniform Interpolation

In Iemhoff (2019b) a modular and constructive method was developed to prove uniform interpolation for certain intermediate and intuitionistic modal logics, which was later generalized to other substructural (modal) logics by Jalali and Tabatabai (2018a, 2018b). In this section we prove that Lax Logic, a logic not covered by these previous methods, has uniform interpolation. Our proof is an adaptation of the proof in Iemhoff (2019b) that the intuitionistic modal logic $\mathsf{iK}_\Box$ has uniform interpolation. For background information and motivation we refer the reader to that paper.

8.6.1 Uniform Interpolants for Formulas

A logic has *uniform interpolation* if for any atom p and any set of atoms $\mathcal{P}$ not containing p, the embedding of $\mathcal{F}(\mathcal{P})$ into $\mathcal{F}(\mathcal{P} \cup \{p\})$ has a right and a left adjoint: For any formula φ and any atom p there exist two formulas, usually denoted by $\forall p\varphi$ and $\exists p\varphi$, in the language of the logic, that do not contain p and such that for all ψ not containing p:

$$\vdash \psi \to \varphi \Leftrightarrow \vdash \psi \to \forall p\varphi \qquad \vdash \varphi \to \psi \Leftrightarrow \vdash \exists p\varphi \to \psi.$$

Given a formula φ, its *universal uniform interpolant with respect to* $p_1 \dots p_n$ is $\forall p_1 \dots p_n\varphi$, which is short for $\forall p_1(\forall p_2(\dots(\forall p_n\varphi)\dots)$, and its *existential uniform interpolant with respect to* $p_1 \dots p_n$ is $\exists p_1 \dots p_n\varphi$, short for $\exists p_1(\exists p_2(\dots(\exists p_n \varphi)\dots)$. The requirements above could be replaced by the following four requirements.

$$\vdash \forall p\varphi \to \varphi \qquad \vdash \psi \to \varphi \Rightarrow \vdash \psi \to \forall p\varphi. \tag{$\forall$}$$

$$\vdash \varphi \to \exists p\varphi \qquad \vdash \varphi \to \psi \Rightarrow \vdash \exists p\varphi \to \psi. \tag{$\exists$}$$

In classical logic one only needs one quantifier, as $\exists p$ can be defined as $\neg\forall p\neg$ and vice versa. Although in the intuitionistic setting $\exists p$ can also be defined in terms of $\forall p$, namely as $\exists p\varphi = \forall q(\forall p(\varphi \to q) \to q)$ for a q not in φ, having it as a separate quantifier is convenient in the proof-theoretic approach presented here (we follow Pitts, 1992, which also uses both quantifiers).

In order to prove that Lax Logic has uniform interpolation, the first task is to adapt the above notions to the setting of sequents and sequent calculus G4iLL.

8.6.2 Reductive Orders and Rank

The set $\mathcal{F}_{\mathsf{ex}}$ is the smallest set of expressions that contains all formulas in the language $\mathcal{L}$, is closed under the connectives and modal operator, and if S is a sequent in $\mathcal{L}$ and p an atom, then $\forall pS$ and $\exists pS$ belong to $\mathcal{F}_{\mathsf{ex}}$. For example, when S is a sequent in $\mathcal{L}$ and φ a propositional formula, then $(\varphi \to \exists pS)$ belongs to $\mathcal{F}_{\mathsf{ex}}$, as does $\bigcirc(\varphi \wedge \forall pS)$, but $\exists p\exists qS$ does not.

The reductive order $\ll$ on sequents defined in Sect. 8.2 determines the following order on formulas in $\mathcal{F}_{\mathsf{ex}}$, denoted $\prec$. For any expression φ in $\mathcal{F}_{\mathsf{ex}}$, let $\mathsf{qf}(\varphi)$ denote the multiset consisting of all occurrences of subformulas of the form QpS in φ, where $Q \in \{\exists, \forall\}$. The order on multisets of the form $\mathsf{qf}(\varphi)$ again is in the style of Dershowitz and Mann (1979): $\mathsf{qf}(\varphi) \prec_{\mathsf{qf}} \mathsf{qf}(\psi)$ iff $\mathsf{qf}(\varphi)$ is the result of replacing one or more formulas of the form QpS in $\mathsf{qf}(\psi)$ by zero or more formulas of the form $Q'qS'$ with $S' \ll S$, where $Q, Q' \in \{\exists, \forall\}$. This order is well–defined since by definition such S are sequents in $\mathcal{L}$ and therefore can be compared via $\ll$. The order on $\mathcal{F}_{ex}$ can now be defined as: if $\varphi, \psi \in \mathcal{F}$, then $\varphi \prec \psi$ iff $(\Rightarrow \varphi) \prec (\Rightarrow \psi)$; if $\varphi \in \mathcal{F}$ and $\psi \notin \mathcal{F}$, then $\varphi \prec \psi$ and not $\psi \prec \varphi$; if $\varphi, \psi \notin \mathcal{F}$, then $\varphi \prec \psi$ if $\mathsf{qf}(\varphi) \prec \mathsf{qf}(\psi)$. When $\varphi \prec \psi$, we say that φ is of *lower rank* than ψ. Clearly, since the order $\ll$ on sequents is well–founded, so is the order $\prec$ on $\mathcal{F}_{ex}$.

8.6.3 Interpolant Assignments

For any given calculus, let $\mathsf{G}_{\mathsf{ins}}$ denote the set of instances of rules in G. A sequent S is *principal* for an instance R of a rule if the conclusion of R is of the form $S' \cdot S$ for some sequent S' and the principal formula of R belongs to S. Otherwise S is *nonprincipal*. For example, suppose R has conclusion $(\Gamma, \varphi \Rightarrow \Delta)$ and φ is the principal formula of R, then any sequent of the form $(\Gamma', \varphi \Rightarrow \Delta')$, where $(\Gamma' \Rightarrow \Delta') \subseteq (\Gamma \Rightarrow \Delta)$, is principal for R.

An *interpolant assignment* ι for G, assigns, for every atom p and sequent S, $\iota\exists p S = \top$ and $\iota\forall p S = \bot$ in case S is empty, and in case S is not empty:

- for every $R \in \mathsf{G}_{\mathsf{ins}}$ with conclusion S, to each of the expressions $\exists \overset{R}{p}\, S$ and $\forall \overset{R}{p}\, S$ a formula in $\mathcal{F}_{ex}$ that is of lower rank than $\exists p S$ (or, equivalently, of lower rank than $\forall p S$), which are denoted by $\iota\exists \overset{R}{p}\, S$ and $\iota\forall \overset{R}{p}\, S$, respectively, and
- for every $\mathcal{R} \in \mathsf{G}$ such that S is nonprincipal for at least one instance of $\mathcal{R}$, to each of the expressions $\exists \overset{\overline{\mathcal{R}}}{p}\, S$ and $\forall \overset{\overline{\mathcal{R}}}{p}\, S$ a formula in $\mathcal{F}_{ex}$ that is of lower rank than $\exists p S$, which are denoted by $\iota\exists \overset{\overline{\mathcal{R}}}{p}\, S$ and $\iota\forall \overset{\overline{\mathcal{R}}}{p}\, S$, respectively.

Given an interpolant assignment we define the following formulas in $\mathcal{F}_{ex}$. Recall that p and q range over atoms.

$$
\begin{aligned}
\forall^+ p S &\equiv_{df} \bigvee\{\iota\forall \overset{R}{p}\, S \mid R \in \mathsf{G}_{\mathsf{ins}}, S \text{ is the conclusion of } R\}\\
\forall^- p S &\equiv_{df} \bigvee\{\iota\forall \overset{\overline{\mathcal{R}}}{p}\, S \mid \mathcal{R} \in \mathsf{G}, S \text{ is nonprincipal for some instance of } \mathcal{R}\}\\
\exists^+ p S &\equiv_{df} \bigwedge\{\iota\exists \overset{R}{p}\, S \mid R \in \mathsf{G}_{\mathsf{ins}}, S \text{ is the conclusion of } R\}\\
\exists^- p S &\equiv_{df} \bigwedge\{\iota\exists \overset{\overline{\mathcal{R}}}{p}\, S \mid \mathcal{R} \in \mathsf{G}, S \text{ is nonprincipal for some instance of } \mathcal{R}\}\\
\forall^{at} p S &\equiv_{df} \bigvee\{q \in S^s \mid q \text{ an atom and } q \neq p, \text{ or } q = \top\} \vee\\
&\qquad \bigvee\{q \wedge \forall p(\varphi, S^a \backslash \{q \to \varphi\} \Rightarrow S^s) \mid (q \to \varphi) \in S^a, q \neq p\}\\
\exists^{at} p S &\equiv_{df} \bigwedge\{q \in S^a \mid q \text{ an atom and } q \neq p, \text{ or } q = \bot\} \wedge\\
&\qquad \bigwedge\{q \to \exists p(\varphi, S^a \backslash \{q \to \varphi\} \Rightarrow S^s) \mid (q \to \varphi) \in S^a, q \neq p\}.
\end{aligned}
$$

Observe that there could be more than one instance of a single rule $\mathcal{R}$ that has S as a conclusion, in which case every instance corresponds to a separate disjunct or conjunct of the interpolant assignment. The definition above is well-defined for reductive calculi, because for such calculi all sets over which the big conjunctions and disjunctions range are finite.

We define a rewrite relation $\rightarrowtail$ on $\mathcal{F}_{\mathsf{ex}}$ that is the smallest relation on $\mathcal{F}_{\mathsf{ex}}$ that preserves the logical operators and satisfies:

$$\forall pS \rightarrowtail \forall^+ pS \vee \forall^- pS \vee \forall^{at} pS \quad \exists pS \rightarrowtail \exists^+ pS \wedge \exists^- pS \wedge \exists^{at} pS.$$

After the definition of uniform interpolation for sequents in Sect. 8.6.4 below, three examples of the rewrite relation are provided.

As G4iLL is a reductive calculus, the following lemma follows from Lemma 3 in Iemhoff (2019b).

Lemma 8.2 *If there exists an interpolant assignment for* G4iLL, *then the relation* $\rightarrowtail$ *on* $\mathcal{F}_{\mathsf{ex}}$ *is confluent and strongly normalizing.*

Note that according to this relation, any formula in $\mathcal{L}$ is in normal form, and any expression in $\mathcal{F}_{\mathsf{ex}}$ that is not a formula in $\mathcal{L}$ is not. Let $\overline{\varphi}$ denote the normal form of an expression φ. For any sequent S in $\mathcal{L}$ we now define its *left* and *right interpolant* to be $\overline{\forall pS}$ and $\overline{\exists pS}$, respectively. We omit the overline most of the time, as it will be clear from the context whether the expression in $\mathcal{F}_{\mathsf{ex}}$ is meant or the actual normal form in $\mathcal{L}$, i.e. the interpolants.

8.6.4 Uniform Interpolants for Sequents

To define uniform interpolants in the setting of sequents, we introduce the notion of a partition, which applies to sequents and to rules. A *partition* of a sequent S is an ordered pair (S^r, S^i) (i for *interpolant*, r for *rest*) such that $S = S^r \cdot S^i$. It is a *p–partition* if p does not occur in S^r. For any sequent S and partition (S^i, S^r) we use the abbreviation:

$$S^r \cdot (\exists pS^i \Rightarrow \forall pS^i \mid \varnothing) \equiv_{df} \begin{cases} S^r \cdot (\exists pS^i \Rightarrow \forall pS^i) & \text{if } S^s \neq \varnothing \text{ and } S^{rs} = \varnothing \\ S^r \cdot (\exists pS^i \Rightarrow\ ) & \text{if } S^s = \varnothing \text{ or } S^{rs} \neq \varnothing. \end{cases}$$

A *(p-)partition* of an instance $R = (S_1 \dots S_n/S_0)$ of a rule is a (p-)partition of the sequents in the rule. Given such a partition and $\star \in \{r, i\}$, let (R^r, R^i) and $R^\star$ respectively denote the expressions

$$\frac{(S_1^r, S_1^i) \quad \dots \quad (S_n^r, S_n^i)}{(S_0^r, S_0^i)}\ (R^r, R^i) \qquad \frac{S_1^\star \quad \dots \quad S_n^\star}{S_0^\star}\ R^\star$$

In Lemma 2 in Iemhoff (2019b) it is shown that given an interpolant assignment an intuitionistic modal logic has uniform interpolation if the following holds:

(∀l) For all p: $\vdash S^a, \forall p S \Rightarrow S^s$;
(∃r) For all p: $\vdash S^a \Rightarrow \exists p S$;
(∀∃) If S is derivable, for all p and all p–partitions (S^r, S^i): $\vdash S^r \cdot (\exists p S^i \Rightarrow \forall p S^i \mid \varnothing)$.

Properties (∀l) and (∃r) are the *independent* (from partitions) *interpolant properties*, and (∀∃) is the *dependent interpolant property*. It is also shown that $\forall p S$ and $\exists p S$ are equivalent to $\forall p I(S)$ and $\exists p (\bigwedge S^a)$, respectively. In particular, $\forall p (\Rightarrow \varphi)$ is equivalent to $\forall p \varphi$ and $\exists p (\varphi \Rightarrow)$ to $\exists p \varphi$.

A partition (S^r, S^i) of S *satisfies* the interpolant properties if, in the case of the independent property, S satisfies them (in which case we also say that S satisfies them), and in case of the dependent property, it holds for that particular partition. A sequent *satisfies* a property if every possible partition of the sequent satisfies it.

To show that Lax Logic has uniform interpolation it thus suffices to show that we can define, for all sequents S, formulas $\forall p S$ and $\exists p S$ in $\mathcal{L}$ in such a way that the three interpolant properties hold. Before we do so, three examples of uniform sequent interpolants are given.

Example 8.1 Suppose the calculus only contains the rule $\mathcal{R}$ for conjunction on the left:

$$\frac{\Gamma, \varphi, \psi \Rightarrow \Delta}{\Gamma, \varphi \wedge \psi \Rightarrow \Delta}$$

We provide the uniform interpolants for $S = (p \wedge q, r, s \Rightarrow t)$. We have

$$\forall p S \rightarrowtail \iota\forall \overset{R}{p} S \vee \iota\forall \overset{\overline{\mathcal{R}}}{p} S \vee \iota\forall^{at} p S \quad \exists p S \rightarrowtail \iota\exists \overset{R}{p} S \wedge \iota\exists \overset{\overline{\mathcal{R}}}{p} S \wedge \iota\exists^{at} p S,$$

where $R = S_1/S$ stands for the instance of $\mathcal{R}$ with $p \wedge q$ as the principal formula. Thus $S_1 = (p, q, r, s \Rightarrow t)$. The standard interpolant assignment introduced below satisfies $\iota\forall \overset{R}{p} S = \forall p S_1$, $\iota\exists \overset{R}{p} S = \exists p S_1$, $\iota\forall \overset{\overline{\mathcal{R}}}{p} S = \bot$, and $\iota\exists \overset{\overline{\mathcal{R}}}{p} S = \top$. This implies that

$$\forall p S \rightarrowtail \forall p S_1 \vee \bot \vee t \quad \exists p S \rightarrowtail \exists p S_1 \wedge \top \wedge r \wedge s.$$

Since S_1 cannot be the conclusion of any rule, we have $\forall p S_1 = \iota\forall \overset{\overline{\mathcal{R}}}{p} S \vee \forall^{at} p S_1 = \bot \vee t$ and $\exists p S_1 = \iota\exists \overset{\overline{\mathcal{R}}}{p} S \wedge \exists^{at} p S_1 = \top \wedge q \wedge r \wedge s$. Therefore

$$\forall p S \rightarrowtail \bot \vee t \vee \bot \vee t \quad \exists p S \rightarrowtail \top \wedge q \wedge r \wedge s \wedge \top \wedge r \wedge s.$$

Indeed, with $\forall p S = t$ and $\exists p S = q \wedge r \wedge s$, the three interpolant properties are satisfied for S.

Example 8.2 For the same calculus as in the previous example, the uniform interpolants of the sequent $S = (p_0 \wedge q_0, p_1 \wedge q_1 \Rightarrow r)$ are computed as follows, where we abbreviate $p_i \wedge q_i$ with ψ_i. For $i = 0, 1$, let R_i stand for the instance of $\mathcal{R}$ with ψ_i as the principal formula, and define sequents $S_0 = (p_0, q_0, \psi_1 \Rightarrow r)$ and

$S_1 = (\psi_0, p_1, q_1 \Rightarrow r)$, which are the premises of these instances. By the above definition,

$$\forall p S \rightarrowtail \iota\forall\overset{R_0}{p} S \vee \iota\forall\overset{R_1}{p} S \vee \iota\forall\overset{\overline{\mathcal{R}}}{p} S \vee \iota\forall^{at} p S \qquad \exists p S \rightarrowtail \iota\exists\overset{R_0}{p} S \wedge \iota\exists\overset{R_1}{p} S \wedge \iota\exists\overset{\overline{\mathcal{R}}}{p} S \wedge \iota\exists^{at} p S.$$

The disjuncts and conjuncts can be computed as in the previous example.

Example 8.3 Suppose the calculus consists only of the rule $\mathcal{R}$ for disjunction on the right:

$$\frac{\Gamma \Rightarrow \varphi_i}{\Gamma \Rightarrow \varphi_0 \vee \varphi_1}\ (i = 0, 1)$$

We provide the uniform interpolants for $S = (r \Rightarrow p \vee q)$. Let R_p, R_q stand for the instances of $\mathcal{R}$ with conclusion S and premises $S_p = (r \Rightarrow p)$ and $S_q = (r \Rightarrow q)$, respectively. By the above definition,

$$\forall p S \rightarrowtail \iota\forall\overset{R_p}{p} S \vee \iota\forall\overset{R_q}{p} S \vee \iota\forall\overset{\overline{\mathcal{R}}}{p} S \vee \iota\forall^{at} p S \quad \exists p S \rightarrowtail \iota\exists\overset{R_p}{p} S \wedge \iota\exists\overset{R_q}{p} S \wedge \iota\exists\overset{\overline{\mathcal{R}}}{p} S \wedge \iota\exists^{at} p S.$$

The standard interpolant assignment introduced below satisfies $\iota\forall\overset{R}{p} S = \forall p S_1$ for any instance $R = S_1/S$ of $\mathcal{R}$, as well as $\iota\forall\overset{\overline{\mathcal{R}}}{p} S = \bot$ and $\iota\exists\overset{\overline{\mathcal{R}}}{p} S = \top$ for any S. This, together with the fact that $\iota\forall^{at} p S = \bot$, implies

$$\forall p S \rightarrowtail \iota\forall p S_p \vee \iota\forall p S_q \vee \bot \vee \bot \quad \exists p S \rightarrowtail \iota\exists p S_p \wedge \iota\exists p S_1 \wedge \top \wedge r.$$

Since S_p and S_q cannot be the conclusion of any rule, we have

$$\forall p S_p = \iota\forall\overset{\overline{\mathcal{R}}}{p} S_p \vee \forall^{at} p S_p = \bot \vee \bot \quad \forall p S_q = \iota\forall\overset{\overline{\mathcal{R}}}{p} S_q \vee \forall^{at} p S_q = \bot \vee q$$
$$\exists p S_p = \iota\exists\overset{\overline{\mathcal{R}}}{p} S_p \wedge \exists^{at} p S_p = \top \wedge r = \iota\exists\overset{\overline{\mathcal{R}}}{p} S_q \wedge \exists^{at} p S_q = \exists p S_q.$$

Therefore $\forall p S \rightarrowtail \bot \vee \bot \vee \bot \vee q \vee \bot \vee \bot$ and $\exists p S \rightarrowtail \top \wedge r \wedge \top \wedge r \wedge \top \wedge r$. Indeed, with $\forall p S = q$ and $\exists p S = r$, the three interpolant properties are satisfied for S.

8.6.5 The Inductive Properties

For a modular development of uniform interpolation, we introduce the following six properties of rules, where $\varnothing \vdash \varphi$ should be read as $\vdash \varphi$. Given an instance $R = (S_1 \dots S_n/S_0)$ of a rule $\mathcal{R}$, we define

$$\mathfrak{I}_R^p \equiv_{df} \{S_j \cdot (\forall p S_j \Rightarrow), (S_j^a \Rightarrow \exists p S_j) \mid 1 \leq j \leq n\} \cup$$
$$\{S^a \Rightarrow \exists p(S^a \Rightarrow) \mid S \subset S_0 \text{ or } \bigcirc S \subseteq S_0 \text{ or } S \subseteq S_j \text{ for some } 1 \leq j \leq n\}$$

$$\mathfrak{D}_R^p \equiv_{df} \bigcup_{j=1}^{n} \{S_j^r \cdot (\exists p S_j^i \Rightarrow \forall p S_j^i \mid \varnothing) \mid (S_j^r, S_j^i) \text{ a } p\text{–partition of } S_j\}.$$

For $\mathfrak{D}_R^p$, note that it contains the sequent $S_j^r \cdot (\exists p S_j^i \Rightarrow \forall p S_j^i \mid \varnothing)$ for *any* possible p-partition (S_j^r, S_j^i) of a premiss S_j of R. And that for S with empty succedent, $S^r \cdot (\exists p S^i \Rightarrow)$ derives $S^r \cdot (\exists p S^i \Rightarrow \forall p S^i)$.

The sets $\mathfrak{I}_R^p$ and $\mathfrak{D}_R^p$ contain the sequents to which, in a proof of the interpolant properties that uses induction along $\prec$, the induction hypothesis could be applied. In such a proof, the assumption that the interpolant properties hold for all sequents below S implies that the sequents in $\mathfrak{I}_R^p$ and $\mathfrak{D}_R^p$ are derivable. Note that in the case that R is an instance of an axiom, the latter set is empty, but the former is not, as it contains all sequents of the form $S^a \Rightarrow \exists p(S^a \Rightarrow)$ for S such that $S \subset S_0$ or $\bigcirc S \subseteq S_0$.

- $(\text{IPP})_{\mathcal{R}}^{\forall}$ $\mathfrak{I}_R^p \vdash S \cdot (\forall \overset{R}{p} S \Rightarrow)$ for every instance R of $\mathcal{R}$ with conclusion S.
- $(\text{IPN})_{\mathcal{R}}^{\forall}$ If S is nonprincipal for some instance of $\mathcal{R}$, then the assumption that all sequents below S satisfy the interpolant properties implies $\vdash S \cdot (\forall \overset{\overline{\mathcal{R}}}{p} S \Rightarrow)$.
- $(\text{IPP})_{\mathcal{R}}^{\exists}$ $\mathfrak{I}_R^p \vdash (S^a \Rightarrow \exists \overset{R}{p} S)$ for every instance R of $\mathcal{R}$ with conclusion S.
- $(\text{IPN})_{\mathcal{R}}^{\exists}$ If S is nonprincipal for some instance of $\mathcal{R}$, then the assumption that all sequents below S satisfy the interpolant properties implies $\vdash (S^a \Rightarrow \exists \overset{\overline{\mathcal{R}}}{p} S)$.
- $(\text{DPP})_{\mathcal{R}}$ For every sequent S that has a derivation of which the last inference is an instance R of $\mathcal{R}$, and for every p–partition (S^r, S^i) such that sequent S^i is principal for R: $\mathfrak{D}_R^p \vdash S^r \cdot (\exists p S^i \Rightarrow \forall p S^i \mid \varnothing)$.
- $(\text{DPN})_{\mathcal{R}}$ For every sequent S that has a derivation of which the last inference is an instance R of $\mathcal{R}$, and for every p–partition (S^r, S^i) such that sequent S^i is nonprincipal for R: if all sequents that are below S satisfy the interpolant properties, then $\vdash S^r \cdot (\exists p S^i \Rightarrow \forall p S^i \mid \varnothing)$.

These six properties are called the *inductive properties* in this paper. "IP" stands for *independent property*, "DP" for *dependent property*, "P" and "N" for *principal* and *nonprincipal*, respectively.

For an intuitionistic modal logic L with calculus G, an interpolant assignment is *sound* for a rule $\mathcal{R}$ in G, if the six inductive properties hold for $\mathcal{R}$, where "$\vdash$" equals "$\vdash_{\mathsf{L}}$". It is *sound* for a calculus if it is sound for all the rules of the calculus.

Remark 8.1 The following observation will be used to prove $(\text{DPP})_{\mathcal{R}}$ and $(\text{DPN})_{\mathcal{R}}$. Consider a sequent S with partition (S^r, S^i), which has a derivation of which the last inference is an instance $R = (S_1 \dots S_n / S)$ of $\mathcal{R}$. To prove $(\text{DPP})_{\mathcal{R}}$, thus in case S^i is principal for R, in order to prove

$$\mathfrak{D}_R^p \vdash S^r \cdot (\exists p S^i \Rightarrow \forall p S^i \mid \varnothing)$$

it suffices to show that

$$\mathfrak{D}_R^p \vdash S^r \cdot (\exists \overset{R^i}{p} S^i \Rightarrow \forall \overset{R^i}{p} S^i \mid \varnothing)$$

for some partition (R^r, R^i) of R with conclusion (S^r, S^i) such that R^i is an instance of $\mathcal{R}$. The reason being, that for such an R^i, $\exists \overset{R^i}{p} S^i$ is a conjunct of $\exists p S^i$ and $\forall \overset{R^i}{p} S^i$ a disjunct of $\forall p S^i$. Likewise, to prove $(\text{DPN})_{\mathcal{R}}$, thus in case S^i is nonprincipal for R, to prove that

$$\vdash S^r \cdot (\exists p S^i \Rightarrow \forall p S^i \mid \varnothing)$$

it suffices to prove that

$$\vdash S^r \cdot (\exists \overset{\overline{\mathcal{R}}}{p} S^i \Rightarrow \forall \overset{\overline{\mathcal{R}}}{p} S^i \mid \varnothing).$$

Corollary 32 and Theorem 31 in Iemhoff (2019b) state the following.

Theorem 8.4 *Any intuitionistic modal logic* L *with a balanced calculus that consists of* $\mathsf{G4iK}_\Box$, *focused rules, and modal focused rules, has uniform interpolation.*

Theorem 8.5 *Any intuitionistic modal logic* L *with a balanced calculus that contains* $\mathsf{G4iK}_\Box$ *and has an interpolant assignment that is sound with respect to all rules that neither are focused, nor modal focused, nor belong to* $\mathsf{G4iK}_\Box$, *has uniform interpolation.*

Given the fact that all rules of G3ip are focused, as shown in Iemhoff (2019b), to prove that PLL has uniform interpolation it suffices to show that the four extra rules of G4iLL are sound with respect to an interpolant assignment that is an extension of the one from Iemhoff (2019b), which is called the *standard interpolant assignment*. This is shown in the next section.

8.6.6 Interpolant Assignment for PLL

Here the interpolant assignment for the four modal rules of G4iLL is introduced. Its soundness is proved in the next section.

For $R = S_1/S$ being an instance of $\mathcal{R} = R\bigcirc$ or $\mathcal{R} = L\bigcirc$, the interpolant assignment is defined as follows:

$$\exists \overset{R}{p} S \equiv_{df} \bigcirc \exists p S_1 \quad \forall \overset{R}{p} S \equiv_{df} \bigcirc \forall p S_1$$
$$\exists \overset{\overline{\mathcal{R}}}{p} S \equiv_{df} \top \qquad \forall \overset{\overline{\mathcal{R}}}{p} S \equiv_{df} \bot$$

For R being an instance

$$\frac{S_1 \quad S_2}{S} = \frac{\Gamma \Rightarrow \varphi \quad \Gamma, \psi \Rightarrow \Delta}{\Gamma, \bigcirc\varphi \to \psi \Rightarrow \Delta}$$

of $\mathcal{R} = R\bigcirc^{\rightarrow}$, the interpolant assignment is defined as follows:

$$\begin{array}{lll} \exists \overset{R}{p} S \equiv_{df} \exists p S_1 \wedge (\forall p S_1 \rightarrow \exists p S_2) & \forall \overset{R}{p} S \equiv_{df} \forall p S_1 \wedge \forall p S_2 & \\ \exists \overset{\overline{\mathcal{R}}}{p} S \equiv_{df} \top & \forall \overset{\overline{\mathcal{R}}}{p} S \equiv_{df} \bot & \text{if } S^s = \varnothing \\ \exists \overset{\overline{\mathcal{R}}}{p} S \equiv_{df} \exists p (S^a \Rightarrow) & \forall \overset{\overline{\mathcal{R}}}{p} S \equiv_{df} \bot & \text{if } S^s \neq \varnothing \end{array}$$

For R being an instance

$$\frac{S_1 \quad S_2}{S} = \frac{\Gamma, \chi \Rightarrow \bigcirc\varphi \quad \Gamma, \bigcirc\chi, \psi \Rightarrow \Delta}{\Gamma, \bigcirc\chi, \bigcirc\varphi \rightarrow \psi \Rightarrow \Delta}$$

of $\mathcal{R} = L\bigcirc^{\rightarrow}$, the interpolant assignment is defined as follows:

$$\exists \overset{R}{p} S \equiv_{df} \bigcirc\exists p S_1 \wedge (\bigcirc\forall p S_1 \rightarrow \exists p S_2) \quad \forall \overset{R}{p} S \equiv_{df} \bigcirc\forall p S_1 \wedge \forall p S_2.$$

For $\exists \overset{\overline{\mathcal{R}}}{p} S$ and $\forall \overset{\overline{\mathcal{R}}}{p} S$, first define for $\gamma = \bigcirc\alpha \rightarrow \beta$:

$$\begin{array}{l} S^{\gamma 0} \equiv_{df} (S^a \backslash \{\gamma\} \rightarrow \bigcirc\alpha) \\ S^{\gamma 1} \equiv_{df} (S^a \backslash \{\gamma\}, \beta \Rightarrow S^s) \\ S^{\gamma} \equiv_{df} \bigwedge\limits_{\bigcirc\alpha \in S^a} \bigcirc\exists p (S^a \backslash \{\bigcirc\alpha\}, \alpha \Rightarrow) \wedge \bigwedge\limits_{\gamma = \bigcirc\alpha \rightarrow \beta \in S^a} \exists p S^{\gamma 0} \wedge (\bigcirc\forall p S^{\gamma 0} \rightarrow \exists p S^{\gamma 1}) \end{array}$$

Then let

$$\exists \overset{\overline{\mathcal{R}}}{p} S \equiv_{df} \begin{cases} S^{\gamma} & \text{if } S^s = \varnothing \\ \exists p (S^a \Rightarrow) \wedge S^{\gamma} & \text{if } S^s \neq \varnothing \end{cases}$$

$$\forall \overset{\overline{\mathcal{R}}}{p} S \equiv_{df} \bigvee\nolimits_{\gamma = \bigcirc\alpha \rightarrow \beta \in S^a} \bigcirc\forall p S^{\gamma 0} \wedge \forall p S^{\gamma 1}$$

8.6.7 *Soundness of the Interpolant Assignment*

Lemma 8.3 *For $\mathcal{R} = R\bigcirc$ all independent inductive properties and (DPP)$_{\mathcal{R}}$ hold.*

Proof That (IPN)$^{\exists}_{\mathcal{R}}$ and (IPN)$^{\forall}_{\mathcal{R}}$ hold is clear. For (IPP)$^{\exists}_{\mathcal{R}}$ and (IPP)$^{\forall}_{\mathcal{R}}$, let R be an instance

$$\frac{S_1}{S} = \frac{\Gamma \Rightarrow \varphi}{\Gamma \Rightarrow \bigcirc\varphi}$$

of $\mathcal{R} = R\bigcirc$, $\exists \overset{R}{p} S$ and $\forall \overset{R}{p} S$ are defined as $\bigcirc\exists p S_1$ and $\bigcirc\forall p S_1$, respectively. Since $(\Gamma \Rightarrow \exists p S_1)$ and $(\Gamma, \forall p S_1 \Rightarrow \varphi)$ belong to $\mathfrak{I}^p_R$, an application of $\mathcal{R}$ shows that $\mathfrak{I}^p_R \vdash \Gamma \Rightarrow \exists \overset{R}{p} S$ holds, and an application of $\mathcal{R}$ followed by an application of $L\bigcirc$ shows that $\mathfrak{I}^p_R \vdash \Gamma, \forall \overset{R}{p} S \Rightarrow \bigcirc\varphi$ holds.

For $(\mathrm{DPP})_{\mathcal{R}}$, in a partition of S in which S^i is principal, S^i has to be of the form $S^i = (\Gamma^i \Rightarrow \bigcirc\varphi)$, where $\Gamma^i \subseteq \Gamma$. We have to show that $\mathfrak{D}_R^p \vdash S^r \cdot (\exists p S^i \Rightarrow \forall p S^i \mid \varnothing)$, which in this case means

$$\mathfrak{D}_R^p \vdash \Gamma^r, \exists p S^i \Rightarrow \forall p S^i. \tag{8.1}$$

Let $S_1^i = (\Gamma^i \Rightarrow \varphi)$. Thus $\Gamma^r, \exists p S_1^i \Rightarrow \forall p S_1^i$ belongs to $\mathfrak{D}_R^p$. An application of $R\bigcirc$ followed by $L\bigcirc$ proves that $\mathfrak{D}_R^p \vdash \Gamma^r, \bigcirc\exists p S_1^i \Rightarrow \bigcirc\forall p_1 S^i$. Since S_1^i/S^i is an instance of $\mathcal{R}$, $\bigcirc\exists p S_1^i$ and $\bigcirc\forall p S_1^i$ are a conjunct and a disjunct of $\exists p S^i$ and $\forall p S^i$, respectively. This implies (8.1). ⊣

Lemma 8.4 *For $\mathcal{R} = L\bigcirc$ all independent inductive properties and $(DPP)_{\mathcal{R}}$ hold.*

Proof That $(\mathrm{IPN})_{\mathcal{R}}^{\exists}$ and $(\mathrm{IPN})_{\mathcal{R}}^{\forall}$ hold is clear. The proofs of $(\mathrm{IPP})_{\mathcal{R}}^{\exists}$ and $(\mathrm{IPP})_{\mathcal{R}}^{\forall}$ are similar to the one for $R\bigcirc$, but we include them nevertheless. For R being an instance

$$\frac{S_1}{S} = \frac{\Gamma, \psi \Rightarrow \bigcirc\varphi}{\Gamma, \bigcirc\psi \Rightarrow \bigcirc\varphi}$$

of $\mathcal{R} = L\bigcirc$, $\exists_p^R S$ and $\forall_p^R S$ are defined as $\bigcirc\exists p S_1$ and $\bigcirc\forall p S_1$, respectively. Thus $(\Gamma, \psi \Rightarrow \exists p S_1)$ and $(\Gamma, \psi, \forall p S_1 \Rightarrow \bigcirc\varphi)$ belong to $\mathfrak{I}_R^p$. An application of $R\bigcirc$ followed by an application of $\mathcal{R}$ to the former shows that $\mathfrak{I}_R^p \vdash \Gamma, \bigcirc\psi \Rightarrow \exists_p^R S$ holds. Two applications of $\mathcal{R}$ to the latter show that $\mathfrak{I}_R^p \vdash \Gamma, \bigcirc\psi, \forall_p^R S \Rightarrow \bigcirc\varphi$ holds.

For $(\mathrm{DPP})_{\mathcal{R}}$, we distinguish the case that S^{is} is empty and that it is not. Note that S^{ia} has to contain $\bigcirc\psi$ because S^i is supposed to be principal for $L\bigcirc$. In the first case, $S^i = (\Gamma^i, \bigcirc\psi \Rightarrow)$ for some $\Gamma^i \subseteq \Gamma$. We have to show that $\mathfrak{D}_R^p \vdash S^r \cdot (\exists p S^i \Rightarrow \forall p S^i \mid \varnothing)$, which in this case means

$$\mathfrak{D}_R^p \vdash \Gamma^r, \exists p S^i \Rightarrow \bigcirc\varphi. \tag{8.2}$$

Consider the partition of S_1 such that $S_1^i = (\Gamma^i, \psi \Rightarrow\ )$. Thus $\Gamma^r, \exists p S_1^i \Rightarrow \bigcirc\varphi$ belongs to $\mathfrak{D}_R^p$ and an application of $L\bigcirc$ gives $\mathfrak{D}_R^p \vdash \Gamma^r, \bigcirc\exists p S_1^i \Rightarrow \bigcirc\varphi$ Since S_1^i/S^i is an instance of $\mathcal{R}$, $\bigcirc\exists p S_1^i$ is a conjunct of $\exists p S^i$. This implies (8.2).

In the second case, $S^i = (\Gamma^i, \bigcirc\psi \Rightarrow \bigcirc\varphi)$ for some $\Gamma^i \subseteq \Gamma$. We have to show that $\mathfrak{D}_R^p \vdash S^r \cdot (\exists p S^i \Rightarrow \forall p S^i \mid \varnothing)$, which in this case means

$$\mathfrak{D}_R^p \vdash \Gamma^r, \exists p S^i \Rightarrow \forall p S^i. \tag{8.3}$$

Consider the partition of S_1 such that $S_1^i = (\Gamma^i, \psi \Rightarrow \bigcirc\varphi)$. Thus $\Gamma^r, \exists p S_1^i \Rightarrow \forall p S_1^i$ belongs to $\mathfrak{D}_R^p$. An application of $R\bigcirc$ followed by $L\bigcirc$ proves $\mathfrak{D}_R^p \vdash \Gamma^r, \bigcirc\exists p S_1^i \Rightarrow \bigcirc\forall p S_1^i$. Since S_1^i/S^i is an instance of $\mathcal{R}$, $\bigcirc\exists p S_1^i$ and $\bigcirc\forall p S_1^i$ are a conjunct and a disjunct of $\exists p S^i$ and $\forall p S^i$, respectively. This implies (8.3). ⊣

Lemma 8.5 *For $\mathcal{R} = R\bigcirc^{\rightarrow}$ all independent inductive properties and $(DPP)_{\mathcal{R}}$ hold.*

Proof That $(\mathrm{IPN})^{\forall}_{\mathcal{R}}$ holds is clear. For $(\mathrm{IPP})^{\exists}_{\mathcal{R}}$ and $(\mathrm{IPP})^{\forall}_{\mathcal{R}}$, let R be an instance

$$\frac{S_1 \quad S_2}{S} = \frac{\Gamma \Rightarrow \varphi \quad \Gamma, \psi \Rightarrow \Delta}{\Gamma, \bigcirc\varphi \to \psi \Rightarrow \Delta}$$

of $\mathcal{R} = R\bigcirc^{\to}$. Since $\exists \overset{R}{p} S$ and $\forall \overset{R}{p} S$ are defined as $\exists p S_1 \wedge (\forall p S_1 \to \exists p S_2)$ and $\forall p S_1 \wedge \forall p S_2$, respectively, we have to show that

$$\mathfrak{I}^p_R \vdash \Gamma, \bigcirc\varphi \to \psi \Rightarrow \exists p S_1 \wedge (\forall p S_1 \to \exists p S_2) \quad \mathfrak{I}^p_R \vdash \Gamma, \forall p S_1 \wedge \forall p S_2, \bigcirc\varphi \to \psi \Rightarrow \Delta.$$

This follows from applications of $\mathcal{R}$ to sequents $(\Gamma \Rightarrow \exists p S_1)$, $(\Gamma, \psi \Rightarrow \exists p S_2)$, $(\Gamma, \forall p S_1 \Rightarrow \varphi)$ and $(\Gamma, \psi, \forall p S_2 \Rightarrow \Delta)$, that all belong to $\mathfrak{I}^p_R$.

For $(\mathrm{IPN})^{\exists}_{\mathcal{R}}$ only the case that Δ is not empty is nontrivial. Thus suppose that S is not principal for R and that all sequents lower than S satisfy the interpolant properties. Let $S_0 = (S^a \Rightarrow)$. Hence $\exists \overset{\overline{\mathcal{R}}}{p} S = \exists p S_0$. Because $S_0 \ll S$, S_0 satisfies the interpolant properties, and thus $S^a_0 \Rightarrow \exists p S_0$ is derivable. Since $S^a = S^a_0$, it follows that $\vdash S^a \Rightarrow \exists \overset{\overline{\mathcal{R}}}{p} S$.

For $(\mathrm{DPP})_{\mathcal{R}}$, in a partition of S in which S^i is principal, S^i has to be of the form $S^i = (\Gamma^i, \bigcirc\varphi \to \psi \Rightarrow \Delta^i)$, where $\Gamma^i \subseteq \Gamma$ and $\Delta^i \subseteq \Delta$. We treat the case that Δ^r is empty, the other case being analogous. We have to show that

$$\mathfrak{D}^p_R \vdash S^r \cdot (\exists p S^i \Rightarrow \forall p S^i). \tag{8.4}$$

Let the partitions of the premisses be given by $S^i_1 = (\Gamma^i \Rightarrow \varphi)$ and $S^i_2 = (\Gamma^i, \psi \Rightarrow \Delta^i)$. Thus $\Gamma^r, \exists p S^i_1 \Rightarrow \forall p S^i_1$ and $S^r_2 \cdot (\exists p S^i_2 \Rightarrow \forall p S^i_2)$ belong to $\mathfrak{D}^p_R$. As $(S^i_1 \ S^i_2 / S^i)$ is an instance of $\mathcal{R}$, $\exists p S^i$ contains a conjunct $\exists p S^i_1 \wedge (\forall p S^i_1 \to \exists p S^i_2)$. Therefore $\mathfrak{D}^p_R$ derives $(\Gamma^r, \exists p S^i \Rightarrow \forall p S^i_1)$ and $(\Gamma^r, \exists p S^i \Rightarrow \forall p S^i_2)$. Since $\forall p S^i$ contains a disjunct $\forall p S^i_1 \wedge \forall p S^i_2$, (8.4) follows. ⊣

Lemma 8.6 *For $\mathcal{R} = L\bigcirc^{\to}$ all independent inductive properties and $(\mathrm{DPP})_{\mathcal{R}}$ hold.*

Proof For $(\mathrm{IPP})^{\exists}_{\mathcal{R}}$ and $(\mathrm{IPP})^{\forall}_{\mathcal{R}}$, let R be an instance

$$\frac{S_1 \quad S_2}{S} = \frac{\Gamma, \chi \Rightarrow \bigcirc\varphi \quad \Gamma, \bigcirc\chi, \psi \Rightarrow \Delta}{\Gamma, \bigcirc\chi, \bigcirc\varphi \to \psi \Rightarrow \Delta}$$

of $\mathcal{R} = L\bigcirc^{\to}$. Given the definition of $\exists \overset{R}{p} S$ and $\forall \overset{R}{p} S$ in this case, we have to show that

$$\mathfrak{I}^p_R \vdash \Gamma, \bigcirc\chi, \bigcirc\varphi \to \psi \Rightarrow \bigcirc\exists p S_1 \wedge (\bigcirc\forall p S_1 \to \exists p S_2)$$
$$\mathfrak{I}^p_R \vdash \Gamma, \bigcirc\chi, \bigcirc\forall p S_1, \forall p S_2, \bigcirc\varphi \to \psi \Rightarrow \Delta.$$

This follows, via applications of $\mathcal{R}$, $R\bigcirc$ and $L\bigcirc$, from the fact that $(\Gamma, \chi \Rightarrow \exists p S_1)$ and $(\Gamma, \bigcirc\chi, \psi \Rightarrow \exists p S_2)$ as well as $(\Gamma, \chi, \forall p S_1 \Rightarrow \bigcirc\varphi)$ and $(\Gamma, \bigcirc\chi, \psi, \forall p S_2 \Rightarrow \Delta)$ belong to $\mathfrak{I}^p_R$.

For $(\mathrm{IPN})^{\exists}_{\mathcal{R}}$ we have to show three things, under the assumption that all sequents lower than S satisfy the interpolant properties:

(1) $\vdash S^a \Rightarrow \exists p(S^a \Rightarrow)$ in case $S^s \neq \varnothing$;
(2) $\vdash S^a \Rightarrow \bigcirc\exists p(S^a\backslash\{\bigcirc\alpha\}, \alpha \Rightarrow)$ for all $\bigcirc\alpha \in S^a$;
(3) $\vdash S^a \Rightarrow \exists p S^{\gamma 0} \wedge (\bigcirc\forall p S^{\gamma 0} \rightarrow \exists p S^{\gamma 1})$ for all $\gamma = \bigcirc\alpha \rightarrow \beta$ in S^a.

(1) follows immediately from the assumption that all sequents lower than S satisfy the interpolant properties and the fact that $(S^a \Rightarrow) \ll S$, which holds if $S^s \neq \varnothing$.
(2) Note that $\big(S^a\backslash\{\bigcirc\alpha\}, \alpha \Rightarrow \exists p(S^a\backslash\{\bigcirc\alpha\}, \alpha \Rightarrow)\big)$ is derivable because all sequents lower than S are assumed to satisfy the interpolant properties. An application of $R\bigcirc$ followed by an application of $L\bigcirc$ gives the desired result.
(3) Assume that $S = (\Pi, \gamma \Rightarrow \Sigma)$, where $\gamma = \bigcirc\alpha \rightarrow \beta$. We use the following abbreviations:

$$S_\alpha \equiv_{df} S^{\gamma 0} = (S^a\backslash\{\gamma\} \Rightarrow \bigcirc\alpha) = (\Pi \Rightarrow \bigcirc\alpha)$$
$$S_\beta \equiv_{df} S^{\gamma 1} = (S^a\backslash\{\gamma\}, \beta \Rightarrow S^s) = (\Pi, \beta \Rightarrow \Sigma).$$

Thus we have to show that $(S^a \Rightarrow \exists p S_\alpha)$ and $(S^a, \bigcirc\forall p S_\alpha \Rightarrow \exists p S_\beta)$ are derivable. The former follows immediately from the fact that S_α and S_β satisfy the interpolant properties, since they are below S. By the same fact, $(S^a_\beta \Rightarrow \exists p S_\beta)$ is derivable. As $\big(S^a_\beta \Rightarrow \exists p S_\beta\big)$ is equal to $(\Pi, \beta \Rightarrow \exists p S_\beta)$, sequent $(\Pi, \beta, \bigcirc\forall p S_\alpha \Rightarrow \exists p S_\beta)$ is derivable too. An application of $L\bigcirc^{\rightarrow}$ to this sequent and $(S^a_\alpha, \forall p S_\alpha \Rightarrow S^s_\alpha) = (\Pi, \forall p S_\alpha \Rightarrow \bigcirc\alpha)$, which by the same fact is derivable, shows that $\vdash \Pi, \gamma, \bigcirc\forall p S_\alpha \Rightarrow \exists p S_\beta$, which is what we had to show.

For property $(\mathrm{IPN})^{\forall}_{\mathcal{R}}$ we use the same notation as in (3) of the previous case and have to show that

$$\vdash S^a, \bigcirc\forall p S_\alpha, \forall p S_\beta \Rightarrow S^s.$$

By the assumption that all sequents lower than S satisfy the interpolant properties, both $(S^a_\alpha, \forall p S_\alpha \Rightarrow S^s_\alpha) = (\Pi, \forall p S_\alpha \Rightarrow \bigcirc\alpha)$ and $(S^a_\beta, \forall p S_\beta \Rightarrow S^s_\beta) = (\Pi, \beta, \forall p S_\beta \Rightarrow \Sigma)$ are derivable. Clearly, applications of weakening, $L\bigcirc$ and $L\bigcirc^{\rightarrow}$ show that $\vdash \Pi, \bigcirc\alpha \rightarrow \beta, \bigcirc\forall p S_\alpha, \forall p S_\beta \Rightarrow \Sigma$, which is what we had to show.

For $(\mathrm{DPP})_{\mathcal{R}}$, in a partition of S in which S^i is principal, S^i has to be of the form $S^i = (\Gamma^i, \bigcirc\chi, \bigcirc\varphi \rightarrow \psi \Rightarrow \Delta^i)$, where $\Gamma^i \subseteq \Gamma$ and $\Delta^i \subseteq \Delta$. We have to show that

$$\mathfrak{D}^p_R \vdash S^r \cdot (\exists p S^i \Rightarrow \forall p S^i \mid \varnothing) \tag{8.5}$$

We treat the case that $\Delta^r = \varnothing$, the other case is analogous. Let the partitions of the premisses be given by $S^i_1 = (\Gamma^i, \chi \Rightarrow \varphi)$ and $S^i_2 = (\Gamma^i, \bigcirc\chi, \psi \Rightarrow \Delta^i)$. Thus $(\Gamma^r, \exists p S^i_1 \Rightarrow \forall p S^i_1)$ and $S^r_2 \cdot (\exists p S^i_2 \Rightarrow \forall p S^i_2)$ belong to $\mathfrak{D}^p_R$. Hence $\mathfrak{D}^p_R$ derives $(\Gamma^r, \bigcirc\exists p S^i_1 \Rightarrow \bigcirc\forall p S^i_1)$. As $(S^i_1 \; S^i_2/S^i)$ is an instance of $\mathcal{R}$, $\exists p S^i$ contains a conjunct $\bigcirc\exists p S^i_1 \wedge (\bigcirc\forall p S^i_1 \rightarrow \exists p S^i_2)$. Therefore $\mathfrak{D}^p_R$ derives $(\Gamma^r, \exists p S^i \Rightarrow \bigcirc\forall p S^i_1)$

and $(\Gamma^r, \exists p S^i \Rightarrow \forall p S^i_2)$. Since $\forall p S^i$ contains a disjunct $\bigcirc \forall p S^i_1 \wedge \forall p S^i_2$, (8.5) follows. ⊣

Lemma 8.7 *For all modal and implication-modal rules $\mathcal{R}$, the interpolant assignment satisfies (DPN)$_{\mathcal{R}}$.*

Proof (DPN)$_{\mathcal{R}}$ is proved with induction to the depth of the derivation $\mathcal{D}$ of S for all rules at the same time. Thus we have to show for any partition (S^r, S^i) of S such that S^i is nonprincipal for the last inference R of $\mathcal{D}$: if all sequents below S satisfy the interpolant properties, then

$$\vdash S^r \cdot (\exists p S^i \Rightarrow \forall p S^i \mid \varnothing). \tag{8.6}$$

Note that for the rules in G4ip (DPN)$_{\mathcal{R}}$ has already been established in Iemhoff (2019b). If $\mathcal{D}$ has depth one, S is an instance of an axiom. If the axiom is $L\bot$, then because S^i is not an instance of it, $\bot \in S^{ra}$. Hence (8.6). If the axiom is Ax, say $S = (\Gamma, q \Rightarrow q, \Delta)$, then if $q \in S^{ra} \cap S^{rs}$, we have (8.6). Therefore assume otherwise: $q \notin S^{ra} \cap S^{rs}$. Because S^i is not an instance of Ax, $q \notin S^{ia} \cap S^{is}$. Thus $q \in S^{ia} \cap S^{rs}$ or $q \in S^{ra} \cap S^{is}$. Since q occurs in S^r, $q \neq p$. Therefore $\vdash q \Rightarrow S^{rs}$ and $q \in S^{ia}$, and thus $\vdash \exists^{at} p S^i \Rightarrow q$, or $\vdash S^{ra} \Rightarrow q$ and $q \in S^{is}$, and thus $\vdash q \Rightarrow \forall^{at} p S^i$. In both cases (8.6) follows, which completes the basis case of the induction.

Suppose the derivation of S has depth bigger than one and ends with an application of $\mathcal{R}$. We show that (DPN)$_{\mathcal{R}}$ holds for S. If $\mathcal{R}$ is not a modal or implication-modal rule, then we already know that (DPN)$_{\mathcal{R}}$ holds for all S. Therefore suppose $\mathcal{R}$ is such a rule. We consider the four rules separately.

If $\mathcal{R}$ is $R\bigcirc$, $S = (\Gamma \Rightarrow \bigcirc\varphi)$ and the premise of the last inference is $S_1 = (\Gamma \Rightarrow \varphi)$. Since we consider (DPN)$_{\mathcal{R}}$, $S^i = (\Gamma^i \Rightarrow)$. We have to show that $\mathfrak{D}^p_R \vdash \Gamma^r, \exists p S^i \Rightarrow \bigcirc\varphi$. Consider the partition of S_1 with $S^i_1 = S^i$. We have $\mathfrak{D}^p_R \vdash \Gamma^r, \exists p S^i_1 \Rightarrow \varphi$ by the induction hypothesis, which implies what we had to show via the consecutive application of $R\bigcirc$ and $L\bigcirc$.

If $\mathcal{R}$ is $L\bigcirc$, $S = (\Gamma, \bigcirc\psi \Rightarrow \bigcirc\varphi)$ and the premise of the last inference is $S_1 = (\Gamma, \psi \Rightarrow \bigcirc\varphi)$. Consider a partition of S such that S^i is not principal for $L\bigcirc$. Thus $S^i = (\Gamma^i \Rightarrow \bigcirc\varphi)$ or $S^i = (\Gamma^i \Rightarrow)$ for some $\Gamma^i \subseteq \Gamma$. We have to show that $\mathfrak{D}^p_R \vdash \Gamma^r, \bigcirc\psi, \exists p S^i \Rightarrow \forall p S^i$ in the first case, and $\mathfrak{D}^p_R \vdash \Gamma^r, \bigcirc\psi, \exists p S^i \Rightarrow \bigcirc\varphi$ in the second case. In both cases we can partition S^i_1 such that $S^i_1 = S^i$. If $S^i = (\Gamma^i \Rightarrow \bigcirc\varphi)$, we have $\mathfrak{D}^p_R \vdash \Gamma^r, \psi, \exists p S^i_1 \Rightarrow \bigcirc\varphi$ by the induction hypothesis. An application of $L\bigcirc$ gives $\mathfrak{D}^p_R \vdash \Gamma^r, \bigcirc\psi, \exists p S^i_1 \Rightarrow \bigcirc\varphi$, and since $S^i_1 = S^i$, this is what we had to show. The case that $S^i = (\Gamma^i \Rightarrow)$ is analogous.

If $\mathcal{R}$ is $R\bigcirc^{\rightarrow}$, the last inference is of the following form:

$$\frac{S_1 \quad S_2}{S} = \frac{\Gamma \Rightarrow \varphi \quad \Gamma, \psi \Rightarrow \Delta}{\Gamma, \bigcirc\varphi \to \psi \Rightarrow \Delta}$$

As $\mathcal{R}$ is not applicable to S^i, $S^i = (\Gamma^i \Rightarrow \Delta^i)$ and $S^r = (\Gamma^r, \bigcirc\varphi \to \psi \Rightarrow \Delta^r)$. We treat the case that $\Delta^r = \varnothing$, the other case is analogous. Thus to show (8.6), we have to show that

$$\vdash \Gamma^r, \bigcirc\varphi \to \psi, \exists p S^i \Rightarrow \forall p S^i. \tag{8.7}$$

Let $S_1^i = (\Gamma^i \Rightarrow\ )$ and $S_2^i = S^i$. By the induction hypothesis,

$$\vdash \Gamma^r, \exists p S_1^i \Rightarrow \varphi \quad \vdash \Gamma^r, \psi, \exists p S_2^i \Rightarrow \forall p S_2^i.$$

Therefore $\vdash \Gamma^r, \bigcirc\varphi \to \psi, \exists p S_1^i, \exists p S_2^i \Rightarrow \forall p S_2^i$. Since $S_2^i = S^i$, for (8.7) it suffices to show that $\exists p S^i$ derives $\exists p S_1^i$. If $S^{is} = \varnothing$, then $S^i = S_1^i$, and if $S^{is} \neq \varnothing$, then $\exists p S_1^i$ is a conjunct of $\exists \overline{p}^{\mathcal{R}} S^i$, which is a conjunct of $\exists p S^i$.

If $\mathcal{R}$ is $L\bigcirc^{\to}$, the last inference is of the following form:

$$\frac{S_1 \quad S_2}{S} = \frac{\Gamma, \chi \Rightarrow \bigcirc\varphi \quad \Gamma, \bigcirc\chi, \psi \Rightarrow \Delta}{\Gamma, \bigcirc\chi, \bigcirc\varphi \to \psi \Rightarrow \Delta}$$

We distinguish three cases.

(1) If $S^i = (\Gamma^i \Rightarrow \Delta^i)$, then let $S_1^i = (\Gamma^i \Rightarrow\ )$ and $S_2^i = S^i$. We treat the case that $\Delta^r \neq \varnothing$, the other case is analogous. Thus to show (8.6), we have to show that

$$\vdash \Gamma^r, \bigcirc\chi, \bigcirc\varphi \to \psi, \exists p S^i \Rightarrow \Delta^r. \tag{8.8}$$

By the induction hypothesis,

$$\vdash \Gamma^r, \chi, \exists p S_1^i \Rightarrow \bigcirc\varphi \quad \vdash \Gamma^r, \bigcirc\chi, \psi, \exists p S_2^i \Rightarrow \Delta^r.$$

Therefore $\vdash \Gamma^r, \bigcirc\chi, \bigcirc\varphi \to \psi, \exists p S_1^i, \exists p S_2^i \Rightarrow \Delta^r$. If $\Delta^i = \varnothing$, then $S^i = S_1^i = S_2^i$, which implies (8.6). Therefore assume that $\Delta^i = S^{is}$ is not empty. Thus $\exists p S_1^i$ is a conjunct of $\exists \overline{p}^{\mathcal{R}} S^i$, which is a conjunct of $\exists p S^i$. Together with $S_2^i = S^i$ this implies (8.8).

(2) If $S^i = (\Gamma^i, \bigcirc\chi \Rightarrow \Delta^i)$, then let $S_1^i = (\Gamma^i, \chi \Rightarrow\ )$ and $S_2^i = S^i$. We treat the case that $\Delta^r = \varnothing$, the other case is analogous. Thus to show (8.6), we have to show that

$$\vdash \Gamma^r, \bigcirc\varphi \to \psi, \exists p S^i \Rightarrow \forall p S^i. \tag{8.9}$$

By the induction hypothesis, $\vdash \Gamma^r, \exists p S_1^i \Rightarrow \bigcirc\varphi$ and $\vdash \Gamma^r, \psi, \exists p S_2^i \Rightarrow \forall p S_2^i$ are derivable. An application of $L\bigcirc^{\to}$ gives

$$\vdash \Gamma^r, \bigcirc\varphi \to \psi, \bigcirc\exists p S_1^i, \exists p S_2^i \Rightarrow \forall p S_2^i.$$

Note that because $S_1^i = (S^i \backslash \{\bigcirc\chi\}, \chi \Rightarrow\ )$, $\bigcirc\exists p S_1^i$ is a conjunct of $\exists \overline{p}^{\mathcal{R}} S^i$, which is a conjunct of $\exists p S^i$. Since also $\exists p S_1^i \Rightarrow \bigcirc\exists p S_1^i$ is derivable and $S_2^i = S^i$, we obtain (8.9).

(3) If $S^i = (\Gamma^i, \bigcirc\varphi \to \psi \Rightarrow \Delta^i)$, then let $S_1^i = (\Gamma^i \Rightarrow \bigcirc\varphi)$ and $S_2^i = (\Gamma^i, \psi \Rightarrow \Delta^i)$. We treat the case that $\Delta^r = \varnothing$, the other case is analogous. Thus to show (8.6), we have to show that

$$\vdash \Gamma^r, \bigcirc\mathcal{X}, \exists p S^i \Rightarrow \forall p S^i. \tag{8.10}$$

By the induction hypothesis, both $(\Gamma^r, \mathcal{X}, \exists p S_1^i \Rightarrow \forall p S_1^i)$ and $S_3 = (\Gamma^r, \bigcirc\mathcal{X}, \exists p S_2^i \Rightarrow \forall p S_2^i)$ are derivable. Thus so are $S_4 = (\Gamma^r, \mathcal{X}, \exists p S_1^i \Rightarrow \bigcirc\forall p S_1^i)$ and $(\Gamma^r, \bigcirc\mathcal{X}, \exists p S_1^i \Rightarrow \bigcirc\forall p S_1^i)$ by applications of $R\bigcirc$ and $L\bigcirc$. An application of $L\bigcirc^{\rightarrow}$ to S_4 and S_3 shows the derivability of $(\Gamma^r, \bigcirc\mathcal{X}, \exists p S_1^i, \bigcirc\forall p S_1^i \rightarrow \exists p S_2^i \Rightarrow \forall p S_2^i)$. Therefore

$$\vdash \Gamma^r, \bigcirc\mathcal{X}, \exists p S_1^i, \bigcirc\forall p S_1^i \rightarrow \exists p S_2^i \Rightarrow \bigcirc\forall p S_1^i \wedge \forall p S_2^i.$$

Let $\gamma = \bigcirc\varphi \rightarrow \psi$. Since $S_1^i = (S^i)^{\gamma 0}$ and $S_2^i = (S^i)^{\gamma 1}$, $\exists p S_1^i$ and $\bigcirc\forall p S_1^i \rightarrow \exists p S_2^i$ are conjuncts of $\exists \overset{\mathcal{R}}{p} S^i$, which is a conjunct of $\exists p S^i$. And since $\bigcirc\forall p S_1^i \wedge \forall p S_2^i$ is a disjunct of $\forall \overset{\mathcal{R}}{p} S^i$ and thus of $\forall p S^i$, we can conclude (8.10). ⊣

8.6.8 Uniform Interpolation for Lax Logic

Theorem 8.6 PLL *has uniform interpolation.*

Proof As the results in Sect. 8.6.7 show, the interpolant assignment for GLL given in Sect. 8.6.6 is sound. Theorem 8.5 then implies the result. ⊣

8.7 Conclusion

In this paper we have developed proof theory for propositional Lax Logic PLL. In particular, we have provided a cut free terminating sequent calculus G4iLL for the logic, which is used to prove that Lax Logic has uniform interpolation. The latter was established by a proof-theoretic method that defines uniform interpolants in terms of the rules of the calculus.

There are many open questions related to uniform interpolation in intuitionistic modal logics. We hope that the methods in this paper and the closely related (Iemhoff, 2019b; Jalali & Tabatabai, 2018b) will be useful in the study of other logics in this class, and perhaps also in the investigation of some of the more general questions in Universal Proof Theory.

Acknowledgements We thank an anonymous referee for the constructive comments on an earlier draft of this paper and Iris van der Giessen for several useful conversations on Lax Logic, as well as Raheleh Jalali and Amir Tabatabai for numerous discussions about proof theory and uniform interpolation.

References

Benton, P. N., Bierman, G. M., & de Paiva, V. C. V. (1993). Computational types from a logical perspective. *Journal of Functional Programming, 1*(1), 1.

Bezhanishvili, G., Bezhanishvili, N., & Ilin, J. (2019). Subframization and stabilization for superintuitionistic logics. *Journal of Logic and Computation, 29*(1), 1–35.

Bílková, M. (2006). *Interpolation in modal logics*. Ph.D. thesis. Charles University.

Buss, S. R. (1998). Chapter 1 of the *Handbook of proof theory*. In S. R. Buss (Ed.) (pp. 1–78). Elsevier.

Curry, H. B. (1957). A theory of formal deducibility. In *Notre dame mathematical lectures*, 2nd edn. (Vol. 6). Notre Dame.

de Paiva, V., & Goré, R., & Mendler, M. (2004). Modalities in constructive logics and type theories. Preface to the special issue on intuitionistic modal logic and application. *Journal of Logic and Computation, 14*(4), 439–446.

Dershowitz, N., & Manna, Z. (1979). Proving termination with multiset orderings. *Communications of the ACM, 22*, 465–476.

Došen, K. (1985). Models for stronger normal intuitionistic modal logics. *Studia Logica, 44*(1), 39–70.

Dyckhoff, R. (1992). Contraction-free sequent calculi for intuitionistic logic. *Journal of Symbolic Logic, 57*(3), 795–807.

Fairtlough, M., & Mendler, M. (1994). An intuitionistic modal logic with applications to the formal verification of hardware. In L. Pacholski & J. Tiuryn (Eds.), *Computer science logic. CSL 1994. Lecture notes in computer science* (Vol. 933, pp. 354–368). Springer.

Fairtlough, M., & Mendler, M. (1997). Propositional lax logic. *Information and Computation, 137*, 1–33.

Fischer-Servi, G. (1977). On modal logic with an intuitionistic base. *Studia Logica, 36*(3), 141–149.

Gabbay, D. M., & Maxsimova, L. L. (2005). *Interpolation and definability*. Clarendon Press.

Ghilardi, S., & Zawadowski, M. (1995). Undefinability of propositional quantifiers in the modal system *S*4. *Studia Logica, 55*, 259–271.

Howe, J. M. (2001). Proof search in lax logic. *Mathematical Structures in Computer Science, 11*, 573–588.

Hudelmaier, J. (1988). A Prolog program for intuitionistic propositional logic. *SNS-Bericht* (pp. 88–28). Tübingen.

Hudelmaier, J. (1992). Bounds for cut elimination in intuitionistic propositional logic. *Archive for Mathematical Logic, 31*, 331–353.

Hudelmaier, J. (1993). An O(n log n)-space decision procedure for intuitionistic propositional logic. *Journal of Logic and Computation, 3*, 63–76.

Iemhoff, R. (2018). Terminating sequent calculi for two intuitionistic modal logics. *Journal of Logic and Computation, 28*(7), 1701–1712.

Iemhoff, R. (2019). Uniform interpolation and the existence of sequent calculi. *Annals of Pure and Applied Logic, 170*(11), 1–37.

Iemhoff, R. (2019a). Uniform interpolation and sequent calculi in modal logic. *Archive for Mathematical Logic, 58*(1–2), 155–181. Online at https://link.springer.com/article/10.1007/s00153-018-0629-0.

Iemhoff, R. (2020). The G4i counterpart of a G3i calculus. *Studia Logica*, accepted for publication. arXiv:2011.11847

Jalali, R., & Tabatabai, A. (2018a) *Universal proof theory: Semi-analytic rules and craig interpolation*. arXiv:1808.06258.

Jalali, R., & Tabatabai, A. (2018b). *Universal proof theory: Semi-analytic rules and uniform interpolation*. arXiv:1808.06258

Litak, T. (2014). Constructive modalities with provability smack. In G. Bezhanishvili (Ed.), *Leo Esakia on duality in modal and intuitionistic logics. Outstanding Contributions to Logic* (Vol. 4, pp. 179–208). Springer.

Litak, T., & Visser, A. (2018). Lewis meets brouwer: Constructive strict implication. *Indagationes Mathematicæ, 29*(1), 36–90.

Maehara, S. (1960). On the interpolation theorem of Crai. *Sûgaku, 12*(4), 235–237 (in Japanese).

Maksimova, L. L. (1977). Craig's theorem in superintuitionistic logics and amalgamated varieties of pseudo-boolean algebras. *Algebra Logika, 16*(6), 643–681.

Melzer, M. (2020). *Canonical Formulas for the lax logic*, MSc Thesis. University of Amsterdam.

Mendler, M. (1993). *A modal logic for handling behavioural constraints in formal hardware verification*. Ph.D. Thesis. University of Edingburgh.

Negri, S. (2011). Proof theory for modal logic. *Philosophy Compass, 6*(8), 523–538.

Ono, H. (1998). Proof-theoretic methods in nonclassical logic: an introduction. In M. Dezani-Ciancaglini, M. Okada, M. Takahashi (Eds.), *Mathematical society of Japan memoirs* (Vol. 2, pp. 207–254).

Pitts, A. (1992). On an interpretation of second order quantification in first order intuitionistic propositional logic. *Journal of Symbolic Logic, 57*(1), 33–52.

Shavrukov, V. Y. (1993). Subalgebras of diagonalizable algebras of theories containing arithmetic. In *Dissertationes Mathematicae* (Vol. 323). Instytut Matematyczny Polskiej Akademi Nauk.

Švejdar, V. (2006). On sequent calculi for intuitionistic propositional logic. *Commentationes Mathematicae Universitatis Carolinae, 47*(1), 159–173.

Troelstra, A. S. (1998). *Corrections to basic proof theory*. http://citeseerx.ist.psu.edu/viewdoc/summary?doi=10.1.1.7.888

Troelstra, A. S., & Schwichtenberg, H. (1996).*Basic proof theory. Cambridge tracts in theoretical computer science* (Vol. 43). Cambridge University Press.

van der Giessen, I., & Iemhoff, R. (2019). Sequent calculi for intuitionistic Gödel-Löb logic, Submitted.

Vorob'ev, N. N. (1970). A new algorithm for derivability in the constructive propositional calculus. *AMS Translations Series, 2*(94), 37–71.

Vorob'ev, N. N. (1952). The derivability problem in the constructive propositional calculus with strong negation. *Doklady Akademii Nauk SSSR, 85*, 689–692.

Wolter, F., & Zakharyaschev, M. (1999). Intuitionistic modal logic. In A. Cantini, E. Casari & P. Minari (Eds.), *Logic and foundations of mathematics. Synthese library* (Vol. 280, pp. 227–238).

Chapter 9
Intermediate Logics in the Setting of Team Semantics

Nick Bezhanishvili and Fan Yang

Abstract Several authors have recently defined intuitionistic logic based on team semantics (tIPC). In this paper we provide two alternative approaches to intermediate logics in the team semantics setting. We do this by modifying tIPC with axioms written with two different versions of disjunction in the logic, a local one and global one. We prove a characterization theorem in the first approach and we introduce a generalized team semantics in the second one.

Keywords Intuitionistic logic · Intermediate logics · Team semantics · de Jongh formulas

9.1 Introduction

One of the important themes in the research of Dick de Jongh has been the study of intuitionistic and other intermediate logics. De Jongh made a number of important contributions in this field and is, in fact, one of the founders of this area (together with Gabbay, Troelstra, Hosoi, Jankov, Kuznetsov and others). Together with Gabbay, de Jongh defined an important class of intermediate logics nowadays called the *Gabbay-de Jongh logics* (Gabbay & de Jongh, 1974). With Troelstra he developed what is currently known as the *discrete duality for Heyting algebras* (de Jongh & Troelstra, 1966). In 1960s, de Jongh proved in his PhD thesis possibly the first De Jongh Theorem,[1] the characterization of intuitionistic propositional calculus IPC among other intermediate logics in terms of the *Kleene Slash* (de Jongh, 1968) (see also Bezhanishvili, 2004). In order to obtain this characterization de Jongh

[1] There are many others, some of them discussed in this volume.

N. Bezhanishvili
Institute for Logic, Language and Computation, University of Amsterdam, Amsterdam, The Netherlands

F. Yang (✉)
Department of Philosophy and Religious Studies, Utrecht University, Utrecht, The Netherlands
e-mail: fan.yang.c@gmail.com

N. Bezhanishvili et al. (eds.), *Dick de Jongh on Intuitionistic and Provability Logics*, Outstanding Contributions to Logic 28, https://doi.org/10.1007/978-3-031-47921-2_9

introduced a type of formulas that we nowadays call *de Jongh formulas*. This is based on special "coloring" of Kripke structures, which was later developed by several authors to a construction known as the n-universal models for intuitionistic logic (see, e.g., Chagrov and Zakharyaschev, 1997 and Bezhanishvili, 2006). Independently, using algebraic techniques, Jankov (1963) introduced a method of, what we now call, *Jankov* or *splitting formulas* for Heyting algebras. This allowed Jankov to construct intermediate logics without the finite model property and to prove that the lattice of intermediate logics has the cardinality that of the continuum (Jankov, 1963, 1968). It turnes out that Jankov and de Jongh formulas are two sides of the same coin. In particular, the Jankov formula of a finite subdirectly irreducible Heyting algebra A is semantically equivalent to the de Jongh formula of the finite rooted Kripke frame dual to A (for details of this connection we refer to Bezhanishvili (2006, Sect. 3.2.2). Because of this, these formulas are also known as *Jankov-de Jongh formulas* (Bezhanishvili, 2006, Remark 3.3.5). An alternative frame-theoretic description of such formulas can be found in Chagrov and Zakharyaschev (1997, Theorem 9.39). From the modern perspective, de Jongh and Jankov formulas axiomatize splittings and their joins in the lattice of intermediate logics. The analogues of de Jongh formulas for transitive modal logics were developed by Fine (1974) and the analogues of Jankov formulas for $\mathsf{K4}$-modal algebras by Rautenberg (1980). For further generalizations of such kind of formulas and their impact on the theory of modal and intermediate logics we refer to Chagrov and Zakharyaschev (1997, Chap. 9) and Bezhanishvili and Bezhanishvili (2022). De Jongh formulas and the method of universal models has become a standard tool in the study of various fragments of intermediate and other non-classical logics (Bezhanishvili et al., 2015, 2019). As a tribute to Dick de Jongh in the first part of this paper we give an overview of the method of de Jongh formulas and sketch some of their new applications.

In the second, main part of the paper, we study intermediate logics in a new domain of team semantics. We present two different approaches. In the first and simpler approach, we provide yet another new application for de Jongh formulas. The second approach is our first attempt to introduce a generalized team semantics in which intermediate logics can be naturally defined. Thus we lift intermediate logics (de Jongh's paradise) to a new setting. We hope that this paper demonstrates again how the ideas developed by Dick de Jongh decades ago could influence new research lines to this day. We now discuss our motivation and contributions in detail.

Team semantics was originally introduced by Hodges (1997a, 1997b) to characterize notions of dependence in Hintikka and Sandu's *independence friendly-logic* (Hintikka, 1998; Hintikka & Sandu, 1989). This framework was later developed by Väänänen in his *dependence logic* (Väänänen, 2007). Independently, Ciardelli and Roelofsen (2011) developed *inquisitive logic*, which turns out to be a logic also based on team semantics[2] and was shown (Ciardelli, 2016; Yang, 2014) to be actually a variant of propositional dependence logic (Yang & Väänänen, 2016). Both dependence

[2] The fact that inquisitive logic essentially also adopts team semantics was first observed by Dick de Jongh and Tadeusz Litak in a private conversation with the second author at the TbiLLC 2011 symposium in Kutaisi, Georgia.

and inquisitive logic can be understood as conservative extensions of classical logic. Recently several authors have defined different intuitionistic logic-based dependence and inquisitive logic (Ciardelli et al., 2020; Holliday, 2020; Punčochář, 2016, 2017). The starting point of this paper is the team-based intuitionistic logic of Ciardelli et al. (2020).

The key idea of team semantics is that formulas are evaluated in a model with respect to *teams*, which, in the intuitionistic Kripke semantics setting, are defined as *sets* $t \subseteq W$ of possible worlds of an intuitionistic Kripke model $\mathfrak{M} = (W, R, V)$. Such sets t correspond to multitudes of state of affairs, or they express certain uncertainly of the current state of affairs. We also extend the standard language of intuitionistic propositional logic (IPC) with a natural disjunction $\mathbin{\vee\!\!\!\vee}$ (called *global disjunction*) on the team level (i.e., the level of $\wp(W)$). The disjunction $\vee$ of IPC is thus referred to as *local disjunction*. We denote our full language as $[\bot, \wedge, \vee, \mathbin{\vee\!\!\!\vee}, \to]$, and the standard language of IPC as $[\bot, \wedge, \vee, \to]$. Formulas in the standard language $[\bot, \wedge, \vee, \to]$ (i.e., $\mathbin{\vee\!\!\!\vee}$-free formulas) are referred to as *standard formulas*. The logical system (denoted by tIPC) as defined in Ciardelli et al. (2020) is a conservative extension of IPC for standard formulas with additional axioms, including IPC axioms for the language with connectives $\bot, \wedge, \mathbin{\vee\!\!\!\vee}, \to$, and the Split axiom

$$\alpha \to (\phi \mathbin{\vee\!\!\!\vee} \psi) \to (\alpha \to \phi) \mathbin{\vee\!\!\!\vee} (\alpha \to \psi), \qquad \text{(Split)}$$

where α is an arbitrary standard formula (thus tIPC is not closed under uniform substitution).

We provide two alternative approaches to intermediate logics in this setting, by modifying tIPC with axioms written with two different disjunctions. Given an intermediate logic $\mathsf{L} = \mathsf{IPC} \oplus \Delta$ with Δ a set of axioms in the standard language, the first approach defines an intermediate team logic tL by adding to tIPC all instances of Δ with propositional variables substituted with standard formulas only, and closing the resulting set under Modus Ponens. This definition has appeared also in a recent work (Quadrellaro, 2022). A similar but subtly different definition was given in Punčochář (2021). We show, by using a disjunctive normal form $\mathbin{\vee\!\!\!\vee}_{i\in I}\alpha_i$ (with each α_i a standard formula) of the logic, that n-universal models of IPC still behave as universal models for tIPC, and thus de Jongh formulas for team-based intuitionistic logic can be defined in the usual manner. Moreover, a large class of intermediate axioms (including de Jongh formulas) for IPC still characterize the same class of frames in the team semantics setting as in the standard (single-world semantics) setting, in the sense that if an intermediate logic L (over the standard language) is complete with respect to a class F of frames, then tL is also complete with respect to F, provided that L has disjunction property or L is canonical.

In the second approach, we seek appropriate ways to obtain logics in which the Split axiom of tIPC can be replaced with other (weaker) axioms (written in the full language $[\bot, \wedge, \vee, \mathbin{\vee\!\!\!\vee}, \to]$). It is fair to say that the Split axiom is the hallmark of team-based propositional logics (as, e.g., it is a crucial axiom for establishing the important disjunctive normal form $\mathbin{\vee\!\!\!\vee}_{i\in I}\alpha_i$ for these logics). Weakening the Split axiom and at the same time keeping certain unique features of team semantics in some meaningful manner is thus not a trivial task. In this paper, we make a first attempt in this direction. We start by exploring a connection between tIPC and intuionistic modal logic IK (Wolter & Zakharyaschev, 1999), given by the observation that every (intuitionistic) Kripke model $\mathfrak{M} = (W, R, V)$ of tIPC gives rise to a (full) powerset model $\mathfrak{M}^\bullet = (\wp(W), \supseteq, R^\circ, \cup, \varnothing, V^\circ)$, where R° and V° are liftings of R and V, and tIPC can thus be given a single-world semantics interpretation over $\mathfrak{M}^\bullet$. In this setting the local disjunction $\vee$ is naturally understood as a binary diamond modality. Such a connection was studied also in the modal dependence logic setting in Yang (2017). Our powerset model construction also corresponds to the set-lifting as discussed in van Benthem (2021), and our local disjunction $\vee$ is essentially the binary diamond $\langle\text{sup}\rangle$ defined in van Benthem (2021). Inspired by these connections, we then introduce the so-called *general intuitionistic team Kripke models* (which generalizes the full powerset models). Over these models we define a generalized team intuitionsitic Kripke semantics in which the Split axiom is no longer sound. The Split axiom essentially enforces the underlying structure of teams (i.e., the part of the full powerset model $\mathfrak{M}^\bullet$ characterizing teams) to be $(\wp(W), \supseteq, \cup, \varnothing)$. In our approach, weakening the Split axiom amounts to changing this structure to $(\wp(W), \succcurlyeq, \Cup, \varnothing)$ with a bounded semilattice $(\wp(W), \Cup, \varnothing)$ and the associated partial order $\preccurlyeq$ satisfying certain constraints, and thus generalizing the standard team semantics. In related work (Punčochář, 2016) generalized the semantics by essentially considering an underlying structure $(I, \supseteq, \cup, \varnothing)$ with a set $I \subseteq \wp(W)$ of teams equipped with certain topology as the underlying structure for teams, resulting in different intermediate team logics from ours (in particular, these logics do not have the local and global disjunction at the same time). Further (Punčochář, 2017) considered a more abstract generalization of team semantics, by treating teams as primitive entities in arbitrary bounded semilattices $(A, \Cup, 0)$. The resulting logic does satisfy the Split axiom in case the semantics is persistent. We will provide more comparisons of our work with Punčochář (2016, 2017) in the concuding section, Sect. 9.5. There are also similarities between our framework and the framework of possibility semantics (Holliday, 2015). In particular, the disjunction of possibility semantics over canonical possibility models coincides with the local disjunction in our setting, as shown in Bezhanishvili and Holliday (2020, Corollary 5.5).

Under the generalized team semantics, two tIPC axioms of the local disjunction $\vee$, the monotonicity and (weak) elimination axiom, turn out to be not sound either. We also study frame conditions under which the soundness of these two axioms as well as the Split axiom can be recovered. We show that distributivity is a sufficient condition for validating all these axioms, and over finite classical frames, distributivity is also a necessary condition for the two axioms of the local disjunction. We leave it for future work to identify in the general case the necessary and sufficient conditions for validating these axioms. We hope that the conditions provided in this paper already give some new insights on the peculiar properties of team semantics and of the logic tIPC.

This paper is organized as follows. In Sect. 9.2, we provide an account of intuitionistic propositional logic IPC, universal models and de Jongh formulas in the standard (single-world Kripke semantics) setting. In Sect. 9.3, we recall from

Ciardelli et al. (2020) intuitionistic propositional logic tIPC based on team semantics, and investigate intermediate team logics through the first approach discussed above. Section 9.4 explores the second approach. We end in Sect. 9.5 by some concluding remarks and open problems.

9.2 Intuitionistic Logic and de Jongh Formulas

In this section, we recall briefly intuitionistic logic and de Jongh formulas in the standard setting. For more detailed discussions, the reader is referred to, e.g., Bezhanishvili (2006), Chagrov and Zakharyaschev (1997).

Fix a set Prop of propositional variables. The language of *intuitionistic propositional logic* is defined as

$$\phi ::= p \mid \bot \mid \phi \wedge \phi \mid \phi \vee \phi \mid \phi \to \phi.$$

We write $\neg\phi := \phi \to \bot$. We consider the usual *intuitionistic propositional calculus* (IPC), which is classical propositional calculus (CPC) without the law of excluded middle ($\phi \vee \neg\phi$) or the double negation elimination axiom ($\neg\neg\phi \to \phi$). We write $\vdash_{\mathsf{IPC}} \phi$ or simply $\vdash \phi$ if ϕ is a theorem.

An *(intuitionistic) Kripke model* is a tuple $\mathfrak{M} = (W, R, V)$ such that W is a non-empty set of *worlds* (or *nodes* or *points*), $R \subseteq W \times W$ is a partial order, and V : $\mathsf{Prop} \to \wp(W)$ is a *persistent* valuation, i.e., $w \in V(p)$ and wRv imply $v \in V(p)$. The underlying pair $\mathfrak{F} = (W, R)$ is called an *(intuitionistic) Kripke frame*. If wRv, then v is called a *successor* of w. An *immediate successor* v of w is a proper successor of w such that there is no other point u with wRu and uRv. Denote by $R\lceil w\rceil$ (or simply $\lceil w\rceil$) the set of all immediate successors of w. A point w is called an *endpoint* if $\lceil w\rceil = \varnothing$. Generated subframes or submodels, and p-morphisms between models or frames are defined as usual. We write $\mathfrak{M}_w$ for the submodel of $\mathfrak{M}$ generated by a point w in $\mathfrak{M}$. The satisfaction relation $\mathfrak{M}, w \models_{\mathsf{IPL}} \phi$ (or simply $\mathfrak{M}, w \models \phi$) is defined inductively in a standard way; in particular,

- $\mathfrak{M}, w \not\models \bot$ always;
- $\mathfrak{M}, w \models \phi \to \psi$ iff for any $v \in W$ with wRv, $\mathfrak{M}, v \models \phi$ implies $\mathfrak{M}, v \models \psi$.

The *persistence* property generalizes easily to arbitrary formulas; that is, $\mathfrak{M}, w \models \phi$ and wRv imply $\mathfrak{M}, v \models \phi$ for arbitrary formulas ϕ.

Given n propositional variables $p_1, \ldots, p_n \in \mathsf{Prop}$. A model $\mathfrak{M} = (W, R, V)$ with the valuation V restricted to $p_1, \ldots, p_n$ is referred to as an n-model. It is often convenient to describe the valuation in an n-model by using *colors*. The *(n-)color* of a point $w \in W$ is defined as a 0-1-sequence $col(w) = c_1 \ldots c_n$ such that $c_i = 1$ iff $w \models p_i$, and $c_i = 0$ iff $w \not\models p_i$. We write $c_1 \ldots c_n \leq c'_1 \ldots c'_n$ if $c_i \leq c'_i$ for each i; and write $c_1 \ldots c_n < c'_1 \ldots c'_n$ if $c_1 \ldots c_n \leq c'_1 \ldots c'_n$ and $c_1 \ldots c_n \neq c'_1 \ldots c'_n$.

De Jongh formulas can be defined over universal models of IPC. For any natural number $n \in \mathbb{N}$, the n-universal model $\mathcal{U}(n) = (W, R, V)$ of IPC is defined inductively in layers as follows:

- The first layer consists of 2^n nodes of distinct n-color.
- Suppose the $\leq k$ layers have been already defined. We define the $(k+1)$th layer as follows:
  - For every node w in the kth layer, and every color $c < col(w)$, add a new node v such that $col(v) = c$ and $\lceil v \rceil = \{w\}$.
  - For every finite set A of pairwise R-incomparable nodes in layers $\leq k$ with at least one element from the kth layer, and for every n-color $c \leq col(w)$ for all $w \in A$, add a new node v such that $col(v) = c$ and $\lceil v \rceil = A$.

Every finite rooted n-model can be found as a p-morphic image of a generated submodel of the n-universal model, as the next lemma shows. We include here also a detailed proof of this well-known fact, proof of which however does not seem to have been well documented in the literature.

Lemma 9.1

(i) Every finite rooted Kripke n-model $\mathfrak{M}$ can be mapped p-morphically onto a unique generated submodel of $\mathcal{U}(n)$.
(ii) Every finite rooted Kripke frame $\mathfrak{F}$ is isomorphic to a generated subframe of $\mathcal{U}(n)$ for some $n \leq |\mathfrak{F}|$.

Proof (i) We prove the claim by induction on the depth of $\mathfrak{M}$ with root r. If $\mathfrak{M}$ is a singleton $\{r\}$, then obviously r can be mapped p-morphically to a unique endpoint w in $\mathcal{U}(n)$ of the same color. Now, suppose $\lceil r \rceil \neq \varnothing$. By induction hypothesis, each $v \in \lceil r \rceil$ can be mapped, via some p-morphism f_v onto a unique submodel of $\mathcal{U}(n)$ generated by some point w_v. Let A be the set of minimal points in $\{w_v \mid v \in \lceil r \rceil\}$.

If A is a singleton $\{w_{v_0}\}$, then we distinguish two cases. If $col(r) = col(w_{v_0})$, then it is easy to see that the function $f = f_{v_0} \cup \{(r, w_{v_0})\}$ is a (unique) onto p-morphism. If $col(r) < col(w_{v_0})$, then by the construction of $\mathcal{U}(n)$, there is a (unique) point w_r in $\mathcal{U}(n)$ with $col(w_r) = col(r)$ and $\lceil w_r \rceil = \{w_{v_0}\}$. It is easy to see that the function $f = f_{v_0} \cup \{(r, w_r)\}$ is a (unique) onto p-morphism.

If A is a set of pairwise incomparable nodes, then by the construction of $\mathcal{U}(n)$, there is a (unique) point w_r in $\mathcal{U}(n)$ with $col(w_r) = col(r)$ and $\lceil w_r \rceil = A$. Thus, the function $f = \{(r, w_r)\} \cup \bigcup_{w_v \in A} f_v$ is a p-morphism. We also have $f_v \restriction \mathcal{U}(n)_u = f_u$ for every $w_v \in A$ and $u \in \lceil r \rceil$ with $w_v R w_u$. Thus, it is not hard to see that f is the unique onto p-morphism.

(ii) Let $\mathfrak{F} = (W, R)$ be a finite rooted frame. Introduce a new propositional variable p_w for each point w in W, and define a valuation V on $\mathfrak{F}$ by setting $V(p_w) = R(w)$. Putting $n = |W|$, by item (i), $(\mathfrak{F}, V_{\mathfrak{F}})$ can be mapped p-morphically via f onto a (unique) generated submodel of $\mathcal{U}(n)$ with root w. Since every point in $(\mathfrak{F}, V_{\mathfrak{F}})$ has a distinct color, the p-morphism f is an isomorphism, namely that $(\mathfrak{F}, V_{\mathfrak{F}})$ is isomorphic to the generated submodel $\mathcal{U}(n)_w$. Note that the underlying frame of $\mathcal{U}(n)_w$ may also be a generated subframe of some $\mathcal{U}(m)$ with $m < n$. □

It follows immediately from the above lemma that universal models are the counter-models for all non-theorems of IPC, as (finite rooted) counter-models of every non-theorem of IPC can be mapped p-morphically onto the universal model.

Theorem 9.1 *For every n-formula* ϕ*, we have that* $\mathcal{U}(n) \models \phi \iff \vdash_{\mathsf{IPC}} \phi$.

We are now ready to recall the definition of de Jongh formulas.

Definition 9.1 (*De Jongh formulas*) Let w be a point in $\mathcal{U}(n)$. Define n-formulas ϕ_w and ψ_w by induction on the depth of w as follows:

- If w is an endpoint, then define

$$\phi_w := \bigwedge_{w \models p_i} p_i \wedge \bigwedge_{w \not\models p_i} \neg p_i \quad \text{and} \quad \psi_w := \neg \phi_w.$$

- If w is not an endpoint (i.e., $\lceil w \rceil \neq \varnothing$), then define

$$\phi_w := \bigwedge_{w \models p_i} p_i \wedge \big(\bigvee \mathsf{np}(w) \vee \bigvee_{v \in \lceil w \rceil} \psi_v \to \bigvee_{v \in \lceil w \rceil} \phi_v\big)$$

$$\text{and} \quad \psi_w := \phi_w \to \bigvee_{v \in \lceil w \rceil} \phi_v,$$

where $\mathsf{np}(w) := \{p_i \mid w \not\models p_i \text{ and } v \models p_i \text{ for all } v \in \lceil w \rceil,\ 1 \leq i \leq n\}$.

The following theorem describes the important property of de Jongh formulas.

Theorem 9.2 *For every points* w, u *in* $\mathcal{U}(n)$*,*

(i) $\mathcal{U}(n), u \models \phi_w \iff wRu$;
(ii) $\mathcal{U}(n), u \not\models \psi_w \iff uRw$.

Proof Item (ii) is a consequence of item (i), since

$$\begin{aligned} u \not\models \psi_w &\iff \exists x : uRx,\ x \models \phi_w \text{ and } \forall v \in \lceil w \rceil : x \not\models \phi_v \\ &\qquad \text{in case } w \text{ is an endpoint, the second part holds trivially} \\ &\iff \exists x : uRx,\ wRx \text{ and } \forall v \in \lceil w \rceil : \neg vRx \qquad \text{by item (i)} \\ &\iff \exists x : uRx \text{ and } w = x, i.e., uRw \qquad \text{since } R \text{ is a partial order} \end{aligned}$$

We now prove item (i) by induction on the depth of w. If w is an endpoint (i.e., w is in the first layer of $\mathcal{U}(n)$), the direction "$\Longleftarrow$" is trivial; for the direction "$\Longrightarrow$", it is easy to see that $\mathcal{U}(n), u \models_{\mathsf{IPL}} \phi_w$ implies that u and all its successors have the same color as w, which can only happen when wRu by the construction of $\mathcal{U}(n)$.

Now, suppose w has proper successors. We first show the direction "$\Longleftarrow$". Suppose wRu. Clearly, $u \models \bigwedge_{w \models p_i} p_i$. If $u = w$, then by induction hypothesis, we have that $w \not\models \bigvee_{v \in \lceil w \rceil} \psi_v$. If u is a proper successor of w, then u is a successor of some point in $\lceil w \rceil$, which by induction hypothesis implies that $u \models \bigvee_{v \in \lceil w \rceil} \phi_v$. Putting everything together entail that $u \models \phi_w$.

For the converse direction "$\Longrightarrow$", assume that $\neg(wRu)$. We will show that $u \not\models \phi_w$. Suppose also that $u \models p_i$ for all p_i with $w \models p_i$, meaning that $col(u) \geq col(w)$. If there exists a successor x of u such that $\neg(wRx)$ and $\neg(xRv_0)$ for some $v_0 \in \lceil w \rceil$, then by induction hypothesis, we have that $x \not\models \bigvee_{v \in \lceil w \rceil} \phi_v$ (for otherwise $wRvRx$ for some $v \in \lceil w \rceil$) and $x \models \bigvee_{v \in \lceil w \rceil} \psi_v$. Thus, $u \not\models \phi_w$.

Now, suppose that no such point exists, namely that

$$\forall x \in R[u]: \text{ either } wRx \text{ (i.e., } w \in R[\lceil w \rceil]) \text{ or } \lceil w \rceil \subseteq R[x].$$

That is, every successor x of u can see all points in $\lceil w \rceil$ in the future, or have seen a point in $\lceil w \rceil$ in the past. By the construction of $\mathcal{U}(n)$, this implies that there exist successors x of u such that $\lceil x \rceil = \lceil w \rceil$. Take such an x with $x \neq w$. Such an $x \neq w$ must exist, for otherwise, by the construction of $\mathcal{U}(n)$, we must have that $\lceil u \rceil$ is a singleton of a predecessor of x and $col(u) < col(w)$, which is a contradiction.

Now, again by the construction of $\mathcal{U}(n)$, we must have that $col(x) \neq col(w)$. Since uRx, we also have $col(x) \geq col(u) \geq col(w)$. Thus $col(x) > col(w)$, meaning that there is some p_j such that $w \not\models p_j$ but $x \models p_j$. For all $v \in \lceil w \rceil$, since xRv, we have that $v \models p_j$. It then follows that $p_j \in \mathsf{np}(w)$ and $x \models \bigvee \mathsf{np}(w)$. On the other hand, by induction hypothesis, we have that $x \not\models \bigvee_{v \in \lceil w \rceil} \phi_v$. Hence, $u \not\models \phi_w$. □

We can now prove the Jankov-de Jongh theorem using de Jongh formulas and universal models. Note that Jankov's proof (Jankov, 1963) was purely algebraic (we refer to Bezhanishvili and Bezhanishvili, 2022 for more details on Jankov's approach). Yet another perspective to this formulas when one introduces new variable for each element of the frame is by Fine (1974) in the modal case and by Zakharyaschev in the intuitionistic one (see Chagrov and Zakharyaschev, 1997, Chap. 9).

Theorem 9.3 (Jankov-de Jongh) *For any finite rooted frame $\mathfrak{F}$, there exists a formula $\chi(\mathfrak{F})$ such that for every frame $\mathfrak{G}$,*

$$\mathfrak{G} \not\models \chi(\mathfrak{F}) \iff \mathfrak{F} \textit{ is a generated subframe of a p-morphic image } \mathfrak{G}.$$

Proof By Lemma 9.1, $\mathfrak{F}$ is isomorphic to a generated subframe of $\mathcal{U}(n)$ with root w and $n \leq \mathfrak{F}$. Let $\chi(\mathfrak{F}) = \psi_w$. "$\Longleftarrow$": Suppose $\mathfrak{F}$ is a generated subframe of a p-morphic image of $\mathfrak{G}$. By Theorem 9.2, $\mathcal{U}(n), w \not\models \psi_w$. Thus $\mathfrak{F} \not\models \psi_w$, which further implies that $\mathfrak{G} \not\models \psi_w$.

"$\Longrightarrow$": Suppose $(\mathfrak{G}, V) \not\models \psi_w$, where $(\mathfrak{G}, V)$ is some n-model on $\mathfrak{G}$. By Lemma 9.1(i), $(\mathfrak{G}, V)$ can be mapped p-morphically onto a generated submodel of $\mathcal{U}(n)$ with root u. Thus, $\mathcal{U}(n), u \not\models \psi_w$. By Theorem 9.2, this means that uRw. Since $\mathfrak{F}$ is isomorphic to the frame of $\mathcal{U}(n)_w$, it follows that $\mathfrak{F}$ is a generated subframe of the underlying frame of $\mathcal{U}(n)_u$, which is a p-morphic image of $\mathfrak{G}$. □

The algebraic counter-part of the next lemma is the *congruence extension property*, which in particular implies that for a Heyting algebra A we have $\mathbf{HS}(A) = \mathbf{SH}(A)$, where **H** and **S** are the operations of taking subalgebras and homomorphic images, respectively (see, e.g., Esakia, 2019).

Lemma 9.2 *For any frames $\mathfrak{F}$ and $\mathfrak{G}$, $\mathfrak{F}$ is a generated subframe of a p-morphic image of $\mathfrak{G}$, iff $\mathfrak{F}$ is a p-morphic image of a generated subframe of $\mathfrak{G}$.*

Proof If $\mathfrak{F}$ is a generated subframe of $f[\mathfrak{G}]$ for some p-morphism f, then $f^{-1}[\mathfrak{F}]$ is a generated subframe of $\mathfrak{G}$ and $f[f^{-1}[\mathfrak{F}]] = \mathfrak{F}$. Conversely, if $\mathfrak{G}_0 = (W_0, R)$ is a generated subframe of $\mathfrak{G} = (W, R)$ and $f[\mathfrak{G}_0] = \mathfrak{F}$ for some p-morphism f, then consider the extension $\mathfrak{H} = (W', R')$ of $\mathfrak{F} = (f[W_0], R')$, defined as

- $W' = f[W_0] \cup (W \setminus W_0)$;
- for any $w, v \in W \setminus W_0$, wRv iff $wR'v$;
- for any $w \in W \setminus W_0$ and $v' \in f[W_0]$, $wR'v'$) iff wRv for some $v \in W_0$ such that $f(v) = v'$.

Let $g : \mathfrak{G} \to \mathfrak{H}$ be an extension of f that is the identity function on $W \setminus W_0$. It is easy to show that g is an onto p-morphism. □

Corollary 9.1 *For any finite rooted frame $\mathfrak{F}$, there exists a formula $\chi(\mathfrak{F})$ such that for every frame $\mathfrak{G}$,*

$$\mathfrak{G} \not\models \chi(\mathfrak{F}) \iff \mathfrak{F} \text{ is a p-morphic image of a generated subframe of } \mathfrak{G}.$$

Given a set Δ of formulas, we write $\mathsf{IPC} \oplus \Delta$ for the set of formulas (or the logic) obtained by closing the set $\mathsf{IPC} \cup \Delta$ under uniform substitution and Modus Ponens. A logic L is called an *intermediate logic* if $\mathsf{IPC} \subseteq \mathsf{L} \subseteq \mathsf{CPC}$. A large class of intermediate logics can be axiomatized by de Jongh formulas or sets of de Jongh formulas. These are exactly the splititng and join splitting logics in the lattice of intermediate logics (Chagrov & Zakharyaschev, 1997, Sect. 10.5). We will now write explicitly some of the axiomatizations of the well-known intermediate logics in terms of de Jongh formulas. Recall that the 1-universal model of IPC is the Rieger-Nishimura ladder drawn in Fig. 9.1 (see e.g., Chagrov and Zakharyaschev, 1997, Sect. 8.7 and Bezhanishvili, 2006, Sect. 3.2). We label it by the elements w_i for $i \in \omega$. Recall also that the *Rieger-Nishimura polynomials* are given by the following recursive definition:

1. $g_0(p) := p$,
2. $g_1(p) := \neg p$,
3. $f_1(p) := p \vee \neg p$,
4. $g_2(p) := \neg\neg p$,
5. $g_3(p) := \neg\neg p \to p$,
6. $g_{n+4}(p) := g_{n+3}(p) \to (g_n(p) \vee g_{n+1}(p))$,
7. $f_{n+2}(p) := g_{n+2}(p) \vee g_{n+1}(p)$.

Then it is well known that the upsets of each w_i are defined by the Rieger-Nishumura poynomials. In particular, For every $k \in \omega$ we have:

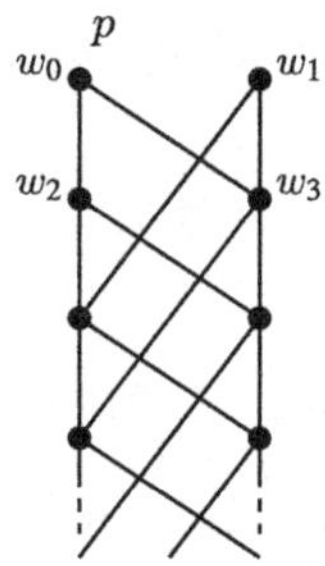

Fig. 9.1 Rieger-Nishimura ladder

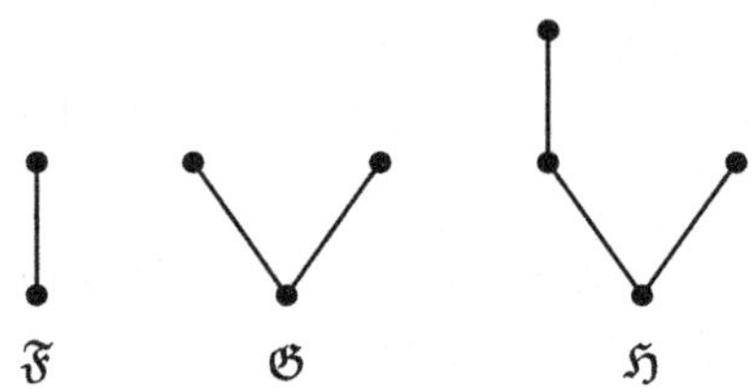

Fig. 9.2 The refutation frames for classical logic, Jankov's logic, and Scott's logic

- $R(w_k) = \{w \in \mathcal{U}(1) : w \models g_k(p)\}$,
- $R(w_k) \cup R(w_{k-1}) = \{w \in \mathcal{U}(1) : w \models f_k(p)\}$.

Recall also that classical logic $\mathsf{CPC} = \mathsf{IPC} \oplus \chi(\mathfrak{F})$, Jankov's logic $\mathsf{KC} = \mathsf{IPC} \oplus \chi(\mathfrak{G})$ and Scott logic $\mathsf{SL} = \mathsf{IPC} \oplus \chi(\mathfrak{H})$, where $\mathfrak{F}$, $\mathfrak{G}$ and $\mathfrak{H}$ are the frames drawn in Fig. 9.2, see, e.g., Chagrov and Zakharyaschev (1997, Chap. 9). Then

$$\mathsf{CPC} = \mathsf{IPC} \oplus \chi(\mathfrak{F}) = \mathsf{IPC} \oplus \psi_{w_2} = \mathsf{IPC} \oplus (\neg\neg p \to p).$$

Jankov's logic

$$\mathsf{KC} = \mathsf{IPC} \oplus \chi(\mathfrak{G}) = \mathsf{IPC} \oplus \psi_{w_3} = \mathsf{IPC} \oplus (\neg\neg p \to p) \to (p \vee \neg p).$$

However, note also that as $\mathfrak{G}$ is a p-morphic image of $\mathfrak{H}$ we have $\mathsf{IPC} \oplus \chi(\mathfrak{G}) =$

$$\mathsf{IPC} \oplus \chi(\mathfrak{G}) \oplus \chi(\mathfrak{H}) = \mathsf{IPC} \oplus \psi_{w_3} \oplus \psi_{w_4} = \mathsf{IPC} \oplus \psi_{w_3} \wedge \psi_{w_4} = \mathsf{IPC} \oplus \neg p \vee \neg\neg p.$$

Scott's logic $\mathsf{SL} =$

$$\begin{aligned} \mathsf{IPC} \oplus \chi(\mathfrak{H}) &= \mathsf{IPC} \oplus \psi_{w_4} = \mathsf{IPC} \oplus g_5(p) \\ &= \mathsf{IPC} \oplus ((\neg\neg p \to p) \to p \vee \neg p) \to \neg p \vee \neg\neg p. \end{aligned}$$

However, not all intermediate logics are axiomatizable by de Jongh formulas (see e.g., Chagrov and Zakharyaschev, 1997, Sect. 9.5). We refer to Chagrov and

Zakharyaschev (1997, Chap. 9) and Bezhanishvili and Bezhanishvili (2022) for a detailed account of generalizations of Jankov-de Jongh formulas that axiomatize all intermediate logics.

Note that given a frame $\mathfrak{F}$, the number of variables used in the de Jongh formula of $\mathfrak{F}$ is the number n such that there is a valuation V on $\mathfrak{F}$ with $(\mathfrak{F}, V)$ being a generated submodel of $\mathcal{U}(n)$. In the Jankov formula of $\mathfrak{F}$ the number of variables coincides with the number of elements of the Heyting algebra of all upsets of $\mathfrak{F}$; and in the Fine-Zakharyaschev approach the number of variables of $\chi(\mathfrak{F})$ equals the cardinality of $\mathfrak{F}$. Thus, in general, de Jongh formulas need fewer varibales than Jankov and Fine-Zakharyaschev formulas. Often the number of variables used in such formulas is not that relevant. After all we are interested in semantic properties of these formulas and what is important is that one can write such formulas and it is less important exactly how. However, the syntactic shape of these and related formulas also plays a role. We refer to Bezhanishvili and de Jongh (2018) and Ilin et al. (2021) for connections of subframe and stable formulas, defined in de Jongh-style, with NNIL and $\mathsf{ONNILLI}$ formulas and to Lauridsen (2019a, 2019b) for their connection with the low levels of the so–called substructural hierarchy. Below we demonstrate that the number of variables used in the Jankov-de Jongh formula is also essential for dealing with some particular problems.

We address the following question: Given a logic L, when does a logic $L' = L \oplus \chi(\mathfrak{F})$ have the finite model property (FMP)? In general, it is still an open problem when $\mathsf{IPC} \oplus \chi(\mathfrak{F})$ or in the modal case $\mathsf{S4} \oplus \chi(\mathfrak{F})$ have the FMP (Kracht, 1993). It is known however that $\mathsf{IPC} \oplus \alpha$ and $\mathsf{S4} \oplus \alpha$ have the FMP if α is a formula in one variable (Chagrov & Zakharyaschev, 1997, Theorem 11.58 and Corollary 11.59). Using the Jankov or Fine-Zakharyaschev method we obtain that the $\chi(\mathfrak{F})$ is in one variable only if $|\mathfrak{F}| \leq 2$. Hence this result cannot be applied to these formulas. However, using de Jongh's method (and its modal analogue) one can show that if a rooted $\mathfrak{F}$ is such that there is V with $(\mathfrak{F}, V)$ being a generated submodel of $\mathcal{U}(1)$ or respectively of $\mathcal{U}_{S4}(1)$ (the 1-universal model of $\mathsf{S4}$) in the modal case, then $\chi(\mathfrak{F})$ is in one variable. Consequently, $\mathsf{IPC} \oplus \chi(\mathfrak{F})$ and $\mathsf{S4} \oplus \chi(\mathfrak{F})$ have the FMP. This is more interesting in the case of $\mathsf{S4}$, since many models of width 2 can be embedded into $\mathcal{U}_{S4}(1)$. For example, the two frames $\mathfrak{F}_1$ and $\mathfrak{F}_2$ drawn in Fig. 9.3 admit valuations such that the corresponding models are (isomorphic to) generated submodels of $\mathcal{U}_{S4}(1)$ (Chagrov & Zakharyaschev, 1997, Chap. 8). Therefore, $\mathsf{S4} \oplus \chi(\mathfrak{F}_1)$ and $\mathsf{S4} \oplus \chi(\mathfrak{F}_2)$ have the FMP.

9.3 Team-Based Intuitionistic Logic and de Jongh Formulas

In the rest of the paper, we investigate intuitionistic and intermediate logics as well as de Jongh formulas in the setting of team semantics. Our starting point is the intuitionistic logic over team semantics (tIPC) introduced in Ciardelli et al. (2020).

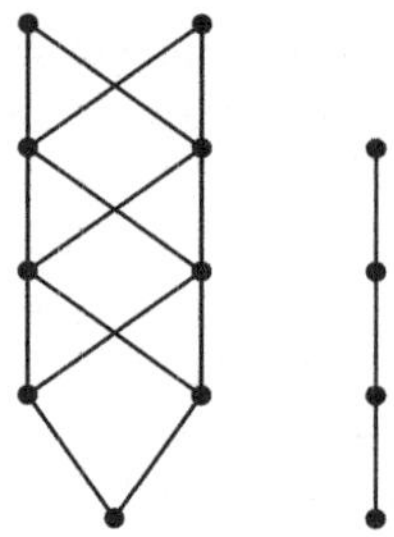

Fig. 9.3 Frames corresponding to generated submodels of $\mathcal{U}_{S4}(1)$

We show in this section, by using a disjunctive normal form of tIPC, that universal models $\mathcal{U}(n)$ for intuitionistic logic IPC over the standard (single-world) semantics behave also as universal models for the intuitionistic logic tIPC over team semantics, and thus de Jongh formulas for tIPC can be defined in the usual manner. Furthermore, we show that a large class of intermediate axioms (including de Jongh formulas) define the same class of Kripke frames in the team-based intuitionistic logic as in the standard (single-world-based) intuitionistic logic. This results in a first approach to obtaining intermediate logics in the team semantics setting. We will explore an alternative approach in the next section.

Let us start by recalling the definition and basic properties of team-based intuitionistic propositional logic as introduced in Ciardelli et al. (2020). The reader is referred to Ciardelli et al. (2020) for further discussions. We extend the syntax of intuitionistic propositional logic with an additional disjunction $\mathbin{\rotatebox[origin=c]{0}{$\vee\!\!\!\vee$}}$, called *global disjunction*; the other disjunction $\vee$ is thus referred to as *local disjunction*. We shall call in this paper the resulting language, denoted as $[\bot, \wedge, \vee, \mathbin{\vee\!\!\!\vee}, \to]$, *team language* or *intuitionistic team logic*. To be precise, formulas of team language are formed by the following grammar:

$$\phi ::= p \mid \bot \mid \phi \wedge \phi \mid \phi \vee \phi \mid \phi \mathbin{\vee\!\!\!\vee} \phi \mid \phi \to \phi.$$

Formulas in the standard language of intuitionistic logic (i.e., $\mathbin{\vee\!\!\!\vee}$-free formulas), denoted as $[\bot, \wedge, \vee, \to]$, are referred to as *standard formulas*. In the rest of the paper, we reserve the first Greek letters $\alpha, \beta, \gamma, \ldots$ for standard formulas, and the last Greek letters $\phi, \psi, \chi, \ldots$ stand for arbitrary formulas in the team language.

Formulas of intuitionistic team logic are evaluated over the usual intuitionistic Kripke models $\mathfrak{M} = (W, R, V)$ but with respect to *teams*. A *team* is a set $t \subseteq W$ of possible worlds, namely $t \in \wp(W)$. We define a natural relation $R^\circ \subseteq \wp(W) \times \wp(W)$ between teams by lifting the relation R on W to the power set of W as:

- $tR^\circ s$ iff $\forall w \in t \exists v \in s : wRv$ (i.e., $t \subseteq R^{-1}[s]$) and $\forall v \in s \exists w \in t : wRv$ (i.e., $s \subseteq R[t]$).

We shall often abuse the notation and write simply tRs for $tR^\circ s$.

Fact 9.1 *Let $\mathfrak{M} = (W, R, V)$ be a model, and t and s teams.*

(i) $\varnothing R \varnothing$, tRt, *and* $tR^{\circ}R[t]$.
(ii) $s \subseteq R[t]$ *iff* $t \supseteq \circ Rs$, *where* $\supseteq \circ R := \{(t, s) \mid \exists r \subseteq W : t \supseteq r \text{ and } rRs\}$ *denotes the composition of* $\supseteq$ *and* R.

Proof Item (i) is immediate. For item (ii), if $s \subseteq R[t]$, let $r = R^{-1}[s] \cap t$. Clearly, $t \supseteq rRs$. Conversely, if $t \supseteq rRs$, then $s \subseteq R[r] \subseteq R[t]$. □

Fact 9.2 *Let $\mathfrak{M} = (W, R, V)$ be a model. We have that $\supseteq \circ R^{\circ}$ is a pre-order.*

Proof Reflexivity follows from the fact that $t \supseteq tRt$. For transitivity, if $t \supseteq \circ Rs$ and $s \supseteq \circ Rr$, it is easy to verify that $t \supseteq (R^{-1}[r] \cap t)R^{\circ}r$. □

Now, we give the definition of the team semantics for formulas in the team language, where our semantic clauses for implication $\to$ and local disjunction $\vee$ deviate slightly from the original definition presented in Ciardelli et al. (2020). The two alternative versions of the semantics are easily seen to be equivalent (by Fact 9.1(ii) above and Corollary 9.2 below).

Definition 9.2 Let $\mathfrak{M} = (W, R, V)$ be an intuitionistic Kripke model and $t \subseteq W$ a team. We define the satisfaction relation $\mathfrak{M}, t \models_{\mathsf{t}} \phi$ (or simply $\mathfrak{M}, t \models \phi$) inductively as follows:

- $\mathfrak{M}, t \models p$ iff $t \subseteq V(p)$
- $\mathfrak{M}, t \models \bot$ iff $t = \varnothing$
- $\mathfrak{M}, t \models \phi \wedge \psi$ iff $\mathfrak{M}, t \models \phi$ and $\mathfrak{M}, t \models \psi$
- $\mathfrak{M}, t \models \phi \vee \psi$ iff there are $s, r \subseteq W$ such that $t \subseteq s \cup r$, $\mathfrak{M}, s \models \phi$ and $\mathfrak{M}, r \models \psi$
- $\mathfrak{M}, t \models \phi \mathbin{\mathbb{v}} \psi$ iff $\mathfrak{M}, t \models \phi$ or $\mathfrak{M}, t \models \psi$
- $\mathfrak{M}, t \models \phi \to \psi$ iff for all $s \subseteq W$ with $t \supseteq \circ Rs$, $\mathfrak{M}, s \models \phi$ implies $\mathfrak{M}, s \models \psi$

We write $\mathfrak{M} \models_{\mathsf{t}} \phi$ if $\mathfrak{M}, t \models \phi$ for all teams $t \subseteq W$, and write $\mathfrak{F} \models_{\mathsf{t}} \phi$ if $(\mathfrak{F}, V) \models_{\mathsf{t}} \phi$ for all valuations V on $\mathfrak{F}$. As usual, for any class F of frames, we write $\mathsf{F} \models_{\mathsf{t}} \phi$ if $\mathfrak{F} \models_{\mathsf{t}} \phi$ for all $\mathfrak{F} \in \mathsf{F}$. We write $\Gamma \models_{\mathsf{t}} \phi$ (or simply $\Gamma \models \phi$) if for all Kripke models $\mathfrak{M}$ and all teams t, $\mathfrak{M}, t \models \Gamma$ implies $\mathfrak{M}, t \models \phi$.

A formula ϕ is said to be *flat*, if for all models $\mathfrak{M}$ and all teams t,

$$\mathfrak{M}, t \models \phi \iff \forall w \in t : \mathfrak{M}, \{w\} \models \phi \qquad \text{(Flatness Property)}$$

For standard formulas the above defined team semantics essentially reduces to the standard (single-world) Kripke semantics, as standard formulas are flat.

Lemma 9.3 (Flatness of standard formulas) *Standard formulas are flat. As a consequence, for any set $\Delta \cup \{\alpha\}$ of standard formulas,*

$$\Delta \models_{\mathsf{t}} \alpha \iff \Delta \vdash_{\mathsf{IPC}} \alpha.$$

Proof The flatness of standard formula is proved by a routine verification by induction. The second part of the lemma follows from completeness of IPC and the observation that

$$\mathfrak{M}, \{w\} \models_{\mathsf{t}} \alpha \iff \mathfrak{M}, w \models \alpha \tag{9.1}$$

holds for all standard formulas α. □

As with the standard intuitionistic logic, formulas of intuitionistic team logic are also persistent, but in a more general sense.

Lemma 9.4 (Persistence) *Let $\mathfrak{M}$ be a model, t and s teams, and ϕ a formula. If $\mathfrak{M}, t \models \phi$ and $t \supseteq \circ Rs$, then $\mathfrak{M}, s \models \phi$. In particular, every formula ϕ is* downward closed, *that is,*

$$\mathfrak{M}, t \models \phi \text{ and } s \subseteq t \Longrightarrow \mathfrak{M}, s \models \phi \qquad \textit{(Downward Closure Property)}$$

Proof By a routine inductive argument. □

As an immediate consequence of the downward closure property, the semantic clause of the local disjunction $\vee$ can be alternatively formulated as follows (which is the original formulation of the clause in Ciardelli et al. (2020)).

Corollary 9.2 *For any model $\mathfrak{M}$, team t and any formulas ϕ and ψ, we have that*

$$\mathfrak{M}, t \models \phi \vee \psi \iff \exists s, r \subseteq t \text{ s.t. } t = s \cup r,\ \mathfrak{M}, s \models \phi \text{ and } \mathfrak{M}, r \models \psi.$$

The empty set $\varnothing$, being the smallest team (with respect to $\subseteq$), plays a special role in intuitionistic team logic. In particular, all formulas have the empty team property.

Lemma 9.5 (Empty team property) *For any model $\mathfrak{M}$ and any formula ϕ, we have that $\mathfrak{M}, \varnothing \models \phi$.*

Proof A routine verification by induction. □

Another closely related property is the union closure property. A formula ϕ is said to be *closed under unions* if for any model $\mathfrak{M}$ and nonempty set T of teams,

$$\mathfrak{M}, t \models \phi \text{ for all } t \in T \Longrightarrow \mathfrak{M}, \bigcup T \models \phi. \qquad \text{(Union Closure Property)}$$

Fact 9.3 *A formula ϕ is flat iff ϕ satisfies the empty team, downward closure and union closure property. In particular, all standard formulas are closed under unions.*

Proof The main claim is easy to verify, and the "in particular" part follows from Lemma 9.3. □

The local disjunction $\vee$ that is inherited from the standard intuitionistic logic has the usual disjunction property in intuitionistic team logic, while the global disjunction $\mathbin{\rotatebox[origin=c]{0}{$\vee$}\!\!\vee}$ turns out to satisfy a stronger disjunction property.

Lemma 9.6 (Disjunction property)*Let ϕ and ψ be two arbitrary formulas.*

(i) If $\models \phi \vee \psi$, then $\models \phi$ or $\models \psi$.
(ii) For any set Δ of standard formulas,

$$\Delta \models \phi \mathbin{\mathbb{v}} \psi \Longrightarrow \Delta \models \phi \text{ or } \Delta \models \psi.$$

Proof We refer to Ciardelli et al. (2020) for a detailed proof. Here we only sketch a proof for item (ii). Suppose $\Delta \not\models \phi$ and $\Delta \not\models \psi$. Then there are models $\mathfrak{M}_1, \mathfrak{M}_2$ and teams t_1, t_2 such that

$$\mathfrak{M}_1, t_1 \models \Delta, \quad \mathfrak{M}_1, t_1 \not\models \phi, \quad \mathfrak{M}_2, t_2 \models \Delta, \quad \text{and} \quad \mathfrak{M}_2, t_2 \not\models \chi.$$

Consider the disjoint union $\mathfrak{M} = \mathfrak{M}_1 \uplus \mathfrak{M}_2$ of $\mathfrak{M}_1$ and $\mathfrak{M}_2$. It is easy to show that $\mathfrak{M}, t_1 \models \Delta$, $\mathfrak{M}, t_1 \not\models \phi$, $\mathfrak{M}, t_2 \models \Delta$ and $\mathfrak{M}, t_2 \not\models \chi$. Since formulas in Δ are flat (by Lemma 9.3), we have also that $\mathfrak{M}, t_1 \cup t_2 \models \Delta$. Moreover, by persistence, we also obtain $\mathfrak{M}, t_1 \cup t_2 \not\models \phi$ and $\mathfrak{M}, t_1 \cup t_2 \not\models \psi$. Thus, $\mathfrak{M}, t_1 \cup t_2 \not\models \phi \mathbin{\mathbb{v}} \psi$. Hence $\Delta \not\models \phi \mathbin{\mathbb{v}} \psi$. □

Closely related to the disjunction property is the fact that for any standard formula α, the following formula (known as the Split *axiom*) is a validity:

$$(\alpha \to \phi \mathbin{\mathbb{v}} \psi) \to (\alpha \to \phi) \mathbin{\mathbb{v}} (\alpha \to \psi). \tag{Split}$$

The Split axiom is often considered the distinguishing axiom of team-based propositional logics. In particular, by using this validity, it is easy to show that every formula in the team language can be transformed into an equivalent formula in disjunctive normal form.

Theorem 9.4 (Disjunctive normal form) *For any formula ϕ, we have that $\phi \equiv \mathbb{V}_{i \in I} \alpha_i$ for some (finite) set $\{\alpha_i \mid i \in I\}$ of standard formulas.*

Proof The proof is by an easy inductive; use $\alpha \to \phi \mathbin{\mathbb{v}} \psi \equiv (\alpha \to \phi) \mathbin{\mathbb{v}} (\alpha \to \psi)$ for the implication case. □

A sound and complete natural deduction system for intuitionistic team logic was introduced in Ciardelli et al. (2020), and the completeness theorem was proved by an argument that makes essential use of the above disjunctive normal form. It is easy to show that the system in Ciardelli et al. (2020) can be presented as the following Hilbert-style system, where the formula α below ranges over standard formulas (i.e., $\mathbb{v}$-free formulas) only. We emphasize that the system below is thus not closed under *uniform substitution*.

Definition 9.3 The Hilbert system for intuitionistic team logic (denoted as tIPC) consists of the following axioms:

(1) $\phi \to \phi \vee \psi$

(2) $(\phi \to \alpha) \to ((\psi \to \alpha) \to (\phi \vee \psi \to \alpha))$
(3) $(\phi \to \chi) \to (\phi \vee \psi \to \chi \vee \psi)$
(4) $\phi \vee \psi \to \psi \vee \phi$
(5) $(\phi \vee \psi) \vee \chi \to \phi \vee (\psi \vee \chi)$
(6) $\phi \vee (\psi \mathbin{\mathbb{v}} \chi) \to (\phi \vee \psi) \mathbin{\mathbb{v}} (\phi \vee \chi)$
(7) All IPC axioms for $\bot, \wedge, \mathbin{\mathbb{v}}, \to$, namely,

- $\phi \to (\psi \to \phi)$
- $\big(\phi \to (\psi \to \chi)\big) \to \big((\phi \to \psi) \to (\phi \to \chi)\big)$
- $(\phi \wedge \psi) \to \phi$, $(\phi \wedge \psi) \to \psi$
- $\phi \to (\psi \to (\phi \wedge \psi))$
- $\phi \to (\phi \mathbin{\mathbb{v}} \psi)$, $\psi \to (\phi \mathbin{\mathbb{v}} \psi)$
- $(\phi \to \chi) \to \big((\psi \to \chi) \to ((\phi \mathbin{\mathbb{v}} \psi) \to \chi)\big)$
- $\bot \to \phi$

(8) $(\alpha \to \phi \mathbin{\mathbb{v}} \psi) \to (\alpha \to \phi) \mathbin{\mathbb{v}} (\alpha \to \psi)$

and the Modus Ponens rule.

Note that the axioms (4)-(6) and (8) can also be written using bi-implications ($\leftrightarrow$) as the main connectives instead of implications ($\to$), since the right to left direction of these implications follows from the other axioms in the system (e.g., $(\phi \vee \psi) \mathbin{\mathbb{v}} (\phi \vee \chi) \to \phi \vee (\psi \mathbin{\mathbb{v}} \chi)$ follows from axiom (3)).

When restricted to standard formulas, the above system tIPC coincides with IPC; in particular, the system tIPC is closed under *standard substitutions*, i.e., substitutions that map propositional variables to standard formulas. The logic tIPC is thus a conservative extension of IPC in the following sense.

Corollary 9.3 *For any set $\Delta \cup \{\alpha\}$ of standard formulas,*

$$\Delta \vdash_{\mathsf{tIPC}} \alpha \iff \Delta \vdash_{\mathsf{IPC}} \alpha.$$

Proof The right to left direction is clear; the other direction follows easily from Lemma 9.3. □

As an interesting application of the disjunctive normal form of intuitionistic team logic, we show next that the universal models $\mathcal{U}(n)$ of IPC behave as universal models also for tIPC.

Proposition 9.1 *For any $n \in \mathbb{N}$, let $\mathcal{U}(n)$ be the n-universal model of IPC. For any formula ϕ, we have that*

$$\mathcal{U}(n) \models_{\mathsf{t}} \phi \iff \vdash_{\mathsf{tIPC}} \phi.$$

Proof It suffices to prove the direction "$\Longrightarrow$". Suppose $\mathcal{U}(n) \models_{\mathsf{t}} \phi$. By Theorem 9.4, we have that $\phi \equiv \alpha_1 \mathbin{\mathbb{v}} \dots \mathbin{\mathbb{v}} \alpha_k$ for some standard formulas $\alpha_1, \dots, \alpha_k$. Thus, $\mathcal{U}(n) \models_{\mathsf{t}} \alpha_i$ for some $1 \leq i \leq k$, which means in particular that $\mathcal{U}(n), W \models_{\mathsf{t}} \alpha_i$ for the domain W of $\mathcal{U}(n)$. Since α_i is a standard formula, by flatness, we have that $\mathcal{U}(n), w \models \alpha_i$ for all $w \in W$. Now, by Theorem 9.1, we obtain that $\vdash_{\mathsf{IPC}} \alpha_i$, which implies by Corollary 9.3 that $\vdash_{\mathsf{tIPC}} \alpha_i$. Hence $\vdash_{\mathsf{tIPC}} \alpha_1 \mathbin{\mathbb{v}} \dots \mathbin{\mathbb{v}} \alpha_k$, thereby $\vdash_{\mathsf{tIPC}} \phi$. □

Having the universal models in place for tIPC, we can consider the de Jongh formulas also in the context of tIPC. Thanks again to the disjunctive normal form of tIPC, a large class of the same de Jongh formulas and intermediate axioms for the standard language of IPC still characterize the same class of frames in the setting of tIPC. Before we present our result, let us first define intermediate logics in this setting. This definition has appeared also in a recent work (Quadrellaro, 2022); cf. also a similar but subtly different definition in Punčochář (2021).

Definition 9.4 Let $\mathsf{IPC} \subseteq \mathsf{L} \subseteq \mathsf{CPC}$ be an intermediate logic with $\mathsf{L} = \mathsf{IPC} \oplus \Delta$ and $\Delta = \{\alpha_i \mid i \in I\}$ a set of axioms written in the standard language (i.e., each α_i is a formula in $[\bot, \wedge, \vee, \to]$). Denote by tL the logic obtained by closing the set

$$\mathsf{tIPC} \cup \{\alpha_i(\overrightarrow{\beta}\,/\,\overrightarrow{p}) \mid i \in I, \text{ and each } \beta_j \text{ is a standard formula}\}$$

under Modus Ponens. We write $\vdash_{\mathsf{tL}} \phi$ if $\phi \in \mathsf{tL}$. Note that $\vdash_{\mathsf{tL}} \phi$ iff $\Delta \vdash_{\mathsf{tIPC}} \phi$.

For instance, consider Jankov's logic $\mathsf{KC} = \mathsf{IPC} \oplus \chi(\mathfrak{G}) = \mathsf{IPC} \oplus \neg p \vee \neg\neg p$, where the frame $\mathfrak{G}$ is as drawn in Fig. 9.2. The intermediate logic tKC is then obtained by closing the set

$$\mathsf{tIPC} \cup \{\neg\alpha \vee \neg\neg\alpha \mid \alpha \in [\bot, \wedge, \vee, \to]\}$$

under Modus Ponens. Note that tKC may not be closed under uniform substitution, as, e.g., an arbitrary formula $\neg\phi \vee \neg\neg\phi$ (with global disjunction $\mathbin{\mathpalette\make@circled\vee}$) is not necessarily in tKC.

The logic tCPC obtained by extending tIPC with the classical axiom $\neg\neg p \to p$ is known in the literature as a variant of *propositional dependence logic* (see Ciardelli, 2016; Yang and Väänänen, 2016), and its fragment with connectives $\bot, \wedge, \mathbin{\vee\!\!\!\vee}, \to$, denoted as $[\bot, \wedge, \mathbin{\vee\!\!\!\vee}, \to]$, is called *inquisitive logic* (Ciardelli & Roelofsen, 2011). We emphasize again that the logic tCPC is not closed under uniform substitution; in particular its $[\bot, \wedge, \mathbin{\vee\!\!\!\vee}, \to]$-fragment, i.e., inquisitive logic (which contains all IPC axioms and the axiom $\neg\neg\alpha \to \alpha$) is not equivalent to classical logic, as an arbitrary formula $\neg\neg\phi \to \phi$ is not necessarily in the logic. It is interesting to note that for inquisitive logic, or the $[\bot, \wedge, \mathbin{\vee\!\!\!\vee}, \to]$-fragment of tCPC, the Split axiom is equivalent to the Kreisel-Putnam axiom

$$(\neg p \to q \mathbin{\vee\!\!\!\vee} r) \to (\neg p \to q) \mathbin{\vee\!\!\!\vee} (\neg p \to r). \tag{KP}$$

The logic can thus be viewed as the intermediate theory (in the language $[\bot, \wedge, \mathbin{\vee\!\!\!\vee}, \to]$) obtained by extending the Kreisel-Putnam logic KP (i.e., $\mathsf{IPC} \oplus \mathsf{KP}$) with the double negation elimination axiom $\neg\neg p \to p$. It was further shown that the logic is equivalent to the so-called *negative variant* of Medvedev's logic ML of finite problems (whose frames are of the form $(\wp(W) \setminus \{\varnothing\}, \supseteq)$ for finite sets W). The reader is referred to Ciardelli and Roelofsen (2011) for detailed discussions. In Sect. 9.4.1 we will explore a related but different power set construction than that of the Medvedev frames.

Now we are ready to show the main result of this section that standard axioms Δ define the same classes of frames in tIPC as in IPC, as long as the intermediate logic $\mathsf{IPC} \oplus \Delta$ has the disjunction property or is canonical.

Theorem 9.5 *Let $\mathsf{L} = \mathsf{IPC} \oplus \Delta$ be an intermediate logic that is complete with respect to a class F_L of L-frames, where Δ is a set of standard formulas. If L has the disjunction property or is canonical, then for every formula ϕ, we have that*

$$\vdash_{\mathsf{tL}} \phi \iff \mathsf{F}_\mathsf{L} \models_\mathsf{t} \phi.$$

Proof By Theorem 9.4 and the completeness theorem of tIPC, we may assume that $\phi = \alpha_1 \vee\!\!\!\vee \ldots \vee\!\!\!\vee \alpha_k$ for some standard formulas $\alpha_1, \ldots, \alpha_k$. Suppose that $\vdash_{\mathsf{tL}} \phi$, i.e., $\Delta \vdash_{\mathsf{tIPC}} \alpha_1 \vee\!\!\!\vee \ldots \vee\!\!\!\vee \alpha_k$. By the disjunction property of $\vee\!\!\!\vee$ (by Lemma 9.6 and completeness of tIPC), we have that $\Delta \vdash_{\mathsf{tIPC}} \alpha_i$ for some $1 \leq i \leq k$. It then follows from Corollary 9.3 that $\Delta \vdash_{\mathsf{IPC}} \alpha_i$, which implies $\mathsf{F}_\mathsf{L} \models \alpha_i$ in the sense of the standard (single world-based) semantics of IPC. By flatness of α_i (Lemma 9.3), we obtain $\mathsf{F}_\mathsf{L} \models_\mathsf{t} \alpha_i$ as well. Hence, $\mathsf{F}_\mathsf{L} \models_\mathsf{t} \alpha_1 \vee\!\!\!\vee \ldots \vee\!\!\!\vee \alpha_k$, namely $\mathsf{F}_\mathsf{L} \models_\mathsf{t} \phi$.

For the converse direction, we first treat the case when L has the disjunction property. Suppose $\mathsf{F}_\mathsf{L} \models_\mathsf{t} \alpha_1 \vee\!\!\!\vee \ldots \vee\!\!\!\vee \alpha_k$. Since $\phi \vee\!\!\!\vee \psi \models_\mathsf{t} \phi \vee \psi$ (by empty team property), we have $\mathsf{F}_\mathsf{L} \models_\mathsf{t} \alpha_1 \vee \cdots \vee \alpha_k$ as well, where the standard formula $\alpha_1 \vee \cdots \vee \alpha_k$ is flat (by Lemma 9.3). Thus, $\mathsf{F}_\mathsf{L} \models \alpha_1 \vee \cdots \vee \alpha_k$ in the sense of the standard (single world-based) semantics of IPC, which implies further that $\Delta \vdash_{\mathsf{IPC}} \alpha_1 \vee \cdots \vee \alpha_k$. Since L has the disjunction property, we have $\Delta \vdash_{\mathsf{IPC}} \alpha_i$ for some $1 \leq i \leq k$. Hence, by Corollary 9.3, we conclude that $\Delta \vdash_{\mathsf{tIPC}} \alpha_i$, which gives $\Delta \vdash_{\mathsf{tIPC}} \alpha_1 \vee\!\!\!\vee \ldots \vee\!\!\!\vee \alpha_k$, i.e., $\vdash_{\mathsf{tL}} \phi$.

Next, we treat the case when L is canonical. Suppose that $\Delta \nvdash_{\mathsf{tIPC}} \alpha_1 \vee\!\!\!\vee \ldots \vee\!\!\!\vee \alpha_k$. Then, for all $1 \leq i \leq k$, we have that $\Delta \nvdash_{\mathsf{tIPC}} \alpha_i$, which implies $\Delta \nvdash_{\mathsf{IPC}} \alpha_i$ by Corollary 9.3. Now, for each $1 \leq i \leq k$, the canonical model $\mathfrak{M}^c_\mathsf{L}$ of L contains a witness w_i satisfying $\mathfrak{M}^c_\mathsf{L}, w_i \not\models \alpha_i$. Thus, $\mathfrak{M}^c_\mathsf{L}, \{w_i\} \not\models_\mathsf{t} \alpha_i$ by Expression (9.1). Let $t = \{w_1, \ldots, w_k\}$. By persistence (Lemma 9.4), we must have that $\mathfrak{M}^c_\mathsf{L}, t \not\models_\mathsf{t} \alpha_i$ for all $1 \leq i \leq k$, and thus $\mathfrak{M}^c_\mathsf{L}, t \not\models_\mathsf{t} \alpha_1 \vee\!\!\!\vee \ldots \vee\!\!\!\vee \alpha_k$. Since L is canonical, we know that $\mathfrak{M}^c_\mathsf{L} \in \mathsf{F}_\mathsf{L}$, and thus $\mathsf{F}_\mathsf{L} \not\models_\mathsf{t} \alpha_1 \vee\!\!\!\vee \ldots \vee\!\!\!\vee \alpha_k$. □

Let us give some examples of the applications of the above theorem. Since Jankov's logic $\mathsf{KC} = \mathsf{IPC} \oplus \chi(\mathfrak{G})$ is canonical, by the above theorem,

$$\vdash_{\mathsf{tKC}} \phi \iff \mathsf{F}_{\mathsf{KC}} \models_\mathsf{t} \phi,$$

where F_{KC} is the class of finite rooted frames with unique top points. Since classical propositional logic $\mathsf{CPC} = \mathsf{IPC} \oplus \chi(\mathfrak{F})$ (where the frame $\mathfrak{F}$ is as drawn in Fig. 9.2) is canonical, the logic tCPC is complete with respect to all classical frames (i.e., frames with dead points only). Note that over classical Kripke models, a team (i.e., a set of possible worlds with no proper successors) can be identified as a set of propositional valuations. In fact, the teams for the logic tCPC (or propositional dependence logic Yang and Väänänen, 2016 and inquisitive logic Ciardelli and Roelofsen, 2011) were originally introduced as such sets of valuations.

Theorem 9.5, however, does not apply for the intermediate logics that neither have the disjunction property nor are canonical. For instance, Scott's logic $\mathsf{SL} = \mathsf{IPC} \oplus \chi(\mathfrak{H})$ (where the frame $\mathfrak{H}$ is as drawn in Fig. 9.2) is one of such examples. Studying the properties of the corresponding intermediate team logics for such logics is left as future work.

9.4 Generalized Team Intuitionistic Kripke Semantics

In the previous section we have defined (in Definition 9.4) intermediate team logics by adding to tIPC a set of standard axioms with the local disjunction $\vee$. As shown in Theorem 9.5, a large class of such axioms characterize, in the usual manner, properties of the underlying Kripke frames. One important feature of such defined intermediate team logics is that they all validate the Split axiom, which characterizes certain unique properties of the underlying structure of the set of teams. Roughly speaking, these logics with two disjunctions (the local one $\vee$ and the global one $\mathbb{v}$) can be considered logics with two layers: The first layer, which we shall call the *core* or *inner* layer, is characterized by the different properties of the underlying Kripke frames. By the result of Theorem 9.5, this layer can be axiomatized by standard formulas, i.e., formulas that contain the local disjunction $\vee$ only, to a large extent (i.e., for axioms that define intermediate logics with disjunction property or are canonical). The second layer, which we shall call the *team* or *outer* layer, is characterized by the properties of the underlying structure of the set of teams. In the setting of Definition 9.4, this team layer is kept fixed by always adopting the Split axiom in the intermediate team logics (as these logics are always extensions of tIPC). Different intermediate logics differ only in the core layer, through validating different standard axioms. See Fig. 9.4 for an illustration of such constructions.

Another natural approach to define intermediate team logics is to allow also the team layer of the logic to vary, by replacing the Split axiom with other axioms in the full language of tIPC (see Fig. 9.5 for an illustration of such constructions). In this section, we make a first attempt in this direction.

Since the Split axiom captures the structure of the teams, defining intermediate team logics that validate other axioms than the Split axiom also amounts to generalizing the original team semantics in some appropriate manner. To this end, in Sect. 9.4.1 we interpret the logic tIPC in the so-called (full) powerset model and compare the logic with a variant of intuitionistic modal logic IK. On the basis of this connection, in Sect. 9.4.2, we define generalized team intuitionistic Kripke semantics and discuss the failure of the Split axiom and two other axioms involving the local disjunction $\vee$ in this new setting. In Sect. 9.4.3, we provide frame conditions validating these three axioms and thus recovering tIPC.

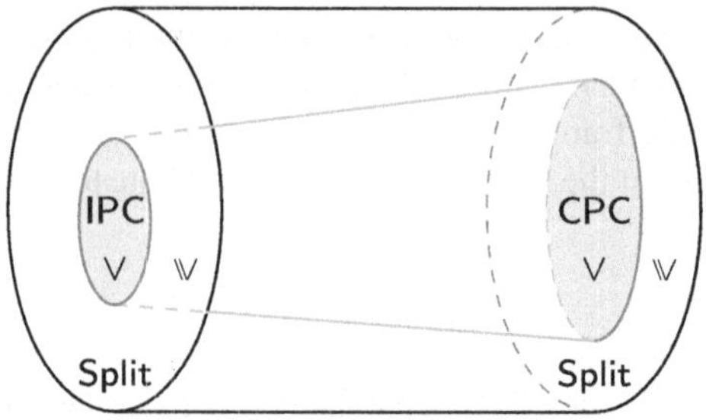

Fig. 9.4 Intermediate team logics obtained by changing the core

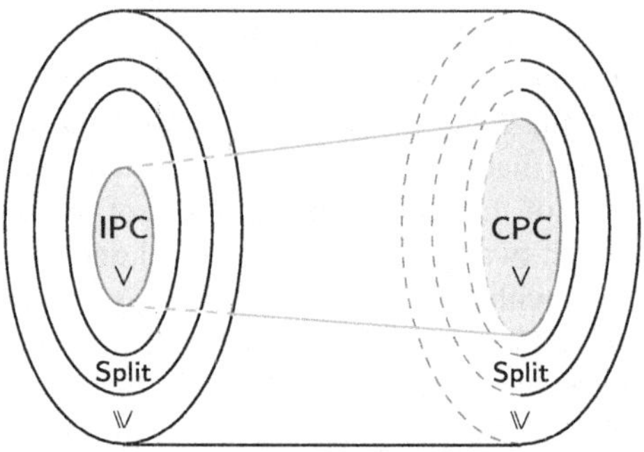

Fig. 9.5 Intermediate team logics obtained by changing the core/team layer

9.4.1 The Power Set Model and Intuitionistic Modal Logic

In this section, we relate tIPC to *inuitionistic modal logic* IK (studied in, e.g., Plotkin and Stirling, 1986, Wolter and Zakharyaschev, 1999) and give a single-world semantics interpretation of the team-based logic tIPC. The key observation for this connection is that the team-based logic tIPC can be given a single-world semantics over the powerset models induced by the underlying intuitionistic Kripke models. We show that the local disjunction $\vee$-free fragment of tIPC can be translated, in a truth-preserving manner, into the diamond-free fragment of intuitionistic modal logic IK over powerset models, while full tIPC corresponds to a variant of IK. Our results are inspired by a similar connection and powerset models discussed in Yang (2017) in the context of modal dependence logics. The logic tIPC can be translated into S4 modal dependence logic via a Gödel-style translation, as shown in Ciardelli et al. (2020).

We first give the translation of the local disjunction $\vee$-free fragment of tIPC into the single-world semantics-based intuitionistic modal logic IK, and then extend this result to full tIPC. Recall that formulas of tIPC are evaluated in an intuitionistic Kripke model $\mathfrak{M} = (W, R, V)$ over *teams* (or sets of worlds), which are elements in the power set of W. Every intuitionistic Kripke model $\mathfrak{M}$ is thus naturally associated

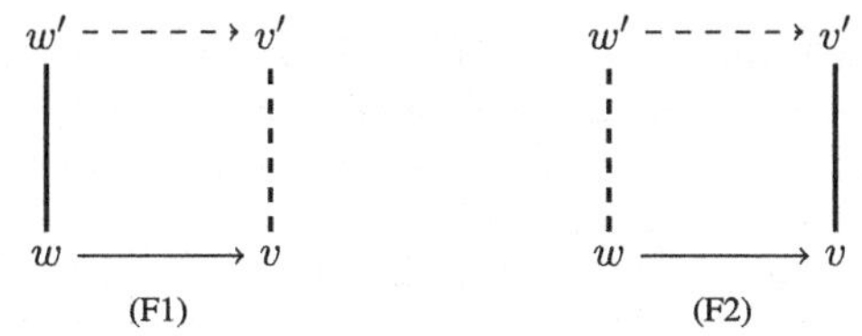

Fig. 9.6 Conditions on frames. The directed lines represent the R relation and the undirected lines represent the $\succcurlyeq$ relation with the nodes positioned above being accessible from the ones positioned below

with a powerset model defined as follows. Similar powerset models were considered also in van Benthem (2021).

Definition 9.5 Let $\mathfrak{M} = (W, R, V)$ be an intuitionistic Kripke model. The *powerset model* $\mathfrak{M}^\circ$ *induced by* $\mathfrak{M}$ is a quadruple $\mathfrak{M}^\circ = (W^\circ, \supseteq, R^\circ, V^\circ)$, where

- $W^\circ = \wp(W) \setminus \{\varnothing\}$, i.e. W° consists of all nonempty teams $t \subseteq W$,
- $\supseteq$ is the superset relation,
- R° is the lifting of R on $\wp(W)$,
- the valuation V° is defined as: $t \in V^\circ(p)$ iff $t \subseteq V(p)$.

If $\mathfrak{M}$ is finite, the reduct $(W^\circ, \supseteq) = (\wp(W) \setminus \{\varnothing\}, \supseteq)$ of the underlying frame of the powerset model $\mathfrak{M}^\circ$ is clearly a frame for Medvedev's logic ML. The whole powerset model $\mathfrak{M}^\circ$ is a special case of a bi-relation intuitionistic Kripke model of intuitionistic modal logic IK. We now recall the definition of such models.

Definition 9.6 A *bi-relation intuitionistic Kripke frame* is a triple $\mathfrak{F} = (W, \succcurlyeq, R)$, where

- W is a nonempty set;
- $\succcurlyeq$ is a partial order on W and R is a binary relation on W satisfying the following conditions (F1) and (F2) (see also Fig. 9.6):

  F1: If $w \succcurlyeq w'$ and wRv, then there exists $v' \in W$ such that vRv' and $w'Rv'$ (i.e., $\preccurlyeq \circ R \subseteq R \circ \preccurlyeq$).

  F2: If wRv and $v \succcurlyeq v'$, then there exists $w' \in W$ such that wRw' and $w'Rv'$ (i.e., $R \circ \succcurlyeq \subseteq \succcurlyeq \circ R$).

A *bi-relation intuitionistic Kripke model* is a pair $\mathfrak{M} = (\mathfrak{F}, V)$ of a bi-relation intuitionistic Kripke frame $\mathfrak{F} = (W, \succcurlyeq, R)$ and a *valuation* $V : \mathsf{Prop} \to \wp(W)$ satisfying *persistence* with respect to $\succcurlyeq$ (i.e., $w \in V(p)$ and $w \succcurlyeq v$ imply $v \in V(p)$).

It is easy to verify that the superset relation $\supseteq$ and the lifting R° in a powerset model $\mathfrak{M}^\circ$ satisfy conditions (F1) and (F2), and the lifting V° of the valuation is persistent with respect to the partial order $\supseteq$. A powerset model $\mathfrak{M}^\circ$ is thus a bi-relation intuitionistic Kripke model. As pointed out already, such type of connection has already been discussed in Yang (2017), where it was shown that modal dependence logic corresponds to a special type of IK and the relation R° is the lifting of an arbitrary relation R in a modal Kripke model. Here in our setting, the relation R° is the lifting of a partial order R of an intuitionistic Kripke model. Observe that the lifting R° is actually a pre-order, or a partial order in case the model is finite.

Fact 9.4 *The lifting R° is a pre-order. If $\mathfrak{M}$ is finite, then R° is a partial order.*

Proof The reflexivity and transitivity of R° follow immediately from the reflexivity and transitivity of R. If $tR^\circ s$, $sR^\circ t$ and s, t are finite, then we must have $t = s$, since R is anti-symmetric. □

Recall that the language of intuitionistic modal logic IK is obtained by enriching the language of intuitionistic propositional logic IPC with modalities. That is, formulas of IK are formed by the following grammar:

$$\phi ::= p \mid \bot \mid \phi \wedge \phi \mid \phi \vee \phi \mid \phi \to \phi \mid \Box\phi \mid \Diamond\phi.$$

Formulas of IK are evaluated over bi-relation intuitionistic Kripke models with respect to single worlds.

Definition 9.7 The *satisfaction* relation $\mathfrak{M}, w \Vdash \phi$ between a bi-relation intuitionistic Kripke model $\mathfrak{M} = (W, \succcurlyeq, R, V)$, a world $w \in W$ and an IK-formula ϕ is defined inductively as follows:

- $\mathfrak{M}, w \Vdash p$ iff $w \in V(p)$
- $\mathfrak{M}, w \nVdash \bot$ always
- $\mathfrak{M}, w \Vdash \phi \wedge \psi$ iff $\mathfrak{M}, w \Vdash \phi$ and $\mathfrak{M}, w \Vdash \psi$
- $\mathfrak{M}, w \Vdash \phi \vee \psi$ iff $\mathfrak{M}, w \Vdash \phi$ or $\mathfrak{M}, w \Vdash \psi$
- $\mathfrak{M}, w \Vdash \phi \to \psi$ iff for all $v \in W$ such that $w \succcurlyeq v$, if $\mathfrak{M}, v \Vdash \phi$, then $\mathfrak{M}, v \Vdash \psi$
- $\mathfrak{M}, w \Vdash \Box\phi$ iff for all $u, v \in W$ such that $w \succcurlyeq u$ and uRv, it holds that $\mathfrak{M}, v \Vdash \phi$
- $\mathfrak{M}, w \Vdash \Diamond\phi$ iff there exists $v \in W$ such that wRv and $\mathfrak{M}, v \Vdash \phi$

It is routine to verify that persistence of propositional variables generalizes to arbitrary IK-formulas, namely, $\mathfrak{M}, w \Vdash \phi$ and $w \succcurlyeq v$ imply $\mathfrak{M}, v \Vdash \phi$.

Formulas in the local disjunction $\vee$-free fragment of tIPC can be translated into the diamond-free fragment of IK via a truth preserving translation defined as follows. A similar translation from modal dependence logic into IK was defined in Yang (2017).

Definition 9.8 Define a translation $(\cdot)^\tau$ from the $\vee$-free fragment of tIPC into (the diamond-free fragment of) IK inductively as:

- $p^\tau := p$
- $\bot^\tau := \bot$
- $(\phi \mathbin{\text{\textbackslash\!/}\!\!\text{/}} \psi)^\tau := \phi^\tau \vee \psi^\tau$
- $(\phi \wedge \psi)^\tau := \phi^\tau \wedge \psi^\tau$
- $(\phi \to \psi)^\tau := \Box(\phi^\tau \to \psi^\tau)$

We now show that the above defined translation preserves truth over powerset models and nonempty teams.

Proposition 9.2 *Let $\mathfrak{M} = (W, R, V)$ be an intuitionistic Kripke model and $t \subseteq W$ a nonempty team. For any $\vee$-free formula ϕ, $\mathfrak{M}, t \models \phi \iff \mathfrak{M}^\circ, t \Vdash \phi^\tau$.*

Proof We prove the lemma by induction on ϕ. If $\phi = \bot$, then for any nonempty team t, we have $\mathfrak{M}, t \not\models \bot$ and $\mathfrak{M}^\circ, t \nVdash \bot$.

If $\phi = p$, then $\mathfrak{M}, t \models p$ iff $t \subseteq V(p)$ iff $\mathfrak{M}^\circ, t \Vdash p$.

If $\phi = \psi \to \chi$, suppose $t \models \psi \to \chi$. We show that $\mathfrak{M}^\circ, t \Vdash \Box(\psi^\tau \to \chi^\tau)$. Let $r, s, u \in \wp(W) \setminus \{\varnothing\}$ be such that $t \supseteq rRs \supseteq u$ and $u \Vdash \psi^\tau$. By induction hypothesis, we have that $u \models \psi$. Since $t \supseteq rRs$ and $t \models \psi \to \chi$, by persistence (Lemma 9.4), we have $s \models \psi \to \chi$. Moreover, since R is reflexive, we have $s \supseteq uRu$. Thus, we obtain $u \models \chi$, from which we conclude $u \Vdash \chi^\tau$ by induction hypothesis.

Conversely, suppose $t \Vdash \Box(\psi^\tau \to \chi^\tau)$. Let $t \supseteq rRs$ be such that $s \models \psi$. If $s = \varnothing$, then $s \models \chi$ by the empty team property. Now, assume that $s \neq \varnothing$. By induction hypothesis, we have that $s \Vdash \psi^\tau$. Moreover, since rRs and $s \neq \varnothing$, we know that $r \neq \varnothing$ and $r \in W^\circ$. Since $t \supseteq rRs \supseteq s$, we obtain that $s \Vdash \chi^\tau$ by assumption. Hence, $s \models \chi$ holds by induction hypothesis.

The other cases are straightforward. $\square$

Now, we turn to full tIPC that has the local disjunction $\vee$ in the language. In the powerset model $\mathfrak{M}^\circ$ as defined in Definition 9.5, we have excluded the empty team from the domain (i.e., $W^\circ = \wp(W) \setminus \{\varnothing\}$), as the empty team behaves as a trivial team in the $\vee$-free fragment of tIPC. When it comes to full tIPC, in the presence of $\vee$ the empty team, however, does play a special and nontrivial role. For this reason, to interpret tIPC in the single-world semantics, we shall instead consider full powerset models $\mathfrak{M}^\bullet$ that has as the domain the full power set $\wp(W)$, as done also in Yang (2017) in the model dependence logic context. The local disjunction $\vee$ will be interpreted essentially as a binary diamond modality $\otimes$ (also called *tensor*) in the single-world semantics over the full powerset models $\mathfrak{M}^\bullet$. The model $\mathfrak{M}^\bullet$ is equipped also with a binary operator, the set-theoretic union $\cup$, which, together with the partial order $\supseteq$, induces a ternary relation for $\otimes$. This treatment of the local disjunction is slightly different from that in Yang (2017), where a ternary relation for $\otimes$ was directly included in the full powerset model.

Definition 9.9 Let $\mathfrak{M} = (W, R, V)$ be an intuitionistic Kripke model. The *full powerset model $\mathfrak{M}^\bullet$ induced by* $\mathfrak{M}$ is a tuple $\mathfrak{M}^\bullet = (W^\bullet, \supseteq, R^\circ, \cup, \varnothing, V^\circ)$, where

- $W^\bullet = \wp(W)$, i.e. $W^\bullet$ consists of all teams $t \subseteq W$,
- $\cup$ is the set-theoretic union,
- and the other components are as defined in the powerset model $\mathfrak{M}^\circ$ (Definition 9.5).

Clearly, $(W^\bullet, \cup, \varnothing)$ forms a bounded join-semilattice (i.e., $\cup$ is associative, commutative and idempotent, and $\varnothing \cup t = t$ for all $t \in W^\bullet$), and $\subseteq$ is the associated partial order (i.e., $\subseteq$ satisfies: $t \subseteq s$ iff $t \cup s = t$). A full powerset model can be viewed as a special case of a bi-relation intuitionistic Kripke model together with an additional binary operator $\Cup$ and a constant 0, which we shall call a *tensored bi-relation intuitionistic Kripke model.*

Definition 9.10 A *tensored bi-relation intuitionistic Kripke frame* is a quadruple $\mathfrak{F} = (W, \succcurlyeq, R, \Cup, 0)$, where

- $(W, \succcurlyeq, R)$ is a bi-relation intuitionistic Kripke frame;
- $(W, \Cup, 0)$ is a bounded join-semilattice;
- $\preccurlyeq$ is the associated partial order of $(W, \Cup, 0)$.

A *tensored bi-relation intuitionistic Kripke model* is a pair $\mathfrak{M} = (\mathfrak{F}, V)$ such that $\mathfrak{F} = (W, \succcurlyeq, R, \Cup, 0)$ is a tensored bi-relation intuitionistic Kripke frame and V is a *valuation* satisfying *persistence* with respect to $\succcurlyeq$.

In order to compare the full logic tIPC with IK, we shall consider a (diamond-free) variant of IK (denoted as $\mathsf{IK}_\otimes$) that has an additional binary diamond, the tensor $\otimes$, and a constant symbol $\mathbf{0}$. We now define the semantics for $\mathsf{IK}_\otimes$ by adapting the standard semantics of IK over bi-relation intuitionistic Kripke models. Such a language was considered already in Yang (2017), where the semantics for tensor $\otimes$ was defined slightly differently and the constant symbol $\mathbf{0}$ was not isolated.

Definition 9.11 The *satisfaction* relation $\mathfrak{M}, w \Vdash \phi$ between a tensored bi-relation intuitionistic Kripke model $\mathfrak{M} = (W, \succcurlyeq, R, \Cup, 0, V)$, a world $w \in W$ and an $\mathsf{IK}_\otimes$-formula ϕ is defined inductively as in Definition 9.7 except for the following cases:

- $\mathfrak{M}, w \Vdash \mathbf{0}$ iff $w = 0$
- $\mathfrak{M}, w \Vdash \phi \otimes \psi$ iff there are $u, v \in W$ such that $w \preccurlyeq u \Cup v$, $\mathfrak{M}, u \Vdash \phi$ and $\mathfrak{M}, v \Vdash \psi$

We now verify that formulas of $\mathsf{IK}_\otimes$ are persistent over tensored bi-relation intuitionistic Kripke models.

Proposition 9.3 (Persistence) *For any $\mathsf{IK}_\otimes$-formula ϕ, if $\mathfrak{M}, w \Vdash \phi$ and $w \succcurlyeq v$, then $\mathfrak{M}, v \Vdash \phi$.*

Proof Let $\mathfrak{M} = (W, \succcurlyeq, R, \Cup, 0, V)$. The proposition is proved by induction. If $\phi = \mathbf{0}$, then $\mathfrak{M}, w \Vdash \mathbf{0}$ and $w \succcurlyeq v$ imply that $w = v = 0$, which then gives $\mathfrak{M}, v \Vdash \mathbf{0}$.

If $\phi = \psi \otimes \chi$, suppose $\mathfrak{M}, w \Vdash \psi \otimes \chi$ and $w \succcurlyeq v$. Then there are $x, y \in W$ such that $w \preccurlyeq x \Cup y$, $\mathfrak{M}, x \Vdash \phi$ and $\mathfrak{M}, y \Vdash \psi$. Since $v \preccurlyeq w \preccurlyeq x \Cup y$, we conclude that $\mathfrak{M}, v \Vdash \psi \otimes \chi$.

The other cases are standard. □

Formulas of tIPC can be translated into $\mathsf{IK}_\otimes$ via a truth preserving translation defined as follows.

Definition 9.12 Define inductively a translation $(\cdot)^\tau$ from tIPC into $\mathsf{IK}_\otimes$ as:

- $\bot^\tau := \mathbf{0}$,
- $(\phi \vee \psi)^\tau := \phi^\tau \otimes \psi^\tau$,
- for the other cases, $\phi^\tau := \phi^\tau$ as in Definition 9.8.

Note that under such defined translation, the negation $\neg\phi(= \phi \to \bot)$ of tIPC is translated into $\mathsf{IK}_\otimes$ as $\phi^\tau \to \mathbf{0}$, which is in general different from the real negation $\phi^\tau \to \bot$. We now show that the translation preserves truth over full powerset models and arbitrary teams.

Proposition 9.4 *Let $\mathfrak{M} = (W, R, V)$ be an intuitionistic Kripke model. For any formula ϕ and any team $t \subseteq W$, $\mathfrak{M}, t \models \phi \iff \mathfrak{M}^\bullet, t \Vdash \phi^\tau$.*

Proof We prove the lemma by induction on ϕ. If $\phi = \bot$, then for any team t, we have $\mathfrak{M}, t \models \bot$ iff $t = \varnothing$ iff $\mathfrak{M}^\bullet, t \Vdash \mathbf{0}$.

The case $\phi = \psi \vee \chi$ follows easily from induction hypothesis. The other cases are proved analogously to the corresponding cases in Proposition 9.2. □

9.4.2 Generalized Team Intuitionistic Kripke Semantics

In this section, we introduce general intuitionistic team Kripke models, which are abstractions of the full powerset models defined in the previous section. We then define generalized team semantics for tIPC-formulas over these models. The Split axiom will not any more be valid under the new semantics. Two other axioms of tIPC, monotonicity and (weak) elimination axiom of the local disjunction $\vee$ (axioms (2) and (3)), turn out to be not valid either. We give counter-examples to illustrate the failure of these three axioms in this section, and we discuss frame conditions validating these axioms in the next section.

Let us now define general intuitionistic team Kripke frames $\mathfrak{F} = (W, R, \succcurlyeq, \Cup)$, which have a partial order $\succcurlyeq$ and a join operation $\Cup$ on $\wp(W)$ resembling and generalizing the superset relation $\supseteq$ and the set-theoretic union $\cup$.

Definition 9.13 A *general intuitionstic team Kripke frame* is a tuple $\mathfrak{F} = (W, R, \succcurlyeq, \Cup)$, where

- (W, R) is an intuitionistic Kripke frame,
- $\Cup$ is a binary operation on $\wp(W)$ such that $(\wp(W), \Cup, \varnothing)$ forms a bounded join-semilattice, and $\preccurlyeq$ is the induced partial order;
- and $\succcurlyeq$, $\Cup$ and the lifting R° of R on $\wp(W)$ satisfy the following conditions:

(a) R° and $\succcurlyeq$ satisfy (F2),

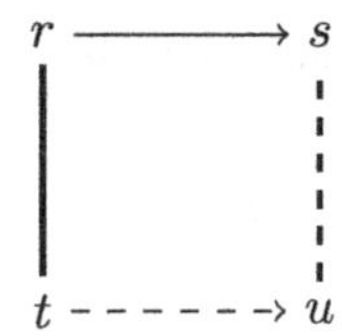

Fig. 9.7 A weaker frame condition

(b) $t \succcurlyeq \circ R^\circ s$ implies $R[t] \succcurlyeq s$ for any $t, s \subseteq W$,
(c) $R[t] \uplus R[s] = R[t \uplus s]$ for any $t, s \subseteq W$.

A *general intuitionistic team Kripke model* $\mathfrak{M} = (W, R, \succcurlyeq, \uplus, V)$ is a general intuitionistic team Kripke frame with a persistent valuation $V : \mathsf{Prop} \to \wp(W)$ on (W, R), i.e., (W, R, V) is an intuitionistic Kripke model.

It follows from the same argument as in Fact 9.4, the lifting R° is a pre-order, and if $\mathfrak{M}$ is finite, R° is a partial order. As before, we sometimes abuse the notation and write simply R for the lifted relation R°. It is easy to verify that composition relation $\succcurlyeq \circ R^\circ$ is a pre-order as well (e.g., transitivity follows from (F2)).

Conditions (a)–(c) in the above definition are imposed to guarantee the persistence of tIPC-formulas under the new semantics (as we will verify in Lemma 9.7). In Condition (a) we do not require R° and $\succcurlyeq$ to satisfy the (F1) condition, which was however also imposed for bi-relation intuitionistic Kripke models (see Definition 9.6). The condition (F1) guarantees the persistence of IK-formulas $\Diamond\phi$ with the diamond modality $\Diamond$. Since the diamond modality does not play a role in our discussion (see the two translations $(\cdot)^\tau$ and $(\cdot)^{\boldsymbol{\tau}}$ in Definitions 9.8 and 9.12), condition (F1) is then not imposed in our models. Every general intuitionistic team Kripke model $\mathfrak{M} = (W, R, \succcurlyeq, \uplus, V)$ is naturally associated with a model $\mathfrak{M}^\bullet = (\wp(W), R^\circ, \succcurlyeq, \uplus, \varnothing, V^\circ)$ whose valuation V° is defined as $t \in V^\circ(p)$ iff $t \preccurlyeq V(p)$. Clearly, this model $\mathfrak{M}^\bullet$ satisfies all the requirements for a tensored bi-relation intuitionistic Kripke model except for (F1)

Condition (b) is a special case of the following weaker condition, introduced by Božić and Došen (1984), on the model $\mathfrak{M}^\bullet$:

- If $t \succcurlyeq r$ and $rR^\circ s$, then there exists $u \in \wp(W)$ such that $tR^\circ u$ and $u \succcurlyeq s$ (i.e., $\succcurlyeq \circ R^\circ \subseteq R^\circ \circ \succcurlyeq$); see also Fig. 9.7.

Condition (b) has an interesting special case: If $t \succcurlyeq s$, then we have $R[t] \succcurlyeq R[s]$, since $t \succcurlyeq sR^\circ R[s]$. This observation implies that the direction $R[t \uplus s] \succcurlyeq R[t] \uplus R[s]$ of Condition (c) is actually always satisfied, as $t \uplus s \succcurlyeq t, s$.

If the partial order $\succcurlyeq$ is taken to be the superset relation $\supseteq$ (or the binary operator $\uplus$ is taken to be the set-theoretic union $\cup$), Conditions (a)–(c) are automatically satisfied regardless of how R is defined. Similarly, if R is taken to be the identity relation $\mathsf{id} = \{(w, w) \mid w \in W\}$ (i.e., the underlying Kripke frame (W, R) is a classical frame), then Conditions (a)–(c) are trivially satisfied regardless of how $\succcurlyeq$ and $\uplus$ are defined. We call such frames *classical frames*. Note that classical-based logics with team semantics, such as propositional dependence logic (Yang & Väänänen, 2016) and inquisitive logic (Ciardelli & Roelofsen, 2011), are defined essentially over frames of the form $\mathfrak{F} = (W, \mathsf{id}, \supseteq, \cup)$, which are very specific types of general intuitionistic team Kripke frames with classical underlying Kripke frames. Below we give a nontrivial example of a general intuitionistic team Kripke frame in which $\succcurlyeq \neq \supseteq$ and $R \neq \mathsf{id}$.

Example 9.1 Consider the frame $\mathfrak{F} = (W, R, \succcurlyeq, \Cup)$, where the underlying intuitionistic Kripke frame (W, R) with $W = \{w, u, v\}$ is depicted as the left figure below, and the bounded semilattice $(\wp(W), \Cup, \varnothing)$ is depicted (upside down) as the right figure below (where the dashed arrows represent the lifting R° with reflexive and transitive arrows omitted).

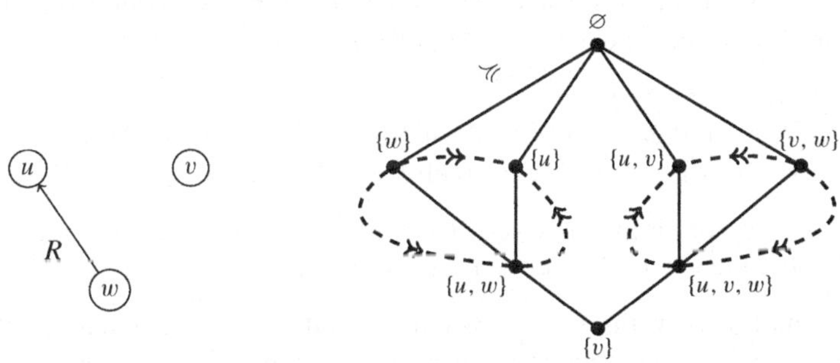

It can be verified that $\mathfrak{F}$ indeed forms a general intuitionistic team Kripke frame. We only illustrate the verification of Conditions (a)-(c) for some nontrivial cases; the remaining cases are left to the reader. Note that the full proof contains many symmetric cases, as the above figure of the semilattice $(\wp(W), \Cup, \varnothing)$ is symmetrical vertically, and the teams in the symmetric positions have similar R°-successors.

For Condition (a), assuming $tRr \succcurlyeq s$ we find t' such that $t \succcurlyeq t'Rs$. We only consider the case $t = \{w\}$ and $r = \{u, w\}$. Then we have that either $s = \varnothing$ or $s \in \{\{u, w\}, \{w\}, \{u\}\}$. In the first case, take $t' = \varnothing = s$ and we have $t \succcurlyeq \varnothing R \varnothing$. In the second case, take $t' = t = \{w\}$ and we have $t \succcurlyeq tRs$.

For Condition (b), assuming $t \succcurlyeq \circ Rs$ we show that $R[t] \succcurlyeq s$. We only consider the case $t = \{w\}$. For any s such that $t \succcurlyeq \circ Rs$, it is easy to observe that $\{u, w\} \succcurlyeq s$. Thus, $R[\{w\}] = \{u, w\} \succcurlyeq s$.

For Condition (c), we only verify $R[t] \Cup R[s] = R[t \Cup s]$ for any $t = \{w\}$ and $s \neq \varnothing, t$. We distinguish three cases according to the value of s. Case 1: $s = \{v\}$: Then

$$R[\{w\}] \Cup R[\{v\}] = \{u, w\} \Cup \{v\} = \{v\} = R[\{v\}] = R[\{w\} \Cup \{v\}].$$

Case 2: $s \in \{\{u\}, \{u, w\}\}$. Since $R[s] = s$ and $R[\{u, w\}] = \{u, w\}$, we have

$$R[\{w\}] \Cup R[s] = \{u, w\} \Cup s = \{u, w\} = R[\{u, w\}] = R[\{w\} \Cup s].$$

Case 3: $s \in \{\{u, v\}, \{v, w\}, \{u, v, w\}\}$: Since $\varnothing \neq R[s] \succcurlyeq \{u, v, w\}$ and $\{w\} \Cup s = \{v\}$, we have

$$R[\{w\}] \Cup R[s] = \{u, w\} \Cup R[s] = \{v\} = R[\{v\}] = R[\{w\} \Cup s].$$

This completes our illustration of the proof.

We now define the generalized team semantics of tIPC-formulas over general intuitionistic team Kripke models. The definition is almost the same as in Definition 9.2, except that we now have the more general partial order $\succcurlyeq$ and join $\Cup$ in place of the superset relation $\supseteq$ and the set-theoretic union $\cup$, respectively.

Definition 9.14 Let $\mathfrak{M} = (W, R, \succcurlyeq, \Cup, V)$ be a general intuitionistic team Kripke model and $t \subseteq W$ a team. The satisfaction relation $\mathfrak{M}, t \models \phi$ is defined inductively exactly as in Definition 9.2 except for the following cases:

- $\mathfrak{M}, t \models p$ iff $t \preccurlyeq V(p)$
- $\mathfrak{M}, t \models \phi \vee \psi$ iff there are $s, r \subseteq W$ such that $t \preccurlyeq s \Cup r$, $\mathfrak{M}, s \models \phi$ and $\mathfrak{M}, r \models \psi$
- $\mathfrak{M}, t \models \phi \to \psi$ iff for all $s \subseteq W$ with $t \succcurlyeq \circ Rs$, $\mathfrak{M}, s \models \phi \Longrightarrow \mathfrak{M}, s \models \psi$

We write $\Gamma \models^{\mathsf{g}} \phi$ (or simply $\Gamma \models \phi$) if for all general intuitionistic team Kripke models $\mathfrak{M}$ and all teams t, $\mathfrak{M}, t \models \psi$ for all $\psi \in \Gamma$ implies $\mathfrak{M}, t \models \phi$.

Let us remark that in the setting of our generalized team semantics, the local disjunction $\vee$ can be understood as a binary diamond modality, for which the corresponding ternary accessibility relation $R_\vee$ is defined as

$$R_\vee(t, s, r) \iff t \preccurlyeq s \Cup r.$$

We now verify that tIPC-formulas are persistent under the new semantics.

Lemma 9.7 (Persistence) *Let $\mathfrak{M}$ be a model, t and s teams, and ϕ a formula. If $\mathfrak{M}, t \models \phi$ and $t \succcurlyeq \circ Rs$, then $\mathfrak{M}, s \models \phi$. In particular, if $\mathfrak{M}, t \models \phi$ and $t \succcurlyeq s$, then $\mathfrak{M}, s \models \phi$.*

Proof We prove the lemma by induction on ϕ. If $\phi = \bot$ and $t \models \bot$, then $t = \varnothing$. For any s such that $\varnothing \succcurlyeq \circ Rs$, since $\varnothing$ is the greatest element for the order $\succcurlyeq$, we must have that $s = \varnothing$. Thus, $s \models \bot$, as required.

If $\phi = p$, suppose $t \models p$ and $t \succcurlyeq rRs$. Then $V(p) \succcurlyeq t \succcurlyeq rRs$, which by Condition (b) and persistence of V implies that $V(p) = R[V(p)] \succcurlyeq s$. Thus, $s \models p$.

If $\phi = \psi \to \chi$, suppose $\mathfrak{M}, t \models \psi \to \chi$ and $t \succcurlyeq rRs$. Suppose also $s \succcurlyeq uRv$ and $v \models \psi$. By (F2), there exists $r' \subseteq W$ such that $r \succcurlyeq r'Ru$. Thus, $t \succcurlyeq r \succcurlyeq r'RuRv$, which by assumption implies that $v \models \chi$.

If $\phi = \psi \vee \chi$, suppose $\mathfrak{M}, t \models \psi \vee \chi$ and $t \succcurlyeq rRs$. Then there exist $t_1, t_2 \subseteq W$ such that $t \preccurlyeq t_1 \Cup t_2$, $t_1 \models \psi$ and $t_2 \models \chi$. Since $t_1 \Cup t_2 \succcurlyeq t \succcurlyeq rRs$, by Conditions (b) and (c), we have that $R[t_1] \Cup R[t_2] = R[t_1 \Cup t_2] \succcurlyeq s$. Moreover, by definition we have $t_1 \succcurlyeq t_1RR[t_1]$ and $t_2 \succcurlyeq t_2RR[t_2]$. Thus, by induction hypothesis, we have that $R[t_1] \models \psi$ and $R[t_2] \models \chi$, from which it follows that $s \models \psi \vee \chi$ follows.

The other cases follow immediately from the induction hypothesis. □

The deduction theorem is an immediate corollary of persistence.

Corollary 9.4 (Deduction theorem) *For any set* $\Gamma \cup \{\phi, \psi\}$ *of formulas, we have that*

$$\Gamma, \phi \models^g \psi \iff \Gamma \models^g \phi \to \psi.$$

Next, we prove that the empty team property holds for tIPC-formulas under the generalized semantics.

Lemma 9.8 (Empty team property) *For any model* $\mathfrak{M}$ *and formula* ϕ*, we have that* $\mathfrak{M}, \varnothing \models \phi$.

Proof The lemma is proved by induction. If $\phi = \bot$, then $\varnothing \models \bot$ by definition. If $\phi = p$, then $\varnothing \models p$ follows from the fact that $\varnothing \preccurlyeq V(p)$ (since $\varnothing$ is the greatest element with respect to $\succcurlyeq$).

If $\phi = \psi \vee \chi$, by induction hypothesis we have $\varnothing \models \psi$ and $\varnothing \models \chi$. Since $\varnothing$ is the identity of the join $\Cup$, we have $\varnothing \preccurlyeq \varnothing = \varnothing \Cup \varnothing$. Hence $\varnothing \models \psi \vee \chi$.

If $\phi = \psi \to \chi$, for any s, r such that $\varnothing \succcurlyeq rRs$ and $s \models \psi$, we must have $r = \varnothing$ and thus $s = \varnothing$. Hence, by induction hypothesis, we conclude $s \models \chi$.

The other cases follow immediately from induction hypothesis. □

The persistence property justifies that our generalized semantics is a well-defined intuitionistic semantics. Together with the empty team property these suggest that our generalized semantics inherits certain nice features of the standard team semantics of tIPC. Let us now examine the logic that the generalized semantics induces.

It is easy to verify that all IPC axioms are sound for the language $[\bot, \wedge, \mathbin{\rotatebox[origin=c]{0}{$\vee$}\!\!\vee}, \to]$ over general intuitionistic team Kripke models (since, in particular, persistence, (F2), and the empty team property hold). Axioms (1) and (4)–(6) of tIPC (see Definition 9.3) also remain sound with respect to the generalized team semantics, as we verify in the following.

Lemma 9.9 *Let* ϕ, ψ *and* χ *be formulas.*

(i) $\phi \vee \psi \equiv^g \psi \vee \phi$ *and* $(\phi \vee \psi) \vee \chi \equiv^g \phi \vee (\psi \vee \chi)$.
(ii) $\phi \models^g \phi \vee \psi$ *and* $\phi \mathbin{\vee\!\!\vee} \psi \models^g \phi \vee \psi$.
(iii) $\phi \vee (\psi \mathbin{\vee\!\!\vee} \chi) \equiv^g (\phi \vee \psi) \mathbin{\vee\!\!\vee} (\phi \vee \chi)$.

Proof The first clause of item (i) follows directly from the commutativity of $\Cup$. For the second clause, we only show $(\phi \vee \psi) \vee \chi \models^g \phi \vee (\psi \vee \chi)$; the other direction is similar. Suppose $\mathfrak{M}, t \models (\phi \vee \psi) \vee \chi$ for some $\mathfrak{M} = (W, R, \succcurlyeq, \Cup, V)$. Then there exist $r, s \subseteq W$ such that $t \preccurlyeq r \Cup s$, $r \models \phi \vee \psi$ and $s \models \chi$. Since $r \models \phi \vee \psi$, there exist $r_0, r_1 \subseteq W$ such that $r \preccurlyeq r_0 \Cup r_1$, $r_0 \models \phi$ and $r_1 \models \psi$. Thus, we have $r_0 \Cup (r_1 \Cup s) \models \phi \vee (\psi \vee \chi)$. To conclude that $t \models \phi \vee (\psi \vee \chi)$, by persistence, it suffices to show that $t \preccurlyeq r_0 \Cup (r_1 \Cup s)$, which reduces to showing $t \preccurlyeq (r_0 \Cup r_1) \Cup s$, as $\Cup$ is associative. Now, since $r \preccurlyeq r_0 \Cup r_1$ and $\preccurlyeq$ is the order of the semilattice $(\wp(W), \Cup, \varnothing)$, we obtain $t \preccurlyeq r \Cup s \preccurlyeq (r_0 \Cup r_1) \Cup s$, as required.

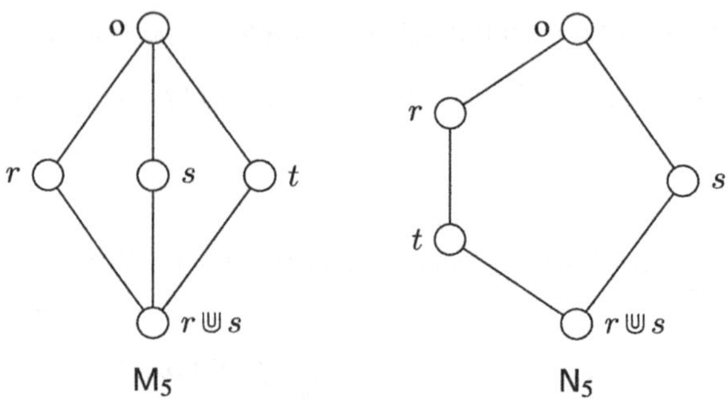

Fig. 9.8 Lattices M_5 and N_5 (placed upside down)

The second clause of item (ii) is a consequence of the first clause. For the first clause of item (ii), suppose $t \models \phi$. By the empty team property (Lemma 9.8), we have $\varnothing \models \psi$. Since $\mathfrak{M}, t \preccurlyeq t = t \Cup \varnothing$, we conclude that $t \models \phi \vee \psi$.

Item (iii) follows directly from definition. □

However, as expected, not all axioms of tIPC are sound in the generalized team semantics. We now give counter-examples to illustrate the failure of the three remaining axioms of tIPC: Axioms (2) and (3), and the Split axiom. For each axiom we provide two counter-models, whose underlying frames are denoted as $\mathfrak{F}_{\mathsf{M}}$ and $\mathfrak{F}_{\mathsf{N}}$. The frame $\mathfrak{F}_{\mathsf{M}} = (W, R, \succcurlyeq, \Cup)$ has as domain an arbitrary finite set W with at least 3 elements, and the partial order $R = \mathsf{id}$ (i.e., the underlying Kripke frame (W, R) is classical, and thus Conditions (a)-(c) in Definition 9.13 are trivially satisfied). The associated (finite) bounded semilattice $(\wp(W), \Cup, \varnothing)$ of $\mathfrak{F}_{\mathsf{M}}$ is thus a lattice (with the meet $\Cap$ defined as $t \Cap s = \bigcup\{r \in W \mid r \preccurlyeq t, s\}$), and we require that it contains the lattice M_5 as shown (upside down) in Fig. 9.8 as a sublattice. The finite frame $\mathfrak{F}_{\mathsf{N}} = (W, R, \succcurlyeq, \Cup)$ is defined the same way as $\mathfrak{F}_{\mathsf{M}}$, except that its bounded (semi)lattice $(\wp(W), \Cup, \varnothing)$ contains instead the lattice N_5 as shown (upside down) in Fig. 9.8 as a sublattice. Recall (see, e.g., Burris and Sankappanavar, 2012) that a lattice is distributive if and only if none of its sublattices is isomorphic to M_5 or N_5 (note that M_5 and N_5 are horizontally symmetric, and thus the direction of the order $\succcurlyeq$ in Fig. 9.8 does not make a substantial difference in the discussion). Thus, the lattices $(\wp(W), \Cup, \Cap, \varnothing)$ for $\mathfrak{F}_{\mathsf{M}}$ and $\mathfrak{F}_{\mathsf{N}}$ are clearly not distributive. The reason for the slightly more involved construction of these counter-models will become clear in the next section, when we discuss distributive frames. Similar counter-examples to our Counter-example 9.2 below can also be found in Punčochář (2017) in a different and algebraic setting.

We now first give our two counter-examples for Axiom (3) $(\phi \to \chi) \to (\phi \vee \psi \to \chi \vee \psi)$, which expresses the monotonicity of the local disjunction $\vee$.

Counter-example 9.1 (Axiom (3) (1). Consider the general intuitionistic team Kripke model $\mathfrak{M} = (\mathfrak{F}_{\mathsf{M}}, V)$, where the frame $\mathfrak{F}_{\mathsf{M}}$ is as described above, the valuation V satisfies

$$V(p) = r,\ \ V(q) = s \text{ and } V(p') = \mathrm{o},$$

and the teams $\mathrm{o}, r, s, t, r \Cup s$ are as shown in M_5 in Fig. 9.8. Note that since $R = \mathsf{id}$, the valuation V is trivially persistent. Moreover, since $\mathrm{o} \models p, q, p'$, the valuation is also legitimate (or consistent with the empty team property) in case $\mathrm{o} = \varnothing$.

We have that $\mathfrak{M}, t \not\models (p \to p') \to (p \vee q \to p' \vee q)$. To see why, first observe that since $R = \mathsf{id}$, a $\succcurlyeq \circ R$-successor of any point is a $\succcurlyeq$-successor, and vice versa. We first show that $t \models p \to p'$. Suppose $t \succcurlyeq t'$ and $t' \models p$. By definition, we have $t' \preccurlyeq V(p) = r$. Since M_5 (with the meet $\Cap_{\mathsf{M}_5}$) is a sublattice of $(\wp(W), \Cup, \Cap, \varnothing)$, $\mathrm{o} = r \Cap_{\mathsf{M}_5} t = r \Cap t \succcurlyeq t'$. Now, since $V(p') = \mathrm{o}$, we obtain $t' \models p'$, as required.

Next, we have $t \models p \vee q$, since $r \models p$, $s \models q$, and $t \preccurlyeq r \Cup s$. Finally, we verify that $t \not\models p' \vee q$, from which we would conclude that $t \not\models p \vee q \to p' \vee q$. For any r', s' such that $r' \models p'$, $s' \models q$, we have, by definition, that $r' \preccurlyeq V(p') = \mathrm{o} \preccurlyeq s$ and $s' \preccurlyeq V(q) = s$, which imply $s \succcurlyeq r' \Cup s'$. Thus, we have that $t \not\preccurlyeq r' \Cup s'$, since otherwise we would have $t \preccurlyeq s$, which is not the case.

(2). Consider the general intuitionistic team Kripke model $\mathfrak{M} = (\mathfrak{F}_{\mathsf{N}}, V)$, where the frame $\mathfrak{F}_{\mathsf{N}}$ is as described above, the valuation V satisfies

$$V(p) = s,\ \ V(p') = \mathrm{o} \text{ and } V(q) = r,$$

and the teams $\mathrm{o}, r, s, t, r \Cup s$ are as shown in N_5 in Fig. 9.8.

We have that $\mathfrak{M}, t \not\models (p \to p') \to (p \vee q \to p' \vee q)$. To see why, we first show that $t \models p \to p'$. For any $t' \preccurlyeq t$ with $t' \models p$, we have $t' \preccurlyeq V(p) = s$. But since N_5 is a sublattice of $(\wp(W), \Cup, \Cap, \varnothing)$, we have that $t' \preccurlyeq t \Cap s = t \Cap_{\mathsf{N}_5} s = \mathrm{o} = V(p')$. Thus, $t' \models p'$.

Next, clearly $t \models p \vee q$, since $t \preccurlyeq r \Cup s$, $r \models q$ and $s \models p$. Finally, we verify that $t \not\models p' \vee q$, from which it would follow that $t \not\models p \vee q \to p' \vee q$. For any r', s' such that $r' \models q$ and $s' \models p'$, we have, by definition, that $r' \preccurlyeq V(q) = r$ and $s' \preccurlyeq V(p') = \mathrm{o} \preccurlyeq r$, which imply that $r' \Cup s' \preccurlyeq r \preccurlyeq t$. Since $r \neq t$, we obtain $t \not\preccurlyeq r' \Cup s'$.

The above counter-examples show the failure of the monotonicity of the local disjunction $\vee$, stated as an implication $(\phi \to \chi) \to (\phi \vee \psi \to \chi \vee \psi)$. Nevertheless, it is straightforward to verify that the following (weaker form of) monotonicity of $\vee$ does hold:

$$\text{If } \phi \models^{\mathsf{g}} \chi, \text{ then } \phi \vee \psi \models^{\mathsf{g}} \chi \vee \psi.$$

In other words, the following deduction rule of $\vee$ is sound:

$$\frac{\phi \vdash \chi}{\phi \vee \psi \vdash \chi \vee \psi}\ \vee\mathsf{Mon}$$

Note that we do not allow any context set Γ to occur on the left-hand side of $\vdash$ in the above rule.

Next, we shall illustrate the failure of Axiom (2) $(\phi \to \alpha) \to ((\psi \to \alpha) \to (\phi \vee \psi \to \alpha))$. We do so by first giving two examples to illustrate the failure of closure under $\Cup$-join for standard formulas (i.e., $\mathbin{\rotatebox[origin=c]{0}{$\vee$}}$-free formulas). Recall that in the standard team semantics, standard formulas are, however, closed under $\cup$-join (or unions). We say that a formula ϕ is *closed under (finite) join* if for any model $\mathfrak{M} = (W, R, \succcurlyeq, \Cup, V)$ and teams $t, s \subseteq W$

$$\mathfrak{M}, t \models \phi \text{ and } \mathfrak{M}, s \models \phi \Longrightarrow \mathfrak{M}, t \Cup s \models \phi. \qquad \text{(Join Closure Property)}$$

Counter-example 9.2 (Failure of join closure for standard formulas) (1). Consider the general intuitionistic team Kripke model $\mathfrak{M} = (\mathfrak{F}_{\mathsf{M}}, V)$, where the valuation V satisfies

$$V(p) = t \text{ and } V(q) = \mathrm{o}.$$

Clearly, $r \Cup s \not\models p \to q$, since for the team t, we have $r \Cup s \succcurlyeq t \models p$, whereas $t \not\models q$. On the other hand, we have $r \models p \to q$ and $s \models p \to q$.

We now give the proof of $r \models p \to q$; the proof of $s \models p \to q$ is similar. Suppose $r' \preccurlyeq r$ and $r' \models p$. Then, by definition, we have that $r' \preccurlyeq V(p) = t$. Since M_5 is a sublattices of $(\wp(W), \Cup, \Cap, \varnothing)$, we have that $r' \preccurlyeq r \Cap t = r \Cap_{\mathsf{M}_5} t = \mathrm{o} = V(q)$, which implies $r' \models q$, as required.

(2). Consider the general intuitionistic team Kripke model $\mathfrak{M} = (\mathfrak{F}_{\mathsf{N}}, V)$, where the valuation V satisfies

$$V(p) = t \text{ and } V(q) = r.$$

Clearly, $r \Cup s \not\models p \to q$, for a similar reason as above. On the other hand, we have $r \models p \to q$ and $s \models p \to q$. The former follows easily from persistence and the fact that $r \models p$ and $r \models q$. The latter is proved by going through a similar argument to that in (1) where the fact $t \Cap s = \mathrm{o}$ is used.

Closure under (finite) joins of a formula ϕ in a model is actually equivalent to validity of the *idempotent law* $(\phi \vee \phi) \leftrightarrow \phi$ in the model, as we now show. Note that $\phi \to \phi \vee \phi$ is always valid (by Lemma 9.9(ii) and the deduction theorem).

Lemma 9.10 *Let $\mathfrak{M} = (W, R, \succcurlyeq, \Cup, V)$ be a general intuitionistic team Kripke model. For any formula ϕ,*

$$\phi \textit{ is closed under (finite) joins in } \mathfrak{M} \iff \mathfrak{M} \models \phi \vee \phi \to \phi.$$

Proof "$\Longrightarrow$": For any team $t \subseteq W$, if $\mathfrak{M}, t \models \phi \vee \phi$, then there exist $r, s \subseteq W$ such that $t \preccurlyeq r \Cup s$, $r \models \phi$ and $s \models \phi$. Since ϕ is closed under joins in $\mathfrak{M}$, we have that $r \Cup s \models \phi$, which then implies $t \models \phi$ by persistence.

"$\Longleftarrow$": Suppose $\mathfrak{M}, s \models \phi$ and $\mathfrak{M}, t \models \phi$ for some $s, t \subseteq W$. By definition $s \Cup t \models \phi \vee \phi$. Now, since $\mathfrak{M} \models \phi \vee \phi \to \phi$, we conclude that $s \Cup t \models \phi$. □

By Corollary 9.3, over the standard team semantics the idempotent law $\alpha \vee \alpha \to \alpha$ for standard formulas α is a theorem in tIPC. By the above lemma, Counter-example 9.2 also shows that $\alpha \vee \alpha \to \alpha$ is no longer sound under the generalized

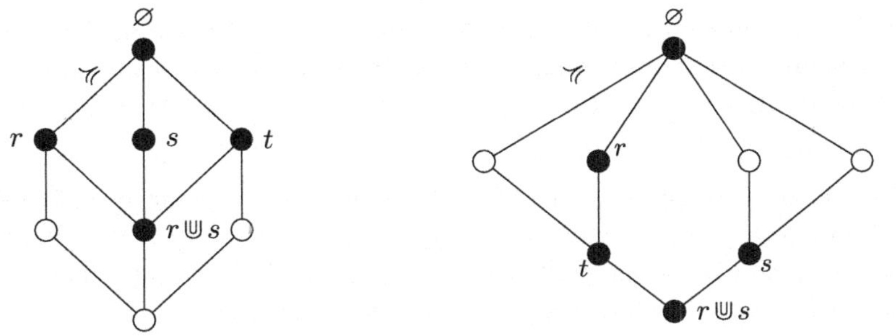

Fig. 9.9 Two bounded join-semilattices $(\wp(W), \Cup, \varnothing)$ (placed upside down) with M_5 and N_5 as a sublattice respectively

semantics. In particular, in the two models of the counter-example we have $r \Cup s \models (p \to q) \vee (p \to q)$, whereas $r \Cup s \not\models p \to q$. The same example also illustrates that the substitution instance of Axiom (2) $(\phi \to \alpha) \to ((\phi \to \alpha) \to (\phi \vee \phi \to \alpha))$ obtained by putting $\phi = \alpha = p \to q$ is not sound, since, e.g., in both models in Counter-example 9.2, we have $r \Cup s \not\models (p \to q) \vee (p \to q) \to (p \to q)$. From this it also follows that $(p \to q) \vee (p \to q) \not\models p \to q$, even though $p \to q \models p \to q$. Therefore, unlike with Axiom (3), the (weaker) rule that corresponds to Axiom (2) "from $\phi \vdash \alpha$ and $\psi \vdash \alpha$, derive $\phi \vee \psi \vdash \alpha$" is not sound either.

Finally, we also construct two counter-models for the Split axiom, whose underlying frames are two concrete instances of the frames $\mathfrak{F}_{\mathsf{M}}$ and $\mathfrak{F}_{\mathsf{N}}$. We, however, did not succeed to construct these counter-models with arbitrary frames $\mathfrak{F}_{\mathsf{M}}$ and $\mathfrak{F}_{\mathsf{N}}$.

Counter-example 9.3 (Split) (1). Consider the general intuitionistic team Kripke model $\mathfrak{M} = (W, R, \succcurlyeq, \Cup, V)$, where W is an arbitrary 3-element set, $R = \mathsf{id}$, the bounded semilattice $(\wp(W), \Cup, \varnothing)$ is the left figure shown (upside down) in Fig. 9.9, and the valuation V satisfies

$$V(p) = t,\ V(q) = \varnothing,\ V(a) = r \text{ and } V(b) = s.$$

Note that the highlighted points in $(\wp(W), \Cup, \varnothing)$ form a sublattice, the lattice M_5. Then $r \Cup s \not\models ((p \to q) \to a \vee b) \to ((p \to q) \to a) \vee ((p \to q) \to b)$. To see why, first observe that $r \Cup s \models (p \to q) \to a \vee b$, since the only teams $t' \preccurlyeq r \Cup s$ such that $t' \models p \to q$ are r, s, and $\varnothing$, on which either a or b is true. But as $s \not\models a$ and $r \not\models b$, we have that $r \Cup s \not\models (p \to q) \to a$ and $r \Cup s \not\models (p \to q) \to b$.

(2). Consider the general intuitionistic team Kripke model $\mathfrak{M} = (W, R, \succcurlyeq, \Cup, V)$, where W is an arbitrary 3-element set, $R = \mathsf{id}$, the bounded semilattice $(\wp(W), \Cup, \varnothing)$ is the right figure shown (upside down) in Fig. 9.9, and the valuation V satisfies

$$V(p) = t,\ V(q) = r = V(a) \text{ and } V(b) = s.$$

Note that the semilattice $(\wp(W), \Cup, \varnothing)$ is the same as the one in Example 9.1, and the highlighted points form a sublattice, the lattice N_5. By a similar argument to the above, it can be verified that $r \Cup s \not\models ((p \to q) \to a \vee b) \to ((p \to q) \to a) \vee ((p \to q) \to b)$.

Consider the logic $\mathsf{gtIPC} := \{\phi \mid \mathsf{F^g} \models \phi\}$ of the class $\mathsf{F^g}$ of all general intuitionistic team Kripke frames. By the observations above, all axioms of tIPC except for Axioms (2) and (3) and the Split axiom belong to gtIPC. Whether these axioms actually axiomatize the logic gtIPC is left as future work.

9.4.3 Recovering the Standard tIPC, *and Distributivity*

In this section, we show that over general intuitionistic team Kripke frames that are distributive, Axioms (2) and (3) and the Split axiom are sound, and thus all tIPC axioms are recovered over such frames; in other words, distributivity is a sufficient condition for validating these axioms. It will then follow also from the counter-examples provided at the end of the previous section that distributivity is actually the necessary and sufficient condition for validating Axioms (2) and (3) over the class of finite classical frames. We also show that over distributive frames standard formulas satisfy the join closure property. Similar results are also found in related work in Punčochář (2017) in a different and simpler setting; we will discuss these connections in more detail in Sect. 9.5.

Let us first define the distributivity condition.

Definition 9.15 A general intuitionistic team Kripke frame $\mathfrak{F} = (W, R, \succcurlyeq, \Cup)$ is called *distributive* if the join-semilattice $(\wp(W), \Cup, \varnothing)$ is distributive, namely

$$t \preccurlyeq r \Cup s \Longrightarrow \exists r', s' \subseteq W : r' \preccurlyeq r,\ s' \preccurlyeq s, \text{ and } t = r' \Cup s'.$$

We denote by DF the class of distributive general intuitionistic team Kripke frames.

A frame $\mathfrak{F} = (W, \supseteq, R, \cup, \varnothing)$ with R a partial order is always distributive, as $(\wp(W), \cup, \varnothing)$ is clearly a distributive semilattice. Let us elaborate on the notion of distributivity. Recall, from e.g., Burris and Sankappanavar (2012), that any distributive join-semilattice in which meets exist is a distributive lattice. Meets of finite join-semilattices always exist. A lattice is distributive if and only if none of its sublattices is isomorphic to M_5 or N_5 shown in Fig. 9.8 (placed upside down with the highlighted nodes r, s, t witnessing the failure of the distributivity condition). We have essentially already pointed out that the two types of finite frames $\mathfrak{F}_\mathsf{M}$ and $\mathfrak{F}_\mathsf{N}$ we constructed in the previous section for Counter-examples 9.1 and 9.2 (as well as the two concrete frames constructed in Counter-example 9.3) are thus not distributive. Furthermore, any finite k-element non-distributive classical frame (i.e., $R = \mathsf{id}$) is thus isomorphic to either $\mathfrak{F}_\mathsf{M}$ or $\mathfrak{F}_\mathsf{N}$ with k elements. Our Counter-examples 9.1 and 9.2 and the discussions afterwards in the previous section thus show that over finite classical frames, distributivity is a necessary condition for validating Axioms (2) and (3), the idempotent law of the local disjunction $\vee$ for standard formulas, and join closure of standard formulas.

Corollary 9.5 *For any finite classical frame $\mathfrak{F}$ that is not in* DF*, we have that Axiom (2), Axiom (3) and the idempotent law $\alpha \vee \alpha \to \alpha$ (with α standard) are not valid on $\mathfrak{F}$, and standard formulas are not closed under joins over $\mathfrak{F}$.*

Whether distributivity is a necessary condition for the same axioms for the local disjunction $\vee$ and join closure property also over non-classical frames (i.e., frames in which $R \neq \mathsf{id}$) is unclear. As for the Split axiom for the global disjunction $\mathbin{\vee\!\!\!\vee}$, we conjecture that distributivity is not a necessary condition for validating the axiom even over finite classical frames. Our Counter-example 9.3 from the previous section was built on two concrete non-distributive frames of $\mathfrak{F}_{\mathsf{M}}$ or $\mathfrak{F}_{\mathsf{N}}$ type; it is not clear how to generalize the argument to arbitrary $\mathfrak{F}_{\mathsf{M}}$ or $\mathfrak{F}_{\mathsf{N}}$ type frames.

We now proceed to show that distributivity is, on the other hand, a sufficient condition for validating all above-mentioned axioms and properties, over arbitrary frames. We start with Axioms (3).

Proposition 9.5 *For any formulas ϕ, ψ and χ, we have that* $\mathsf{DF} \vdash (\phi \to \chi) \to (\phi \vee \psi \to \chi \vee \psi)$.

Proof Let $\mathfrak{M} = (W, R, \succcurlyeq, \Cup, V)$ be a model with a distributive underlying frame. Suppose $\mathfrak{M}, t \models \phi \to \chi$ for some $t \subseteq W$. Suppose also that $t \succcurlyeq \circ Rs$ and $s \models \phi \vee \psi$. Then there are $r_0, r_1 \subseteq W$ such that $s \preccurlyeq r_0 \Cup r_1$, $r_0 \models \phi$ and $r_1 \models \psi$. By distributivity, there exist $r_0', r_1' \subseteq W$ such that $r_0' \preccurlyeq r_0$, $r_1' \preccurlyeq r_1$, and $s = r_0' \Cup r_1'$. Since $t \succcurlyeq \circ Rs \succcurlyeq r_0' \preccurlyeq r_0$, by persistence, we have $r_0' \models \phi \to \chi$ and $r_0' \models \phi$, which imply $r_0' \models \chi$. On the other hand, since $r_1 \succcurlyeq r_1'$, we also have $r_1' \models \psi$ by persistence. Hence, we conclude that $s = r_0' \Cup r_1' \models \chi \vee \psi$, as required. □

Next, we show that over general intuitionistic team Kripke models whose underlying frame are distributive, standard formulas are closed under finite joins in general, and under arbitrary joins given certain additional conditions.

Lemma 9.11 (Join closure of standard formulas) *Let $\mathfrak{M} = (W, R, \succcurlyeq, \Cup, V)$ be a model with a distributive underlying frame. For any standard formula α, any teams $t_0, t_1 \subseteq W$,*

$$\mathfrak{M}, t_0 \models \alpha \text{ and } \mathfrak{M}, t_1 \models \alpha \Longrightarrow \mathfrak{M}, t_0 \Cup t_1 \models \alpha.$$

Moreover, if $(\wp(W), \Cup, \varnothing)$ forms a distributive complete semilattice, and distributivity and Condition (c) hold for arbitrary $\Cup$-joins, then

$$\mathfrak{M}, t \models \alpha \text{ for every } t \in T \Longrightarrow \mathfrak{M}, \biguplus T \models \alpha.$$

Proof We only give the proof for the finite join case. The arbitrary join case follows from essentially the same argument, in which all the steps with finite joins can be easily adapted to the infinite case given the completeness assumption.

We proceed by induction on α. If $\alpha = \bot$, then $\mathfrak{M}, t_0 \models \bot$ and $\mathfrak{M}, t_1 \models \bot$ imply that $t_0 = \varnothing = t_1$. Thus, $t_0 \Cup t_1 = \varnothing \models \bot$.

If $\alpha = p$, then $t_0, t_1 \models p$ implies $t_0, t_1 \preccurlyeq V(p)$. Since $(\wp(W), \Cup, \varnothing)$ is a lattice, $t_0 \Cup t_1 \preccurlyeq V(p)$ and so $t_0 \Cup t_1 \models p$.

Case $\alpha = \beta \vee \gamma$. Suppose $t_0, t_1 \models \beta \vee \gamma$. Then there are $r_0, s_0, r_1, s_1 \in \wp(W)$ such that $t_0 \preccurlyeq r_0 \Cup s_0$, $t_1 \preccurlyeq r_1 \Cup s_1$, $r_0 \models \beta$, $s_0 \models \gamma$, $r_1 \models \beta$, and $s_1 \models \gamma$. By induction hypothesis, we have that $r_0 \Cup r_1 \models \beta$ and $s_0 \Cup s_1 \models \gamma$. Since $(\wp(W), \Cup, \varnothing)$ is a lattice,

$$t_0 \Cup t_1 \preccurlyeq (r_0 \Cup s_0) \Cup (r_1 \Cup s_1) = (r_0 \Cup r_1) \Cup (s_0 \Cup s_1)$$

Hence we conclude that $t_0 \Cup t_1 \models \beta \vee \gamma$.

Case $\alpha = \beta \to \gamma$. Suppose $t_0, t_1 \models \beta \to \gamma$. Suppose also that $t_0 \Cup t_1 \succcurlyeq \circ Rs$ and $s \models \beta$. Then, by Conditions (b) and (c), we have $R[t_0] \Cup R[t_1] = R[t_0 \Cup t_1] \succcurlyeq s$. By distributivity, there exist $r_0, r_1 \subseteq W$ such that $r_0 \preccurlyeq R[t_0]$, $r_1 \preccurlyeq R[t_1]$, and $s = r_0 \Cup r_1$. Since $t_0 R^\circ R[t_0] \succcurlyeq r_0 \preccurlyeq s$ and $t_1 R^\circ R[t_1] \succcurlyeq r_1 \preccurlyeq s$, by persistence, we have $r_0, r_1 \models \beta \to \gamma$ and $r_0, r_1 \models \beta$. Thus, $r_0, r_1 \models \gamma$. Hence, by induction hypothesis, we conclude that $s = r_0 \Cup r_1 \models \gamma$.

The case $\alpha = \beta \wedge \gamma$ is straightforward. □

Let us remark that in the above proof, only the implication case requires the distributivity assumption. This means that for arbitrary general intuitionistic team Kripke model, all $[\bot, \wedge, \vee]$-formulas are closed under finite joins.

An immediate corollary of the above lemma is that the idempotent law for $\vee$ and Axiom (2) are sound over distributive frames.

Corollary 9.6 *For any standard formula α, and any formulas ϕ and ψ, we have that* $\mathsf{DF} \models \alpha \vee \alpha \to \alpha$ *and* $\mathsf{DF} \models (\phi \to \alpha) \to ((\psi \to \alpha) \to (\phi \vee \psi \to \alpha))$.

Proof The soundness of $\alpha \vee \alpha \to \alpha$ follows from Lemmas 9.11 and 9.10. The soundness of Axiom (2) follows from the observation that given Axiom (3), Axiom (2) is equivalent to $\alpha \vee \alpha \to \alpha$. □

As a consequence of Corollaries 9.5–9.6, distributivity is a necessary and sufficient condition for validating Axiom (2), Axiom (3), idempotent law of $\vee$ for standard formulas and closure under join for standard formulas over finite classical frames.

Finally, we show that over distributive frames the Split axiom is also sound, and thus all tIPC axioms are recovered in DF or $\mathsf{DF} \models \mathsf{tIPC}$. We first prove the following lemma.

Lemma 9.12 *Let $\mathfrak{F} = (W, R, \succcurlyeq, \Cup)$ be a general frame. For any teams $t, r, s \subseteq W$, if $t \succcurlyeq \circ Rr$ and $t \succcurlyeq \circ Rs$, then $R[t] \succcurlyeq r \Cup s$.*

Proof By condition (b) we have that $R[t] \succcurlyeq r$ and $R[t] \succcurlyeq s$, which then imply that $R[t] \succcurlyeq r \Cup s$. □

Proposition 9.6 *For any standard formula α, and any formulas ϕ and ψ, we have that* $\mathsf{DF} \models (\alpha \to \phi \mathbin{\vee\!\!\!\vee} \psi) \to (\alpha \to \phi) \mathbin{\vee\!\!\!\vee} (\alpha \to \psi)$.

Proof Let $\mathfrak{M} = (W, R, \succcurlyeq, \Cup, V)$ be an arbitrary model with a distributive underlying frame. Suppose $\mathfrak{M}, t \not\models (\alpha \to \phi) \vee (\alpha \to \psi)$. We show that $\mathfrak{M}, t \not\models \alpha \to \phi \vee \psi$. By assumption, there are teams $r, s \subseteq W$ such that $t \succcurlyeq \circ Rr, t \succcurlyeq \circ Rs, r \models \alpha, r \not\models \phi$, $s \models \alpha$ and $s \not\models \psi$. Consider the team $r \Cup s$. By Lemma 9.11, we have $r \Cup s \models \alpha$. Since $r, s \preccurlyeq r \Cup s$, by persistence, we have $r \Cup s \not\models \phi$ and $r \Cup s \not\models \psi$, meaning $r \Cup s \not\models \phi \vee \psi$. Now, by Lemma 9.12, we have $R[t] \succcurlyeq r \Cup s$, from which we conclude that $R[t] \not\models \alpha \to \phi \vee \psi$. Finally, since $t R^\circ R[t]$, we obtain $t \not\models \alpha \to \phi \vee \psi$ by persistence. □

Hence, we have shown that DF $\models$ tIPC. Whether the logic of DF is exactly tIPC, or what the logic of DF is is left as future work.

9.5 Concluding Remarks and Open Problems

In this paper, we have studied intermediate logics in the setting of team semantics. Starting from the team-based intuitionistic logic tIPC introduced in Ciardelli et al. (2020), we have explored two alternative approaches to define intermediate logics for team semantics. In the first approach, we extend tIPC with axioms written in the standard language only. We showed that universal models for IPC behave also as universal models for tIPC, and a large class of intermediate standard axioms (including some de Jongh-formulas) still define the same class of Kripke frames. In the second approach, we made a first attempt to define team-based logics in which the Split axiom is not sound. This is done by generalizing the standard team semantics through general intuitionistic team Kripke frames $\mathfrak{F} = (W, R, \succcurlyeq, \Cup)$, where $(\wp(W), \succcurlyeq, \Cup, \varnothing)$ is an abstraction of the underlying structure $(\wp(W), \supseteq, \cup, \varnothing)$ of teams in the standard team semantics.

A few remarks on our choice of definitions in the second approach are in order, especially in comparison with two related approaches considered by Punčochář (2016, 2017) (as sketched in the introduction section). Instead of modifying the underlying structure of teams to $(\wp(W), \succcurlyeq, \Cup, \varnothing)$, it also makes sense to consider the structure $(I, \supseteq, \cup, \varnothing)$, where $I \subseteq \wp(W)$ is a set of "admissible" teams, as it was done in Punčochář (2016). Another appealing alternative generalization is to treat teams simply as elements of the domain W in Definition 9.13; that is, instead of "lifting" the order R to the power set level, one "flattens" teams to the single world level. Or, as considered in Punčochář (2017), teams are viewed as primitive entities in an arbitrary bounded semilattice $(A, \Cup, 0)$ (i.e., a team is an element in the domain A). This approach, however, does not seem to be in line with the fundamental idea of team semantics itself, at least not directly. Team semantics was introduced by Hodges (1997a, 1997b) to characterize notions of dependence, which, as Hodges observed, can only make sense in the presence of *multitudes* or *sets* of possibles worlds (i.e., teams), rather than in single worlds. We thus choose to maintain this most distinguishing feature of team semantics, and define teams only as elements in the power set of the domain W. The benefit of this team semantics approach will become apparent when we investigate notions of dependence in the intermediate logic framework laid down in this paper, which we leave as future work.

There is another main difference between our approach and the approaches in Punčochář (2016, 2017). Our general intuitionistic team Kripke frames $\mathfrak{F} = (W, R, \succcurlyeq, \Cup)$ are, in a sense, always two-layered—the base layer consists of a usual intuitionistic Kripke frame (W, R), and the structure $(\wp(W), \succcurlyeq, \Cup, \varnothing)$ belongs to the team layer. The "intermediateness" can come from variations in both layers as well as their interactions. In particular, our first approach (Sect. 9.3) can be viewed as keeping the team layer fixed to $(\wp(W), \supseteq, \cup, \varnothing)$, while varying the underlying Kripke frame (W, R). The interaction between the two layers (as reflected in the composition $\succcurlyeq \circ R^{\circ}$) also affects the behavior of implication $\rightarrow$ (see Definition 9.14). On the other hand, the approach in either (Punčochář, 2016) or (Punčochář, 2017) is essentially single-layered. The focus there is only on the team layer, being either $(I, \supseteq, \cup, \varnothing)$ or $(A, \succcurlyeq, \Cup, 0)$ (where $\preccurlyeq$ is the partial order associated with the join $\Cup$), and the implication $\rightarrow$ is thus defined by using the partial order $\supseteq$ or $\succcurlyeq$ alone as the successor relation. In particular, putting $A = \wp(W)$, the setting in Punčochář (2017) essentially corresponds to the case $R = \mathsf{id}$ in our setting. It is interesting to note, though, that in Punčochář (2017), it was shown that over the frames $(A, \succcurlyeq, \Cup, 0)$ with the semilattice $(A, \Cup, 0)$ being distributive, the logic in the standard language (with local disjunction $\vee$) is exactly IPC. Distributivity is also shown to be a necessary and sufficient condition for join closure in the setting. Moreover, Punčochář (2017) also introduced a sound and complete system of natural deduction for the language of tIPC extended with the necessity $\Box$ modality, which is roughly tIPC axioms/rules together with rules of IK. These results correspond to our results in Sect. 9.4.3. Whether the arguments for the completeness theorem in Punčochář (2017) can be adapted in our context to show that distributivity is actually the characterization condition for tIPC axioms is open.

A subtle but important technical difference between our setting and the setting in Punčochář (2017) also deserves comments. The semantics of the local disjunction $\vee$ in Punčochář (2017) is defined with the same clause as in our Definition 9.14 except that our condition "$t \preccurlyeq s \Cup r$" is changed to "$t = s \Cup r$" in Punčochář (2017). While the two definitions for $\vee$ are actually equivalent in the standard team semantics setting (when $\succcurlyeq = \supseteq$ and $\Cup = \cup$, see Corollary 9.2), our definition turns out to be more general in the generalized team semantics setting. In particular, the persistence property in Punčochář (2017) (especially for the $\vee$ case in the proof) also requires the assumption that the underlying semilattice in question is distributive (in which case Split axiom is also sound), whereas the persistence property in our setting (Lemma 9.7) does not require the distributivity assumption (and thus the Split axiom is in general not sound in our setting).

We end the paper by mentioning some more open problems. Section 9.4 is a first step towards a generalized team semantics and many further directions deserve investigation. For instance, in Definition 9.13 of general intuitionistic team Kripke frames, whether it is possible to weaken conditions (b) and (c) and still preserve the persistence property of the logic needs to be further explored. It also makes sense to weaken the assumption that $\preccurlyeq$ is the partial order of the join-semilattice

$(\wp(W), \Cup, \varnothing)$. While this lattice order assumption was used heavily in Sect. 9.4.3 to recover all tIPC axioms, in Sect. 9.4.2 the assumption is used actually only in establishing Lemma 9.9(i) (the associativity of the local disjunction $\vee$), where the minimal condition seems to be that "$t \preccurlyeq s$ implies $t \Cup r \preccurlyeq s \Cup r$ for all teams r, s, t". Weakening the lattice order assumption to this alternative assumption will give rise to a different logic that requires further investigation. How to axiomatize the logic of the class of all general intuitionistic team Kripke frames, what is the logic of the class DF of all distributive such frames, and do these logics have the disjunction property are also topics for future research. Providing appropriate definitions of universal models and de Jongh formulas in the generalized team semantics setting is also an interesting direction. Related work on universal models for fragments of intuitonistic modal logic has been studied recently in Bezhanishvili et al. (2021). Another very recent work (Quadrellaro, 2022) has provided algebraic semantics for tIPC and the team-based intermediate logics obtained from our first approach (as defined in Definition 9.4). Some other work on algebraic semantics of inquisitive and other relevant logics can be found in Bezhanishvili et al. (2022), Grilletti and Quadrellaro (2019). It is natural to study also the algebraic semantics for the logics obtained from our second approach. A recent Master's thesis (Dmitrieva, 2021) introduced a related type of general team semantics for a positive (modal) logic by considering complete (modal) lattices (where the distributivity of the lattices also plays a related subtle role). Studying modal extensions of the language tIPC is also an interesting direction for future work. Different variants of modal dependence and inquisitive logic have been studied by many authors, see, e.g., Kontinen et al. (2015), Sevenster (2009), Väänänen (2008), Yang (2017).

Acknowledgements The authors are thankful to the referee for useful comments. They would also like to thank Davide Quadrellaro for interesting discussion related to the topic of this paper. The second author was supported by grant 330525 of the Academy of Finland, and Research Funds of the University of Helsinki.

References

Bezhanishvili, N. (2004). *De Jongh's characterization of intuitionistic propositional calculus*. Liber Amicorum Dick de Jongh. University of Amsterdam.

Bezhanishvili, N. (2006). *Lattices of intermediate and cylindric modal logics*. Ph.D. thesis, University of Amsterdam.

Bezhanishvili, G., & Bezhanishvili, N. (2022). Jankov formulas and axiomatization techniques for intermediate logics. In A. Citkin, & I. M. Vandoulakis (Eds.), *V. A. Yankov on Non-Classical Logics, History and Philosophy of Mathematics, Outstanding contributions to Logic*, volume 24 (pp. 71–124), Springer.

Bezhanishvili, G., Bezhanishvili, N., Carai, L., Gabelaia, D., Ghilardi, S., & Jibladze, M. (2021). Diego's theorem for nuclear implicative semilattices. *Indagationes Mathematicae, 32*(2), 498–535.

Bezhanishvili, N., Colacito, A., & de Jongh, D. (2017). A study of subminimal logics of negation and their modal companions. In A. Silva, S. Staton, P. Sutton & C. Umbach (Eds.), *Language,*

logic, and computation—12th international tbilisi symposium, TbiLLC 2017, volume 11456 of Lecture notes in computer science (pp. 21–41). Springer.
Bezhanishvili, N., de Jongh, D., Tzimoulis, A., & Zhao, Z. (2015). Universal models for the positive fragment of intuitionistic logic. In H. Hvid Hansen, S. Murray, M. Sadrzadeh & H. Zeevat (Eds.), *Logic, Language, and Computation—11th international tbilisi symposium, TbiLLC 2015, volume 10148 of Lecture notes in computer science* (pp. 229–250). Springer.
Bezhanishvili, N., Grilletti, G., & Quadrellaro, D. (2022). An algebraic approach to inquisitive and DNA-logics. *The Review of Symbolic Logic, 15*(4), 950–990.
Bezhanishvili, N., & Holliday, W. H. (2020). Choice-free stone duality. *Journal of Symbolic Logic, 85*(1), 109–148.
Bezhanishvili, N., & de Jongh, D. (2018). Stable formulas in intuitionistic logic. *Notre Dame Journal of Formal Logic, 59*(3), 307–324.
Božić, M., & Došen, K. (1984). Models for normal intuitionistic modal logics. *Studia Logica, 43*(3), 217–245.
Burris, S., & Sankappanavar, H. P. (2012). *A course in universal algebra (The Millennium Edition)*. Springer.
Chagrov, A., & Zakharyaschev, M. (1997). *Modal logic*. Oxford University Press.
Ciardelli, I. (2016). Dependency as question entailment. In S. Abramsky, J. Kontinen, J. Väänänen & H. Vollmer (Eds.), *Dependence logic: Theory and application, progress in computer science and applied logic* (pp 129–182). Birkhauser.
Ciardelli, I., Iemhoff, R., & Yang, F. (2020). Questions and dependency in intuitionistic logic. *Notre Dame Journal of Formal Logic, 61*(1), 75–115.
Ciardelli, I., & Roelofsen, F. (2011). Inquisitive logic. *Journal of Philosophical Logic, 40*(1), 55–94.
de Jongh, D., & Troelstra, A. (1966). On the connection of partially ordered sets with some pseudo-boolean algebras. *Indagationes Mathematicae, 28*(3), 317–329.
de Jongh, D. (1968). *Investigations on the intuitionistic propositional calculus*. Ph.D. thesis, The University of Wisconsin-Madison.
Dmitrieva, A. (2021). *Positive modal logic beyond distributivity: duality, preservation and completeness*. Master's thesis, University of Amsterdam.
Esakia, L. (2019). Heyting algebras: Duality theory. In G. Bezhanishvili & W. H. Holliday (Eds.), *Trends in logic—studia logica library* (Vol. 50). Translated from the Russian edition by A. Evseev. Springer.
Fine, K. (1974). Logics containing $K4$ Part I. *Journal of Symbolic Logic, 39*, 31–42.
Gabbay, D. M., & de Jongh, D. (1974). A sequence of decidable finitely axiomatizable intermediate logics with the disjunction property. *Journal of Symbolic Logic, 39*(1), 67–78.
Grilletti, G., & Quadrellaro, D. (2019). Lattices of intermediate theories via Ruitenburg's theorem. In *Tbilisi symposium on language, logic and computation (TbiLLC 2019)*, Lecture notes in computer science. Springer, forthcoming.
Hintikka, J. (1998). *The principles of mathematics revisited*. Cambridge University Press.
Hintikka, J., & Sandu, G. (1989). Informational independence as a semantical phenomenon. In R. Hilpinen J. E. Fenstad, I. T. Frolov (Eds.), *Logic, methodology and philosophy of science* (pp. 571–589). Elsevier.
Hodges, W. (1997b). Some strange quantifiers. In J. Mycielski, G. Rozenberg & A. Salomaa (Eds.), *Structures in logic and computer science: A selection of essays in honor of A. Ehrenfeucht, volume 1261 of Lecture notes in computer science* (pp. 51–65). Springer.
Hodges, W. (1997). Compositional semantics for a language of imperfect information. *Logic Journal of the IGPL, 5*, 539–563.
Holliday, W. (2015). *Possibility frames and forcing for modal logic*.
Holliday, W. (2020). Inquisitive intuitionistic logic. In N. Olivetti & R. Verbrugge (Eds.), *Advances in modal logic* (Vol. 13, pp. 329–348). College Publications.
Ilin, J., de Jongh, D., & Yang, F. (2021). NNIL-formulas revisited: Universal models and finite model property. *Journal of Logic and Computation, 31*(2), 573–596.

Jankov, V. A. (1963). On the relation between deducibility in intuitionistic propositional calculus and finite implicative structures. *Doklady Akademii Nauk SSSR, 151*, 1293–1294. (In Russian).

Jankov, V. A. (1968). The construction of a sequence of strongly independent superintuitionistic propositional calculi. *Soviet Mathematics Doklady, 9*, 806–807.

Kontinen, J., Müller, J., Schnoor, H., & Vollmer, H. (2015). A van Benthem theorem for modal team semantics. In S. Kreutzer (Ed.), *24th EACSL annual conference on computer science logic (CSL 2015)* (pp. 277–291).

Kracht, M. (1993). Splittings and the finite model property. *Journal of Symbolic Logic, 58*(1), 139–157.

Lauridsen, F. M. (2019). *Cuts and completions: Algebraic aspects of structural proof theory*. Ph.D. thesis, ILLC, University of Amsterdam.

Lauridsen, F. M. (2019). Intermediate logics admitting a structural hypersequent calculus. *Studia Logica, 107*(2), 247–282.

Plotkin, G. D., & Stirling, C. P. (Eds.), A framework for intuitionistic modal logic. In J. Y. Halpern (Ed.), *Theoretical aspects of reasoning about knowledge* (pp. 399–406).

Punčochář, V. (2016). A generalization of inquisitive semantics. *Journal of Philosophical Logic, 45*, 399–428.

Punčochář, V. (2017). Algebras of information states. *Journal and Logic and Computation, 23*, 1643–1675.

Punčochář, V. (2021). Inquisitive Heyting algebras. *Studia Logica, 109*, 995–1017.

Quadrellaro, D. (2022). On intermediate inquisitive and dependence logics: An algebraic study. Special Issue on Logics of Dependence and Independence. *Annals of Pure and Applied Logic, 173*(10), 103143.

Rautenberg, W. (1980). Splitting lattices of logics. *Archive for Mathematical Logic Logik und Grundlagenforschung, 20*(3–4), 155–159.

Sevenster, M. (2009). Model-theoretic and computational properties of modal dependence logic. *Journal of Logic and Computation, 19*(6), 1157–1173.

Simpson, A. K. (1994). *The proof theory and semantics of intuitionistic modal logic*. Ph.D. thesis, University of Edinburgh.

Väänänen, J. (2007). *Dependence logic: A new approach to independence friendly logic*. Cambridge University Press.

Väänänen, J. (2008). Modal dependence logic. In K. Apt & R. van Rooij (Eds.), *New perspectives on games and interaction, volume 4 of Texts in logic and games* (pp. 237–254). University Press.

van Benthem, J. (2021). Relational patterns, partiality, and set lifting in modal semantics. Technical report, PP-2021-02, https://eprints.illc.uva.nl/id/eprint/1773

Wolter, F., & Zakharyaschev, M. (1999). Intuitionistic modal logic. In E. Casari A. Cantini & P. Minari (Eds.), *Logic and foundations of mathematics* (pp. 227–238). Synthese Library.

Yang, F. (2014). *On extensions and variants of dependence logic*. Ph.D. thesis, University of Helsinki.

Yang, F. (2017). Modal dependence logics: Axiomatizations and model-theoretic properties. *Logic Journal of the IGPL, 25*(5), 773–805.

Yang, F., & Väänänen, J. (2016). Propositional logics of dependence. *Annals of Pure and Applied Logic, 167*(7), 557–589.

Chapter 10
Well Partial Orders

Andreas Weiermann

Abstract This article provides a (personal) survey on well partial orders. We mainly concentrate on aspects which we find most interesting and apologize for omitting many other important results. Starting with recalling basic facts from De Jongh and Parikh's fundamental paper on maximal order types we survey some related key results by Diana Schmidt. Then we discuss generalized trees, their embeddability relation and their associated maximal order types. These results can be condensed neatly into the formula $o(T(W)) = \vartheta(o(W(\Omega)))$. At the end we cover Friedman style miniaturizations of Kruskal's theorem and their associated phase transitions.

Keywords Well partial orders · Well quasi orders · Finite trees · Higman's lemma · Kruskal's tree theorem · Maximal order types · Friedman style miniaturization

10.1 Introduction

Well partial orders are partial orders which play an important role in logic, computer science, algebra and other branches of mathematics. Modulo minor details, i.e. after dropping the antisymmetry condition, these orders are equivalent to well-quasi orders, and for the purposes this article, we can mainly identify these two concepts. Well partial orders are significant for proving termination in various contexts. By now, famous applications of well partial orders have been provided in the study of substructural logics by Kripke's lemma (Kripke, 1959), in the study of Skolem's problem by Ehrenfeucht (1973), in the study of Gröbner bases by Baader and Nipkow (1998), in the proof of the graph minor theorem via Friedman's tree theorem

A. Weiermann (✉)
Vakgroep Wiskunde: Analysis, Logic and Discrete Mathematics, Krijgslaan 281 Gebouw S8, 9000 Ghent, Belgium
e-mail: Andreas.Weiermann@UGent.be

N. Bezhanishvili et al. (eds.), *Dick de Jongh on Intuitionistic and Provability Logics*, Outstanding Contributions to Logic 28, https://doi.org/10.1007/978-3-031-47921-2_10

(Friedman et al., 1987), in minor testing by the graph minor theorem of Robertson et al. (2004), in proving termination for rewrite systems by Dershowitz (1982), and in the theory of transition systems (Abdula, 2010). In fact well partial orderings are ubiquitous, and so one article by Kruskal carries the title *The Theory of Well quasi orderings; a frequently discovered concept*.

For an overview about essential aspects of state of the art we allow ourselves to recommend the recently published book (Schuster et al., 2020). In this article, I will survey parts of the theory of well partial orders from my personal perspective. There will be numerous omissions, and I apologize for skipping so many interesting results, which I should have included. The focus of this article will be mainly on well partial orders on strings and (generalized) trees. There will be no discussion of well partial orderings on graphs, and we refer to Schuster et al. (2020) for the graph minor theorem. There are also profound results about well partial orderings related to Carlson's patterns of resemblance. We refer to the recent papers by Carlson (2017) and Wilken (2018). We also omit the discussion of better quasi orderings and again we refer to Schuster et al. (2020).

10.2 My Personal View on Well Partial Orderings

This section is a purely subjective description of how I came into the subject, how I have been influenced by it and what I personally still consider intriguing open problems. The exact definitions are postponed to later sections.

In the beginning of the nineties I fell in love with the subject after having read the article *Well partial orderings and hierarchies* by De Jongh and Parikh (1977) and the Habilitation thesis *Well partial orderings and their maximal order types* by Schmidt (1979). However, at that time, it wasn't easy to access Schmidt's work, and I was very grateful to Gert Müller, who sent her thesis to the Institute for Mathematical Logic in Münster, where I was based at that time. Fortunately, the thesis has been reprinted in Schuster et al. (2020) and is now widely available.

Around that period, a constructive proof for Higman's lemma was published in Murthy and Russel (1990), and a challenge to find a constructive proof of Kruskal's theorem was offered. It turns out that the key elements to these constructive proofs were already contained in the two seminal contributions by De Jongh and Parikh, and Schmidt.

In fact, the work of De Jongh and Parikh could be used to provide (by a clever explicit construction) a constructive proof of Higman's Lemma as shown by Simpson in Simpson (1988).

Another seminal result by De Jongh (unpublished) is that the maximal order type of the set of binary trees under homeomorphic embedding equals ε_0. This result has the beautiful consequence that the order type of the closure of $\{0\}$ under a monotonic increasing binary ordinal valued function is at most ε_0. Diana Schmidt was previously able to calculate sharp bounds on the order type of the closure of $\{0\}$ under an n-ary monotonic increasing function in terms of Schütte's Klammer term functions

(Schmidt, 1975). Still, her calculations were quite complicated and were only for $n > 2$.

Quite surprisingly, by clever and beautiful applications of the De Jongh and Parikh theory, Schmidt proved stronger results with easier proofs in her habilitation thesis. She calculated several maximal order types of well partial orderings related to trees and showed that the maximal order type operation for tree classes behaves functorial in a certain way. Her results included in particular the De Jongh result for binary trees, so a proof of this important result (and many more) is now in published form available via Schmidt's contribution to Schuster et al. (2020). (H. Friedman arrived independently at similar results about maximal order types, but his notes are still unpublished.)

Harvey Friedman pioneered the application of well partial orders to proof theory and he became famous for proving that ATR_0 does not prove Kruskal's tree theorem. In that context he also pioneered corresponding miniaturizations, which provide examples for concrete combinatorial statements which are independent of strong theories.

This type of material attracted my attention due to its beauty and elegance and since it is related to the fascinating and fundamental natural well ordering problem of Feferman (Crossley & Kister, 1986; Feferman, 1987). Therefore, it was an obvious idea to investigate in particular possible extensions of Schmidt's results.

In a first step, motivated by Schütte (1954) and Veblen (1908), I studied the order type of the closure of $\{0\}$ under a monotonic increasing ordinal valued function of varying countable arity and proved a corresponding result for generalized trees (Weiermann, 1993). Afterwards, I continued with studying the order type of the closure of $\{0\}$ under an essentially increasing monotonic increasing function (Weiermann, 1993). However, I got stuck for a few years without being able to provide further extensions.

After some time, Michael Rathjen and I exploited the work of De Jongh and Parikh, Schmidt, and Simpson to give a constructive proof of Kruskal's theorem (Rathjen & Weiermann, 1993). We were able to show how to transform in a primitive recursive way a given bad sequence of finite trees into a descending sequence of ordinals below the small Veblen ordinal. (Results of comparable strengths have been obtained by Hasegawa, 1994 and Friedman (unpublished)). The details in our calculation were quite involved, but the seminal works by De Jongh and Parikh as well as Schmidt mentioned before provided the necessary red thread in the calculations. Furthermore, these calculations allowed for extracting subrecursive bounds for Kruskal's theorem (Weiermann, 1994) using (Buchholz et al., 1994).

At that time, the extensions of Kruskal's theorem (based on the gap condition) by Harvey Friedman (Simpson, 1985) remained a complete mystery. It was completely unclear to me how the maximal order types for Friedman's embeddability relation could be analyzed in any way, but we will return to this in a minute.

I also got interested in proving termination for term rewriting systems in that period. By a nice coincidence, well partial orderings played a central role here via Dershowitz's simplification orderings (Dershowitz, 1982). In Weiermann (1992), I calculated maximal order types for the multiset ordering and some other orderings. I introduced relativized ordinal notation systems, defined by adjoining a given partial order X into a given system of notations for a proof-theoretic ordinal. It turned out that many order-theoretic properties of X transferred over to the enlarged system $T(X)$.

During my presentation at the CSL91 conference in Bern, I formulated a general hypothesis over the relation between proof-theoretic ordinals and well partial orders. Although speculative in character, this hypothesis has formed a fruitful red thread through my later investigations on maximal order types. The hypothesis goes as follows:

Let us assume that $\langle T, \leq_T \rangle$ is a notation system for a proof-theoretic ordinal. Let $\trianglelefteq$ be the restriction of $\leq_T$ such that, for $\alpha, \beta \in T$ we have $\alpha \trianglelefteq_T \beta$ iff $\alpha \leq_T \beta$ holds for graph theoretic reasons, e.g. a subterm or a monotonicity reason. Then $\langle T, \trianglelefteq_T \rangle$ is a well partial order and the maximal order type of $\langle T, \trianglelefteq_T \rangle$ is equal to the order type of $\langle T, \leq_T \rangle$.

A typical application of this principle concerns the notations for ordinals below ε_0 and the set of binary trees. However, there are many more situations in which this hypothesis is true, as can be seen already by the results of Schmidt. Moreover, the principle even allows to form reasonable conjectures about newly formed partial orders. So, for example, the ordering studied in Weiermann (1993) emerged from this hypothesis.

In the term rewriting discipline, I was fascinated by Dieter Hofbauer's result on primitive recursive derivation lengths for the recursive path ordering and Cichon's grand idea to connect derivation lengths with the slow growing hierarchy along the order type of the underlying termination ordering. Although Cichon's exposition Cichon (1992) contained some technical errors, I still find it an interesting research problem to figure out precisely the side conditions that make his principle true.

For some time, I worked without success on the hypothesis that the derivation lengths resulting from termination proofs via a simplification ordering (these are orderings extending the homeomorphic embeddability relation and therefore are well founded orderings) were multiple recursive. To my surprise, this hypothesis was proved wrong by Hèléne Touzet in Touzet (2002). On the positive side, in Weiermann (1998), I could prove the hypothesis in case that the rewrite relation could be mapped into Buchholz's $<_k$ relation (Buchholz, 1994). Still today, I would find it very interesting to explore where the threshold for slow growing derivation length is situated between simplification orderings and the $<_k$-relation. Another route for approaching the problem was explored in Moser and Weiermann (2003).

A second problem which intrigued me in the nineties was determining the critical constant for Friedman's miniaturization of Kruskal's theorem. By applying compactness (in the form of König's lemma) to Kruskal's theorem one can obtain for any number-theoretic function f a true combinatorial $\forall\exists$ statement FFF_f, the Friedman finite form of Kruskal's theorem with respect to f. Friedman showed that for f the identity function FFF_f was unprovable in predicative analysis ATR_0. Matousek and Loebl could show in Loebl and Matoušek (1985) that for the function $f_r(i) := \lceil r \cdot \log_2(i) \rceil$, first order Peano arithmetic proves $\mathrm{FFF}_{f_{0.5}}$, but that FFF_{f_4} is true but independent from the Peano axioms.

A student of Dick De Jongh, namely Niklas Engsner, brought the problem around 1995 to my attention, and Niklas intended to work on narrowing the gap between 0.5 and 4 as part of his PhD project. His idea was to extract better bounds from the Rathjen Weiermann analysis of Kruskal's theorem. Unfortunately, he left academia at some moment.

In 2000 while visiting Tosiyasu Arai in Hiroshima, I reconsidered this problem. I could narrow the phase transition interval down to one single point consisting of a primitive recursive real number (which might be transcendental) (Weiermann, 2003). In the following years, I considered this phase transition phenomenon further and developed a general research program on phase transitions in logic and combinatorics. It turned out that similar phase transitions hold for independent statements based on well orders, Ramseyan statements and well partial orders (Weiermann, 2004, 2007, 2009).

Phase transitions for well partial orders are related to the analytic combinatorics of strings and trees on which Flajolet and Sedgewick have written a seminal and encyclopedic textbook (Flajolet & Sedgewick, 2009). The magical formula for tree counting is that $t_n \sim C \cdot \rho^n / n^{3/2}$ where C, ρ are constants and t_n is the number of trees having n nodes. This formula holds for various tree classes and even more refined results are available when needed.

Based on this background there emerged a general recipe on guessing phase transition thresholds related to well partial orders, and this recipe has yielded by now a plethora of results. The critical constant for unprovability turns out quite often to be $\frac{1}{\log_2(\rho)}$. The proof details vary from situation to situation and come with a varying degree of difficulty. Over the years, Lev Gordeev and I proved some of the strongest results in this direction, introducing scaling windows around the critical threshold number (as in Gordeev and Weiermann, 2012). Later in this article I will sketch some elements of the proof related to the phase transition for Kruskal's theorem.

The De Jongh Parikh formula for the maximal order type shows that this order type is equal to the strict supremum of the maximal order types of the left sets in question. Interestingly, this formula can often be translated into a convergence statement on a sequence of real numbers that converge from above against the critical constant for the phase transition under consideration. The sequence of real numbers controls the counting of the left sets in question, whereas the critical constant controls the counting of the full sets in question.

From 2007 onward, I worked with Michael Rathjen on a project funded by the Royal Society. Inspired by previous results of Friedman, we worked on well-ordering principles in reverse mathematics, concrete numerical lower bounds on provable Σ_1-sentences and generalizing Kruskal's theorem.

A major advance in this research was the formula

$$o(T(W(\cdot)) = \vartheta(o(W(\Omega))$$

which predicts the maximal order type of tree classes. These tree classes are defined inductively relative to a mapping W which maps (countable) well partial orders to *(countable) well partial orders. Additionally, this mapping is used to assign a finite combinatorial structure to the set of immediate subtrees of a tree.*

Here o refers to maximal order type, ϑ to the collapsing function from Rathjen and Weiermann (1993) and Ω to the first uncountable ordinal. In case of finite rooted structured trees, W would be the finite sequence operation $X \mapsto X^* \setminus \{\emptyset\}$ since a tree different from the root comes from connecting the root with a finite non-empty sequence of immediate subtrees. The formula predicted precisely several extensions of Kruskal's theorem.

We have explored these ideas further in several joint articles with Jeroen Van der Meeren and Michael Rathjen. For example, if $W(X) = M(X \times X)$ forming finite multisets of pairs of elements from X then $o(T(W(\cdot)) = \vartheta(o(\Omega^{\Omega}))$ (Van der Meeren, 2015) generalizing (Weiermann, 1993). Here $M(X)$ consists of the multisets over a finite set with respect to injective mappings. The maximal order type for $M(X)$ has been determined only recently in Weiermann (2009), Van der Meeren (2015). It turned out that this maximal order type has a much more complicated structure than the maximal order type of multisets over X with respect to the multiset ordering. Moreover, if $W(X) = (X \times X)^*$ forming finite sequences of pairs of elements from X then $o(T(W(\cdot)) = \vartheta(o(\Omega^{\Omega^{\Omega}})))$ (Van der Meeren, 2015). If $W(X)$ is the set of binary trees with leaf labels in X then $o(T(W(\cdot)) = \vartheta(\varepsilon_{\Omega+1})$ (Van der Meeren, 2017) strongly generalizing (Weiermann, 1993). These results give a nice graph-theoretic description of the set of ordinals below the Bachmann Howard ordinal.

In his seminal PhD thesis (Van der Meeren, 2015), Jeroen Van der Meeren brought the formula $o(T(W(\cdot)) = \vartheta(o(W(\Omega))$ into full blossom. Moreover, Van der Meeren even calculated the maximal order types related to Friedman's Kruskal theorem with gap condition for two labels. For the first time, it was possible to make non-trivial progress on analyzing the Friedman embeddability relation with gap condition. In a current research project, we try to obtain a result regarding the maximal order types for embeddability relations for any fixed finite label n, which comes down to analysing trees defined by an n-fold nesting of tree constructors. But the technical problems seem to be demanding.

Applications of well partial order theory to algebra have been investigated in Simpson (1988). In this paper, Simpson covers the underlying complexity of Dickson's Lemma and Higman's Lemma in great detail. Aschenbrenner and Pong (2004) recently made a fascinating contribution to partial order theory from the algebraic point of view. In joint work with Pelupessy, I could provide refinements on the computational complexity of lengths of effectively given chains of monomial ideals, and we could prove that even two variables already lead to Ackermannian complexity (Pelupessy & Weiermann, 2012). This result contrasts the situation with Hilbert's basis theorem, where n variables usually lead to the nth branch of the Ackermann function (Moreno Socias, 1992). Aschenbrenner and Pong posed the question on the maximal order types of downward closed sets, and jointly with Harry Altman, I recently obtained a satisfactory solution in Altman and Weiermann (2024).

One beautiful aspect of well partial orders is the technique used in the proofs. For example, a short and elegant proof of Kruskal's theorem can be given by using the minimal bad sequence argument of Nash Williams (See, for example, Simpson,

1985 or later in this text for such a proof). Quite recently, in joint work with Anton Freund and Michael Rathjen, I determined the reverse mathematics strength of the minimal bad sequence argument in terms of a uniform Kruskal theorem (Freund et al., 2022). I hope this category-theoretic framework will help to analyze the strength of Friedman's Kruskal theorem with gap conditions and its extensions.

There are numerous open problems left regarding well partial orders. Besides the ones mentioned above, my favourite ones include (among others):

1. Determine the proof-theoretic strength of Fraïsse's order type conjecture (Montalbán, 2017).
2. Determine the maximal order type of the embeddability relation in Kriz's theorem when the labels are taken from a fixed infinite ordinal α (Kriz, 1989).
3. Determine the maximal order types emerging from α-indexed sequences with finite range contained in a well quasi order (Nash-Williams, 1965).
4. Find a functorial version of the formula $o(T(W(\cdot)) = \vartheta(o(W(\Omega)))$.

10.3 Well Partial Orderings: Basic Material

As usual, small Greek letters range over ordinals and small Latin letters range over finite ordinals. ω stands for the first limit ordinal. We follow the standard von Neumann convention that $\alpha = \{\beta : \beta < \alpha\}$.

In this article, a *partial order* X is a set together with a binary relation $\leq$ which is reflexive, transitive and antisymmetric. A *quasi order* X is a set together with a binary relation $\leq$ which is reflexive and transitive. Obviously, a quasi order can be transformed into a partial order by factoring out via the equivalence relation $x \equiv y \iff x \leq y \wedge y \leq x$. In this article, if not mentioned otherwise, the notions partial order and quasi order can (by this identification) be interchanged freely without affecting the results. We write $x < y$ for $x \leq y$ and $\neg y \leq x$. If different partial orders are involved, we write $x \leq_X y$ or $x \leq y \mod X$ to indicate $x \leq y$ when $x, y \in X$. We also write $x \geq y$ for $y \leq x$.

Definition 10.1 A partial order X is a well partial order iff for every ω-sequence (x_i) of elements in X there exist $i, j < \omega$ such that $i < j$ and $x_i \leq x_j \mod X$.

Let us reformulate this definition. For $\alpha \leq \omega$ let us call a sequence $(x_i : i < \alpha)$ of elements in a partial order *bad* if there do not exist i, j with $i < j < \alpha$ and $x_i \leq x_j \mod X$. If for a given partial order X there are no bad ω-sequences of elements in X then X is a well partial order and vice versa.

From now on, we abbreviate well partial order by wpo as this is a standard abbreviation.

Clearly, a well order is a wpo and every wpo is well-founded. Moreover, every non empty subset of a wpo has at least one minimal element and at most finitely many minimal elements. Also, every finite set is a wpo with respect to any partial ordering on it. Nevertheless, it is not yet clear that wpo's are useful at all. In passing, let us note that it can be proved by first order compactness that wpo'ness like well foundedness is not a first order property.

10.3.1 Operations on Well Partial Orders

To get a better feeling for wpo's it is useful to have a basic stock of operations which transform wpo's into wpo's. Let us first gather some obvious equivalences.

Lemma 10.1 *Let X be a partial order. Then the following are equivalent*

1. *X is a wpo.*
2. *Every non empty subset of X has at least one and at most finitely many minimal elements (with respect to the given partial ordering on X).*
3. *Every relation $\leq'$ on X which extends $\leq$ to a linear ordering is a well ordering.*

More basic material on wpo's can be found in De Jongh and Parikh's classic (de Jongh, 1977) and Higman's (also seminal) article (Higman, 1952).

In a next step, let us consider forming *disjoint unions*. For two given sets X_0, X_1 let $X_0 \oplus X_1 = \{(x, 0) : x \in X_0\} \cup \{(y, 1) : y \in X_1\}$. If X_0 and X_1 are partial orders then $X_0 \oplus X_1$ is naturally equipped with the following partial order. $(x, i) \leq (y, j)$ mod $X_0 \oplus X_1 \iff i = j \wedge x \leq y \mod X_i$. By a routine verification, one can prove that $X_0 \oplus X_1$ is a partial order if X_0 and X_1 are partial orders. The following first closure property is also easy to verify.

Lemma 10.2 *$X_0 \oplus X_1$ is a wpo if X_0 and X_1 are wpos.*

Proof. By the infinite pigeonhole principle. □.

Another basic operation concerns forming *cartesian products*.

For two given sets X_0, X_1 let $X_0 \otimes X_1 = \{(x, y) : x \in X_0 \wedge y \in X_1\}$.

If X_0 and X_1 are partial orders then $X_0 \otimes X_1$ is naturally equipped with the following partial order.

$(x, y) \leq (x', y') \mod X_0 \otimes X_1 :\iff x \leq x' \mod X_0 \wedge y \leq y' \mod X_1$. By routine verification one can prove that $X_0 \otimes X_1$ is a partial order if X_0 and X_1 are partial orders.

Lemma 10.3 *$X_0 \otimes X_1$ is a wpo if X_0 and X_1 are wpos.*

Proof This is already somewhat tricky and requires a non trivial argument. A direct argument can be found, for example, in Baader and Nipkow (1998). We here recall a standard simple proof using Ramsey's theorem. Assume that we are given an ω-sequence $((x_i, y_i))$ of pairs (x_i, y_i) in $X_0 \otimes X_1$. Now let us zoom into the first coordinate and so we obtain an ω-sequence (x_i). Define a partition of pairs (two element sets) of integers as follows. $P(\{i, j\}_<) := 1$ if $x_i \leq x_j \mod X_0$ and $P(\{i, j\}_<) := 0$ otherwise. Then there exists an infinite monochromatic set I for P. If the colour of I is 0 then $(x_i : i \in I)$ forms an infinite bad sequence of elements in X_0 which is excluded since X_0 is a wpo. So the colour of I is 1 and $(x_i : i \in I)$ forms (with respect to the ordering on X_0 a weakly increasing infinite sequence. Let us now switch to the second coordinate. Since X_1 is a wpo we find for the sequence $(y_i : i \in I)$ indices $i, j \in I$ with $i < j$ and $y_i \leq y_j \mod X_1$. This means that $(x_i, y_i) \leq (x_j, y_j) \mod X_0 \otimes X_1$. □

It is this type of nice mathematics which makes the theory of wpo's so attractive.

A third basic operation for wpos concerns the *finite sequence operation.*

For a given set X let $X^* = \{(x_0, \ldots, x_{n-1}) : n < \omega \wedge (\forall i < n)[x_i \in X]\}$. If X is a partial order then X^* is naturally equipped with the following partial order. $(x_i : i < m) \leq (y_j : j < n) \mod X^* :\iff (\exists i_0, \ldots i_{m-1})[i_0 < \ldots < i_{m-1} < n \wedge (\forall l < m)[x_l \leq y_{i_l} \mod X]]$. By a routine verification one can prove that X^* is a partial order if X is a partial order.

Lemma 10.4 *X^* is a wpo if X is a wpo.*

Proof Already at this basic level a proof is rather difficult to find. The conceptually simplest proof (which we recall here) goes back to Nash-Williams (1963) and it goes nowadays under the minimal bad sequence argument or even the Nash Williams argument.

Assume that there exists an infinite bad sequence of finite sequences with elements in X. Among all these bad sequences there exists one sequence (s_i^0) such that the length of s_0^0 is minimal possible. Among all bad sequences there exists one sequence (s_i^1) such that $s_0^0 = s_0^1$ and such that the length of s_1^1 is minimal possible. Then iterate the construction for all $i < \omega$. Finally put $s_i := s_i^i$. This sequence (s_i), the so called *minimal bad sequence*, is bad by construction. Indeed, if $s_i \leq s_j$ for some i, j with $i < j$ then $s_i^{j+1} \leq s_j^{j+1}$ which is excluded. We see that no finite sequence s_i of elements in X has length zero since a sequence of length zero is embeddable into any finite sequence with respect to the ordering on X^*.

So we can write $s_i = \langle x_i \rangle \star t_i$ where $\star$ denotes concatenation and x_i is an element of X and t_i is an element of X^* which has smaller length than s_i. We claim that the set $S := \{t_i : i < \omega\}$ is well partially ordered. Otherwise we could pick an infinite bad sequence (t_{j_l}) of elements in this set. Here j_l refers to some numbering of the indices $j < \omega$ and this numbering is yet not required to be strictly monotone. But by passing to a subsequence we may well assume that j_l is strictly monotone. The reason is that when an index j_l is fixed by the pigeonhole principle only finitely many elements in the tail $(t_j)_{j>l}$ can have indices smaller than j_l and we can leave these out recursively by passing to a still bad subsequence. Now consider the sequence $s_0, \ldots, s_{j_0-1}, t_{j_0}, t_{j_1}, \ldots$. We claim that this is a bad sequence. Otherwise we find two indices p, q such that the former element with index p is $\leq \mod X^*$ the latter with index q. If $p < q < j_0$ then this contradicts the badness of (s_j). If $j_0 \leq p < q$ then this contradicts the badness of (t_{j_l}). But if $q < j_0 \leq p$ then we would obtain $(\mathrm{mod}\ X^*)\ s_p \leq t_q \leq \langle x_q \rangle \star t_q = s_q$ which contradicts the badness of (s_i). So S is a wpo.

Now consider the ω-sequence (t_j) of elements in S. Since S is a wpo we can assume by Ramsey's theorem that this sequence contains a weakly increasing (modulo X^*)

subsequence (t_{j_l}). Consider the sequence (x_{j_l}). Since X is a wpo we find l, l' such that $l < l'$ and $x_{j_l} \leq x_{j_{l'}} \mod X$. But then ($\mod X^*$) $s_{j_l} = \langle x_{j_l}\rangle \star t_{j_l} \leq \langle x_{j_{l'}}\rangle \star t_{j_{l'}} = s_{j_{l'}}$ so after all the sequence s_i was not bad. Contradiction. □

10.4 Finite Trees

A fourth operation for forming wpo's concerns finite trees (with or without labels). Here one has te be very specific about the nature of trees. We have to agree on whether there are roots and whether there is an ordering on the immediate subtrees of a given tree in our discussion. We assume in all cases that trees have a root.

Let us first consider the following inductive definition of the set of *structured trees* $\mathbb{T}^s(X)$ with labels in X.

If $n \geq 0$ and $t_0, \ldots, t_{n-1} \in \mathbb{T}^s(X)$ and $x \in X$ then $\bullet_x(t_0, \ldots, t_{n-1}) \in \mathbb{T}^s(X)$. We say that the number of nodes of $\bullet_x(t_0, \ldots, t_{n-1})$ is equal to 1 plus the sum over the number of nodes of all the t_q for $q < n$.

Note that $\bullet_x()$ is a tree with labels in X. This tree is considered as having just a root with label in X. In this way we identify trees with varyadic terms over a signature indexed by elements of X.

If X is a partial order then $\mathbb{T}^s(X)$ is naturally equipped with the following partial order $\trianglelefteq$. $s = \bullet_x(s_0, \ldots, s_{m-1}) \trianglelefteq t = \bullet_y(t_0, \ldots, t_{n-1})$ if there exists a $q < n$ such that $s \trianglelefteq t_q$, or $x \leq y \mod X$ and $(\exists i_0, \ldots i_{m-1})[i_0 < \ldots < i_{m-1} < n \wedge (\forall l < m)[s_l \trianglelefteq t_{i_l}]]$. By routine verification one can prove that $\langle \mathbb{T}^s(X), \trianglelefteq\rangle$ is a partial order if X is a partial order.

If X is a singleton we write $\mathbb{T}^s$ for $\mathbb{T}^s(X)$.

Lemma 10.5 *$\mathbb{T}^s(X)$ is a wpo if X is a wpo.*

Proof Having seen the Nash William's proof of Higman's Lemma before this is now relatively straight forward.

Assume that there exists an infinite bad sequence of finite trees with labels in X. Among all these there exist one sequence which we write as (s_{0i}) such that the number of nodes of s_{00} is minimal possible. Among all bad sequences there exist one sequence which we write as (s_{1i}) such that $s_{00} = s_{10}$ and such that the number of nodes of s_{11} is minimal possible. Then iterate the construction. Finally put $s_i := s_{ii}$. This (minimal bad) sequence is bad by construction.

Now we can write $s_i = \bullet_{x_i}(s_i^0, \ldots, s_i^{n_i-1})$ and x_i is an element of X for every $i \in \omega$ and the s_i^l are the immediate subtrees of s_i. We claim that the set $S := \{s_i^l : i < \omega \wedge l < n_i\}$ is well partially ordered. Otherwise we could pick an infinite bad sequence $s_{i_p}^{l_p}$ of elements in this set. Here i_p refers to some numbering of the indices $i < \omega$ and this numbering is yet not required to be strictly monotone. But by passing to a subsequence we may well assume that the sequence i_p is monotone. The reason is that when an index i_p is fixed by the pigeonhole principle only finitely many elements in the tail can have indices smaller than i_p and we can leave these out recursively by passing to a subsequence. Now consider the sequence $s_0, \ldots, s_{i_0-1}, s_{i_0}^{l_0}, s_{i_1}^{l_1}, \ldots$. We claim that this is a bad sequence. Otherwise we find two indices p, q that the the former element with index p is $\trianglelefteq$ the latter element with index q. If $p < q < j_0$ then this contradicts the badness of s_j. If $i_0 \leq p < q$ then this contradicts the badness of $s_{i_p}^{l_p}$. But if $p < i_0 \leq q = i_r$ then we would obtain $s_p \leq s_{i_r}^{l_r} \trianglelefteq s_{i_r} = s_q$ which contradicts the badness of (s_i). So S is a wpo and therefore S^* is a wpo. Now consider the sequence $(s_j^0, \ldots, s_j^{n_j-1})$ of elements in S^*. Since S^* is a wpo by Ramsey's theorem we can assume that this sequence contains a weakly increasing (with respect to $\trianglelefteq$) subsequence $(s_{i_p}^0, \ldots, s_{i_p}^{n_{i_p}-1})$. Consider the sequence (x_{i_p}). Since X is a wpo we find p, q such that $p < q$ and $x_{i_p} \leq x_{i_q} \mod X$. But then $s_{i_p} = x_{i_p}(s_{i_p}^0, \ldots, s_{i_p}^{n_{i_p}-1}) \trianglelefteq x_{i_q}(s_{i_q}^0, \ldots, s_{i_q}^{n_{i_q}-1}) = s_q$. So after all the minimal bad sequence (s_i) was not bad. Contradiction. □

Trees come also in the version of unstructured (or non planar) trees. These can be defined graph-theoretically or as equivalence classes of structured trees. Let us consider the graph-theoretic notion first.

An *unstructured tree* (tree for short) is a finite partial order $\langle T, \leq_T \rangle$ such that T has a unique minimal element with respect to $\leq_T$ (the root of T) and such that for every $s \in T$ the set $\{t \in T : t \leq_T s\}$ is linearly ordered by $\leq_T$. If T is a tree and $s, t \in T$ then there exists a maximal element in the set $\{r \in T : r \leq s\} \cap \{r \in T : r \leq t\}$ which is called the *meet* (or *infimum*) of s and t, written as $s \wedge_T t$. We say that a tree S *embeds* into a tree T if there exists a one to one mapping $h : S \to T$ such that for all $s, t \in S$ we have $h(s \wedge_S t) = h(s) \wedge_T h(t)$. We denote this embeddability (again) by $\trianglelefteq$. Let $\mathbb{T}$ be the set of unstructured trees. Then $\trianglelefteq$ is a partial order on $\mathbb{T}$.

If X is a partial order then a labeled tree with labels in X is a pair (T, l) consisting of a tree together with a labeling function $l : T \to X$. Let $\mathbb{T}(X)$ be the set of unstructured trees with labels in X. We say that a labeled trees (S, l) can be embedded into a labeled tree (T, l') if there exists a tree embedding $h : S \to T$ such that for all $s \in S$ $l(s) \leq l'(h(s)) \mod X$. We also write $(S, l) \trianglelefteq (T, l')$ in this case. Then this embeddability is a partial order on $\mathbb{T}(X)$.

The relation between $\mathbb{T}^s(X)$ and $\mathbb{T}(X)$ can be understood as follows.

If X is a partial order then $\mathbb{T}^s(X)$ is naturally equipped with the following quasi order (recall that this means a transitive and reflexive order). $s = \bullet_x(s_0, \ldots, s_{m-1}) \trianglelefteq^u t = \bullet_y(t_0, \ldots, t_{n-1})$ if there exists a $q < n$ such that $s \leq t_q$, or $x \leq y \mod X$ and there is a one to one function $h : m \to n$ such that $s_l \trianglelefteq^u t_{h(l)} \mod X$ for all $l < m$.

Then $\trianglelefteq^u$ is a quasi order on $\mathbb{T}^s(X)$. If we factorize $\mathbb{T}^s(X)$ through the equivalence relation $\equiv^u$ which is induced by $\trianglelefteq^u$ then the equivalence class of a structured labeled tree corresponds in a bijective fashion to an unstructured labeled tree.

Lemma 10.6 (Kruskal's Theorem) *$\langle \mathbb{T}(X), \trianglelefteq \rangle$ is a wpo if X is a wpo. Moreover $\mathbb{T}$ is a wpo.*

Proof This follows because $\trianglelefteq$ is a wpo on $\mathbb{T}^s(X)$, since $\trianglelefteq \subseteq \trianglelefteq^u$ and since $\langle \mathbb{T}^s(X), \trianglelefteq^u \rangle$ is isomorphic with $\langle \mathbb{T}(X), \trianglelefteq \rangle$. □

If T is a tree and $s \in T$ then $t \in T$ is called a direct successor of s if $s \leq_T t$ and $s \neq t$ and there is no $r \in T$ such that $s <_T r$ and $r <_T t$. Let $\mathbb{T}_n$ be the set of finite (unstructured) trees with labels in $n = \{0, \ldots, n-1\}$. Let us now consider the famous Friedman gap embeddability relation on $\mathbb{T}_n$. For $T_0, T_1 \in \mathbb{T}_n$ with labeling functions l_0, l_1 we define $(T_0, l_0) \trianglelefteq^w_{gap} (T_1, l_1)$ if there exists a one to one mapping $h : T_0 \to T_1$ such that for all $s, t \in T_0$ $h(s \wedge_{T_0} t) = h(s) \wedge_{T_1} h(t)$ and such that moreover for all $s \in T_0 : l_0(t) = l_1(h(t))$ and such for all $s \in T_0$ and all direct successors $s' \in T_0$ of s and all $t \in T_1$ with $h(s) < t < h(s')$ *mod* T_1 we have that $l_1(t) \geq l_1 h(s') = l_0(s') \mod T_1$.

This is the so called *weak gap embeddability relation.* If we demand in addition that for all $t < h(root(T_0)) \mod 4T_1$ we have $l_1(t) \geq l_0(root(T_0))$ then we say that T_0 is strongly gap embeddable into T_1 and we write $(T_0, l_0) \trianglelefteq^s_{gap} (T_1, l_1)$.

Lemma 10.7 (Friedman's Kruskal Theorem) *For all n* $\langle \mathbb{T}_n, \trianglelefteq^s_{gap} \rangle$ *is a wpo. In particular* $\langle \mathbb{T}_n, \trianglelefteq^w_{gap} \rangle$ *is a wpo.*

Proof This is very difficult to prove. See e.g. Simpson (1985). The proof is based on an ingenious nested minimal bad sequence argument.

10.5 Well Partial Orderings and Their Maximal Order Types

In this section we recall the basic theory of maximal order types from De Jongh and Parikh (1977). To some extent this theory has been redeveloped more recently in Blass and Gurevich (2008), Kriz and Thomas (1990).

If $\langle X, \leq \rangle$ is a wpo then

$$o(\langle X, \leq \rangle) := \sup\{otype(\langle X, \leq' \rangle) : \leq' \text{ is a well order on } X \text{ extending } \leq\}.$$

If $\leq$ is clear from the context we simply write $o(X)$ instead of $o(\langle X, \leq \rangle)$.

As it turns out this supremum is actually a maximum and this non trivial fact is a fundamental and far reaching result.

Let us fix a partial ordering $\leq$ on a set X. We define $L_X(x) := \{y \in X : \neg x \leq y\}$ which is the set of y which possibly are to the left of x in a linear extension of $\leq$. Moreover we put $U_X(x) := X \setminus L_X(x)$. $U_X(x)$ is the upper cone determined by x. Furthermore we put $l_X(x) := o(L_X(x))$ and $u_X(x) := o(U_X(x))$. If X is clear from the context we drop the subscript X.

Lemma 10.8 *If* $\langle Y, \leq \rangle$ *is a wpo and* $Z \subseteq Y$, *then* $\langle Z, \leq \rangle$ *is a wpo and* $o(\langle Z, \leq \rangle) \leq o(\langle Y, \leq \rangle)$. *Hence* $l_X(x) \leq o(X)$ *and* $u_X(x) \leq o(X)$.

Let us put this lemma into a more general context. Let X and Y be partial orders. We call a mapping $e : X \to Y$ a quasi-embedding if for all $x, x' \in X$: If $e(x) \leq e(x')$ mod Y then $x \leq x' \mod X$.

Lemma 10.9 *If X is a partial order and if Y is a wpo and if $e : X \to Y$ is a quasi-embedding then X is a wpo and $o(X) \leq o(Y)$.*

Lemma 10.10 *If $o(X)$ is a limit then $o(X) = \sup\{l(x) : x \in X\}$.*

In developing their theory De Jongh and Parikh introduced the following technical notions. Until Theorem 10.6 let us assume that $\langle X, \leq \rangle$ denotes a wpo.

An element $x \in X$ is called *superfluous* if $l(x) = o(X)$.

Lemma 10.11 *There exists a $Y \subseteq X$ such that $\langle Y, \leq \rangle$ does not contain superfluous elements and so that $o(X) = o(Y)$.*

Lemma 10.12 *If $o(X) = \omega^{\alpha_1} + \cdots + \omega^{\alpha_n}$ is written in Cantor normal form, $x \in X$ and $u(x) < \omega^{\alpha_n}$ then x is superfluous.*

Theorem 10.1 *There exists a well ordering $\leq'$ on X extending $\leq$ such that otype $(\leq') = o(X)$.*

Lemma 10.13 *For each $x \in X$, $l(x) + u(x) \leq o(X))$ and hence $l(x) < o(X)$ and therefore superfluous elements do not exist.*

Lemma 10.14 *If $o(X) = \omega^{\alpha_1} + \cdots + \omega^{\alpha_n}$ is written in Cantor normal form then there exists an $x \in X$ with $u(x) = \omega^{\alpha_n}$ and $\omega^{\alpha_1} + \cdots + \omega^{\alpha_{n-1}} \leq l(x)$.*

For calculating the maximal order types of cartesian products the following lemma is very useful.

Lemma 10.15 *If $o(X) = \omega^{\alpha_1} + \cdots + \omega^{\alpha_n}$ is written in Cantor normal form then for some m there exist $x_1, \ldots x_m \in X$ such that $o(U(x_1) \cup \ldots U(x_m)) = \omega^{\alpha_n}$ and $o(L(x_1) \cap \ldots \cap L(x_m)) = \omega^{\alpha_1} + \cdots + \omega^{\alpha_{n-1}}$.*

Lemma 10.16 *The ordinal $o(X)$ is a successor ordinal iff there exists an $x \in X$ such that $U_X(x) = \{x\}$ in which case $o(X) = l(x) + 1$.*

The following theorem is of key importance for calculating maximal order types. It follows from the previous lemmas.

Theorem 10.2 $o(X) = \sup\{(l(x) + 1 : x \in X\}$.

Let $\oplus$ and $\otimes$ denote the (commutative) Hessenberg sum and product of ordinals. The following Lemmas are a highlight of de Jongh (1977). Somewhat weaker results have been obtained before by Carruth (1942).

In the following two theorems let Y be a wpo.

Theorem 10.3 $o(X \oplus Y) = o(X) \oplus o(Y)$.

Theorem 10.4 $o(X \otimes Y) = o(X) \otimes o(Y)$.

Theorem 10.5 *Let $X := \{0, \ldots, n-1\}$ partially ordered by $=$. Then $o(X^*) = \omega^{\omega^{n-1}}$.*

Theorem 10.6 *If $o(X) = \varepsilon + n$ for some epsilon number ε then $o(X^*) = \omega^{\omega^{\varepsilon+n+1}}$. If $o(X) = \alpha$ is infinite and if α has not the form $\varepsilon + n$ for some epsilon number ε then $o(X^*) = \omega^{\omega^{\alpha}}$.*

Proof A proof for the upper bounds can be found in Schmidt's contribution to Schuster et al. (2020). The proof of $\geq$ is easy but maybe not so well known and so we give it here. Let $\underline{\omega}^\alpha := \omega^{\alpha+1}$ if $\alpha = \varepsilon + n$ for some epsilon number ε and let $\underline{\omega}^\alpha := \omega^\alpha$ otherwise. Our claim is that for infinite α we have $o(X^*) = \omega^{\underline{\omega}^\alpha}$. We show $o(X^*) \geq \omega^{\underline{\omega}^\alpha}$. The proof goes in two steps. First assume that α has not the form $\varepsilon + n$ with an epsilon number ε. Then $\underline{\omega}^\alpha = \omega^\alpha$. We construct a quasi-embedding $h_\alpha : \omega^{\omega^\alpha} \to \alpha^*$ by recursion on α and assume by recursion that $\beta < \alpha$ yields that h_β is the restriction of h_α to ω^{ω^β} for $\beta < \alpha$.

If $\alpha = \beta + 1$ and $\gamma < \omega^{\omega^\alpha}$ then $\gamma = (\omega^{\omega^\beta})^l \cdot \gamma_l + \cdots (\omega^{\omega^\beta})^0 \cdot \gamma_0$ with $\gamma_i < \omega^{\omega^\beta}$ and $\gamma_l > 0$. We can assume that there is a quasi-embedding h_β of $\omega^{\underline{\omega}^\beta}$ into β^*. We put $h_\alpha(\gamma) := h_\beta(\gamma_l) \star \langle\beta\rangle \star \cdots \star \langle\beta\rangle \star h_\beta(\gamma_0)$.

This will be a quasi-embedding and moreover h_β is the desired restriction of h_α for which the condition on restrictions is valid by the induction hypothesis. Assume $h_\alpha(\gamma) \leq h_\alpha(\gamma') \mod \alpha^*$. Then the number of occurrences of β in $h_\alpha(\gamma)$ is less than or equal to the number of occurrences of β in $h_\alpha(\gamma')$. If we have strict inequality then $\gamma < \gamma'$. Otherwise the $h_\beta(\gamma_i)$ map into the corresponding $h_\beta(\gamma_i')$ so that $\gamma_i \leq \gamma_i'$ (for all $i \leq l$) ensuring $\gamma \leq \gamma'$.

If α is a limit and $\gamma < \omega^{\omega^\alpha}$ then $\gamma < \omega^{\omega^\beta}$ for some $\beta < \alpha$ and we can assume that $h_\beta(\gamma)$ is defined and we may put $h_\alpha(\gamma) := h_\beta(\gamma)$. Then h_α is well defined by the condition on restrictions and h_α is a quasi-embedding.

Now consider the case that α has the form $\varepsilon + n$ with an epsilon number ε. If $n = 0$ then define a quasi embedding $h_\alpha : \omega^{\omega^{\varepsilon+1}} \to \varepsilon$ as follows. Let $\gamma = \varepsilon^l \alpha_l + \cdots + \varepsilon^0 \alpha_0$ where $\alpha_l \neq 0$ and $\alpha_i < \varepsilon$. Then put $h_\alpha(\gamma) := (\alpha_l, \alpha_l + \alpha_{l-1}, \ldots, \alpha_l + \alpha_{l-1} + \cdots + \alpha_0)$. This is a quasi-embedding. If $n > 0$ we can use the same argument as before in the case $\alpha = \beta + 1$. □

An interesting unpublished result by De Jongh states that set of structured binary trees has maximal order type ε_0. A published proof for the upper bound can be found in Schmidt's contribution in Schuster et al. (2020). The lower bound is easy to see since ε_0 is the closure of $\{0\}$ under $\alpha, \beta \mapsto \omega^\alpha + \beta$. Quite remarkably this lower bound even holds for unstructured binary trees. In Schmidt et al. (2020) it is shown that in general the lower bound for maximal order types for unstructured trees matches quite often the lower bound for maximal order types for structured trees.

Let us now consider the construction of $\mathbb{T}(W)$ of generalized trees for a wpo operator W and let us discuss its relation to the ϑ function. To explain the context let us recall briefly the definition of the collapsing function ϑ which can, for example, be used for a proof-theoretic analysis of ID_1, the first order theory of non iterated inductive definitions. Let Ω denote the first uncountable ordinal and $\varepsilon_{\Omega+1}$ the first epsilon number above Ω. First note that any non zero ordinal $\alpha < \varepsilon_{\Omega+1}$ can be described uniquely in terms of its Cantor normal form:

$$\alpha = \Omega^{\alpha_1}\beta + \cdots + \Omega^{\alpha_n} \cdot \beta_n$$

where $\alpha_1 > \ldots > \alpha_n$ and $0 < \beta_1, \ldots, \beta_n < \Omega$. In this situation we define the countable subterms $K\alpha$ of α recursively via

$$K\alpha := K\alpha_1 \cup \ldots \cup K\alpha_n \cup \{\beta_1, \ldots, \beta_n\}$$

where $K0 := \{0\}$. Let $AP = \{\omega^\delta : \delta \in ON\}$ be the class of additive principal numbers. We can then put by recursion on α

$$\vartheta\alpha := \min\{\beta \in AP : \beta \geq \max K\alpha \wedge \forall\gamma < \alpha(K\gamma < \beta \rightarrow \vartheta\gamma < \beta)\}.$$

One easily verifies $\vartheta\alpha < \Omega$ by induction on α using a cardinality argument. It is easy to verify that then $\varepsilon_0 = \vartheta\Omega$ and $\Gamma_0 = \vartheta\Omega^2$.

To explain the expected formula concerning wpo's let us consider a given explicit operator W which maps a (countable) wpo X to a (countable) wpo $W(X)$ so that the elements of $W(X)$ can be described as generalized terms in which the variables are replaced by constants for the elements of X. We assume that the ordering between elements of $W(X)$ is induced effectively by the ordering from X. (This resembles Feferman's notion of relative categoricity Feferman, 1968.) In concrete situations W may for example stand for an iterated application of basic constructions like disjoint union and cartesian product, the set of finite sequences construction, the multiset construction, or a tree constructor and the like. We assume that for W we have an explicit knowledge of $o(W(X))$ such that $o(W(X)) = o(W(o(X)))$ and such that this equality can be proved using an effective reification (An example for this technique is, for example, given in Rathjen and Weiermann, 1993).

Using W we then build the set of W-constructor trees $T(W(\cdot))$ as follows:

1. $\bullet \in T(W(\cdot))$.
2. If (s_i) is a sequence of elements in $T(W(\cdot))$ and $w((x_i))$ is a term from $W(X)$ then $\bullet(w((s_i))) \in T(W(\cdot))$.

The embeddability relation $\trianglelefteq$ on $T(W(\cdot))$ is defined recursively as follows:

1. $\bullet \trianglelefteq t$.
2. If $s \trianglelefteq t_i$ for some i then $s \trianglelefteq \bullet(w((t_i)))$
3. If $w((s_i)) \leq w'((t_j))$ mod W is induced recursively by $\trianglelefteq$ then $\bullet(w((s_i))) \trianglelefteq \bullet(w'((t_j)))$.

The general principle is now that

$$T(W(\cdot))) \text{ is a wpo}$$

and

$$o(\langle T(W(\cdot))), \trianglelefteq\rangle) \leq \vartheta o(W(\Omega)) \tag{10.1}$$

for $o(W(\Omega)) \in dom(\vartheta)$ with $o(W(\Omega)) \geq \Omega^3$. Moreover, the reverse inequality follows in many cases by direct inspection. (For $o(W(\Omega)) \leq \Omega^3$ the formula and the corresponding reverse inequality are still morally true but it might require some extra modifications regarding the values of ϑ.)

The formula (10.1) is even true for several natural examples which appear as suborderings of Friedman's Kruskal theorem with gap condition for n labels (Simpson, 1985) (provided that the domain of ϑ is suitably extended, but discussing such matters is beyond the scope of the present article). We believe that the formula will be the key property in finally analyzing Friedman's theorem on the embeddability with gap condition in full and at UGent we have already obtained promising approximations.

In general the formula (10.1) can be proved along the following general outline. (This outline applies to all cases which we considered so far.)

Proof ("Proof outline" for (10.1)) The inequality is proved by induction on $o(W(\Omega))$. Let $t = w((t_j)) \in T(W(\cdot))$. We claim $o(L_{T(W(\cdot)))}(t)) < \vartheta o(W(\Omega))$ and may assume by induction hypothesis that

$$o(L_{T(W(\cdot)))}(t_j)) < \vartheta o(W(\Omega)).$$

If now $s \in L_{T(W(\cdot)))}(t)$ then there will be a natural quasi-embedding putting s into a wpo $T(W'(\cdot, (t_i)))$ such that

$$o(W'(\Omega, (t_i))) < o(W(\Omega))$$

and such that

$$K(o(W'(\Omega, (t_i)))) \leq \max K(o(W(\Omega)) \cup o(L_{T(W(\cdot))}(t_i)).$$

This step uses the assumption that the maximal order type resulting from W can be computed by an effective reification a la (Rathjen & Weiermann, 1993) or (Simpson, 1988). Therefore the definition of ϑ can be used to show

$$\vartheta(o(W'(\Omega, (t_i))))) < \vartheta(o(W(\Omega))).$$

By induction hypothesis

$$o(L_{T(W(\cdot))}(t)) \leq o(T(W'(\Omega, (t_i)))) \leq \vartheta(o(W'(\Omega, (t_i)))) < \vartheta(o(W(\Omega)))$$

and we are done.

This proof outline can e.g. be used to prove (rigorously) the main results of the Habilitationsschrift of Diana Schmidt in a short and uniform way. Let us give a concrete example.

Theorem 10.7 *Let Y_i be countable wpos. Let $W(X) := Y_0 \times X^0 \oplus \cdots \oplus Y_n \times X^n$. Then $o(T(W(\cdot))) \leq \vartheta(o(W(\Omega)) = \vartheta(\Omega^n o(Y_n) + \cdots + \Omega^0 o(Y_0))$.*

Proof Let $\vartheta := \vartheta(\Omega^n o(Y_n) + \cdots + \Omega^0 o(Y_0))$. Let $s \in T(W(\cdot))$. We are looking for an wpo operator W' such that there exists a quasi-embedding $e : L(s) \to T(W'(\cdot))$ and such that $o(T(W'(\cdot))) < \vartheta$.

The construction of W' emerges from an analysis of $L(s)$ and is basically contained in Schmidt (1979). Let $s = \bullet(y_i, \langle s_0, \dots, s_{i-1}\rangle \in T(W(\cdot))$. We identy s with a finite tree where the root is labeled by y_i and where $\langle s_0, \dots, s_{i-1}\rangle$ is the sequence of direct subtrees of s. So we may also write $s = \bullet_{y_i}(s_0, \dots, s_{i-1})$. We want to prove $o(L(s)) < \vartheta$. This is done by induction on the number of nodes of s. Assume that $t \in L(s)$ and that $t = \bullet_{z_j}(t_0, \dots, t_{j-1})$. We have $\neg s \trianglelefteq t$.

Case 1. $j \neq i$.

Case 2. $j = i$ and $z_j = z_i \in L(y_i)$.

Case 3. $j = i$ and $y_i \leq z_i \mod Y_i$. Then there is a minimal $k < j$ such that $s_l \trianglelefteq t_l$ for all $l < k$ and $\neg s_k \trianglelefteq t_k$. We assume by recursion hypothesis that $e_{s_k} : L(s_k) \to T(W_{s_k}(\cdot))$ is a quasi-embedding with $o(T(W_{s_k}(\cdot))) < \vartheta$.

In this case we consider t as $\bullet_{(z_j, t_k)}(t_0, \dots, t_{k-1}, t_{k+1}, \dots, t_{j-1})$.

Let $W'(X) := \bigcup_{k<j} Y_0 X^0 \oplus \cdots \oplus (Y_{k-1} \oplus (Y_k \otimes T(W_{s_k}(\cdot))) \otimes X^{k-1}) \oplus (L(y_k) \otimes X^k) \oplus \cdots \oplus Y_n \otimes X^n)$.

We define the quasi embedding e as follows. We assume that $e(t_i)$ are defined and that $e(t_i) \in T(W'(\cdot))$.

In Case 1 and 2 we put $e(t) = \bullet_{z_j}(e(t_0), \dots, e(t_{j-1}))$. Then $e(t) \in T(W'(\cdot))$. In Case 3 we put $e(t) = \bullet_{(z_j, e_{s_k}(t_k))}(e(t_0), \dots, e(t_{k-1}), e(t_{k+1}), \dots, e(t_{j-1}))$.

Then $e(t) \trianglelefteq e(t') \mod T(W'(\cdot))$ yields $t \trianglelefteq t' \mod T(W(\cdot))$ for $t, t' \in L(s)$.

The induction hypothesis yields $o(T(W'(\cdot)))) \leq \vartheta(\oplus_{k \leq j-1}(\Omega o(Y_n) + \cdots + \Omega^{k-1}(o(Y_{k-1}) \oplus (o(Y_k) \otimes o(T(W_{s_k}(\cdot))) + \Omega^k o(L(y_k)) + \cdots + \Omega^0 o(Y_0)) =: \vartheta'$. Since $o(T(W_{s_k}(\cdot))) < \vartheta$ we have $K(\Omega^n o(Y_n) + \cdots + \Omega^{k-1}(o(Y_{k-1}) \oplus (o(Y_k) \otimes o(T(W_{s_k}(\cdot))) + \Omega^k o(L(y_k)) + \cdots + \Omega^0 o(Y_0)) < \vartheta$. Since $\Omega^n o(Y_n) + \cdots + \Omega^{k-1}(o(Y_{k-1}) \oplus (o(Y_k) \otimes o(T(W_{s_k}(\cdot))) + \Omega^k o(L(y_k)) + \cdots + \Omega^0 o(Y_0) < \Omega^n o(Y_n) + \cdots + \Omega^0 o(Y_0)$ we find $o(L(s)) \leq \vartheta' < \vartheta$. □

To make this proof even more constructive we observe that we can define $W_s := W'$ in a (primitive recursive) way so that $L(s)$ quasi-embeds into a recursively defined wpo $T(W_s(\cdot))$ with $o(T(W_s(\cdot))) < \vartheta$. If a finite bad sequence $s_0, \dots, s_m \in T(W(\cdot))$ is given, then $s_1, \dots, s_m \in L(s_0)$. Let us assume that we have available a quasi-embedding $e_{s_0} : L(s_0) \to T(W_{s_0}(\cdot))$. Then $e_{s_0}(s_1), \dots, e_{s_0}(s_m) \in T(W_{s_0}(\cdot))$ is a bad squence. Then $e_{s_0}(s_2), \dots, e_{s_0}(s_m) \in L(e_{s_0}(s_1))$ and we can fine a quasi embedding $L(e_{s_0}(s_1)) \to T(W_{e_{s_0}(s_1)}(\cdot))$ and we can iterate. We associate $o(T(W))$ to the empty sequence, $o(T(W_{s_0}(\cdot)))$ to the sequence $\langle s_0\rangle$ and $o(T(W_{e_{s_0}(s_1)}(\cdot)))$ to the sequence $\langle s_0, s_1\rangle$, etcetera, and we move the bad sequence primitive recursively into a strictly descending sequence of ordinals. This allows for constructive well partial orderedness proofs from appropriate well orderedness assumptions. Moreover we can read off uniform subrecursive bounds for the lengths of effectively given (finite) bad sequences.

Further examples for applications of the formula (10.1) arise e.g. from the *multiset construction.* Let $M'(X)$ be the set of finite multisets over X ordered by $m \ll m' \iff (\forall x \in m \setminus (m \cap m'))(\exists y \in m' \setminus (m \cap m'))[x < y]$. Moreover let

$B(X)$ be the set of binary stuctured trees with only the leaves labeled with elements from X ordered under homeomorphic embeddability.

Example 10.1 (Van der Meeren, 2015, 2017)

1. If $W(X) = X^*$ then $o(\langle T(W(\cdot)), \trianglelefteq \rangle) = \vartheta(\Omega^\omega)$ (since $\omega^{\omega^{\Omega+1}} = \Omega^\omega$).
2. If $W(X) = \bigotimes_{i<n} X$ then $o(\langle T(W(\cdot)), \trianglelefteq \rangle) = \vartheta(\Omega^n)$ (since $\bigotimes_{i<n} \Omega = \Omega^n$).
3. If $W(X) = M'(X \otimes X)$ then $o(\langle T(W(\cdot)), \trianglelefteq \rangle) = \vartheta(\Omega^\Omega)$ (since $\omega^{\Omega \otimes \Omega} = \Omega^\Omega$).
4. If $W(X) = (X \otimes X)^*$ then $o(\langle T(W(\cdot)), \trianglelefteq \rangle) = \vartheta(\Omega^{\Omega^\Omega})$ (since $o((\Omega \times \Omega)^*) = \omega^{\omega^{\Omega^2}} = \Omega^{\Omega^\Omega}$).
5. If $W(X) = (X^*)^*$ then $o(\langle T(W(\cdot)), \trianglelefteq \rangle) = \vartheta(\Omega^{\Omega^{\Omega^\omega}})$. (since $o(((\Omega)^*)^*) = \omega^{\omega^{\omega^{\omega^{\Omega+1}}}} = \Omega^{\Omega^{\Omega^\omega}}$).
6. If $W(X) = B(X)$ then $o(\langle T(W(\cdot)), \trianglelefteq \rangle) = \vartheta(\varepsilon_{\Omega+1})$ (since $o(B(\Omega)) = \varepsilon_{\Omega+1}$.

The most far reaching applications of (10.1) can be found in the thesis of Van der Meeren (2015).

10.6 Friedman Style Miniaturizations and Their Phase Transitions

A famous result of Harvey M. Friedman states that ATR_0 does not prove Kruskal's theorem (Simpson, 1985). Moreover Friedman showed that his finite form of Kruskals theorem (which is of complexity Π^0_2) is also true but unprovable in ATR_0 (Simpson, 1985). In the last section we deal with phase transitions related to well partial orders and we indicate how to find the critical constant and its scaling window for Friedman's finite form of Kruskal's theorem. We work with unstructured trees and denote the embeddability relation as before with $\trianglelefteq$.

Let $f : \mathbb{N} \to \mathbb{N}$ be a number-theoretic function. FFF_f, Friedman's finite form of Kruskal's theorem with respect to f, is the assertion: For every number K there exists a number M which is so big that for all sequences $T_0, \ldots, T_M$ of finite trees, if they satisfy the condition that for all $i \leq M$ the number of nodes in T_i is less than or equal to $K + f(i)$, then there exist indices $i, j < \omega$ such that $i < j \leq M$ and $T_i \trianglelefteq T_j$.

Theorem 10.8 *The assertion* FFF_f *is true for any choice of* f.

Proof This follows by a routine application of the compactness theorem. Let us still indicate the details.

Assume that FFF_f is false. Then there exists a K such that for all M there exist a finite bad sequence $T_0, \ldots, T_M$ such that the number of nodes in T_i is less than or equal to $K + f(i)$. Let $\mathcal{T} := \{(T_0, \ldots, T_M) : \forall i \leq M[|T_i| \leq K + f(i)] \wedge \neg\exists i, j[i < j \leq M \wedge T_i \trianglelefteq T_j]\}$. Then $\mathcal{T}$ is an infinite finitely branching tree and by compactness $\mathcal{T}$ will have an infinite path of the form $()$, (T_0), (T_0, T_1), (T_0, T_1, T_2), ... This yields an infinite bad sequence of finite trees in contradiction to Kruskal's theorem. □

It is quite remarkable that Friedman showed that FFF_f is unprovable in ATR_0 for $f(i) = i$.

A further refinement is provided by Friedman's theorem on internal tree embeddings where finite sequences of finite trees are replaced by truncations of just one big tree (Friedman, 1998). There also exists a similar finite form for Higman's Lemma, the Friedman theorem on long finite sequences (Friedman, 2001), but we do not discuss these results here.

In this section we treat one of the most advanced phase transitions which has been investigated so far, the phase transition for Friedman's finite form of Kruskal's theorem.

To obtain good phase transition results for FFF_f we need some sophisticated results from analytic combinatorics (Flajolet & Sedgewick, 2009). Let $B_{1,k}$ be the tree which consists precisely of one root and k direct successor nodes of the root. Let $B_{d+1,k}$ be the tree which consists of a root and k immediate subtrees of the form $B_{d,k}$. For a given tree B, let $L(B)$ be the set of trees in which B is not embeddable.

Let t_n be the number of trees which have n nodes. Let $t_{d,k,n}$ be the number of finite trees in $L(B_{d,k})$ which have n nodes. Note that $t_{1,k,n}$ is the number of finite trees which have n nodes and have at every node less than k direct successors.

We are interested in the asymptotics for the numbers t_n and $t_{d,k,n}$. As usual we write $a_n \sim b_n$ if $\lim_{n\to\infty} \frac{a_n}{b_n} = 1$.

Lemma 10.17 *There exists strictly positive real number constants β, ρ, $\beta_{d,k}$ and $\rho_{d,k}$ such that (for n going to infinity):*

1. $t_n \sim \beta \cdot \frac{\rho^n}{n^{\frac{3}{2}}}$.
2. $t_{d,k,n} \sim \beta_{d,k} \cdot \frac{(\rho_{d,k})^n}{n^{\frac{3}{2}}}$.

Moreover, there exist two uniform strictly positive real number constants C_1 and C_2 such that for all d and k we have that

1. $\rho - \rho_{1,k} \leq 2^{-C_1 \cdot k}$.
2. $\rho - \rho_{d,k} \geq 2^{2^{-C_2 \cdot d \cdot k}}$.

Proof A detailed proof would be beyond the scope of this exposition. We confine ourselves with describing the basic set up for the proof. Let $T(z) := \sum_{i=0}^{\infty} t_n z^n$ be the generating function for counting finite trees. Then by Polya theory the power series T fulfills the following functional equation. $T(z) = z \cdot \exp(\sum_{i=1}^{\infty} \frac{T(z^i)}{i})$. Let η be the radius of convergence of T. Then $\rho = \frac{1}{\eta}$. Using complex analysis Otter showed in Otter (1948) the asymptotic result for t_n which is mentioned in the first assertion.

For analyzing the number $t_{d,k,n}$, we define polynomials P_k as follows. Expand $\exp(\sum_{i=1}^{\infty} \frac{X^i}{i})$ into a formal power series, and let P_k be the polynomial consisting of the sum of the monomials $X_1^{a_1} \cdot \ldots \cdot X_m^{a_m}$ satisfying $\sum i \cdot a_i \leq k$. For the generating functions $T_{d,k}(z) := \sum_{i=0}^{\infty} t_{d,k,n} z^n$ one obtains the following functional equations

$$T_{1,k}(z) = z \cdot P_k(T_{1,k}(z), T_{1,k}(z^2), \ldots, T_{1,k}(z^k))$$

and recursively

$$T_{d+1,k}(z) = z \cdot \exp(\sum_{i=1}^{\infty} \frac{T_{d,k}(z^i)}{z^i}) \cdot P_k(T_{d+1,k}(z), T_{d+1,k}(z^2), \ldots, T_{d+1,k}(z^k)).$$

Let $\eta_{d,k}$ be the radius of convergence of $T_{d,k}$. Then $\rho_{d,k} = \frac{1}{\eta_{d,k}}$, and by Pólya theory one obtains the asymptotic estimate for $\rho_{d,k}$. By a highly non-trivial analysis of the power series involved, one can finally extract the information regarding the speed of the convergence of $\rho_{d,k}$ against ρ. The case $d = 1$ is already rather difficult. Full details are given in Goh (1994). The case $d > 1$ can also be treated by another rather sophisticated analysis and has been carried out by Gordeev and Weiermann around 2014 (unpublished). The bound on the convergence speed of $\rho_{d,k}$ to ρ in mentioned in Lemma 10.17 can actually be sharpened considerably.

10.6.1 Proof of the Upper Bound

We are now going to prove Friedman's miniaturization theorem (for PA) by reducing it to the unprovability of the termination of the hydra game. For this we assume that the reader is familiar with the standard assignments of the fundamental sequences for the ordinals below ε_0. Details can e.g. be found in Buchholz et al. (1994). What we basically need is that $\text{PA} \nvdash \forall K \exists i \omega_K[0][1] \ldots [i-1] = 0$, which is the unprovability of the termination of the hydra game in PA. Here $\omega_1 := \omega$ and $\omega_{K+1} = \omega^{\omega_K}$. Moreover $\alpha[x]$ denotes the x-th member of the fundamental sequence for α with the convention that $(\alpha + 1)[x] = \alpha$ and $0[x] := 0$. We also assume that the reader is aware of the fact that the length of a hydra game can be measured by the Hardy functions $(H_\alpha)_{\alpha \leq \varepsilon_0}$ (see, e.g. for more background about this hierarchy Buchholz et al., 1994).

The first approximation for the phase transition for FFF is as follows.

Lemma 10.18 *If* $f(i) = 2^{i^2}$ *then* $\text{PA} \nvdash \text{FFF}_f$

Proof Assume for a contradiction that the assertion is provable in PA, and let us argue further in PA. Fix $K > 0$.

Let $\alpha_i := \omega_K[0] \ldots [i-1]$. We show that we can find (provably in PA) an M such that $\alpha_M = 0$. In this situation we have a contradiction with the PA-unprovability of the hydra game. For this purpose define a mapping T from the ordinals below ε_0 to the set of finite trees as follows. $T(0)$ is the singleton tree consisting of a root. If $\beta = \omega^{\beta_1} + \cdots + \omega^{\beta_n}$ (written in Cantor normal form), then $T(\beta)$ is the tree with a new root and immediate subtrees formed by $T(\beta_1), \ldots, T(\beta_n)$. Let $T_i := T(\alpha_i)$. Then a simple calculation shows that the number of nodes in T_i is bounded by $K \cdot i! \leq K^2 + 2^{i^2}$ (where $i!$ refers to the factorial function). By assumption we can pick for K^2 an M so big that there exist $i < j$ such that $T_i \trianglelefteq T_j$. Then a moment's reflection shows $\alpha_i \leq \alpha_j$ and hence $\alpha_j = 0$ so that also $\alpha_M = 0$. As explained above this yields a contradiction.

Now we have to compress bounds of the form $K + 2^{i^2}$ to smaller bounds. This will be achieved by iterating a certain compression technique which is taken from Weiermann (2003). This technique has features which are similar to the renormalization method which is used to classify phase transitions in physics. The compression operator lowers the growth rate control function up to a certain point when a kind of fixed point is reached. In the sequel we denote by $|i|$ the binary length of i. (For our purposes we could switch between $\log_2$ and $|i|$ without affecting the provability and unprovability results for PA.)

Lemma 10.19 *1. Let* $f(i) := 2^i$. *Then* $\mathrm{PA} \nvdash \mathrm{FFF}_f$.
2. Let $f(i) := i^3$. *Then* $\mathrm{PA} \nvdash \mathrm{FFF}_f$.

Proof We prove only the second assertion (using the first assertion). The first assertion can be proved similarly (using Lemma 10.18) but its proof is easier since the bounding functions are more coarse.

We argue in PA. Assume that $\mathrm{PA} \vdash \mathrm{FFF}_f$. Let K be given. (We may assume w.l.o.g. that K is large enough to fulfil some basic asymptotic properties needed later.) Let M be maximal so that there exists a bad sequence of trees $(T_i)_{i \le M}$ such that $|T_i| \le K + 2^i$. Then by the first assertion M as a function of K has PA-unprovable growth and in fact the proof of the first assertion shows that M is at least as big as $H_{\varepsilon_0}(K)$. We are going to show within PA that M is a lower bound for the maximal number M' for which there exists a bad sequence of trees $(T_i')_{i \le M'}$ such that $|T_i'| \le (4K + 4) + i^3$. $\mathrm{PA} \vdash \mathrm{FFF}_f$ then would yield that M' as a function (of $4K + 4$ hence) of K is provably total in PA. This would lead to a contradiction with the first assertion.

For $i \le M$ let us construct a tree T_i' as follows. For i large enough (this can be assumed without much loss of generality so that the asymptotic bounds apply) let T_i' have two immediate subtrees U_i, V_i. The tree U_i consists of two immediate subtrees W_i, X_i. The tree X_i consists of a tree consisting of a root and $K + 2$ immediate singleton subtrees. The tree W_i is equal to $T_{|i|}$. The tree V_i is the $(2^{|i|} - i)$th element (in some prescribed linear ordering which extends the tree embeddability relation) of the set M_i where M_i is defined by

$$M_i := \{T \in L(B_{1,K+1}) : |T| \le 2^{|i| \cdot 2}\}.$$

(For i small we can let W_i be equal to T_0 and let V_i be a linear tree whose number of nodes decreases with i. We hereby assume that V_0 starts with a sufficiently big number of nodes.)

By construction the sequence T_i' is bad. This is mainly due to the facts that trees of the form X_i have to be mapped onto trees of the same form, that the original sequence of the T_i is bad, and that, for small ranges where $|i|$ is constant, the enumeration of M_i produces a bad sequence.

By the exponential growth of $t_{1,K+1,n}$ as a function of n we have $|M_i| \ge 2^{|i|}$ so that T_i' is well defined.

Moreover we obtain

$$\begin{aligned} |T_i'| &\leq K + 2^{|i|} + 2K + 3 + 1 + i^2 \\ &\leq K' + i^3 \end{aligned}$$

where $K' = 4K + 4$. Since K' depends of course provably recursively on K we have derived a contradiction with $\mathrm{PA} \vdash \mathrm{FFF}_f$.

The previous proof does not really exploit the power of the counting argument and we can easily sharpen it to obtain the following result.

Lemma 10.20 (Loebl and Matoušek, 1985)

1. *Let* $f(i) = |i|^4$. *Then* $\mathrm{PA} \nvdash \mathrm{FFF}_f$.
2. *Let* $f(i) = 4 \cdot |i|$. *Then* $\mathrm{PA} \nvdash \mathrm{FFF}_f$.

Proof The proof strategy will be the same as for the proof of Lemma 10.19. We prove only the second assertion (using the first assertion). The first assertion can be proved similarly but its proof is easier since the bounding functions are more coarse.

Let K be given. Let M be of order $H_{\varepsilon_0}(K)$ so that there exists a sequence of trees $(T_i)_{i \leq M}$ such that $|T_i| \leq K + |i|^4$ and such that there are no indices $i < j \leq M$ such that $T_i \trianglelefteq T_j$. For $i \leq M$ construct a tree T_i' as follows. For i large enough let T_i' have two immediate subtrees U_i, V_i. The tree U_i consists of two immediate subtrees W_i, X_i. The tree X_i consists of a tree consisting of a root and $K + 2$ immediate singleton subtrees. The tree W_i is equal to $T_{|i|}$. The tree V_i is the $(2^{|i|} - i)$th element (in some prescribed linear ordering which extends the embeddability relation) of the set M_i where M_i is defined by

$$M_i := \{T \in L(B_{1,K+1}) : |T| = |i|\}.$$

By construction the sequence T_i' is bad. This is mainly due to the facts that trees of the form X_i have to be mapped onto trees of the same form, that the original sequence of the T_i is bad, and that, for short intervals on which $|i|$ is a constant function of i, the enumeration of M_i produces a bad sequence.

We need to show that $|M_i| \geq 2^{|i|}$ so that T_i' is well defined. To this end we employ an asymptotic bound on the number of trees. Recall that

$$t_{K+1,n} = |\{T \in L(B_{1,K+1}) : |T| = n\}|$$

and that

$$t_{K+1,n} \sim \beta_{K+1} \cdot \frac{\rho_{K+1}^n}{n^{\frac{3}{2}}}.$$

Moreover ρ_K converges to $\rho = 2.95576...$ which is bigger than 2. So for K large enough (and for which a lower bound can be found effectively) the desired estimate $|M_i| \geq 2^{|i|}$ will hold.

Moreover we obtain

$$\begin{aligned}|T_i'| &\leq K + ||i||^4 + K + 3 + 1 + |i| \\ &\leq K' + 4 \cdot |i|,\end{aligned}$$

where $K' = 2K + 4$.

This yields a contradiction as in the proof of Theorem 10.19.

Theorem 10.9 *Let* $r > \frac{1}{\log_2(\rho)}$ *and* $f(i) = r \cdot |i|$. *Then* PA $\nvdash$ FFF_f.

Proof This follows by another compression step. We will soon give a proof of a stronger result, so we omit the proof of this result. A proof can also be found in Weiermann (2003).

Theorem 10.10 *Let* $r \leq \frac{1}{\log_2(\rho)}$ *and* $f(i) = r \cdot |i|$. *Then* PA $\vdash$ FFF_f.

Proof This follows by an easy counting argument using the asymptotic estimate for t_n. Given K, we can take for M roughly 2^{2^K}. Details can be found in Weiermann (2003). (A stronger result is proved in Theorem 10.11.)

The previous two theorems describe the static version of the phase transition for Kruskal's theorem. We are now going to describe the dynamic version by describing the phase transition window around the critical constant $\frac{1}{\log_2(\rho)}$.

For an unbounded non decreasing function $f : \mathbb{N} \to \mathbb{N}$ let $f^{-1}(m)$ be the least n such that $f(n) \geq m$.

Theorem 10.11 *Let* $f_{\varepsilon_0}(i) := (\frac{1}{\log_2(\rho)} + \frac{1}{H_{\varepsilon_0}^{-1}(i)}) \cdot |i|$. *Then* PA $\nvdash$ $\mathrm{FFF}_{f_{\varepsilon_0}}$.

Proof Let K be given. For M of order $H_{\varepsilon_0}(K')$ (where K' depends linearly on K modulo a small factor) there exists a bad sequence of finite trees $(T_i)_{i \leq M}$ such that $|T_i| \leq K' + 4 \cdot |i|$.

Construct a tree T_i' as follows. For i large enough let T_i' have two immediate subtrees U_i, V_i. The tree U_i consists of two immediate subtrees W_i, X_i. The tree X_i consists of a tree consisting of a root and $K' + 2$ immediate singleton subtrees. The tree W_i is equal to $T_{|i|}$. The tree V_i is the $2^{|i|} - i^{\text{th}}$ element (in some prescribed linear ordering which extends the embeddability relation) of the set M_i, where M_i is defined by

$$M_i := \{T \in L(B_{1,K'+1}) : |T| = \lceil (\frac{1}{\log_2(\rho)} + \frac{1}{K'+1}) \cdot |i| \rceil\}.$$

By construction the sequence T_i' is bad. This is mainly due to the facts that trees of the form X_i have to be mapped onto trees of the same form, that the original sequence of the T_i is bad, and that, on short intervals on which $|i|$ is constant, the enumeration of M_i produces a bad sequence.

Regarding the number of elements of M_i we make the following side calculation

$$\begin{aligned}|M_i| &\geq \rho_K'^{(\frac{1}{\log_2(\rho)}+\frac{1}{K'+2})\cdot|i|}\\ &\geq 2^{\log_2(\rho_K')\cdot(\frac{1}{\log_2(\rho)}+\frac{1}{K'+2})\cdot|i|)}\\ &\geq 2^{\log_2(\rho-2^{-C_1\cdot K'})\cdot(\frac{1}{\log_2(\rho)}+\frac{1}{K'+2})\cdot|i|)}\\ &\geq 2^{|i|},\end{aligned}$$

provided that $\log_2(\rho - 2^{-C_1\cdot K'})\cdot(\frac{1}{\log_2(\rho)}+\frac{1}{K'+2}) > 1$ for K' being large enough. But this is a simple exercise in calculus since, say by l'Hospital's rule, the fraction $\frac{\log_2(\rho-2^{-C_1\cdot K'})-\log_2(\rho)}{2^{-C_1\cdot K'}}$ converges to some constant and so $\log_2(\rho - 2^{-C_1\cdot K'})$ converges exponentially fast to $\log_2(\rho)$ from below whereas $\frac{1}{\log_2(\rho)}+\frac{1}{K'+2}$ converges linearly slow from above against $\frac{1}{\log_2(\rho)}$.

Moreover, since $i \leq M = H_{\varepsilon_0}(K')$ we obtain $H_{\varepsilon_0}^{-1}(i) \leq K'$ for $i \leq M$, and therefore

$$\begin{aligned}|T_i'| &\leq K' + 4\cdot||i|| + K' + 3 + 1 + \left(\frac{1}{\log_2(\rho)} + \frac{1}{K'+1}\cdot|i|\right)\\ &\leq K'' + \left(\frac{1}{\log_2(\rho)} + \frac{1}{K'}\right)\cdot|i|\\ &\leq K'' + \left(\frac{1}{\log_2(\rho)} + \frac{1}{H_{\varepsilon_0}^{-1}(i)}\right)\cdot|i|,\end{aligned}$$

where $K'' = 2K' + 4$. This shows the claim since K'' depends provably recursively on K' hence on K.

10.6.2 Proof of the Lower Bound

Theorem 10.12 *Let* $\alpha < \varepsilon_0$ *and let* $f_\alpha(i) := (\frac{1}{\log_2(\rho)} + \frac{1}{H_\alpha^{-1}(i)})\cdot|i|$. *Then* $\mathrm{PA} \vdash \mathrm{FFF}_{f_\alpha}$.

Proof We argue in PA and may use the totality of H_α. For a given large enough K, we may thus put $M := 2^{H_\alpha(2^{2^{2^K}})} - 1$. We show that this M is big enough to find the desired indices i and j. First note that $\frac{M}{2} > H_\alpha(2^{2^{2^K}})$. Assume for a contradiction that $T_0, \ldots, T_M$ is a given bad sequence of finite trees such that

$$|T_i| \leq K + \left(\frac{1}{\log_2(\rho)} + \frac{1}{H_\alpha^{-1}(i)}\right)\cdot|i|$$

for $0 \leq i \leq M$. For $H_\alpha(2^{2^{2^K}}) \leq i \leq M$ we find that

$$|T_i| \leq K + \left(\frac{1}{\log_2(\rho)} + \frac{1}{H_\alpha^{-1}(H_\alpha(2^{2^{2^K}}))}\right) \cdot |M|$$
$$= K + \left(\frac{1}{\log_2(\rho)} + \frac{1}{2^{2^{2^K}}}\right) \cdot H_\alpha(2^{2^{2^K}})$$

Let $N := K + \left(\frac{1}{\log_2(\rho)} + \frac{1}{2^{2^{2^K}}}\right) \cdot H_\alpha(2^{2^{2^K}})$. Recall that $B_{1,K}$ is the tree with one root and K immediate singleton subtrees and that $B_{d+1,K}$ is the tree with one root and K immediate identical subtrees of the form $B_{d,K}$. Since $|T_0| \leq K$, we see that $T_0 \trianglelefteq B_{K,K}$.

Therefore we see that $T_1, \ldots, T_M \in L(B_{K,K})$. By analytic combinatorics and Lemma 10.17 there is a real number constant C_1 such that

$$\sum_{i \leq n} t_{k,k,i} = |\{T \in L(B_{K,K}) : |T| \leq n\}| \leq C_1 \cdot (\rho_{K,K})^n$$

for all n, and there is a constant C_2 such that

$$\rho - \frac{1}{2^{2^{K^2 \cdot C_2}}} \geq \rho_{K,K}.$$

On the one hand we see that

$$M \leq |\{T \in L(B_{K,K}) : |T| \leq N\}| \leq C_1 \cdot (\rho_{K,K})^N$$

On the other hand, we now show that

$$C_1 \cdot (\rho_{K,K})^N < \frac{M}{2}$$

for K large enough, which yields a contradiction.

To see this, note that on the one hand $\frac{1}{\log_2(\rho)} + \frac{1}{2^{2^{2^K}}}$ converges from above triple exponentially fast to $\frac{1}{\log_2(\rho)}$. On the other hand, we have $\log_2(\rho_{K,K}) \leq \log_2(\rho - \frac{C_2}{2^{2^{K^2}}})$, which converges not faster than poly-double exponentially fast to $\log_2(\rho)$. This means that the product $\log_2(\rho_{K,K}) \cdot (\frac{1}{\log_2(\rho)} + \frac{1}{2^{2^{2^K}}})$ converges not faster than poly-double exponentially from below to 1 and so is eventually strictly smaller than 1. This yields after some elementary calculations that for K large enough we find

$$
\begin{aligned}
M \le C_1 \cdot {\rho_{K,K}}^N &\le C_1 \cdot (\rho_{K,K})^{K+(\frac{1}{\log_2(\rho)}+\frac{1}{2^{2^K}})\cdot H_\alpha(2^{2^{2^K}})} \\
&= C_1 \cdot 2^{\log_2(\rho_{K,K})\cdot(K+(\frac{1}{\log_2(\rho)}+\frac{1}{2^{2^K}})\cdot H_\alpha(2^{2^{2^K}}))} \\
&\le C_1 \cdot 2^{\log_2(\rho-\frac{C_2}{2^{2^{K^2}}})\cdot(K+(\frac{1}{\log_2(\rho)}+\frac{1}{2^{2^K}})\cdot H_\alpha(2^{2^{2^K}}))} \\
&< \frac{M}{2}.
\end{aligned}
$$

Note that the threshold functions basically appear as functional inverses of the tree counting function, so the functional inverse of the function covering the required finite combinatorics of trees. The examples we have at hand show that this is a quite general phenomenon in analyzing phase transitions for Gödel's incompleteness theorem (Weiermann, 2009).

Let us finally remark that in Theorems 10.11 and 10.12 the theory PA can be replaced by the theory $ACA_0 + (\Pi^1_2 - BI)_0$ when ε_0 is replaced by the small Veblen ordinal.

Acknowledgements I would like to thank Oriola Gjetaj for her careful proof reading which helped in improving this article.

References

Abdula, P. A. (2010). Well (and better) quasi-ordered transition systems. *The Bulletin of Symbolic Logic, 16*(4), 457–515.

Altman, H., & Weiermann, A. (2024). Maximum linearizations of lower sets in $\mathbb{N}^m$ with application to monomial ideals. https://arxiv.org/abs/1909.06719

Aschenbrenner, M., & Pong, W. Y. (2004). Orderings of monomial ideals. *Fundamenta Mathematicae, 181*(1), 27–74.

Baader, F., Nipkow, T. (1998). *Term rewriting and all that* (pp. xii+301) Cambridge University Press.

Blass, A., Gurevich, Y. (2008). Program termination and well partial orderings. *ACM Transactions on Computational Logic, 9*(3)Art. 18, 1–26.

Buchholz, W., Cichon, A., & Weiermann, A. (1994). A uniform approach to fundamental sequences and hierarchies. *Mathematical Logic Quarterly, 40*(2), 273–286.

Buchholz, W. (1995). Proof-theoretic analysis of termination proofs. Proof theory, provability logic, and computation (Berne, 1994). *Annals of Pure and Applied Logic, 75*(1–2), 57–65.

Carlson, T. (2017). Analysis of a double Kruskal theorem. *Transactions of the American Mathematical Society, 369*(4), 2897–2916.

Carruth, P. W. (1942). Arithmetic of ordinals with application to the theory of ordered abelian groups. *Bulletin of the American Mathematical Society, 48*, 262–271.

Cichon, E. A. (1992). *Termination orderings and complexity characterisations. Proof theory* (pp. 171–193). Cambridge University Press.

Crossley, J., Kister, J. B. (1986/1987). Natural well-orderings. *Archive for Mathematical Logik Grundlag, 26*(1–2), 57–76.

de Jongh, D. H. J., & Parikh, R. (1977). Well-partial orderings and hierarchies. In *Proceedings of the Koninklijke Nederlandse Akademie van Wetenschappen, Series A* (Vol. 80 (Indigationes Math. 39), pp. 195–207).

Dershowitz, N. (1982). Orderings for term-rewriting systems. *Theoretical Computer Science, 17*(3), 279–301.

Ehrenfeucht, A. (1973). Polynomial functions with exponentiation are well ordered. *Algebra Universalis, 3*, 261–262.

Feferman, S. (1968). Systems of predicative analysis, II: Representations of ordinals. *The Journal of Symbolic Logic, 33*(2), 193–220.

Feferman, S. (1987). Proof theory: a personal report. In G. Takeuti (Ed.), *Proof theory* (pp. 447–481). North-Holland.

Flajolet, P., Sedgewick, R. (2009). *Analytic combinatorics* (pp. xiv+810). Cambridge University Press. ISBN: 978-0-521-89806-5

Freund, A., Rathjen, M., & Weiermann, A. (2022). Minimal bad sequences are necessary for a uniform Kruskal theorem. https://arxiv.org/abs/2001.06380. The online version appeared in Advances in Mathematics 400 (2022) 108265, 1–44.

Friedman, H. M., Robertson, N., & Seymour, P. (1987). The metamathematics of the graph minor theorem. In *Logic and combinatorics* (Arcata, Calif., 1985) (Vol. 65, pp. 229–261). Contemporary Mathematics, American Mathematical Society Providence.

Friedman, H. (2001). Long finite sequences. *Journal of Combinatorial Theory, Series A, 95*(1), 102–144.

Friedman, H. (2002). Internal finite tree embeddings. In *Reflections on the foundations of mathematics (Stanford, CA, 1998), Lecture Notes login* (Vol. 15, pp. 60–91). The Association for Symbolic Logic.

Goh, W., & Schmutz, E. (1994). Unlabeled trees: distribution of the maximum degree. *Random Structures Algorithms, 5*(3), 411–440.

Gordeev, L., & Weiermann, A. (2012). Phase transitions of iterated Higman-style well-partial-orderings. *Archive for Mathematical Logic, 51*(1–2), 127–161.

Hasegawa, R. (1994). Well-ordering of algebras and Kruskal's theorem. Lecture Notes in Computer Science *Logic, language and computation* (Vol. 792, pp. 133–172). Springer.

Higman, G. (1952). Ordering by divisibility in abstract algebras. *Proceedings of the London Mathematical Society, 3*(2), 326–336.

Kripke, S. (1959). The problem of entailment (Abstract). *The Journal of Symbolic Logic, 24*, 324.

Kriz, I. (1989). Well-quasiordering finite trees with gap-condition. Proof of Harvey Friedman's conjecture. *Annals of Mathematics, 2*(130), 215–226.

Kriz, I., & Thomas, R. (1990). Ordinal types in Ramsey theory and well-partial-ordering theory. In *Mathematics of Ramsey theory, algorithms combinations* (Vol. 5, pp. 57–95). Springer.

Loebl, M., Matoušek, J. (1987). On undecidability of the weakened Kruskal theorem. In *Logic and combinatorics (Arcata, Calif., 1985), Contemporary mathematics* (Vol. 65, pp. 275–280) American Mathematical Society Providence.

Montalbán, A. (2017). Fraïssé's conjecture in Π^1_1-comprehension. *Journal of Mathematical Logic, 17*(2), 1750006.

Moreno Socias, G. (1992). Length of polynomial ascending chains and primitive recursiveness. *Mathematica Scandinavica, 71*(2), 181–205.

Moser, G., & Weiermann, A. (2003). Relating derivation lengths with the slow-growing hierarchy directly. Lecture Notes in Computer Science *Rewriting techniques and applications* (Vol. 2706, pp. 296–310). Springer.

Murthy, C. R., Russell, J. R. (1990). A constructive proof of Higman's lemma. In Fifth annual IEEE symposium on logic in computer science (pp. 257–267). IEEE Comput. Soc. Press.

Nash-Williams, C. S. T. J. A. (1963). On well-quasi-ordering finite trees. *Mathematical Proceedings of the Cambridge Philosophical Society, 59*, 833–835.

Nash-Williams, C. S. T. J. A. (1965). On well-quasi-ordering transfinite sequences. *Mathematical Proceedings of the Cambridge Philosophical Society, 61*, 33–39.

Otter, R. (1948). The number of trees. *Annals of Mathematics, 2*(49), 583–599.

Pelupessy, F., & Weiermann, A. (2012). On the lengths of bad sequences of monomial ideals over polynomial rings. *Fundamenta Mathematicae, 216*(2), 101–108.

Rathjen, M., & Weiermann, A. (1993). Proof-theoretic investigations on Kruskal's theorem. *Annals of Pure and Applied Logic, 60*, 49–88.

Robertson, N., & Seymour, P. (2004). Graph Minors. XX. Wagner's conjecture. *Journal of Combinatorial Theory, Series B, 92*(2), 325–357.

Schmidt, D. (1979). *Well-partial orderings and their maximal order types*. Habilitationsschrift. This thesis appeared finally in [46].

Schmidt, D. (1975). Bounds for the closure ordinals of replete monotonic increasing functions. *Journal of Symbolic Logic, 40*(3), 305–316.

Schmidt, D., Van der Meerenen, J., & Schmidt, D. (2020). Calculating maximal order types for finite rooted unstructured labeled trees. In *The legacy of Kurt Schütte* (pp. 253–264). Springer.

Schütte, K. (1954). Kennzeichnung von Ordnungszahlen durch rekursiv erklärte Funktionen. *Mathematische Annalen, 127*, 15–32.

Schuster, P., Seisenberger, M., & Weiermann, A. (Eds.). (2020). Well-quasi orders in computation, logic, language and reasoning. a unifying concept of proof theory, automata theory, formal languages and descriptive set theory

Simpson, S. G. (1985). Nonprovability of certain combinatorial properties of finite trees. In *Friedman's research on the foundations of mathematics* (pp. 87–117). North-Holland.

Simpson, S. G. (1988). Ordinal numbers and the Hilbert basis theorem. *Journal of Symbolic Logic, 53*(3), 961–974.

Touzet, H. (2002). A characterisation of multiply recursive functions with Higman's lemma. RTA '99 (Trento). *Information and Computer Security, 178*(2), 534–544.

Van der Meeren, J. (2015). Connecting the Two Worlds: Well-partial orders and Ordinal Notation Systems. Ph.D. thesis, Ghent.

Van der Meeren, J., Rathjen, M., & Weiermann, A. (2015). Well-partial-orderings and the big Veblen number. *Archive for Mathematical Logic, 54*(1–2), 193–230.

Van der Meeren, J., Rathjen, M., & Weiermann, A. (2017). An order-theoretic characterization of the Howard-Bachmann-hierarchy. *Archive for Mathematical Logic, 56*(1–2), 79–118.

Veblen, O. (1908). Continuous increasing functions of finite and transfinite ordinals. *Transactions of the American Mathematical Society, 9*(3), 280–292.

Weiermann, A. (1992). Proving termination for term rewriting systems. Lecture notes in computer science *Computer science logic (Berne, 1991)* (Vol. 626, pp. 419–428). Springer.

Weiermann, A. (1993). An order-theoretic characterization of the Schütte-Veblen-hierarchy. *Mathematical Logic Quarterly, 39*(3), 367–383.

Weiermann, A. (1993). Bounds for the closure ordinals of essentially monotonic increasing functions. *Journal of Symbolic Logic, 58*(2), 664–671.

Weiermann, A. (1994). Complexity bounds for some finite forms of Kruskal's theorem. *Journal of Symbolic Logic, 18*(5), 463–488.

Weiermann, A. (2003). An application of graphical enumeration to PA. *Journal of Symbolic Logic, 68*(1), 5–16.

Weiermann, A. (1998). Bounding derivation lengths with functions from the slow growing hierarchy. Logic Colloquium '95 (Haifa). *Archive for Mathematical Logic, 37*(5–6), 427–441.

Weiermann, A. (2004). A classification of rapidly growing Ramsey functions. *Proceedings of the American Mathematical Society, 132*(2), 553–561.

Weiermann, A. (2007). Phase transition thresholds for some Friedman-style independence results. *MLQ Mathematical Logic Quarterly, 53*(1), 4–18.

Weiermann, A. (2009). Phase transitions for Gödel incompleteness. *Annals of Pure and Applied Logic, 157*(2–3), 281–296.

Weiermann, A. (2009). A computation of the maximal order type of the term ordering on finite multisets. Lecture notes in computer science *Mathematical theory and computational practice* (Vol. 5635, pp. 488–498). Springer.

Wilken, G. (2018). Tracking chains revisited. Lecture notes series, institute for mathematical sciences, national university of Singapore *Sets and computations* (Vol. 33, pp. 183–220). World Sci. Publ.

Chapter 11
Learning to Act and Observe in Partially Observable Domains

Thomas Bolander, Nina Gierasimczuk, and Andrés Occhipinti Liberman

Abstract We consider a learning agent in a partially observable environment, with which the agent has never interacted before, and about which it learns both what it can observe and how its actions affect the environment. The agent can learn about this domain from experience gathered by taking actions in the domain and observing their results. We present learning algorithms capable of learning as much as possible (in a well-defined sense) both about what is directly observable and about what actions do in the domain, given the learner's observational constraints. We differentiate the level of domain knowledge attained by each algorithm, and characterize the type of observations required to reach it. The algorithms use dynamic epistemic logic (DEL) to represent the learned domain information symbolically. Our work continues that of Bolander and Gierasimczuk (2015), which developed DEL-based learning algorithms based to learn domain information in fully observable domains.

Keywords Artificial intelligence · Dynamic epistemic logic · Partial observability · Action model learning · Behavioural correctness · Learnability

11.1 Introduction

This paper explores a learning task which we refer to as *domain learning*. Domain learning involves learning a representation of the dynamics of a domain (sometimes called an environment), from the experience gathered by performing actions in this domain and observing their results. We are concerned with domain learning in what we call unknown, partially observable domains. We introduce the learning task and the main results informally in this section, leaving formal details for later.

Consider an agent inhabiting an unknown, partially observable domain. By an *unknown* domain, we mean one that the agent hasn't interacted with before, so it

---

T. Bolander · N. Gierasimczuk (✉)
Danmarks Tekniske Universitet, Kongens Lyngby, Denmark
e-mail: nigi@dtu.dk

A. Occhipinti Liberman
DTIC, Universitat Pompeu Fabra, Barcelona, Spain

N. Bezhanishvili et al. (eds.), *Dick de Jongh on Intuitionistic and Provability Logics*, Outstanding Contributions to Logic 28, https://doi.org/10.1007/978-3-031-47921-2_11

doesn't know how its actions may affect the domain. For example, a river would be an unknown domain to an agent that has never seen or been to a swimming pool, the sea, a river, or any other body of water. The agent would indeed not know how its actions (moving the limbs, breathing, etc.) would interact with this domain. By a *partially observable* domain, we mean one in which the agent may get to see only a part of the domain at any point in time. For example, the agent in the river may be able to see underwater only within two meters of distance, because the water is murky. Although the agent starts without any knowledge about how it can act in this domain, and what it can or cannot observe about it, we assume that it has access to experiences of interaction with the domain, gathered by trying out actions and observing what happens as a result. The goal of the agent is to *learn a correct representation of what actions do, and what it can and cannot observe, from such experience.* We present learning algorithms that enable the agent to learn "as much as possible" about the domain, given the observational limits, provided that the learner is given a sufficient number of interactions to learn from. Of course, the exact meanings of "learning as much as possible", "sufficient number of interactions", etc., are important here, and will be made precise in the coming sections.

We are interested in learning representations of domains that rely on *dynamic epistemic logic* (DEL). We therefore introduce DEL in Sect. 11.2, where we also motivate our choice of this representation. Section 11.3 presents domains and their properties formally. Sections 11.4–11.6 contain the main results of the paper. In Sect. 11.4, we first introduce a distinction between what we call learning *explicit* domain knowledge and learning *implicit* domain knowledge. We define learning explicit domain knowledge as learning to know what will be directly observed when actions are executed. We contrast this with implicit knowledge, which includes what might additionally be inferred from the history of earlier actions and from the general experience with the domain. Section 11.4 focuses on learning explicit domain knowledge. We motivate this learning goal, formalise it, present a learner that achieves explicit domain knowledge, and we characterise the observations required to do so. Section 11.5 extends learning beyond explicit knowledge, illustrating situations in which an agent can learn more about the domain than what can be directly observed in each state. Section 11.6 is focused on learning implicit domain knowledge. We formalise the goal of learning implicit knowledge, present a learner that attains this goal, and characterise the type of observations that are sufficient for doing so. Section 11.7 presents related work and Sect. 11.8 concludes with final remarks and possible avenues for future research.

11.2 Dynamic Epistemic Logic (DEL)

Given a finite set P of propositional symbols (atomic propositions), we define the (single-agent) *dynamic epistemic language* $\mathcal{L}_{\mathrm{DEL}}(P)$ over P by the following BNF:

$$\phi ::= p \mid \neg\phi \mid \phi \wedge \phi \mid K\phi \mid [\mathcal{E}]\phi,$$

where $p \in P$ and $\mathcal{E}$ denotes an event model as defined below. We read $K\phi$ as "it is known that ϕ" and $[\mathcal{E}]\phi$ as "executing the action $\mathcal{E}$ necessarily leads to an epistemic model where ϕ holds". By means of the standard abbreviations we introduce the additional symbols $\rightarrow$, $\vee$, $\leftrightarrow$, $\bot$, and $\top$. The modalities $[\mathcal{E}]$ are called *dynamic modalities*. Formulas without dynamic modalities are called *static formulas*.

Definition 11.1 An *epistemic model* over a set of propositional symbols P is $\mathcal{M} = (W, R, V)$, where

- W is a finite set of *worlds*
- $R \subseteq W \times W$ is an equivalence relation called the *indistinguishability relation*
- $V : W \rightarrow 2^P$ is a *valuation function* (or *labelling function*, as it maps worlds to valuations)

Dynamic epistemic logic (DEL) introduces the concept of an event model for modelling the changes to states brought about by executions of actions (Baltag & Moss, 2004). We here use a variant that includes (boolean) postconditions (Ditmarsch & Kooi, 2008), which means that actions can have both ontic effects (changing the factual states of affairs) and epistemic effects (changing the beliefs of agents).

Definition 11.2 An *event model* over P is $\mathcal{E} = (E, Q, pre, post)$, where

- E is a finite set of *events*
- $Q \subseteq E \times E$ is an equivalence relation called the *indistinguishability relation*
- $pre : E \rightarrow \mathcal{L}_{\text{DEL}}(P)$ assigns to each event a *precondition*
- $post : E \rightarrow (P \rightarrow \mathcal{L}_{\text{DEL}}(P))$ assigns to each event a *postcondition mapping* mapping each event e into a *postcondition* $post(e)$. In this paper, postconditions are boolean, meaning that for each event e and proposition p, $post(e)(p) \in \{\top, \bot, p\}$ (p is set true, false or unchanged).

Intuitively, events correspond to the ways in which an action changes the epistemic model, and the indistinguishability relation codes (an agent's) inability to recognize the difference between those different ways. In an event e, $pre(e)$ specifies what conditions have to be satisfied for it to take effect, and $post(e)$ specifies its outcome.

Example 11.1 Consider the action of tossing a coin. It can be represented by the following event model over $P = \{h\}$, where h means that the coin is facing heads up:

$$\mathcal{E} = \quad \overset{\bullet}{e_1 : \langle\top, h\rangle} \qquad\qquad \overset{\bullet}{e_2 : \langle\top, \neg h\rangle}$$

We label each event e by a pair whose first argument is the event's precondition while the second is its postcondition represented compactly as a sequence $l_1 \ldots l_n$ of propositional literals given by: if $post(e)(p) = \top$ then p is one of the l_i; if $post(e)(p) = \bot$, then $\neg p$ is one of the l_i; if $post(e)(p) = p$ then p doesn't occur in $l_1 \ldots l_n$. Hence, formally we have $\mathcal{E} = (E, Q, pre, post)$ with $E = \{e_1, e_2\}$, Q is the identity on E, $pre(e_1) = pre(e_2) = \top$, $post(e_1)(h) = \top$ and $post(e_2)(h) = \bot$. The event model encodes that tossing the coin will either make h true (e_1) or h false (e_2).

Given an epistemic model $\mathcal{M} = (W, R, V)$ and a world $w \in W$, truth of an epistemic formula ϕ in w of $\mathcal{M}$ is defined as follows:

$$
\begin{array}{lll}
\mathcal{M}, w \models p & \text{iff} & p \in V(w) \\
\mathcal{M}, w \models \neg\phi & \text{iff} & \mathcal{M}, w \not\models \phi \\
\mathcal{M}, w \models \phi \wedge \psi & \text{iff} & \mathcal{M}, w \models \phi \text{ and } \mathcal{M}, w \models \psi \\
\mathcal{M}, w \models K\phi & \text{iff} & \text{for all } v \in W, \text{if } wRv \text{ then } \mathcal{M}, v \models \phi \\
\mathcal{M}, w \models [\mathcal{E}]\phi & \text{iff} & \text{for all events } e \text{ in } \mathcal{E}, \\
 & & \text{if } \mathcal{M}, w \models pre(e) \text{ then } \mathcal{M} \otimes \mathcal{E}, (w, e) \models \phi
\end{array}
$$

When $\mathcal{M}, w \models \phi$ for all $w \in W$, we write $\mathcal{M} \models \phi$. Our syntax and semantics of the dynamic modality, $[\mathcal{E}]$, is a bit non-standard. Normally, one considers *pointed* event models $(\mathcal{E}, e)$, where e is an event of $\mathcal{E}$ called the *actual event*, and then provides a semantics for a dynamic modality of the form $[\mathcal{E}, e]$. In our setting, event models are going to be used by agents to represent their uncertainty about the dynamics of an action they are trying to learn. For instance, an agent could use an event model with two events e_1 and e_2 to represent that it doesn't know whether the execution of a particular action a results in the occurrence of event e_1 or event e_2. In this case, of course the agent cannot point out an actual event among the two. So it makes sense to only consider non-pointed event models (or, equivalently, multi-pointed event models where all events are designated). Technically speaking, we could alternatively just have introduced the standard syntax and semantics and then introduced the notation $[\mathcal{E}]\phi$ as an abbreviation of $\bigwedge_{e \in E}[\mathcal{E}, e]\phi$, for all $\mathcal{E} = (E, Q, pre, post)$ and all formulas ϕ.

Definition 11.3 (*Product update*) Let $\mathcal{M} = (W, R, V)$ be an epistemic model over P and $\mathcal{E} = (E, Q, pre, post)$ an event model over P. The *product update* of $\mathcal{M}$ with $\mathcal{E}$ is the epistemic model $\mathcal{M} \otimes \mathcal{E} = (W', R', V')$, where

- $W' = \{(w, e) \in W \times E \mid (\mathcal{M}, w) \models pre(e)\}$
- $R' = \{((w, e), (v, f)) \in W' \times W' \mid wRv \text{ and } eQf\}$
- $V'((w, e)) = \{p \in P \mid \mathcal{M}, w \models post(e)(p)\}$

The product update $\mathcal{M} \otimes \mathcal{E}$ represents the result of executing the action $\mathcal{E}$ in the epistemic model represented by $\mathcal{M}$. It is well known that for finite models, two epistemic models are modally equivalent (satisfy the same formulas) iff they are bisimilar (Blackburn et al., 2001). For single-agent models as we consider here, bisimilarity of two models $\mathcal{M}$ and $\mathcal{M}'$ reduces to checking whether for each connected component of $\mathcal{M}$ (each equivalence class of worlds wrt R in $\mathcal{M}$) there exists a connected component of $\mathcal{M}'$ containing the same valuations, and vice versa. Each epistemic model can easily be replaced by its minimal bisimilar representation (its *bisimulation contraction*) achieved by only preserving one world for each set of worlds with identical valuations within a connected component. In the following, we will systematically assume each epistemic model achieved through a product update to be replaced by its bisimulation contraction, and we will generally identify isomorphic models. In this way, we consider bisimilar models to be identical.

Example 11.2 Continuing Example 11.1, consider a situation of an agent seeing a coin lying heads up. It can be represented by the epistemic model $\mathcal{M} = (\{w\}, \{(w, w)\}, V)$ with $V(w) = \{h\}$. Let us now calculate the result of executing the coin toss in this model;

$$\mathcal{M} \otimes \mathcal{E} = \quad \overset{\bullet}{(w_1, e_1) \colon h} \qquad\qquad \overset{\bullet}{(w_1, e_2) \colon}$$

Here, each world is labelled by its valuation, i.e. the atomic propositions true at the world. In $\mathcal{M}$, the agent knows that the coin is facing heads up, that tossing it is neither guaranteed to lead to heads nor tails, but after the coin has been tossed, the agent will know. Those facts are encoded by the following:

$$\mathcal{M}, w \models K(h \wedge \neg[\mathcal{E}]h \wedge \neg[\mathcal{E}]\neg h \wedge [\mathcal{E}](Kh \vee K\neg h))$$

In this paper, we only consider single-agent DEL, as we consider a single agent trying to learn the dynamics of an environment. It might at first seem excessive to introduce all the machinery of DEL to only consider the single-agent version. It is well known that for single-agent epistemic logic, it is sufficient to represent epistemic models as sets of propositional states (often called *belief states*). However, framing our results in the general setting of DEL is relevant for (at least) four reasons. The first is that it gives us the ability to use the event models of DEL to provide compact representations of actions (more compact than just describing actions as a set of possible transitions). Of course, other languages for action descriptions exist, in particular languages like STRIPS (Fikes & Nilsson, 1971) for describing planning domains, but those languages rarely support compact representations of partially observable actions. The second reason is that DEL integrates dynamic modalities in the logical language, so that our agents/learners can explicitly formulate their knowledge about action consequences, as was illustrated in Example 11.2. The third is that the ultimate goal of the line of research introduced here is to be able to generalise to the multi-agent case, where a learner might learn not only what an action does and from what the agent observes, but also from what other agents observe. The fourth reason is that we also intend to integrate our learning algorithms into epistemic planning robots based on DEL (Dissing & Bolander, 2020). The goal is that the robots can not only do planning based on known actions, but can also learn new actions and new environment dynamics.

11.3 Transition Systems and Partially Observable Domains

A domain typically consists of a set of states, a set of actions, and a state transition function mapping pairs of states and actions into the possible successor states (Geffner & Bonet, 2013). When domains are partially observable, we also need to specify what is observed, e.g., with the use of an observation function. A domain can be deterministic or non-deterministic depending on whether an action can have one or more possible outcomes (one or more possible successor states). In this paper, we are only concerned with deterministic actions. However, a learner might during learning consider several possible outcomes of a given action, and hence we also consider non-deterministic transition systems. The set of states of a domain are typically specified as a subset of 2^P for some finite set of propositional symbols P. In this paper, our learners only try to learn domains of this type, but they might still represent their knowledge of such domains by domains of a more general type. Hence our definition of a domain will be more general.

Consider a domain where the states are subsets of P for some set of propositional symbols P. Observation functions often map into so-called 'observation tokens' that can be completely separate from the language P used to describe states. In our setting, however, we will assume that what is being observed in a state is the truth-value of a subset of the propositional symbols, hence directly connecting observations to the state descriptions themselves. For instance, if l_r is a propositional symbol denoting that the light is on in room r, and s is a state in which an agent is present in room r, the agent would be observing the truth-value of l_r in state s. Given these assumptions, we now first define (labelled) transition systems (Sangiorgi, 2011, Chap. 1) and then our partially observable domains.

Definition 11.4 A *transition system* is a tuple $\mathcal{T} = (S, A, T, s_0)$ where

- S is a finite set of *states*
- A is a finite set of *actions*
- $T : S \times A \to 2^S$ is a *transition function*
- $s_0 \in S$ is the *initial state*

A transition system is called *deterministic* if for all $s \in S$ and $a \in A$, $|T(s, a)| \leq 1$. An action $a \in A$ is called *universally applicable* if for all $s \in S$, $|T(s, a)| \geq 1$. In a deterministic transition system where all actions are universally applicable, we hence have that $T(s, a)$ is a singleton for all s and a. In that case, we often write $T(s, a) = s'$ instead of $T(s, a) = \{s'\}$, that is, we take T to be a mapping $T : S \times A \to S$. A transition system over a finite set of propositional symbols P is a transition system where $S \subseteq 2^P$.

In this paper, we are concerned with learning actions of transition systems that are deterministic and in which every action is universally applicable. It might not be natural for any action to be applicable in any state, e.g., "open door" might not be applicable in a state where the door is locked. However, we can replace any such action a by a "try a" action a' that is universally applicable by simply letting

$T(s, a') = s$ for the states s in which a is not applicable (we can always attempt to open the door even if locked, but then it will simply stay closed). We will also restrict attention to transition systems that are 'generated' by their initial state s_0, that is, where any state $s \in S$ can be reached from s_0 by some action sequence. This simplifies things by making it clear that the learner always starts in the same state, and has the possibility to reach any state of the system.

Definition 11.5 A *(partially observable) domain* is a tuple $\mathcal{D} = (\mathcal{T}, \Omega, Obs)$ where

- $\mathcal{T} = (S, A, T, s_0)$ is a transition system in which every action is universally applicable and in which every state can be reached by an action sequence applied to s_0
- Ω is a set of *observations*
- $Obs : S \to 2^{\Omega}$ is an *observation function* mapping each state s into the set of observations that are possible to receive in s

An observation function Obs is *deterministic* if $|Obs(s)| = 1$. In that case we often write $Obs(s) = o$ instead of $Obs(s) = \{o\}$, that is, we take Obs to be a mapping $S \to \Omega$. In this case, an agent entering state s will always receive the same observation $Obs(s)$. A domain $\mathcal{D} = ((S, A, T, s_0), \Omega, Obs)$ is *deterministic* if both T and Obs are deterministic. In this paper, we are going to assume that the observation function is deterministic.

A domain *over a set of propositional symbols P* is $\mathcal{D} = ((S, A, T, s_0), \Omega, Obs)$ where $S \subseteq 2^P$ and $\Omega = 2^P \times 2^P$. An observation $(o^+, o^-) \in Obs(s)$ is split into a set o^+ of the *propositions observed to be true* and o^- of *propositions observed to be false*. We will assume observation functions to be *noiseless*, that is, every proposition in o^+ is true in s, and every proposition in o^- is false in s. Mostly, our domains over P will be deterministic, and any state s then always produces the same observation $Obs(s) = (o^+, o^-)$.

When $\mathcal{D} = ((S, A, T, s_0), \Omega, Obs)$, we will sometimes use $s \in \mathcal{D}$ as an abbreviation for $s \in S$.

Even though we will only consider learning deterministic domains, such domains could still appear non-deterministic to a learner. For instance, an action a might produce distinct outcomes when applied in two states s and t that are *observationally indistinguishable*, that is, for which $Obs(s) = Obs(t)$. In such cases, a learner might (provisionally) decide to represent what has been learned as a non-deterministic transition system, even if the learner actually knows that the underlying transition system must be deterministic.

In the following, we will identify a transition function $T : S \times A \to 2^S$ with its induced relation on $S \times A \times S$, given by: $(s, a, s') \in T$ iff $s' \in T(s, a)$. Given a state s in a transition system over P, we define $s^+ := s$ and $s^- := P - s$. In other words, s^+ is the set of atomic propositions true in s, and s^- is the set of atomic propositions false in s. Given a domain $\mathcal{D} = ((S, A, T, s_0), \Omega, Obs)$, we define the set of *possible observations* in $\mathcal{D}$ to be $\{o \in Obs(s) \mid s \in S\}$. For deterministic domains over P, the set of possible observations is $\{(x, y) \in 2^P \times 2^P \mid Obs(s) =$

(x, y) for some $s \in S\}$. For a possible observation $o = (x, y)$, we often use o^+ to denote the set of positively observed propositions x and o^- the set of negatively observed propositions y. Similarly, for an observation $Obs(s) = (x, y)$, we often write $Obs^+(s)$ for x and $Obs^-(s)$ for y.

To improve readability, we will often represent states by the sequence of literals true in the state, so, e.g., if $P = \{p, q, r\}$, the state $s = \{p, r\}$ will be represented as $p\neg qr$. More generally, for any sequence of propositional literals $l_1, \ldots, l_n$ containing exactly one occurrence of each propositional symbol in P, the state s satisfying all the l_i is denoted $l_1 \ldots l_n$. When the sequence $l_1, \ldots, l_n$ contains *at most* one occurrence of each propositional symbol in P, we use the term $l_1 \ldots l_n$-*state* to denote any state s satisfying all the l_i, i.e., any state s with $s \models l_1 \wedge \cdots \wedge l_n$. So for instance if $P = \{p, q, r\}$, then the pq-states are pqr and $pq\neg r$. We will also sometimes use this type of notation for observations: Given the observation $(Obs^+(s), Obs^-(s))$ received in s, if $Obs^+(s) = \{p_1, \ldots, p_n\}$ and $Obs^-(s) = \{q_1, \ldots, q_m\}$, we will represent the observation compactly as $p_1 \ldots p_n \neg q_1 \ldots \neg q_n$.

Example 11.3 Consider an environment with two rooms, the left room containing a toggle switch and the right room containing a lamp, see Fig. 11.1. The switch controls the lamp, but the status of the light in the right room cannot be observed when being in the left room—and the status of the switch cannot be observed when being in the right room. We can model the environment as a domain over $P = \{l, r, s\}$, where l means that the light is on, r means that the agent is in the right room, and s means that the switch is on. The initial state is $s_0 = \neg l \neg r \neg s$: the light and the switch are both off and the agent is in the left room. There are two actions available to the agent, *flip* and *move*. Executing the *flip* action flips the switch, that is, flips the truth value of both s and l. Executing the *move* action means moving to the other room, that is, flipping the truth value of r. Hence $A = \{$*flip*, *move*$\}$, and the transition system (S, A, T, s_0) underlying the domain is then the system illustrated in Fig. 11.1. We have no outgoing edges for the *flip* action in s_2 and s_3, since the switch cannot be operated when being in the right room. To get a transition system in which every action is universally applicable, we simply assume there to be a reflexive loop for the *flip* action in s_2 and s_3 (the *flip* action is replaced by a "try *flip*" action). We often leave such reflexive edges implicit.

Note that the transition system is deterministic. Letting $S = \{s_0, s_1, s_2, s_3\}$, it is also a transition system in which each state is reachable from the initial state s_0.

The environment can now be described as the domain $\mathcal{D} = ((S, A, T, s_0), \Omega, Obs)$ over P with $Obs(s_0) = \neg r \neg s$, $Obs(s_1) = \neg rs$, $Obs(s_2) = lr$, and $Obs(s_3) = \neg lr$. Note that we could have equivalently defined Obs by saying that the truth value of r (location of agent) is observed in any state, that the truth value of s (the position of the switch) is observed only in the left room, and that the truth value of l (the status of the lamp) is only observed in the right room. In Fig. 11.1, we have underlined the observed literals in each state.

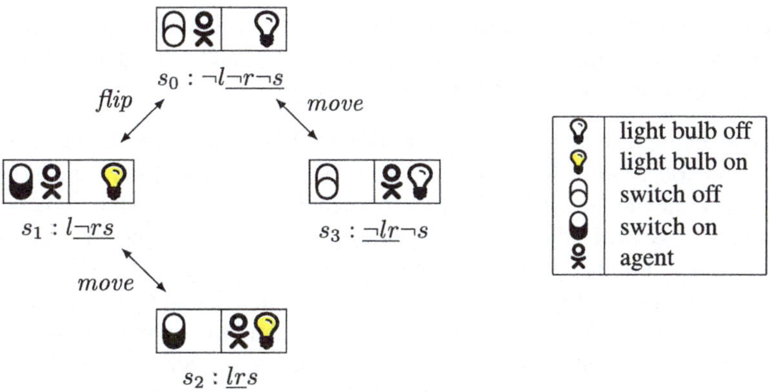

Fig. 11.1 The transition system for the light switch domain

11.4 Learning Explicit Domain Knowledge

When learning fully observable domains, the learning goal is to learn the full underlying transition system (what each action does). This is possible under some reasonable assumptions (Walsh & Littman, 2008). However, for partially observable domains, the goal of learning the full underlying transition system will in general not be attainable. An agent that can in no state observe the truth value of p, will never learn how its actions affect p. So we have to revise the learning goal to be that the learner learns whatever is possible given its observational limitations. The challenge is then generalising the learning from the case of full observability to the case of partial observability. Indeed, even the earlier claim—that an agent that can never observe the truth value of p, can never learn how its actions affect p—is not always true.

Example 11.4 Consider a domain over two propositions p and q. There is a single action a. The truth value of q is always observed and the truth value of p is never observed. Suppose that whenever a sequence of executions of a is performed in the initial state, the learner receives the following sequence of observations: $q, q, \neg q, q, q, \neg q, \ldots$, that is, a sequence of two q-states followed by a $\neg q$-state, then two q-states again, etc. Hence, in some of the q-states, the action a produces another q-state, and in other q-states it produces a $\neg q$-state. Since the underlying transition system is assumed to be deterministic, then if we assume our learners to know this (which we are generally going to do), a learner receiving such observation sequences should hence be able to infer that there must be two distinct kinds of q-worlds. Assume furthermore that our learners know the underlying language of the domain, that is, the available propositional symbols (another assumption that we are generally going to make in this paper). Then a learner can infer that the two distinct q-worlds must be distinct by assigning different truth values to p. In other words, the learner can infer that the a action affects p. The learner can never know exactly *how*, i.e., will never be able in any state to infer the real truth value of p. But the learner can still learn something about the relation between the action a and the proposition p, e.g., that p does not always have the same truth value, but will have the same truth value every third time a is executed.

We can distinguish between two types of knowledge that the learner can achieve, *implicit* and *explicit* knowledge. There is a rich literature in epistemic logic on explicit and implicit knowledge, and with different meanings assigned to the two concepts. In this paper, we will take explicit knowledge to be what is known because it is directly observed in the current state, and implicit knowledge to be whatever might additionally be inferred from the history of earlier actions and general experience with the environment (domain). In Example 11.4 the only explicit knowledge the learner can gain is that action a sometimes makes a q-state into another q-state and sometimes into a $\neg q$-state; and that it always makes a $\neg q$-state into a q-state. Learning that every second time a is executed in a q-state it produces a $\neg q$-state is not something we will consider to be explicit knowledge, as it can never be explicitly known how many times a has been executed before (it is not directly observed in any state).

At first, we will only focus on learning explicit knowledge, that is, learners that learn to know what will be directly observed when actions are executed. Even if an agent is able to learn more than this, it is still interesting to have learners that can identify the explicit knowledge resulting from action executions. Consider for instance a person, Agnes, in a room with a switch that controls the light in the room next door, similar to the scenario of Example 11.3. Suppose the door to the other room is closed, so that it is not possible to observe whether the light in there is on or not. Suppose further that initially Agnes doesn't know what the light switch does, since she is in a house that she hasn't been in before, a summer house that she borrowed from a friend. If Agnes decides to explore the environment (the summer house), she might learn that the light switch controls the light in the other room, e.g., by opening the door or peeking through the key hole. Hence, potentially, she might in this case be able to learn the full underlying transition system of the domain consisting of the light switch and the lamp in the other room. In this case, her implicit knowledge about the domain will be complete. However, identifying the implicit knowledge and the full underlying transition system is not the only thing that's relevant. Consider for instance that after the end of her stay, another friend, Bertram, comes to stay in the summer house. Bertram also never stayed there before. When Bertram arrives, the door to the other room might be closed, and the light in there is on. In this case, it would clearly be relevant for Agnes to know that Bertram now is in the same information state as she was initially, and then he will not be able to observe the light in the other room. So she might tell him: "The light in the other room is on. This switch controls the light. Please make sure to turn off the light before you leave." In other words, at least in multi-agent scenarios, it is relevant to learn not only as much as we can possibly come to know about the environment, but also what we are able to directly see and not see in this environment.

11.4.1 Compatibility Domain

To define a learner for explicit knowledge, we first need some additional technical definitions. Given a deterministic domain $\mathcal{D} = ((S, A, T, s_0), \Omega, Obs)$ over some set P, suppose an execution of an action $a \in A$ produces a transition (s, t) (that is, we have $T(s, a) = t$). Then the learner observing the action execution doesn't get to see the state transition from s to t itself, but only observes a transition from the observation $(Obs^+(s), Obs^-(s))$ to the observation $(Obs^+(t), Obs^-(t))$. Generally, given a true state s, the learner observes only $(Obs^+(s), Obs^-(s))$. What is the set of states t that the learner thinks could have produced the observation? If the learner has no additional information about the domain, it can only conclude that the underlying state t must be among the states consistent with what has been observed, that is, a state t satisfying $Obs^+(s) \subseteq t^+$ and $Obs^-(s) \subseteq t^-$. More formally, given a state s, we define the states *observationally compatible* with s as the set of states $comp(s) := \{t \in 2^P \mid Obs^+(s) \subseteq t^+ \text{ and } Obs^-(s) \subseteq t^-\}$. We can extend this notion to observations $(x, y) \in 2^P \times 2^P$ in the obvious way: $comp((x, y)) = \{t \in 2^P \mid x \subseteq t^+ \text{ and } y \subseteq t^-\}$. We then get that $comp(s) = comp((Obs^+(s), Obs^-(s))$. Note that for any states $s, t \in S$, we have $comp(s) = comp(t)$ iff $Obs(s) = Obs(t)$.

Definition 11.6 Let $\mathcal{D} = ((S, A, T, s_0), \Omega, Obs)$ be a deterministic domain over P. The *compatibility domain* $((S', A', T', s_0'), \Omega', Obs')$ induced by $\mathcal{D}$ is given by:

- $S' = \{comp(s) \mid s \in S\}$
- $A' = A$
- $T' = \{(comp(s), a, comp(t)) \mid (s, a, t) \in T\}$
- $s_0' = comp(s_0)$
- $\Omega' = 2^P \times 2^P$
- $Obs'(comp(s)) = Obs(s)$.[1]

The compatibility domain $\mathcal{D}'$ induced by a domain $\mathcal{D}$ is the image of $\mathcal{D}$ under the compatibility mapping *comp*. It hence encodes the original transition systems as seen through the lens of the observation function. It encodes what is directly observed by the agent when actions are executed in the domain. Note that $\mathcal{D}'$ is not necessarily deterministic even if $\mathcal{D}$ is: a compatibility state $comp(s)$ might contain distinct states for which a given action a produces distinct outcomes, and where these distinct outcomes can be observationally distinguished.

Example 11.5 Consider again the light switch domain $\mathcal{D}$ of Example 11.3. Figure 11.2 shows the compatibility domain $\mathcal{D}'$ induced by $\mathcal{D}$. In the figure, each state is marked by which compatibility state, $comp(s_i)$, it is, and additionally by the observation that defines it (the observation received in s_i). Note that $\mathcal{D}'$ is still deterministic. In fact, the underlying transition system of $\mathcal{D}$ is isomorphic to the underlying transition system of $\mathcal{D}'$. At first one might think that this implies that a

[1] This is well-defined since $comp(s) = comp(t)$ iff $Obs(s) = Obs(t)$.

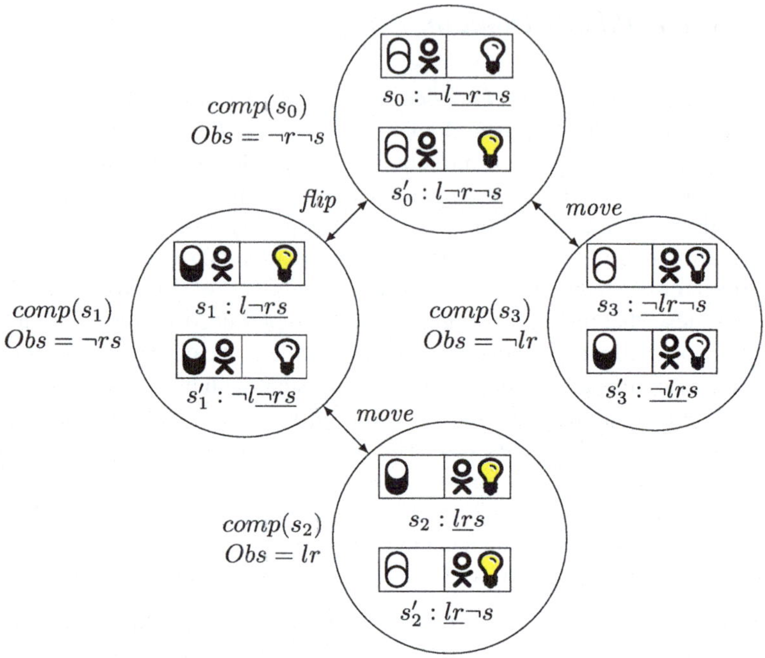

Fig. 11.2 The compatibility domain $\mathcal{D}'$ induced by the light switch domain $\mathcal{D}$ of Example 11.3

learner can learn to identify the full underlying system, despite its observational limitations. This, however, is not so. Even if a learner can use its observational powers to construct $\mathcal{D}'$ from observing action executions, there are things it can never learn. For instance, while the learner will be able to infer that, from the initial state, first flipping the switch and then moving to the other room will lead to a state where the light is on (the compatibility state $comp(s_2) = \{s_2, s_2'\}$), it will never be able to know whether it is the action of flipping the switch itself, or only the combined action of first flipping the switch and then moving, that made the light go on. This is due to the effect of the *flip* action applied to $comp(s_0)$ being $comp(s_1) = \{s_1, s_1'\}$, in which there is both an l-state and a $\neg l$-state. In other words, the agent will never be able to distinguish the real domain from one in which the electrical circuit controlling the light has a serial connection containing both the light switch and a movement detector, so that the light only goes on when both the switch is on and the agent is in that room. Since this distinction is not learnable, we of course need a notion of *learnability* or *behavioural correctness* that takes these observational limitations into account, so that a learning agent can still be considered "complete" even if it doesn't manage to identify the underlying transition system, but only identifies it up to its observational limitations.

Definition 11.7 An *isomorphism* between two domains $\mathcal{D} = ((S, A, T, s_0), \Omega, Obs)$ and $\mathcal{D}' = ((S', A, T', s_0), \Omega, Obs')$ is a bijection $f : S \to S'$ satisfying:

1. $f(s_0) = s'_0$
2. for any $s, t \in S$, $(s, a, t) \in T$ iff $(f(s), a, f(t)) \in T'$
3. for all $s \in S$, $Obs(s) = Obs'(f(s))$.

Two domains are called *isomorphic* if there exists an isomorphism between them.

This is the natural generalisation of isomorphisms between labelled transition systems (Gorrieri & Versari, 2015) to domains. Let $\mathcal{D}'$ denote the compatibility domain of a domain $\mathcal{D} = ((S, A, T, s_0), \Omega, Obs)$. An alternative definition of the compatibility domain induced by $\mathcal{D}$ could be the domain $\mathcal{D}'' = ((S'', A, T'', s''_0), \Omega, Obs'')$ given by $S' = \{Obs(s) \mid s \in S\}$, $T' = \{(Obs(s), a, Obs(t)) \mid (s, a, t) \in T\}$, $s''_0 = Obs(s_0)$, and $Obs''(s) = s$. Since we have, for all $s, t \in S$, $comp(s) = comp(t)$ iff $Obs(s) = Obs(t)$, we can define a bijection $f : S' \to S''$ by $f(comp(s)) = Obs(s)$. It is easily verified that f is an isomorphism between $\mathcal{D}'$ and $\mathcal{D}''$, so the two definitions of compatibility domains are equivalent (up to isomorphism). It means that we can also think of the compatibility domain to be the domain of observations with the transitions on observations induced by the real underlying transitions in the obvious way.

In a state s of a domain $\mathcal{D}$, we consider the explicit knowledge of the agent to be what is true in all of $comp(s)$, which is of course simply what follows logically from the observation made in s, since the following are equivalent for all propositional formulas ϕ (where $\models$ denotes the standard entailment relation in propositional logic),

- for all $t \in comp(s)$, $t \models \phi$
- for all t with $Obs^+(s) \subseteq t^+$ and $Obs^-(s) \subseteq t^-$, $t \models \phi$
- $\models (\bigwedge_{p \in Obs^+(s)} p \wedge \bigwedge_{p \in Obs^-(s)} \neg p) \to \phi$

To make the notion of explicit knowledge precise in an epistemic setting, we identify any set of states S over a set of propositional symbols P with an *induced* epistemic model $\mathcal{M}_S = (W, R, V)$ over P with $W = S$, $R = W \times W$ and $V(w) = w$. By this identification, we can evaluate (static) epistemic formulas in sets of states, e.g., we then have $comp(s) \models K\phi$ iff ϕ is true in every world (state) w in $comp(s)$, which again holds iff ϕ follows from the observation received in s. The modal operator K is here interpreted as the operator for explicit knowledge. In the following, we will generally identify sets of states S with their induced epistemic models $\mathcal{M}_S$ without notice. It will be clear from the context whether a set of states should be considered as just that, or as the induced epistemic model.

By the identification of sets of states with epistemic models, we can see the compatibility domain as a domain on epistemic models. In dynamic epistemic logic, any set of event models similarly defines a domain on epistemic models.

Definition 11.8 Let $(\mathcal{E}_a)_{a \in A}$ be a collection of event models over a set of propositions P, and let $\mathcal{M}_0$ be an epistemic model over P. The domain $((S', A', T', s'_0), \Omega, Obs)$ induced by $(\mathcal{E}_a)_{a \in A}$ and $\mathcal{M}_0$ is given by:

- $S' = \{\mathcal{M}_0 \otimes \mathcal{E}_{a_1} \otimes \cdots \otimes \mathcal{E}_{a_n} \mid a_1, \ldots, a_n \in A\}$
- $A' = A$
- $T' = \{(\mathcal{M}, a, \mathcal{M}') \mid \mathcal{M}'$ is a connected component of $\mathcal{M} \otimes \mathcal{E}_a\}$
- $s'_0 = \mathcal{M}_0$
- $\Omega = 2^P \times 2^P$
- $Obs(\mathcal{M}) = (\{p \in P \mid \mathcal{M} \models p\}, \{p \in P \mid \mathcal{M} \models \neg p\})$

11.4.2 Behavioural Correctness and Learners

We can use isomorphisms between domains to define (strong) equivalence between them. We could of course also consider a weaker notion of equivalence between domains given by bisimulation, and indeed we will later do so when defining learners for implicit knowledge. However, for now it suffices to consider isomorphisms.

If the domain induced by a collection of event models is isomorphic to the compatibility domain, it means that the event models represent exactly what is explicitly knowable about the domain. A learner that identifies such event models will be called *behaviourally correct with respect to explicit knowledge*. It will be a learner that identifies the underlying transition system up to the limitations of its distinguishing powers defined by the compatibility mapping (defined by what is explicitly observable), as follows.

Definition 11.9 Let $\mathcal{D} = ((S, A, T, s_0), \Omega, Obs)$ and $\mathcal{D}' = ((S', A, T', s'_0), \Omega', Obs')$ be domains. We say that $\mathcal{D}'$ is *behaviourally correct with respect to explicit knowledge* about $\mathcal{D}$ if $\mathcal{D}'$ is isomorphic to the compatibility domain induced by $\mathcal{D}$. A collection of event models $(\mathcal{E}_a)_{a \in A}$ is *behaviourally correct with respect to explicit knowledge* about $\mathcal{D}$ if the domain induced by $(\mathcal{E}_a)_{a \in A}$ and $comp(s_0)$ is behaviourally correct with respect to explicit knowledge about $\mathcal{D}$.

Learners learn from observing the execution of actions. Executions of actions are represented as transitions $(s, a, t) \in T$. As mentioned earlier, when a transition $(s, a, t) \in T$ occurs, the learner only observes $(Obs(s), a, Obs(t))$. We call $(Obs(s), a, Obs(t))$ the *observed transition* of (s, a, t). In general, for a domain $\mathcal{D} = ((S, A, T, s_0), \Omega, Obs)$, an *observed transition* is any $(Obs(s), a, Obs(t))$ with $(s, a, t) \in T$. A learner takes as input a set of observed transitions of a domain, and attempts to provide a representation of what it has learned about the actions of the domain. There are many possible ways a learner could represent its learned actions, one of them being through event models. In that setting, a learner can be seen as an algorithm that takes as input a set σ of observed transitions of actions A over P and produces a collection of event models $\mathcal{E}_a$, one for each action $a \in A$. The goal for the learner is then to produce a collection of event models that is behaviourally correct with respect to explicit knowledge. If the learner hasn't been exposed to all possible transitions of the system, it cannot of course in general be expected to produce behaviourally correct event models. If a set of observed transitions contains observations of all possible transitions of the system, we call it *sound and complete*. More precisely, a set of observed transitions σ is called *sound and complete* for a domain $\mathcal{D} = ((S, A, T, s_0), \Omega, Obs)$ if for all $(s, a, t) \in T$, $(Obs(s), a, Obs(t)) \in \sigma$.

11.4.3 A Behaviourally Correct Learner of Explicit Knowledge

We now present our first learner (learning algorithm) and prove that it achieves its goal (producing a behaviourally correct set of event models) when presented with a sound and complete set of observed transitions. The algorithm is included as Algorithm 1. It relies on the following additional definition. For any observation o in a domain over P, we define the epistemic formula ϕ_o representing the explicit knowledge that an agent observing o has, where we use $Kw\ \phi$ as shorthand for $K\phi \vee K\neg\phi$ (knowing whether ϕ):

$$\phi_o := K\left(\bigwedge_{p\in o^+} p \wedge \bigwedge_{p\in o^-} \neg p\right) \wedge \bigwedge_{p\in P-(o^+\cup o^-)} \neg Kw\ p$$

Consider for instance the initial state $s_0 = \neg l \neg r \neg s$ of the light switch domain (Example 11.3). In this state, the agent receives the observation $Obs(s_0) = \neg r \neg s$ and hence $\phi_{Obs(s_0)}$ is the formula $K(\neg r \wedge \neg s) \wedge \neg Kw\ l$ representing that the agent knows that the switch is off and that the agent itself is in the left room, but not knowing whether the light is on or off.

Theorem 11.1 *The learning algorithm* LEARNER(P, A, σ) *(Algorithm 1) applied to a sound and complete set of observed transitions of a domain $\mathcal{D}$ outputs a collection of event models that are behaviourally correct with respect to explicit knowledge about $\mathcal{D}$.*

Proof Let σ be a sound and complete set of observed transitions for a domain $\mathcal{D} = ((S, A, T, s_0), \Omega, Obs)$, and let $(\mathcal{E}_a)_{a\in A}$ denote the output produced by Algorithm 1. We need to show that the domain induced by $(\mathcal{E}_a)_{a\in A}$ and $comp(s_0)$ is isomorphic to the compatibility domain induced by $\mathcal{D}$. So let $\mathcal{D}' = ((S', A, T'), x_0), \Omega, Obs')$ denote the compatibility domain of $\mathcal{D}$ and let $\mathcal{D}'' = ((S'', A, T'', \mathcal{M}_0), \Omega, Obs'')$ denote the domain induced by $(\mathcal{E}_a)_{a\in A}$ and $\mathcal{M}_0 = comp(s_0)$, where $\Omega = 2^P \times 2^P$. We need to find a bijection $f : S' \to S''$ such that $f(x_0) = \mathcal{M}_0$, for any $x, y \in S'$ we have $(x, a, y) \in T'$ iff $(f(x), a, f(y)) \in T''$, and for all $x \in S'$, $Obs'(x) = Obs''(f(x))$.

Claim 1. If $(x_i, a_i, x_{i+1}) \in T'$ for all $0 \le i \le n$, then $(\mathcal{M}_i, a_i, \mathcal{M}_{i+1}) \in T''$ for all $0 \le i \le n$ where $\mathcal{M}_i$ is the epistemic model induced by x_i.

Algorithm 1: LEARNER(P, A, σ)

Input : P (propositional symbols), A (actions), σ (observed transitions)
Output : $(\mathcal{E}_a)_{a \in A}$ (event models)

for *each* $a \in A$ **do**
  let $\mathcal{E}_a$ be an empty event model;
  for *each observation* o **do**
    $O := \{o' \mid (o', a, o) \in \sigma\}$;
    if $O \neq \emptyset$ **then**
      $E := \{\langle \bigvee_{o' \in O} \phi_{o'}, \mathsf{post} \rangle \mid \mathsf{post}(p) = \top$ for all $p \in o^+$ and $\mathsf{post}(p) = \bot$ for all $p \in o^-\}$;
      Add to $\mathcal{E}_a$ the events E and make all events in E mutually indistinguishable, but distinguishable from all other events in $\mathcal{E}_a$;
    end
  end
end
return $(\mathcal{E}_a)_{a \in A}$;

Proof of Claim 1. The proof is by induction on n. Note that by definition, $\mathcal{M}_0$ is the epistemic model induced by the state $x_0 = comp(s_0)$, which covers the base case (action sequences of length 0). For the induction step, suppose $(x_i, a_i, x_{i+1}) \in T'$ for all $0 \leq i \leq n$ and suppose $(\mathcal{M}_i, a_i, \mathcal{M}_{i+1}) \in T''$ for all $0 \leq i \leq n-1$ where $\mathcal{M}_i$ for all $i \leq n$ is the epistemic model induced by x_i. We need to show $(\mathcal{M}_n, a_n, \mathcal{M}_{n+1}) \in T''$ where $\mathcal{M}_{n+1}$ is the epistemic model induced by x_{n+1}. Since $(x_n, a_n, x_{n+1}) \in T'$, there exist $s_n, s_{n+1} \in T$ such that $x_n = comp(s_n)$, $x_{n+1} = comp(s_{n+1})$ and $(s_n, a_a, s_{n+1}) \in T$. By soundness and completeness of σ, $(Obs(s_n), a_n, Obs(s_{n+1})) \in \sigma$. This implies that $\mathcal{E}_{a_n}$ contains a subset E of all events of the form $\langle \bigvee_{(o', a_n, Obs(s_{n+1})) \in \sigma} \phi_{o'}, \mathsf{post} \rangle$ where $\mathsf{post}(p) = \top$ for all $p \in Obs^+(s_{n+1})$ and $\mathsf{post}(p) = \bot$ for all $p \in Obs^-(s_{n+1})$. Furthermore, the events in E are all mutually indistinguishable, but distinguishable from all other events of $\mathcal{E}_{a_n}$. Since $\mathcal{M}_n$ is induced by $x_n = comp(s_n)$, the valuations occurring in the worlds of $\mathcal{M}_n$ are exactly the ones represented by the states in $comp(s_n)$. That is, $\mathcal{M}_n$ has a world w with valuation $V(w) = P'$ iff $Obs^+(s_n) \subseteq P'$ and $Obs^-(s_n) \cap P' = \emptyset$. From this it follows that $\mathcal{M}_n \models \phi_{Obs(s_n)}$. Since $(Obs(s_n), a_n, Obs(s_{n+1})) \in \sigma$, any world satisfying $\phi_{Obs(s_n)}$ will satisfy the precondition of each of the events in E. In particular, each world of $\mathcal{M}_n$ will satisfy the precondition of each event in E. For any state $t \in x_{n+1} = comp(s_{n+1})$ we must have $Obs^+(s_{n+1}) \subseteq t^+$ and $Obs^-(s_{n+1}) \subseteq t^-$. Hence for any state $t \in x_{n+1}$, there must exist an event $e_t \in E$ with $\mathsf{post}(e_t)(p) = \top$ for all $p \in t$ and $\mathsf{post}(e_t)(p) = \bot$ for all $p \notin t$. Conversely, for any event $e \in E$, from the conditions $\mathsf{post}(e)(p) = \top$ for all $p \in Obs^+(s_{n+1})$ and $\mathsf{post}(e)(p) = \bot$ for all $p \in Obs^-(s_{n+1})$ we get the existence of a state t observationally compatible with s_{n+1} such that $\mathsf{post}(e) = \mathsf{post}(e_t)$. This implies that $\mathcal{M}_n \otimes \mathcal{E}_{a_n}$ contains a connected component in which the world valuations are exactly the states $t \in x_{n+1}$. This proves that $\mathcal{M}_n \otimes \mathcal{E}_{a_n}$ contains $\mathcal{M}_{n+1}$ (the epistemic model induced by x_{n+1}) as a connected component, and hence $(\mathcal{M}_n, a_n, \mathcal{M}_{n+1}) \in T''$, as required (recall the models are identified under bisimilarity, so it is irrelevant that the connected component might contain several worlds with the same valuation). This completes the proof of the claim.

Claim 2. If $(\mathcal{M}_i, a_i, \mathcal{M}_{i+1}) \in T''$ for all $0 \leq i \leq n$, then there exists a state sequence $x_1, \ldots, x_{n+1} \in S'$ such that for all i, $(x_i, a_i, x_{i+1}) \in T'$ and $\mathcal{M}_i$ is the epistemic model induced by x_i.

Proof of Claim 2. The base case is as for Claim 1. For the induction step, suppose $(\mathcal{M}_i, a_i, \mathcal{M}_{i+1}) \in T''$ for all $0 \leq i \leq n$ and suppose $(x_i, a_i, x_{i+1}) \in T'$ for all $0 \leq i \leq n-1$ where $\mathcal{M}_i$ for all $i \leq n$ is the epistemic model induced by x_i. We need to show $(x_n, a_n, x_{n+1}) \in T'$ and that $\mathcal{M}_{n+1}$ is the epistemic model induced by x_{n+1}. Since $(\mathcal{M}_n, a_n, \mathcal{M}_{n+1}) \in T''$, $\mathcal{M}_{n+1}$ is one of the connected components of $\mathcal{M}_n \otimes \mathcal{E}_{a_n}$. Since $\mathcal{M}_n$ is induced by x_n, it must be connected, and hence there must exist a connected component E of events in $\mathcal{E}_{a_n}$ such that $\mathcal{M}_{n+1}$ is the product update of $\mathcal{M}_n$ with those events. By the definition of the algorithm, there must exist an observation o such that $(o', a_n, o) \in \sigma$ for some o' and such that E must be the set of all events of the form $\langle \bigvee_{(o',a_n,o)\in\sigma} \phi_{o'}, \mathsf{post} \rangle$ where $\mathsf{post}(p) = \top$ for all $p \in o^+$ and $\mathsf{post}(p) = \bot$ for all $p \in o^-$. Since $\mathcal{M}_{n+1}$ is the product update of $\mathcal{M}_n$ with the events in E, and since $\mathcal{M}_{n+1}$ is non-empty, at least one of the events in E must have its precondition satisfied in at least one of the worlds of $\mathcal{M}_n$. Since all events in E have the same precondition, *all* events have their precondition satisfied in at least one world of $\mathcal{M}_n$. In other words, there must exist o' such that $(o', a_n, o) \in \sigma$ and such that $\phi_{o'}$ is satisfied in at least one of the worlds of $\mathcal{M}_n$. From $(o', a_n, o) \in \sigma$ we get the existence of $s_n, s_{n+1} \in S$ with $(s_n, a_n, s_{n+1}) \in T$, $Obs(s_n) = o'$ and $Obs(s_{n+1}) = o$. By the definition of $\phi_{o'}$ and since $\phi_{o'}$ holds in a world of $\mathcal{M}_n$, all propositions in o'^+ are true in all worlds of $\mathcal{M}_n$, all propositions in o'^- are false in all worlds of $\mathcal{M}_n$, and all other propositions are true in some worlds of $\mathcal{M}_n$ and false in others. Since $\mathcal{M}_n$ is the epistemic model induced by x_n, we then get that $x_n = \{t \in 2^P \mid o'^- \subseteq t^- \text{ and } o'^+ \subseteq t^-\} = \{t \mid Obs^-(s_n) \subseteq t^- \text{ and } Obs^+(s_n) \subseteq t^+\} = comp(s_n)$. Since $(s_n, a_n, s_{n+1}) \in T$, we must have $(comp(s_n), a_n, comp(s_{n+1})) \in T'$. It now suffices to prove that $\mathcal{M}_{n+1}$ is the epistemic model induced by $comp(s_{n+1})$. Since $Obs(s_{n+1}) = o$, E is the set of all events of the form $\langle \bigvee_{(o',a_n,Obs(s_{n+1}))\in\sigma} \phi_{o'}, \mathsf{post} \rangle$ where $\mathsf{post}(p) = \top$ for all $p \in Obs^+(s_{n+1})$ and $\mathsf{post}(p) = \bot$ for all $p \in Obs^-(s_{n+1})$. This implies that the update of $\mathcal{M}_n$ with the events in E must be all states in $comp(s_{n+1})$, and hence $\mathcal{M}_{n+1}$ is induced by $comp(s_{n+1})$, as required. This completes the proof of the claim.

We can now construct an isomorphism f from $\mathcal{D}'$ to $\mathcal{D}''$ by, for all $x \in S'$, letting $f(x)$ be the epistemic model induced by x. Note that Claim 2 guarantees that all states of S'' are induced epistemic models of states in S', hence guaranteeing that the mapping correctly maps elements of S' into elements of in S''. We now first get $f(x_0) = \mathcal{M}_0$, as required. Suppose then $(x, a, y) \in T'$. Since each state S is reachable from s_0, there must then exist an action sequence $a_0, \ldots, a_n \in A$ and a state sequence $x_0, \ldots, x_{n+1} \in S'$ such that for all $i \leq n$ we have $(x_i, a_i, x_{i+1}) \in T'$ and where $x = x_n$, $a = a_n$ and $y = x_{n+1}$. It follows from Claim 1 that $(\mathcal{M}_n, a_n, \mathcal{M}_{n+1}) \in T''$, where $\mathcal{M}_n$ is the epistemic model induced by x_n and $\mathcal{M}_{n+1}$ is the epistemic model induced by x_{n+1}. We then get $f(x) = f(x_n) = \mathcal{M}_n$ and $f(y) = f(x_{n+1}) = \mathcal{M}_{n+1}$ and hence $(f(x), a, f(y)) \in T''$, as required. Suppose instead that $(f(x), a, f(y)) \in T''$. Then we need to prove that $(x, a, y) \in T'$. There must exist an action sequence $a_0, \ldots, a_n \in A$ and a state sequence $\mathcal{M}_0, \ldots, \mathcal{M}_{n+1} \in S''$ such that for all $i \leq n$ we have $(\mathcal{M}_i, a_i, \mathcal{M}_{i+1}) \in T''$ and where $f(x) = \mathcal{M}_n$, $a = a_n$ and $f(y) = \mathcal{M}_{n+1}$. From Claim 2 we now get that $(x_n, a_n, x_{n+1}) \in T'$ where $\mathcal{M}_n$ is the epistemic model induced by x_n and $\mathcal{M}_{n+1}$ is the epistemic model induced by x_{n+1}. It follows that $f(x_n) = f(x)$ and $f(x_{n+1}) = f(y)$, and hence $x_n = x$ and $x_{n+1} = y$, and thus finally $(x, a, y) \in T'$, as required. For the observation functions, it finally follows that for all $comp(s) \in S'$ we have $Obs'(comp(s)) = Obs''(f(comp(s)))$, since $f(comp(s))$ is the induced epistemic model of $comp(s)$, and hence $Obs''(f(comp(s))) = (\{p \mid f(comp(s)) \models p\}, \{p \mid f(comp(s)) \models \neg p\}) = (\{p \mid p \in t \text{ for all } t \in comp(s)\}, \{p \mid p \notin t \text{ for all } t \in comp(s)\}) = (Obs^+(s), Obs^-(s)) = Obs(s) = Obs'(comp(s))$. □

Example 11.6 Consider again the light switch domain $\mathcal{D}$ of Example 11.3. Letting $P = \{l, r, s\}$, $A = \{move, flip\}$ and σ be a sound and complete set of observed transitions for $\mathcal{D}$, LEARNER(P, A, σ) will for the *move* action produce the event model $\mathcal{E}_{move}$ presented in Fig. 11.3 and for the *flip* action produce the event model $\mathcal{E}_{flip}$ presented in Fig. 11.4. Note that the event model $\mathcal{E}_{move}$ has four connected components, corresponding to whether the move goes from the left to the right room (the first and the third connected components of Fig. 11.3) or from the right to the left room (the second and the fourth connected components), and whether the move starts in a state where the switch/light is off (first and second component) or on (third and fourth). In total, the two event models have 6 connected components corresponding to the 6 transitions of the compatibility domain induced by $\mathcal{D}$. Since each connected com-

Fig. 11.3 The event model $\mathcal{E}_{move}$ provided as output of Algorithm 1 running on the light switch domain

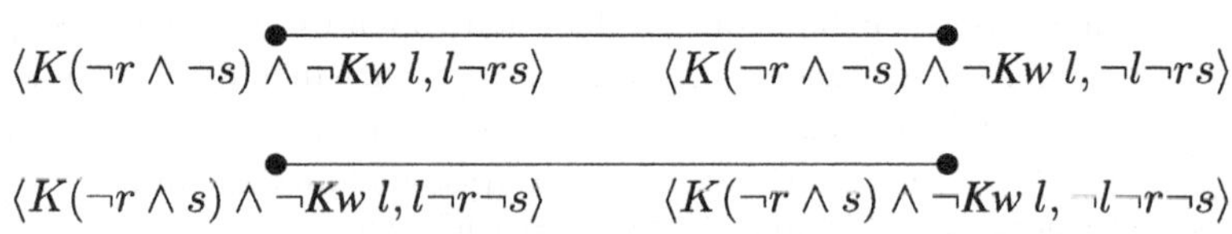

Fig. 11.4 The event model $\mathcal{E}_{flip}$ provided as output of Algorithm 1 running on the light switch domain

ponent has 2 events, this appears to be an even less compact representation than the compatibility transition system itself, but this is due to the event models representing action outcomes as being nondeterministic rather than representing them as deterministically mapping into a set of states (a "belief state"). The induced compatibility domain of course represents the exact same information as the set of event models, only in a different way. It would probably be possible to represent the event models more compactly, e.g. using a three-valued version of DEL where having an event e with $post(e)(p) = \textit{undefined}$ would be a compact representation of the two events $\langle pre(e), post(e) \cup \{p \mapsto \top\}\rangle$ and $\langle pre(e), post(e) \cup \{p \mapsto \bot\}\rangle$. Such a representation would reduce the event models of this example to only contain 6 events in total. We leave the detailed consideration of such three-valued versions of DEL to a future paper. Another possible optimisation of the action representation would be to use non-boolean postconditions, that is, postcondition mappings where $post(e)(p)$ can be an arbitrary formula. Actually, such postconditions are the most standard in DEL, however, for simplicity we decided here to only work with boolean postconditions. As an example, consider the second and fourth component of Fig. 11.3. Using non-boolean postconditions (Ditmarsch & Kooi, 2008), we could represent the two left events of the two components by a single event $\langle Kr \wedge \neg Kw\, s, \mathsf{post}\rangle$ with $\mathsf{post}(l) = \top$, $\mathsf{post}(r) = \bot$ and $\mathsf{post}(s) = l$, and similarly for the two right events. Event preconditions can also sometimes be simplified, e.g. the aforementioned event could in this context be simplified to $\langle Kr, \mathsf{post}\rangle$. In future work it will be relevant to consider extended learning algorithms including methods for simplifying event models (ensuring that they still induce the same domain).

The learned event models of Figs. 11.3 and 11.4 represent the explicit knowledge the agent can gain about the *dynamics* of the domain. Earlier, we explained how to evaluate static formulas in $comp(s_0)$, but given the learned event models, we can now also evaluate dynamic formulas including the modalities $[\mathcal{E}_{flip}]$ and $[\mathcal{E}_{move}]$. This allows the agent to reason about the dynamics of its explicit knowledge, for instance we get:

$$comp(s_0) \models \neg Kw\, l \wedge [\mathcal{E}_{flip}]\neg Kw\, l \wedge [\mathcal{E}_{flip}][\mathcal{E}_{move}]Kl$$

Thus a learner having used Algorithm 1 to learn the dynamics of the domain and produced the event models of the two figures, would be able to conclude that it initially doesn't explicitly know whether the light is on, that flipping the switch doesn't change this, but both flipping the switch and moving to the other room guarantees explicitly knowledge of the light being on. This can potentially be used in combination with epistemic planning based on DEL (Bolander & Andersen, 2011). For instance, if the learner was given the planning goal Kl (turning on the light and explicitly knowing to have done so), it would produce the plan (*flip*, *move*), since this is the shortest action sequence $(a_1, \ldots, a_n)$ satisfying $comp(s) \models [a_1] \cdots [a_n]Kl$. Note that if the learner was given the simpler goal l ("turn on the light"), the plan would still be the same, since $comp(s) \not\models [\mathcal{E}_{flip}]l$ and $comp(s) \not\models [\mathcal{E}_{move}]l$.

11.5 Extending Learning Beyond Explicit Knowledge

We have presented a learning algorithm for learning explicit knowledge, and illustrated how it works on the light switch domain. How about implicit learning in that domain? Is there more a learner would be able to learn—or deduce—about the domain than what can be directly observed in each state? In this domain, there isn't. The learner of explicit knowledge has already identified the underlying transition system up to isomorphism, so it is only the identification of which propositions are true in each state that are missing. Taking another look at Fig. 11.2, it can be seen that this additional insight can not be achieved. Consider for any choice of states $s_i'' \in comp(s_i), i = 1, \dots, 3$, the domain with states $\{s_1'', \dots, s_3''\}$ and with transition function and observation function induced by the compatibility domain of Fig. 11.2. No sequence of observations of action executions can make the agent distinguish this new domain from the real domain (the two domains are bisimilar, a notion to be formally introduced in the next section). Take for instance $s_0'' = s_0'$, $s_1'' = s_1'$, $s_2'' = s_2$ and $s_3'' = s_3$. In the domain induced by these states, the light is initially turned on (s_0'), and flipping the switch makes it turn off (s_1'). So the function of the switch has been reversed, at least when the agent is in the left room. However, if moving into the other room with the switch up, the light will actually go on again (s_2), and if moving with the switch down, the light will turn off (s_3). So moving from one room to the other reverses the function of the switch (making you perhaps reconsider which electrician to call for your next electrical wiring job). For each choice of the s_i'', we can make a similar description of a domain that would make (some) sense and that would be consistent with any observations that can be made in the domain (bisimilar with the real domain). One could consider learners with an inductive bias that would make them choose certain domains over others. An inductive bias could for instance be the simplicity (size) of the produced set of event models, a kind of basic Ockham's razor principle. However, it can be shown that in the light switch domain, such an inductive bias will still not allow the learner to uniquely settle on the correct domain. To sum up, the light switch domain doesn't allow us to see any difference between learning explicit or implicit knowledge. Let us know consider another domain that does.

Example 11.7 Consider the domain $\mathcal{D} = ((S, A, T, s_0), \Omega, Obs)$ over $P = \{p\}$ with $S = \{s_0, s_1\}$, $s_0 = \emptyset$, $s_1 = \{p\}$, $A = \{\mathit{flip}\}$, $T = \{(s_0, \mathit{flip}, s_1), (s_1, \mathit{flip}, s_0)\}$, $Obs(s_0) = \top$, and $Obs(s_1) = p$ (that is, $Obs(s_0) = (\emptyset, \emptyset)$ and $Obs(s_1) = (\{p\}, \emptyset)$). So the *flip* action flips the truth value of p, but the truth value is only observable when true. It sounds perhaps a bit esoteric, since if p is observed when true, why don't we also observe it when false? We can however think of a concrete example of this type. Consider a box that can either be empty ($\neg p$) or full (p), and *flip* is the action of emptying it if full, and making it full if empty. Suppose further that when the box is empty its walls are completely opaque, but when filled, the pressure on the bottom of the box activates a switch that turns on a light inside the box so that its walls become transparent and it becomes visible that it is full. In this case, when it is empty nothing is observed, $Obs(s_0) = \top$, but when full, it is observed to be full, $Obs(s_1) = p$.

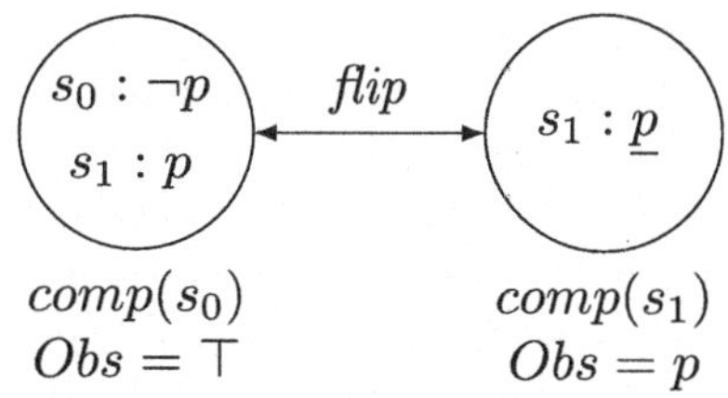

Fig. 11.5 The compatibility domain $\mathcal{D}'$ induced by the box domain $\mathcal{D}$ of Example 11.7

A learner that is initially unfamiliar with the electronics inside the box and has only been exposed to the state where it is empty and hence opaque, will of course both consider it possible that it is empty and that it is full. So there should be no explicit knowledge about p in that state. However, if the learner has also been exposed to the state where it is full, and has observed p in that state, the learner should be able to conclude that when nothing is observed, p is false (the learner is supposed to be aware that both the transition and observation functions are deterministic). So it should be possible for a learner to come to implicitly know the full dynamics of the domain.

Let us try to make these things a bit more formal. Consider first the induced compatibility domain $\mathcal{D}' = ((S', A, T', s_0'), \Omega, Obs')$ of $\mathcal{D}$, shown in Fig. 11.5. Now note that both states of the compatibility domain contain the state s_1. A learner having identified $\mathcal{D}'$ from its interactions with the domain will then conclude that two distinct possible observations are possible in the state s_1: the empty observation $\top$ (from the occurrence of the state s_1 in $comp(s_0)$) and the observation p (from the occurrence of the state s_1 in $comp(s_1)$). Assuming the agent knows the observation function to be deterministic, this is clearly a contradiction. A learner should be able to use the additional knowledge of the observation function being deterministic to refine its representation of the domain.

We define an *observation determinisation* of a compatibility domain to be any domain $\mathcal{D}''$ defined as follows. The domain of $\mathcal{D}''$ contains for each $x \in S'$ a state $g(x) \subseteq x$ such that the set of states $\{g(x) \mid x \in S'\}$ form a partition of S. The relations and functions on $\mathcal{D}''$ are then inherited from $\mathcal{D}'$ in the canonical way. The only way $\mathcal{D}''$ differs from $\mathcal{D}'$ is that we have removed elements of the states of $\mathcal{D}'$ in order to ensure that each original state $s \in S$ appears in exactly one state of $\mathcal{D}''$. Since each pair of states of $\mathcal{D}'$, and hence $\mathcal{D}''$, have distinct observations, this ensures that $\mathcal{D}''$ only specifies a single observation for each original state $s \in S$. Considering the domain above as an example, there only exists a single observation determinisation given by $g(comp(s_0)) = \{s_0\}$ and $g(comp(s_1)) = \{s_1\}$. The corresponding domain is shown in Fig. 11.6. We can now of course also define a notion of behavioural

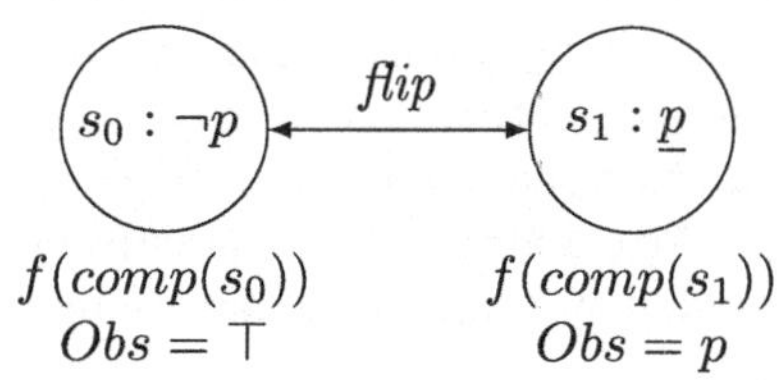

Fig. 11.6 The unique observation determinisation of the compatibility domain $\mathcal{D}'$ shown Fig. 11.5

correctness with respect to a observation determinisation, and use that to define a notion of implicit knowledge. We will not introduce the technical details here, but just mention that it would be possible to learn event models with two relations R_e and R_i, corresponding to two modalities, K_e for explicit knowledge and K_i for implicit knowledge. We could require that explicit knowledge is behaviourally correct with respect to the compatibility domain, and implicit knowledge with respect to its observation determinisation. In that case, a learner would be able to express facts such as:

$$comp(s_0) \models K_i \neg p \wedge \neg K_e \neg p$$

expressing that in the initial state it is implicitly known that the box is empty, but it is not explicitly known. We previously gave examples of the usefulness of being able to express both kinds of knowledge, e.g., in multi-agent settings.

The domain considered here is of course very simple. It only has a single observation determinisation, and that determinisation identifies the full underlying transition system. In general, domains can have many observation determinisations, and then we can not expect there to be a single domain to define behavioural correctness with respect to. Behavioural correctness would then have to be defined in terms of behavioural equivalence with respect to a single or all observation determinisations. We leave these consideration for the next section focusing on learning algorithms for implicit knowledge.

The example above considered a notion of refinement of compatibility domains based on knowing that the observation function is deterministic. We will now give an example of a possible way to refine compatibility domains based on knowing that the transition function is deterministic.

Example 11.8 Consider again the earlier sketched domain over $P = \{p, q\}$ with a single action a that produces the following sequence of observations: $q, q, \neg q, q, q, \neg q, \ldots$. For instance, it could be a domain where a is an action of knocking on a door that is initially closed (q), and only when knocking twice will it open ($\neg q$). Knocking once when open then closes it again. Formally, it could be represented as a domain $\mathcal{D} = ((S, A, T, s_0), Obs, \Omega)$ with $S = \{s_0, s_1, s_2\}$, $s_0 = \{p, q\}$, $s_1 = \{q\}$, $s_2 = \emptyset$, $A = \{a\}$, $T = \{(s_0, a, s_1), (s_1, a, s_2), (s_2, a, s_0)\}$ and $Obs(s_0) = Obs(s_1) = q$ and $Obs(s_2) = \neg q$. Note that, as mentioned earlier, the truth value of q is always observed, and the truth value of p is never observed. The proposition p being true encodes that the door hasn't been knocked at since it was last closed. The domain $\mathcal{D}$ is presented in Fig. 11.7 and its compatibility domain in Fig. 11.8. Note that the compatibility domain is non-deterministic, since the state $comp(s_0)$ has two outgoing a-edges. If the learner knows the underlying domain to be deterministic, it should be able to refine $\mathcal{D}'$ into a deterministic domain. In this case, we cannot refine the compatibility domain by simply modifying its existing states. There are simply too few (compatibility) states to make the transition function deterministic.

We could probably alternatively define ways to "unfold" compatibility domains. Note that in this particular example, the compatibility domain is obtained simply

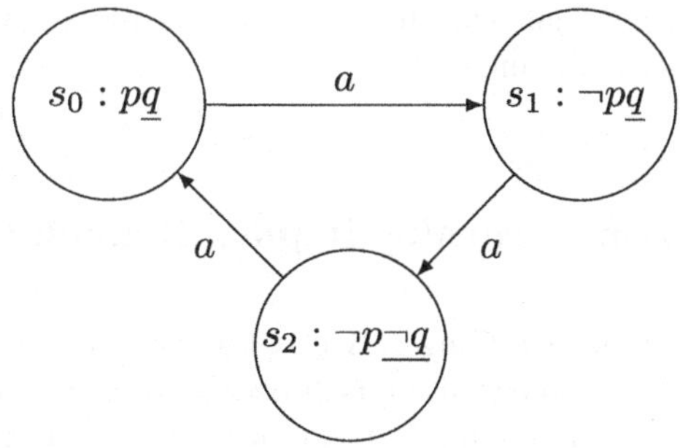

Fig. 11.7 The door knocking domain $\mathcal{D}$ of Example 11.8

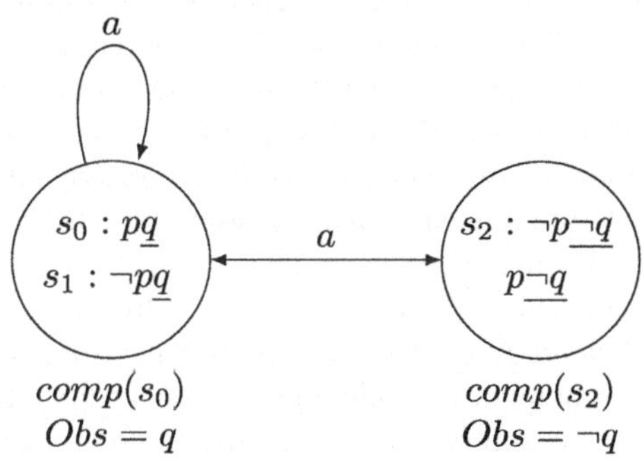

Fig. 11.8 The compatibility domain $\mathcal{D}'$ induced by the door knocking domain $\mathcal{D}$ shown Fig. 11.7. Note that $comp(s_2)$ contains the state $\{p\}$ (denoted as $p\underline{\neg q}$), which is not contained in the state space of the domain $\mathcal{D}$

by identifying the two upper states of Fig. 11.7 (but in general the compatibility domain is not defined as a simple quotient in this way, as the compatibility function doesn't always induce an equivalence relation on states as the previous example showed). As for observation determinisation, we would in general then get many distinct ways of unfolding a compatibility domain. In this example, one unfolding would produce the real domain, and another would produce the one with p swapped by $\neg p$ everywhere. We are not going to pursue the technical details of defining such unfoldings in this paper. To define them in detail, we would need to take into account that whether an unfolding is consistent with the observations received is not only a matter of which individual transitions have been observed, but also the order of transitions. Unfolding the domain of Fig. 11.8 can clearly be done in many different ways, also in ways in which the frequency of observing $\neg q$ is different, e.g. only observing $\neg q$ every fourth time (though that would require a bigger language P). Observing streams of action executions is here needed to determine which domain unfoldings are consistent with the actual underlying domain. Furthermore, to produce all deterministic domains consistent with a given compatibility domain, we would need to consider both observation determinisations (as in the previous example) and unfoldings, potentially even interleaved. Defining algorithms for producing these domains, and then proving them to have the expected properties, is probably possible, but also non-trivial. Instead, we are in the next section going to define a new algorithm for learning implicit knowledge that in a more direct way builds all the domains consistent with the observed transitions, also taking the order of action execution into account.

11.6 Learning Implicit Domain Knowledge

At the beginning of Sect. 11.4, we informally described explicit knowledge as what is known because it is directly observed in the current state, and implicit knowledge as whatever might additionally be inferred from the history of earlier actions and from the general experience with the domain. To formalise the notion of explicit knowledge, we introduced the notion of a *compatibility domain*: a domain that captures what is explicitly knowable about the real domain, and whose states are sets of states over the alphabet of the domain. A learner was then deemed behaviourally correct with respect to explicit knowledge if the learner outputs a set of event models whose induced domain is isomorphic to the compatibility domain.

In this section, we proceed in a similar way with respect to implicit domain knowledge. To capture it, we introduce the notion of a *behavioural equivalence domain*. A learner is deemed *behaviourally correct with respect to implicit knowledge* if it outputs a domain that is isomorphic to the behavioural equivalence domain, and not, as before, to the compatibility domain.

11.6.1 Implicit Knowledge and Behavioural Equivalence

To formalise the notion that implicit knowledge is whatever can be inferred from a history of earlier actions and general experience with the domain, we resort to the concept of *behavioural equivalence*. Many different notions of *behavioural equivalence* have been proposed in the literature on labelled transition systems (Gorrieri & Versari, 2015, Chap. 2). Behavioural equivalence relations seek to establish in which cases two transition systems offer similar interaction capabilities. The intuition is that two systems should be equivalent if they cannot be distinguished by interacting with them. If the initial states of two systems are behaviourally equivalent, then they cannot be distinguished by any experience gathered from these initial states, through any sequence of actions.

In this section, we introduce notions of behavioural equivalence for domains, based on what the agent can observe about the underlying transition system. Two domains will be deemed equivalent if they cannot be distinguished, *from observations*, by interacting with them. In other words, we will formalise implicit knowledge as knowledge up to behavioural equivalence. This is in fact as much knowledge as a learner can possibly acquire. Learning proceeds by making "experiments" with, or "testing" the domain, i.e. by taking certain sequences of actions and observing the results. But, in two behaviourally equivalent domains, any such experiment yields the same observations. As a result, the agent can only come to know with certainty the information that holds in every domain that's equivalent to the real one.

To define the behavioural equivalence relations used in this section, we need to fix some notions. Let $\mathcal{D} = ((S, A, T, s_0), \Omega, Obs)$ be a deterministic domain over P. Let A^* denote the set of all finite sequences of elements of A. An element of A^* is called an action sequence or *action trace* (Gorrieri & Versari, 2015). Each such trace defines an interaction with the system, e.g., the trace $(move, move, flip, move)$ is an interaction in the light switch domain of Example 11.3. We denote by $\mathrm{Tr}(\mathcal{D})$ the set of traces in $\mathcal{D}$. The *execution trace of* $(a_i)_{0\leq i\leq n}$ *from* s is the sequence $(t_0, a_0, t_1, a_1, \ldots, t_n, a_n, t_{n+1})$ where $t_0 = s$ and $(t_i, a_i, t_{i+1}) \in T$ for all $0 \leq i \leq n$. That is, an execution trace is an alternating sequence of states and actions ending with a state. We denote the set of executions traces from s by $\mathrm{ExTr}(\mathcal{D}, s) := \{\varepsilon \mid \varepsilon \text{ is the execution trace from} s \text{of some} \alpha \in \mathrm{Tr}(\mathcal{D})\}$. The *observation trace for* an execution trace

$$(s_0, a_0, s_1, \ldots, a_n, s_{n+1})$$

is the sequence

$$(Obs(s_0), a_0, Obs(s_1), \ldots, a_n, Obs(s_{n+1})).$$

We denote the set of observation traces from s by

$$\mathrm{ObsTr}(\mathcal{D}, s) := \{\tau \mid \tau \text{ is the observation trace for } \varepsilon, \text{ for some } \varepsilon \in \mathrm{ExTr}(\mathcal{D}, s)\}$$

The first behavioural equivalence we discuss is *trace equivalence*, based on equivalence of traces up to what can be observed.

Definition 11.10 Let $\mathcal{D} = ((S, A, T, s_0), \Omega, Obs)$ and $\mathcal{D}' = ((S', A', T', s_0'), \Omega', Obs')$ be two deterministic domains over P. Two states $s \in S$ and $s' \in S'$ are called *(observationally) trace equivalent* if $\mathrm{ObsTr}(\mathcal{D}, s) = \mathrm{ObsTr}(\mathcal{D}', s')$. $\mathcal{D}$ and $\mathcal{D}'$ are called *(observationally) trace equivalent* if their initial states s_0 and s_0' are observationally trace equivalent.

For transition systems, the classical alternative to trace equivalence is bisimilarity. We introduce a version of bisimulation for domains which relates two states s and t when s and t are observationally indistinguishable, and such that if the same action is executed in both states, the resulting states are again observationally indistinguishable.

Definition 11.11 Let $\mathcal{D} = ((S, A, T, s_0), \Omega, Obs)$ and $\mathcal{D}' = ((S', A, T', s_0'), \Omega, Obs')$ be two deterministic domains over P. A relation $Z \subseteq S \times S'$ is called a *bisimulation* between S and S' if for every $s \in S$ and $s' \in S'$, the following conditions hold:

Observational indistinguishability: if sZs', then $Obs(s) = Obs'(s)$.
Forth: if sZs' and $(s, a, t) \in T$, then there exists a $t' \in S'$ s.t. $(s', a, t') \in T'$ and tZt'.
Back: if sZs' and $(s', a, t') \in T'$, then there exists a $t \in S$ s.t. $(s, a, t) \in T$ and tZt'.

Two states $s \in S$ and $s' \in S$ are called *(observationally) bisimilar*, denoted $s \underline{\leftrightarrow} s'$, if there is a bisimulation Z such that sZs'. The domains $\mathcal{D}$ and $\mathcal{D}'$ are called *(observationally) bisimilar*, denoted $\mathcal{D} \underline{\leftrightarrow} \mathcal{D}'$, if their starting states are observationally bisimilar, i.e. $s_0 \underline{\leftrightarrow} s_0'$.

For classical, deterministic transition systems, trace equivalence and bisimilarity coincide (Gorrieri & Versari, 2015), trace equivalence simply means that the same action traces are possible (the same action sequences are applicable). In our case, for deterministic domains, trace equivalence and bisimilarity are based on a different notion of equivalence under observability. However, we still get that (observational) trace equivalence and (observational) bisimilarity coincide.

Lemma 11.1 *Two deterministic domains $\mathcal{D}$ and $\mathcal{D}'$ over P are trace equivalent iff they are bisimilar.*

Proof Let $\mathcal{D} = ((S, A, T, s_0), \Omega, Obs)$ and $\mathcal{D}' = ((S', A, T', s_0'), \Omega, Obs')$ be two deterministic domains over P.

($\Rightarrow$) Suppose that $\mathcal{D}$ and $\mathcal{D}$ are trace equivalent. Then $\text{ObsTr}(\mathcal{D}, s_0) = \text{ObsTr}(\mathcal{D}', s_0')$. Define a relation Z by sZs' iff $\text{ObsTr}(\mathcal{D}, s) = \text{ObsTr}(\mathcal{D}', s')$. Thus, Z relates s_0 and s_0' by definition. To show that Z is a bisimulation, we show conditions (i)–(iii) for bisimulations. Suppose that sZs'. For condition (i), take any observation trace $(o, \dots) \in ObsTr(D, s)$. Then $Obs(s) = o$. Since sZs', we have $ObsTr(D, s) = ObsTr(D', s')$ and hence $(o, \dots) \in ObsTr(D', s')$. This implies $Obs'(s') = o$, and hence $Obs'(s') = o = Obs(s)$, as required. For condition (ii), suppose that sZs' and $(s, a, t) \in T$. Since $(s, a, t) \in T$, we get $(s, a, t, \dots) \in \text{ExTr}(\mathcal{D}, s)$. Hence, $(Obs(s), a, Obs(t), \dots) \in \text{ObsTr}(\mathcal{D}, s)$, and since $\text{ObsTr}(\mathcal{D}, s) = \text{ObsTr}(\mathcal{D}', s')$, we get $(Obs'(s'), a, Obs'(t'), \dots) \in \text{ObsTr}(\mathcal{D}', s')$, for some $t' \in S$. Since $Obs(s) = Obs'(s')$, $Obs(t) = Obs'(t')$ and $\text{ObsTr}(\mathcal{D}, s) = \text{ObsTr}(\mathcal{D}', s')$, we get

$$
\begin{aligned}
\text{ObsTr}(\mathcal{D}, t) &= \{(Obs(t), \dots) \mid (Obs(s), a, Obs(t), \dots) \in \text{ObsTr}(\mathcal{D}, s)\} \\
&= \{(Obs'(t'), \dots) \mid (Obs'(s'), a, Obs'(t'), \dots) \in \text{ObsTr}(\mathcal{D}', s')\} \\
&= \text{ObsTr}(\mathcal{D}', t').
\end{aligned}
$$

Condition (iii) is symmetric.

($\Leftarrow$) Suppose that $\mathcal{D} \underline{\leftrightarrow} \mathcal{D}'$, i.e. there is a bisimulation Z such that $s_0 Z s_0'$. To show $\text{ObsTr}(\mathcal{D}, s_0) = \text{ObsTr}(\mathcal{D}', s_0')$, we prove inclusion both ways. Let $(Obs(s_0), a_0, Obs(s_1), \dots, a_n, Obs(s_{n+1})) \in \text{ObsTr}(\mathcal{D}, s_0)$. Then $(s_0, a_0, s_1, \dots, a_n, s_{n+1}) \in \text{ExTr}(\mathcal{D}, s_0)$. By the forth condition, there are $s_1', \dots, s_{n+1}' \in S'$ such that $(s_i', a_i, s_{i+1}') \in T'$, $s_i Z s_i'$, for all $0 \leq i \leq n$. Since $s_i Z s_i'$, $Obs(s_i) = Obs'(s_i')$ for all $0 \leq i \leq n$. Thus, $(Obs(s_0), a_0, Obs(s_1), \dots, a_n, Obs(s_{n+1})) \in \text{ObsTr}(\mathcal{D}', s_0')$. The proof for $\text{ObsTr}(\mathcal{D}', s_0') \subseteq \text{ObsTr}(\mathcal{D}, s_0)$ is analogous to the one just given, except for using the back condition instead of the forth one. □

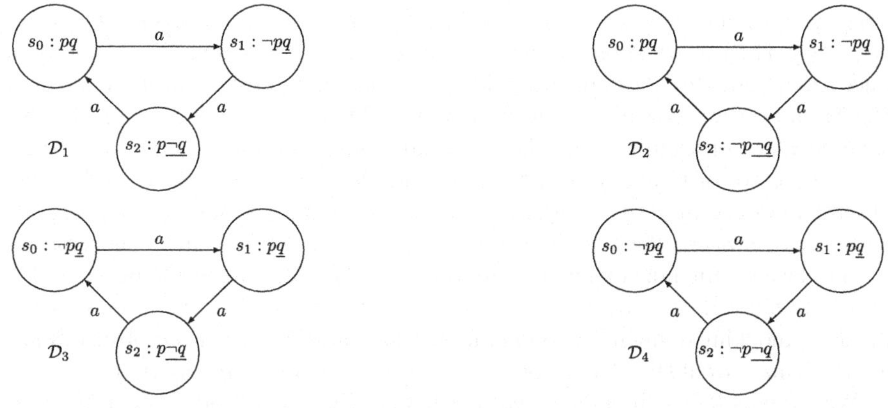

Fig. 11.9 The domains $\mathcal{D}_1, \ldots, \mathcal{D}_4$ over $P = \{p, q\}$ that are bisimilar to the door knocking domain of Example 11.8. The real domain is denoted here as $\mathcal{D}_2$

Example 11.9 There are four domains that are bisimilar, i.e. trace equivalent, to the door knocking domain of Example 11.8. Figure 11.9 shows them, where the real domain is $\mathcal{D}_2$. The two top domains, $\mathcal{D}_1$ and $\mathcal{D}_2$, differ from the two bottom ones, $\mathcal{D}_3$ and $\mathcal{D}_4$, in the initial state. The left domains differ from the right ones in the bottom state. To see that these are in fact the only bisimilar or trace equivalent domains, recall first that, by definition, domains over a set of propositional atoms P have a state space $S \subseteq 2^P$. This means that any domain over P that is trace equivalent to the original one will have at most four states. (In general, since the set of actions in a domain $\mathcal{D}$ over P is finite, and the set of states is finite as well, the set of domains bisimilar to $\mathcal{D}$ is finite).

Note that a domain with a single state s_0^* cannot be trace equivalent to the original domain $\mathcal{D}_2$, since it must be able to produce the observation trace $(q, a, q, a, \neg q)$, which would require that both q and $\neg q$ are observable in s_0^*. Now consider any domain with two states, s_0^* and s_1^*, and suppose it was bisimilar to the real one. By the observational indistinguishability condition of bisimulations, and the fact that in the initial state of the door knocking q is observed as true, s_0^* must also be a state in which q is observed as true. Since the domain must be able to produce the observation trace $(q, a, q, a, \neg q)$, there must be a state in which $\neg q$ is observed. This must be s_1^*. In order to produce the observation trace (q, a, q), this domain would need to have (s_0^*, a, s_0^*) in its transition function. But then, in order to produce the observation trace $(q, a, q, a, \neg q)$, the domain would need to have (s_0^*, a, s_1^*) in its transition function. The transition function would then be non-deterministic. But bisimilarity is only defined for deterministic domains over the same set of propositional variables P.

Next, consider domains with three states, s_0^*, s_1^* and s_2^*. It is straightforward to check that the domains in Fig. 11.9 are bisimilar to the real domain; the relation linking s_i to s_i^* is a bisimulation. Now take any three-state domain not listed in Fig. 11.9 and suppose it is bisimilar to the real domain. In order to produce the observation trace (q, a, q), the a-successor of s_0^* must be a q-state. Suppose the a-successor of s_0^* is s_0^* itself. Then, as with domains with two states, in order to produce the observation trace $(q, a, q, a, \neg q)$, the domain would need a non-deterministic transition function. Hence, the a-successor of s_0^* must be a q-state different from s_0^*. W.l.o.g., let this state be s_1^*. In order to produce the observation trace $(q, a, q, a, \neg q)$, s_2^* must be the a-successor of s_1^* and it must be a $\neg q$-state. Moreover, in order to produce the observation trace $(q, a, q, a, \neg q, a, q)$, the a-successor of s_2^* must be a q-state. If the successor was s_0^*, the domain would be listed in Fig. 11.9. So the a-successor of s_2^* must be a s_1^*. But note that then, in this domain, the following would be a possible observation trace: $(q, a, q, a, \neg q, a, q, a, \neg q)$. This trace is however impossible in the door knocking domain; $\neg q$ observations occur always after *two* q observations. Such a domain would break the repeated observation pattern $q, q, \neg q, q, q, \neg q, \ldots$[2]

We are thus left with domains with four states: s_0^*, s_1^*, s_2^* and s_3^*. Reasoning as before, we can show that the a-successor of s_0^* must be a q-state different from s_0^*. W.l.o.g., let's say that this state is s_1^*. In order to produce the observation trace $(q, a, q, a, \neg q)$, the a-successor of s_1^* must be a $\neg q$-state. W.l.o.g., let this state be s_2^*. Again, in order to produce the observation trace $(q, a, q, a, \neg q, a, q)$, the a-successor of s_2^* must be a q-state. If this successor is s_1^*, we get a non-bisimilar domain. If this successor is s_0^*, then s_3^* is not reachable from s_0^*, which cannot be, by definition of domains. And if this successor is s_4^*, then the domain produces the observation trace $(q, a, q, a, \neg q, a, \neg q)$, which is not a possible observation trace for the real domain.

11.6.2 Behavioural Equivalence Domain

We now have all prerequisite notions to define the behavioural equivalence domain. Before defining the domain formally, we provide some intuition about it. Consider an agent situated in the initial state s_0 of a domain $\mathcal{D}$. Since the agent only has access to observations of states, it cannot distinguish s_0 from the initial state s'_0 of any other bisimilar domain $\mathcal{D}'$. The agent therefore considers s'_0 as a possible world. That is, the agent's initial uncertainty can be represented by the set W_0 of initial states from all domains that are bisimilar to $\mathcal{D}$. Now, the agent is unable to tell, just from its observations, whether it is interacting with the real domain, or a bisimilar alternative. Thus, after executing any action a in s_0, the agent won't be able to distinguish the a-successor s_1 of s_0, from the a-successor s'_1 of the bisimilar state s'_0. That is, after a is executed, the agent's new state of uncertainty can be represented by the set of a-successors of the states in W_0. The definition of the behavioural equivalence domain follows this intuition.

[2] Note that an observation trace re-visiting states is needed to notice that the domain under consideration is not bisimilar to the real one. Later on, in this section, we will bound the length of observation traces required to tell with certainty whether a domain is bisimilar or not to the one being explored.

Definition 11.12 Let $\mathbb{D} = \{((S^i, A, T^i, s_0^i), \Omega, Obs^i) \mid 1 \leq i \leq n\}$ be a finite set of domains, all bisimilar. The *synchronous composition* of (the domains in) $\mathbb{D}$ is the domain $\mathcal{D}' = ((S', A, T', s_0'), \Omega, Obs')$ given by

- $(s_0^i)_{1 \leq i \leq n} \in S'$ and for all $a \in A$: if $(s^i)_{1 \leq i \leq n} \in S'$ then $T^i(s^i, a)_{1 \leq i \leq n} \in S'$.
- $T'((s^i)_{1 \leq i \leq n}, a) = (T^i(s^i, a))_{1 \leq i \leq n}$.
- $s_0' = (s_0^i)_{1 \leq i \leq n} \in S'$.
- $Obs'((s^i)_{1 \leq i \leq n}) = Obs^1(s^1)$.

When $\mathbb{D}$ is the set of all domains bisimilar to a given domain $\mathcal{D}$, we call $\mathcal{D}'$ the *behavioural equivalence domain* induced by $\mathcal{D}$.

Our notion of synchronous composition of bisimilar domains is the natural counterpart of the notion of synchronous composition of finite automata (Cassandras & Lafortune, 2009): To get from the synchronous compositions of automata to domains, we just need to take the observation functions into account and restrict to the states reachable by synchronous traces from the initial state. We refer to the states of the behavioural equivalence domain as *global states* and to its transitions as *global transitions*. The state space of the behavioural equivalence domain is built recursively. The tuple $(s_0^i)_{1 \leq i \leq n}$, which has the initial states of all bisimilar domains as components, is the initial global state. Then for each global state $(s^i)_{1 \leq i \leq n} \in S'$ and each action $a \in A$, we add to the set of global states the tuple $T^i(s^i, a)_{1 \leq i \leq n}$, which has as i-th component the a-successor of s^i. The global transition function maps the global state $(s^i)_{1 \leq i \leq n} \in S'$ and the action a to the global state that has as i-th component the a-successor of s^i. The observation function Obs' assigns to the global state $(s^i)_{1 \leq i \leq n} \in S'$ the same as the observation that each Obs^i assigns to its i-th component.[3] Note that the behavioural equivalence may have two non-identical global states $(s^i)_{1 \leq i \leq n} \in S'$ and $(t^i)_{1 \leq i \leq n} \in S'$ with the same underlying sets, i.e. $\{s^i \mid 1 \leq i \leq n\} = \{t^i \mid 1 \leq i \leq n\}$.

We will now use the behavioural equivalence domain to define our notion of implicit knowledge. The definition is completely symmetric to the definition of explicit knowledge, except based on the behavioural equivalence domain instead of the compatibility domain. In a global state of the behavioural equivalence domain, we consider the implicit knowledge of the agent to be what is true in *all* its component states. To make this notion precise in an epistemic setting, we identify each global state $s = (s^i)_{1 \leq i \leq n}$ with an *induced* epistemic model $\mathcal{M}_s = (W, R, V)$ with $W = \{s^i \mid 1 \leq i \leq n\}$, $R = W \times W$ and $V(w) = w$. By this identification, we can evaluate (static) epistemic formulas in global states, e.g. we then have $(s^1, \ldots, s^n) \models K\phi$ iff ϕ is true in every world $s^1, \ldots, s^n$. The modal operator K is here interpreted as the operator for implicit knowledge. In the following, we will generally identify global states s with their induced epistemic models $\mathcal{M}_s$ without notice. It will be clear from the context whether a global state should be considered as just that or as the induced epistemic model.

---

[3] The observation function is well-defined, as the components of a global state are all bisimilar (else, the forth condition would be violated) and thus all receive the same observation.

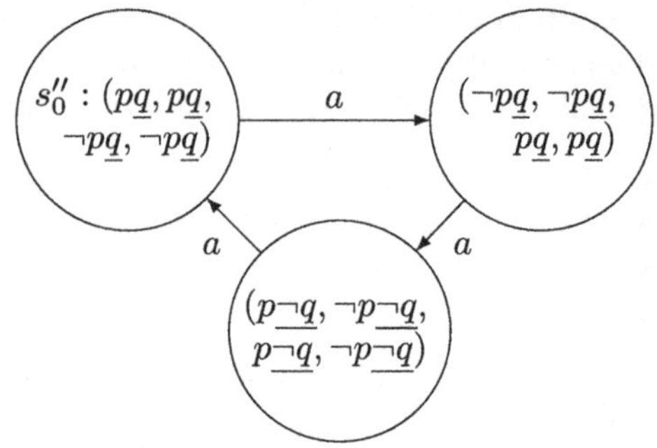

Fig. 11.10 The behavioural equivalence domain $\mathcal{D}''$ induced by the door knocking domain $\mathcal{D}$ of Fig. 11.7

11.6.3 Behavioural Correctness and Learnability

A domain $\mathcal{D}$ that is isomorphic to the behavioural equivalence domain represents all that is implicitly knowable about the domain. A learner that identifies such a domain will be called *behaviourally correct with respect to implicit knowledge.*

Definition 11.13 Let $\mathcal{D} = ((S, A, T, s_0), \Omega, Obs)$ and $\mathcal{D}' = ((S', A, T', s_0'), \Omega', Obs')$ be domains. We say that $\mathcal{D}'$ is *behaviourally correct with respect to implicit knowledge* about $\mathcal{D}$ if $\mathcal{D}'$ is isomorphic to the behavioural equivalence domain induced by $\mathcal{D}$. A collection of event models $(\mathcal{E}_a)_{a \in A}$ is *behaviourally correct with respect to implicit knowledge* about $\mathcal{D}$ if the domain induced by $(\mathcal{E}_a)_{a \in A}$ and the initial state of the behavioural equivalence domain of $\mathcal{D}$ is behaviourally correct with respect to implicit knowledge about $\mathcal{D}$.

Example 11.10 Figure 11.10 depicts the behavioural equivalence domain $\mathcal{D}''$ induced by the door knocking domain $\mathcal{D}$ from Example 11.8. Example 11.9 proves that there are 4 domains bisimilar to $\mathcal{D}$ (including $\mathcal{D}$ itself), hence the states of $\mathcal{D}''$ are quadruples of states over $P = \{p, q\}$ (the 4 bisimilar domains were provided in Fig. 11.9). Hence, more precisely, the states of $\mathcal{D}''$ are quadruples $(s^1, \ldots, s^4)$ where $s^i \in \mathcal{D}_i$, $i = 1, \ldots, 4$, using the domain names introduced in Fig. 11.9. In particular, the initial state s_0'' of $\mathcal{D}''$ is the quadruple of the initial states of the 4 domains $\mathcal{D}_1, \ldots, \mathcal{D}_4$, i.e., $s_0'' = (pq, pq, \neg pq, \neg pq)$. Note that $\mathcal{D}''$ is isomorphic to the original domain $\mathcal{D}$. This is different from the situation with the compatibility domain $\mathcal{D}'$ of $\mathcal{D}$, shown in Fig. 11.8. The compatibility domain $\mathcal{D}'$ only has two states and represents a as a non-deterministic action, hence is not isomorphic to $\mathcal{D}$. In other words, for this particular example, the behavioural equivalence domain gives a perfect representation of the original domain, whereas the compatibility domain does not.

As mentioned earlier, the compatibility domain captures explicit knowledge and the behavioural equivalence domain captures implicit knowledge. In the door knocking domain, the explicit knowledge is very limited. Consider a learner that learns an event model $\mathcal{E}_a$ that is behaviourally correct with respect to explicit knowledge about $\mathcal{D}$. This means that the domain induced by $\mathcal{E}_a$ and $s_0 = pq$ is isomorphic to the compatibility domain $\mathcal{D}'$. Hence, consulting Fig. 11.8, we would get $comp(s_0) \models Kq \wedge \neg K[\mathcal{E}_a]q \wedge \neg K[\mathcal{E}_a]\neg q$. So, initially the agent explicitly knows q, but doesn't explicitly know whether knocking on the door will lead to another q-state or to a $\neg q$-state.

Since the behavioural equivalence domain is isomorphic to the true domain, this confusion would not arise in the implicit knowledge of the agent. Consider an event model $\mathcal{E}'_a$ that is behaviourally correct with respect to implicit knowledge about $\mathcal{D}$. Then the domain induced by $\mathcal{E}'_a$ and $s_0 = pq$ is isomorphic to the behavioural equivalence domain $\mathcal{D}''$ of $\mathcal{D}$. Consulting Fig. 11.10, we can then conclude $s''_0 \models Kq \wedge K[\mathcal{E}'_a]q \wedge K[\mathcal{E}'_a][\mathcal{E}'_a]\neg q$. In other words, the agent implicitly knows q, and implicitly knows that knocking once doesn't change the truth value of q, but knocking twice does.

We mentioned earlier that an agent would never be able to learn exactly how the knocking action affects p. However, it *is* able to learn the possible ways it *might* affect p and still be consistent with all observation traces. This is encoded in the behavioural equivalence domain. Recall that, intuitively, p is needed to keep track of how many times the door has been knocked at since it was last closed. Of course it is only relevant to distinguish between whether it was knocked at zero times or once since last closed, as after two knocks it is open again. This implies that we can use p to encode the necessary information in four different ways: (1) p becomes false when the door has been knocked at once since last closed; (2) p becomes true when the door hasn't been knocked at since last closed; (3) p becomes false when the door hasn't been knocked at since last closed; (4) p becomes true when the door has been knocked at once since last closed. Consulting again Fig. 11.10, we can see that these four options are exactly the ones encoded by each of the elements of the quadruples: the first elements encode option (1), the second elements encode option (2), etc. In other words, the behavioural equivalence domain encodes exactly all the possible underlying dynamics of the domain (of which there are 4, corresponding to $\mathcal{D}_1, \ldots, \mathcal{D}_4$). Furthermore, it represents these compactly as a single domain, by identifying the dynamics that are not distinguishable by any observation traces (how exactly p encodes the book-keeping of number of knocks at the door, we can never observe).

We can now also explain why the states of the behavioural equivalence domain are represented as tuples instead of sets of states. Consider what would happen if we replaced each of the tuples in Fig. 11.10 by the set of elements contained in the tuple. Then both of the two upper states would become the set $\{pq, \neg pq\}$. In other words, these two states would become identified, and we would get back to the compatibility domain! Tuples are needed to be able to keep track of which trace we are currently following, and hence to be able to distinguish states that are observationally indistinguishable, but not observationally trace equivalent.

In the next subsection, we will present a learning algorithm for implicit knowledge and prove that it produces a domain that is isomorphic to the behavioural equivalence domain, when presented with a certain set of observation traces for the initial state of the real domain. But before looking at this specific learner, let us discuss the issue of whether behavioural correctness with respect to implicit knowledge is achievable, in general, when presented with a *finite* set of observation traces for the initial state of the real domain. Note that the set of observation traces for an initial state s_0 may be infinite. For example, an unknown domain with a single state s_0 and a single 'loopy' action a, i.e. $(s_0, a, s_0) \in T$ induces infinitely many observation traces. In fact, since we assume universal applicability of actions, *every* domain has an infinite set of observation traces. However, does the learner *need* to see all possible observation traces to achieve behavioural correctness? As we will soon show, the answer is no. A learner that wants to achieve behavioural correctness, i.e. to produce a domain that is isomorphic to the behavioural equivalence domain of the real domain, can do so in finite time, from a finite set of observation traces. Borrowing some notions from formal learning theory, we can then say that the behavioural equivalence domain is *finitely identifiable* (Lange & Zeugmann, 1992; Mukouchi, 1992): the agent can conclusively learn it in finite time (up to isomorphism), given an appropriate set of observation traces. This is in contrast with *identifiability in the limit* (Gold, 1967), which holds for any learning process in which the learner converges to the right solution after seeing finitely many inputs, but does not necessarily know at which point convergence takes place. In formal learning theory, a set that enables finite identification is called a *definite finite tell-tale set* (DFTT for short, see Mukouchi, 1992; Gierasimczuk and de Jongh, 2010, 2012, for an application in action learning see Bolander and Gierasimczuk, 2017). We adapt this notion to our setting.

Definition 11.14 Let $\mathcal{D} = ((S, A, T, s_0), \Omega, Obs)$ be a domain over P. A set $\Sigma \subseteq \text{ObsTr}(\mathcal{D}, s_0)$ is called a *definite finite tell-tale (DFTT) for* $\mathcal{D}$ if

(i) Σ is finite;
(ii) for any domain $\mathcal{D}' = ((S', A, T', s_0'), \Omega, Obs')$ over P, if $\Sigma \subseteq \text{ObsTr}(\mathcal{D}', s_0')$ then $\mathcal{D} \underline{\leftrightarrow} \mathcal{D}'$.

As we show next, every domain $\mathcal{D}$ over P has a DFTT, and such a DFTT is sufficient for learning a behaviourally correct domain with respect to $\mathcal{D}$ in finite time.

Proposition 11.1 *Every domain over P has a DFTT.*

Proof Let $\mathcal{D} = ((S, A, T, s_0), \Omega, Obs)$ be a domain over P. Note that in condition (ii) for DFTTs, we quantify over domains over P with the same set of actions and observations as $\mathcal{D}$, as bisimilarity is defined for domains with the same sets of actions and observations. As P, A and Ω are finite, this set of domains is finite. Let $\mathcal{D}_1, \ldots, \mathcal{D}_n$ be an enumeration of all domains over P with actions A and observations Ω, and for each $\mathcal{D}_i$, define a function $Q_i : A^* \to \{0, 1\}$ by

$$Q_i(x) = \begin{cases} 1 & \text{if } \mathcal{D} \text{ and } \mathcal{D}_i \text{ produce the same observation trace based on } x \text{ from their initial states} \\ 0 & \text{otherwise} \end{cases}$$

Every domain $\mathcal{D}_i$ which is not bisimilar to $\mathcal{D}$ is also not trace equivalent to it (Lemma 11.1). Hence, the function Q_i outputs 0 at least on one of its inputs. For each $1 \leq i \leq n$ for which Q_i outputs 0 at least once, pick one x_i with $Q_i(x_i) = 0$, and let τ_i be the observation trace produced by $\mathcal{D}$ based on x_i. Define the set Σ consisting of all such τ_i. We claim that Σ is a DFTT. For condition (i): Σ is clearly finite, since the τ_i range over $1 \leq i < n$. For condition (ii), take a domain $\mathcal{D}'$ and suppose that $F \subseteq ObsTr(\mathcal{D}', s_0')$. For a contradiction, suppose that $\mathcal{D}'$ is not bisimilar to $\mathcal{D}$. $\mathcal{D}' = \mathcal{D}_i$, for some $i \in \{1, \ldots, n\}$. Since $\mathcal{D}_i$ is not bisimilar to $\mathcal{D}$, by construction of Σ, $\tau_i \notin ObsTr(\mathcal{D}_i, s_0')$. Hence $\Sigma \not\subseteq ObsTr(\mathcal{D}_i, s_0')$, which gives a contradiction. □

Proposition 11.1 shows that a DFTT exists for each domain $\mathcal{D}$ over P. But we can in fact do better and *bound* the length of the observation traces in a DFTT.

Proposition 11.2 *Every domain over P has a DFTT consisting of all observation traces that have $2^{2|P|}$ actions.*

Proof Let $\mathcal{D}_1 = ((S^1, A, T^1, s_0^1), \Omega, Obs^1)$ be a domain over P. Note that in condition (ii) for DFTTs, we quantify over domains over P with the same set of actions and observations as $\mathcal{D}_1$, as bisimilarity is defined for domains with the same sets of actions and observations. For each domain over P of the form $\mathcal{D}_2 = ((S^2, A, T^2, s_0^2), \Omega, Obs^2)$, we can construct a *product* domain $\mathcal{D} = ((S, A, T, s_0), \Omega, Obs)$ where $S = S^1 \times S^2$, for all $(s, s') \in S$ and $a \in A$, $T((s, s'), a) = (T^1(s, a), T^2(s', a))$, $Obs((s, s')) = (Obs(s), Obs(s'))$ and $s_0 = (s_0^1, s_0^2)$. If $\mathcal{D}_1$ and $\mathcal{D}_2$ are trace inequivalent, there will be a trace in $\mathcal{D}$ leading to a state (s, s') with $Obs(s) \neq Obs(s')$. Now note that the number of states in $\mathcal{D}$ is $|S^1| \times |S^2| \leq 2^{|P|} \cdot 2^{|P|} = 2^{2|P|}$. If (s, s') is reachable, then it is reachable by a simple path, i.e. a sequence of alternating states and actions that does not visit the same state twice (Chartrand, 2012, Theorem 6). Hence, we need not consider paths longer than that—if a state (s, s') with $Obs(s) \neq Obs(s')$ is reachable, it will be reachable by a simple path, and hence a path of length at most $2^{2|P|}$. □

Having established the existence of DFTTs of a bounded finite size, we can now show that domains can be identified, up to bisimilarity, in finite time.

Proposition 11.3 *If there is a DFTT for $\mathcal{D}$ over P, then it is possible to learn a domain that is behaviourally correct with respect to implicit knowledge in finite time.*

Proof Let Σ be a DFTT for $\mathcal{D}$.
Claim 1. Σ features all actions and observations in $\mathcal{D}$.
Proof of Claim 1. For contradiction, suppose that there is an action $a \in A$ which does not occur in any observation trace of Σ. Define a domain $\mathcal{D}'$ with $S' = S$, $A' = A$, $s_0 = s_0'$ and $T'(s, a') = T(s, a')$ for all $a \in A - \{a\}$. Let $T(s_0, a) = t$, and choose an observation for t in $\mathcal{D}'$ so that $Obs'(t) \neq Obs(T(s_0, a))$. Then $\mathcal{D}$ and $\mathcal{D}'$ have different observation traces, and thus are not bisimilar, but $\Sigma \subseteq \mathrm{ObsTr}(\mathcal{D}', s_0')$. This gives contradiction. Now suppose that there is some observation $o \in \Omega$ not featured in Σ. Let $s \in S$ be some state with $Obs(s) = o$. Define $\mathcal{D}'$ as $\mathcal{D}$ except for

$Obs(s) \neq Obs'(s)$. Since o does not occur in Σ, $\Sigma \subseteq \text{ObsTr}(\mathcal{D}', s_0')$, which gives contradiction. This completes the proof of Claim 1.

From Claim 1, we know that all actions and observations from $\mathcal{D}$ appear in Σ. Let $\mathcal{D}_1, \ldots, \mathcal{D}_n$ be an enumeration of all domains over P with actions A and observations Ω. For each domain $\mathcal{D}'$ over P, check if $\Sigma \subseteq \{\tau \in \text{ObsTr}(\mathcal{D}', s_0') \mid \tau$ has length at most $2^{2|P|}\}$. Any domain passing this check is, by definition, bisimilar to $\mathcal{D}$. As the set of domains over P with actions A and observations Ω is finite, and Σ is a finite set of finite sequences (Proposition 11.1), this check can be done in finite time. Once we have computed the set $\mathbb{D}$ of domains passing the check, we can construct the synchronous composition of the domains in $\mathbb{D}$, following Definition 11.12. As $\mathbb{D}$ is the set of domains bisimilar to $\mathcal{D}$, and the behavioural equivalence domain induced by $\mathcal{D}$ is simply the synchronous composition of the domains in $\mathbb{D}$, we will have constructed, from $\mathbb{D}$, the behavioural equivalence domain induced by $\mathcal{D}$. □

Using the bound for the size of a DFTT established in Proposition 11.2, we can define a notion of a *sound and complete* set of observation traces for a domain.

Definition 11.15 Let $\mathcal{D} = ((S, A, T, s_0), \Omega, Obs)$. A *sound and complete* set of observation traces Σ for $\mathcal{D}$ is a subset of $\text{ObsTr}(\mathcal{D}, s_0)$ such that every observation trace from s_0 with $2^{2|P|}$ actions is in Σ.

11.6.4 A Behaviourally Correct Learner of Implicit Knowledge

We now present our second learner of the paper and prove that it achieves its goal (producing a behaviourally correct domain) when presented with a sound and complete set of observation traces from the initial state of the real domain. The learner first computes, in an incremental fashion, the set $\mathbb{D}$ of domains bisimilar to the real domain $\mathcal{D}$. It then computes and outputs the synchronous composition of the domains in $\mathbb{D}$, which is, by definition, the same as the behavioural equivalence domain induced by $\mathcal{D}$.

To describe the learner, we need the following additional technical definition. A *history for observation trace* $(o_0, a_0, o_1, \ldots, a_n, o_{n+1})$ is a sequence $h = (\bar{s}_0, o_0, a_1, \bar{s}_1, o_1, \ldots, a_n, s_{n+1}, o_{n+1})$ where $\bar{s}_i \in comp(o_i)$ for $0 \leq i \leq n+1$. Intuitively, h *hypothesises* the state $\bar{s}_i$ that gave rise to the observations o_i. Since observations are non-noisy, it is only relevant to consider hypothesised states that are compatible with the observations.

By choosing such states to explain $(o_0, a_0, o_1, \ldots, a_n, o_{n+1})$, the history h induces a possible domain that could have generated it.

Definition 11.16 Let $h = (\bar{s}_0, o_0, a_1, \bar{s}_1, o_1, \ldots, a_n, s_{n+1}, o_{n+1})$ be a history (for some observation trace). We then define the following notations

- $S^h = \{\bar{s}_i \mid i = 0, \ldots, n+1\}$
- $A^h = \{a_i \mid i = 0, \ldots, n\}$
- $T^h = \{(\bar{s}_i, a_i, \bar{s}_{i+1}) \mid i = 0, \ldots, n\}$
- $s_0^h = \bar{s}_0$
- $\Omega^h = \{o_i \mid i = 0, \ldots, n+1\}$
- $Obs^h(\bar{s}_i) = o_i$, for $i = 0, \ldots, n+1$.

The *domain induced by* h is the domain $\mathcal{D}^h$ given by $\mathcal{D}^h = ((S^h, A^h, T^h, s_0^h), \Omega^h, Obs^h)$.

The learner has three related components, described in Algorithms 2 and 3, and in Theorem 11.2. The procedure in Theorem 11.2 builds on the output of Algorithm 3, which in turn builds on the output of Algorithm 2. Algorithm 2, called HISTORIES(P, τ), generates a set of histories from a set of proposition symbols P and a single observation trace τ. The algorithm generates the set of all histories for τ, assuming the domain is described by propositional symbols P. Each history h produced by this algorithm on input τ induces a domain $\mathcal{D}^h$ over P that mimics the interaction seen in τ. However, as τ may not exhibit *every* possible interaction with the real domain $\mathcal{D}$, $\mathcal{D}^h$ provides a partial description of $\mathcal{D}$. Intuitively, such partial descriptions could be carefully 'stitched together' to generate complete domain descriptions bisimilar to $\mathcal{D}$. Algorithm 3 presents such a procedure, DOMAINS(P, Σ). Given a set of observation traces $\Sigma = \{\tau_1, \ldots, \tau_n\}$, it iteratively calls HISTORIES(P, τ_i) for $1 \leq i \leq n$. Each call produces a set of histories $\mathcal{H}_i$ for τ_i. DOMAINS(P, Σ) then checks each set of histories of the form $H = \{h_1, \ldots, h_n \mid h_i = (s, \ldots), h_i \in \mathcal{H}_i, 1 \leq i \leq n,$ for some state $s\}$, and constructs a domain $\mathcal{D}^H$ by taking the union of the components of the domains $\mathcal{D}^{h_i}$, component-wise. If $\mathcal{D}^H$ is deterministic, the domain is added to a set $\mathbb{D}$. The set of domains $\mathbb{D}$ is the output of DOMAINS(P, Σ). Intuitively, each $\mathcal{D}^H \in \mathbb{D}$ is a domain that mimics all observation traces in the input, without breaking the condition of determinism for domains. Lemma 11.2 shows that DOMAINS(P, Σ) produces all and only the domains that are bisimilar to the real domain $\mathcal{D}$, when Σ is a sound and complete set of observation traces for $\mathcal{D}$. The overall learner, described in Theorem 11.2, then returns the synchronous composition of the domains $\mathbb{D}$ produced by DOMAINS(P, Σ). Since $\mathbb{D}$ is the set of domains bisimilar to $\mathcal{D}$, the learner will return the behavioural equivalence domain induced by $\mathcal{D}$, and hence be behaviourally correct with respect to implicit knowledge about $\mathcal{D}$.

Example 11.11 Figure 11.11 shows the behaviour of HISTORIES(P, τ) (Algorithm 2) when executed with $P = \{p, q\}$ and the observation trace $\tau = (q, a, q, a, \neg q, a, q, a, q)$ from the door knocking domain. The algorithm iterates over $i = 0, 1, 2, 3$. At step 0, it generates the set $Con_0 = \{(\bar{s}_0, \bar{s}_1) \mid \bar{s}_0, \bar{s}_1 \in comp(q)\} = \{(pq, pq), (pq, \neg pq), (\neg pq, pq), (\neg pq, \neg pq)\}$. The algorithm starts a history $(\bar{s}_0, q, a, \bar{s}_1, q)$ for each such pair, as they all induce deterministic domains. Each path of length 1 starting at the root of a tree corresponds to one such history. At step 1, the algorithm computes Con_1. For each pair $(\bar{s}_1, \bar{s}_2) \in Con_1$, it considers each history from step 0 whose last state is $\bar{s}_1$ and tries to extend it with

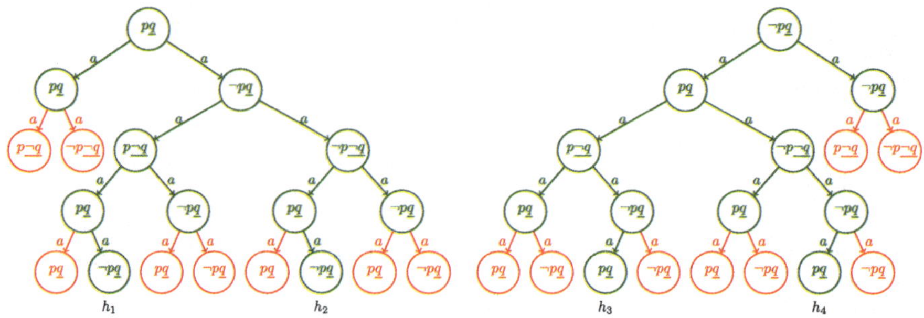

Fig. 11.11 A visualisation of the execution of HISTORIES(P, τ) (Algorithm 2) with input $P = \{p, q\}$ and the observation trace $\tau = (q, a, q, a, \neg q, a, q, a, q)$ from the door knocking domain. Each branch of each tree represents an initial segment of a history for τ. The leftmost branch of the leftmost tree represents the history $(pq, q, a, pq, q, a, p\neg q, \neg q)$: The root is the state pq in which q is observed; then a is executed leading again to the state pq where q is observed; finally, a is executed again, this time leading to the state $p\neg q$ where $\neg q$ is observed

---

Algorithm 2: HISTORIES(P, τ)

Input : P (propositional symbols), $\tau = (o_0, a_0, o_1, \ldots, o_n, a_n, o_{n+1})$ (observation trace).
Output : H_n (histories).

for $i = 0, \ldots, n$ **do**
  $Con_i := \{(\bar{s}_i, \bar{s}_{i+1}) \in 2^P \times 2^P \mid \bar{s}_i \in comp(o_i), \bar{s}_{i+1} \in comp(o_{i+1})\}$;
  $H_i = \emptyset$;
  for $(\bar{s}_i, \bar{s}_{i+1}) \in Con_i$ **do**
    if $i = 0$ **then**
      Start a history $h = (\bar{s}_0, o_0, a_0, \bar{s}_1, o_1)$;
      if $\mathcal{D}^h$ *is deterministic* **then**
        $H_0 := H_0 \cup \{h\}$;
      end
    end
    else
      for $h = (\bar{s}_0, o_0, \ldots, \bar{s}_i, o_i) \in H_{i-1}$ **do**
        $h' := (\bar{s}_0, o_0, \ldots, \bar{s}_i, o_i, a_i, \bar{s}_{i+1}, o_{i+1})$;
        if $\mathcal{D}^{h'}$ *is deterministic* **then**
          $H_i := H_i \cup \{h'\}$;
        end
      end
    end
  end
end
return H_n;

Algorithm 3: DOMAINS(P, Σ)

Input : P (propositional symbols), $\Sigma = \{\tau_1, \ldots, \tau_n\}$ (set of observation traces).
Output : $\mathbb{D}$ (set of domains over P).

$\mathbb{D} := \emptyset$;
for $i = 1, \ldots, n$ **do**
  $\mathcal{H}_i :=$ HISTORIES(P, τ_i);
end
for $h_1 \in \mathcal{H}_1, \ldots, h_n \in \mathcal{H}_n$ **do**
  if $h_1, \ldots, h_n$ *all start in the same state* s_0 **then**
    $H := \{h_1, \ldots, h_n\}$;
    $s_0^H := s_0$;
    $\mathcal{D}^H := ((\bigcup_{i=1}^k S^{h_i}, \bigcup_{i=1}^k A^{h_i}, \bigcup_{i=1}^k T^{h_i}, s_0^H), \bigcup_{i=1}^k \Omega^{h_i}, \bigcup_{i=1}^k Obs^{h_i})$;
    if $\mathcal{D}^H$ *is deterministic* **then**
      $\mathbb{D} := \mathbb{D} \cup \{\mathcal{D}^H\}$
    end
  end
end
return $\mathbb{D}$;

$(a, \bar{s}_2, \neg q)$. From the left child of the root of the left tree, two history extensions are tried and discarded (shown in red). The first one would extend (pq, q, a, pq, q) into $(pq, q, a, pq, q, a, p\neg q, \neg q)$. This extension is discarded because it would induce the non-deterministic transition function $T(pq, a) = \{pq, p\neg q\}$. The second one would extend (pq, q, a, pq, q) into $(pq, q, a, pq, q, a, \neg p\neg q, \neg q)$. This is discarded as well, as it would induce the non-deterministic transition function $T(pq, a) = \{pq, \neg p\neg q\}$. On the right tree, the history extensions marked in red are also discarded, as they would similarly induce non-deterministic transition functions. The remaining extensions, marked in green, are accepted. In step 2, the histories generated in step 1 are again extended. In step 3, several history extensions are discarded, since they would yield non-deterministic domains. The algorithm then outputs the set of histories $\{h_1, \ldots, h_4\}$, each corresponding to a path from one root of a tree to a leaf of that tree.

Figure 11.12 shows the output of DOMAINS(P, Σ) (Algorithm 3) when executed with $P = \{p, q\}$ and $\Sigma = \{\tau\}$. The algorithm first calls HISTORIES(P, τ), which as just seen, generates the set of histories $\{h_1, \ldots, h_4\}$. The algorithm then generates one domain $\mathcal{D}^{\{h_i\}}$ for each $1 \leq i \leq n$. For instance $\mathcal{D}^{\{h_1\}}$ is generated by the history $h_1 = (pq, q, a, \neg pq, q, a, p\neg q, \neg q, a, pq, q, a, \neg pq, \neg p)$ from Fig. 11.11. The domain $\mathcal{D}^{h_1}$ is simply read off the history h_1: the history tells us that the a-successor of the state pq is the state $\neg pq$, that the successor of $\neg pq$ is $p\neg q$, and that, finally, the successor of $p\neg q$ is the original state pq. This immediately produces the domain $\mathcal{D}^{\{h_1\}}$ shown in the top left of Fig. 11.12, where we also underlined the observations. These observations are similarly read directly off h_1. Note that, for all i, $\mathcal{D}^{\{h_i\}}$ is the same domain as $\mathcal{D}_i$ from Fig. 11.9, which depicted all domains bisimilar to the door knocking domain. The algorithm has thus produced exactly the set of domains bisimilar to the door knocking domain.

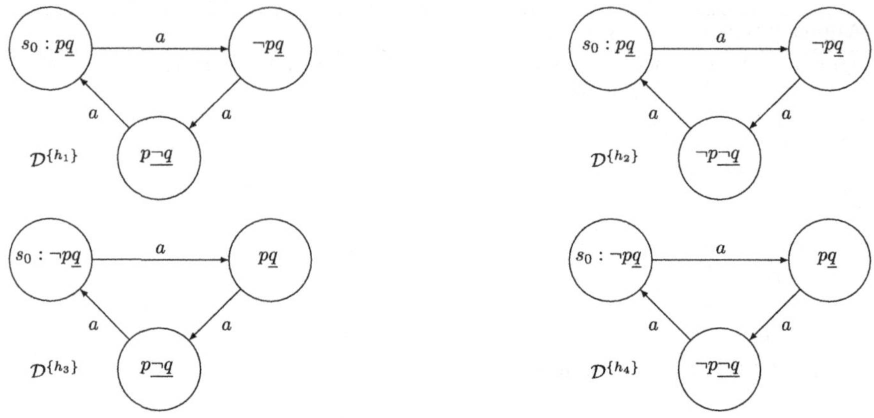

Fig. 11.12 The domains $\mathcal{D}^{\{h_1\}}, \dots, \mathcal{D}^{\{h_4\}}$ outputted by DOMAINS$(P, \{\tau\})$ (Algorithm 3) with input $P = \{p, q\}$ and the observation trace $\tau = (q, a, q, a, \neg q, a, q, a, q)$ from the door knocking domain

Finally, the overall learner, on input $P = \{p, q\}$ and $\Sigma = \{\tau\}$, takes the result of DOMAINS$(P, \{\tau\})$, i.e. $\{\mathcal{D}^{\{h_1\}}, \dots, \mathcal{D}^{\{h_4\}}\}$, and returns the synchronous composition of $\{\mathcal{D}^{\{h_1\}}, \dots, \mathcal{D}^{\{h_4\}}\}$. As $\{\mathcal{D}^{\{h_1\}}, \dots, \mathcal{D}^{\{h_4\}}\} = \{\mathcal{D}_1, \dots, \mathcal{D}_4\}$, the output of the learner is in fact the behavioural equivalence domain induced by the door knocking domain.

Lemma 11.2 *With input P and a sound and complete set Σ of observation traces for $\mathcal{D}$, algorithm* DOMAINS(P, Σ) *(Algorithm 3) returns the set of all domains over P that are bisimilar to $\mathcal{D}$.*

Proof Throughout the proof, let $\tau_1, \dots, \tau_k$ be an enumeration of Σ, let $\mathcal{H}_i$ denote the output of HISTORIES(P, τ_i), and denote each τ_i as follows:

$$\tau_i = (Obs(s_0^i), a_0^i, Obs(s_1^i), \dots, a_n^i, Obs(s_{n+1}^i)).$$

The execution trace associated with τ_i will be denoted ε_i.

($\subseteq$) We show first that DOMAINS$(P, \Sigma) \subseteq \{\mathcal{D}' \mid \mathcal{D} \underline{\leftrightarrow} \mathcal{D}'\}$. Let $\mathcal{D}^H \in$ DOMAINS(P, Σ), $H = \{h_1, \dots, h_k\}$. We refer to the components of D^H as S^H, A^H, etc. We show that $\mathcal{D}^H \underline{\leftrightarrow} \mathcal{D}$. Note that each $h_i \in H$ is a history for τ_i, i.e. $h_i \in \mathcal{H}_i$. Note that h_i can thus be written as follows:

$$h_i = (\bar{s}_0^i, Obs(s_0^i), a_0^i, \bar{s}_1^i, Obs(s_1^i), \dots, a_n^i, \bar{s}_{n+1}^i, Obs(s_{n+1}^i))$$

where each $\bar{s}_j^i$ is some state from $comp(o_j^i)$. Define a relation $Z \subseteq S \times S^H$ by

$$sZs' \quad \text{iff } s = s_j^i \quad \text{and} \quad s' = \bar{s}_j^i \quad \text{for some} \quad i \in \{1, \dots, k\} \quad \text{and some } j \in \{0, \dots, n\}.$$

We claim that Z is a bisimulation between $\mathcal{D}$ and $\mathcal{D}^h$, i.e., Z satisfies (i) $s_0 Z s_0^H$, (ii) Observational indistinguishability, (iii) Forth, and (iv) Back.

(i) Since each $\tau_i \in \text{ObsTr}(\mathcal{D}, s_0)$, $s_0^1 = s_0^2 = \cdots = s_0^k = s_0$. By definition of $\mathcal{D}^H$, $\bar{s}_0^1 = \bar{s}_0^2 = \cdots = \bar{s}_0^k = s_0^H$ (line 8, Algorithm 3). Thus, by definition of Z, $s_0 Z s_0^H$.

(ii) Let $s_j^i Z \bar{s}_j^i$. From lines 6 and 11 of Algorithm 2 and the definition of $\mathcal{D}^{h_i}$, it follows that $Obs(s_j^i) = Obs^{h_i}(\bar{s}_j^i)$. Since $\mathcal{D}^H$ is deterministic (line 10, Algorithm 3) $Obs^H(\bar{s}_j^i) = Obs^{h_i}(\bar{s}_j^i)$, since $Obs^H = \bigcup_i Obs^{h_i}$ (line 9, Algorithm 3). Hence $Obs(s_j^i) = Obs^H(\bar{s}_j^i)$.

(iii) Suppose that sZs' and $T(s, a) = t$. As sZs', $s = s_j^i$ and $s' = \bar{s}_j^i$ for some $i \in \{1, \ldots, k\}$ and some $j \in \{0, \ldots, n\}$. Consider an execution trace of the form

$$\varepsilon = (s_0, a_0^i, \ldots, a_{j-1}^i, s, a, t, \ldots)$$

with n actions. Note that $\varepsilon \in \text{Tr}(\mathcal{D}, s_0)$, so there is some $\ell \in \{1, \ldots, k\}$ such that $\varepsilon = \varepsilon_\ell$. It thus follows that $t = s_{j+1}^\ell$ and that $\tau_\ell \in \text{ObsTr}(\mathcal{D}, s_0)$.
Claim 1. $\bar{s}_m^i = \bar{s}_m^\ell$ implies $\bar{s}_{m+1}^i = \bar{s}_{m+1}^\ell$, for $0 \le m \le j - 1$.
Suppose for contradiction that for some m, $\bar{s}_m^i = \bar{s}_m^\ell$ but $\bar{s}_{m+1}^i \neq \bar{s}_{m+1}^\ell$. From lines 6 and 11 of Algorithm 2, we can see that then $(\bar{s}_m^i, a_m^i, \bar{s}_{m+1}^i) \in T^{h_i}$ and $(\bar{s}_m^\ell, a_m^i, \bar{s}_{m+1}^\ell) \in T^{h_\ell}$. By definition of $\mathcal{D}^H$ (line 9, Algorithm 3), $(\bar{s}_m^i, a_m^i, \bar{s}_{m+1}^i), (\bar{s}_m^\ell, a_m^i, \bar{s}_{m+1}^\ell) \in T^H$. But then, since $\bar{s}_{m+1}^i \neq \bar{s}_{m+1}^\ell$, T^H is not deterministic, which gives contradiction. This completes the proof of Claim 1.
Note that $\bar{s}_0^i = \bar{s}_0^\ell = s_0^H$, by definition of $\mathcal{D}^H$. From Claim 1, we then get $\bar{s}_j^i = \bar{s}_j^\ell$. From lines 6 and 11 of Algorithm 2, we get that $T^{h_\ell}(\bar{s}_j^i, a) = \bar{s}_{j+1}^\ell$. By definition of $\mathcal{D}^H$, $T^H(\bar{s}_j^i, a) = \bar{s}_{j+1}^\ell$. As $s' = \bar{s}_j^i$ and $t = s_{j+1}^\ell$, letting $t' = \bar{s}_{j+1}^\ell$ we get: there is a $t' \in S^H$ s.t. $T^H(s', a) = t'$ and tZt'.

(iv) Let sZs' and $T^H(s', a) = t'$. As sZs', $s = s_j^i$ and $s' = \bar{s}_j^i$ for some $i \in \{1, \ldots, k\}$ and some $j \in \{0, \ldots, n\}$.
From $T^H(s', a) = t'$, by definition of T^H, it follows that there is some T^{h_ℓ} such that $s' = \bar{s}_m^\ell$, $t' = \bar{s}_{m+1}^\ell$ and $T^{h_\ell}(s', a) = t'$. As $\bar{s}_j^i = s' = \bar{s}_m^\ell$, we get $T^H(\bar{s}_j^i, a) = \bar{s}_{m+1}^\ell$. As actions are universally applicable, there is some $t \in S$ such that $T(s, a) = t$. Consider an execution trace of the form

$$\varepsilon = (s_0, a_0^i, \ldots, a_{j-1}^i, s, a, t, \ldots)$$

with n actions. Note that $\varepsilon \in \text{Tr}(\mathcal{D}, s_0)$, so there is some $p \in \{1, \ldots, k\}$ such that $\varepsilon = \varepsilon_p$. It thus follows that $t = s_{j+1}^p$ and $\tau_p \in \text{ObsTr}(\mathcal{D}, s_0)$. From Claim 1, we get $\bar{s}_j^i = \bar{s}_j^p$, so from lines 6 and 11 of Algorithm 2, we get $T^{h_p}(\bar{s}_j^i, a) = \bar{s}_{j+1}^p$. By definition of T^H, we get $T^H(\bar{s}_j^i, a) = \bar{s}_{j+1}^p$. But since we already know that $T^H(\bar{s}_j^i, a) = \bar{s}_{m+1}^\ell$ and that T_H is deterministic, we get $\bar{s}_{j+1}^p = \bar{s}_{m+1}^\ell$. Since $s = s_j^i$, $t = s_{j+1}^p$ and$t' = \bar{s}_{m+1}^\ell$, we get: $T(s, a) = t$ and tZt'.

($\supseteq$) We show now that $\text{DOMAINS}(P, \Sigma) \supseteq \{\mathcal{D}' \mid \mathcal{D} \underline{\leftrightarrow} \mathcal{D}'\}$. By Lemma 11.1, this is equivalent to showing $\text{DOMAINS}(P, \Sigma) \supseteq \{((S', A', T', s_0'), \Omega', Obs') \mid \text{ObsTr}(\mathcal{D}, s_0) = \text{ObsTr}(\mathcal{D}', t_0)\}$, which is what we will do. Assume that $((S', A', T', t_0), \Omega, Obs)$ satisfies $\text{ObsTr}(\mathcal{D}, s_0) = \text{ObsTr}(\mathcal{D}', t_0)$. Let $\tau_i \in \Sigma$. Since

$\tau_i \in \text{ObsTr}(\mathcal{D}, s_0)$,$\tau_i \in \text{ObsTr}(\mathcal{D}', s_0')$. Since $s_0^i \in comp(s_0^i)$ and $s_1^i \in comp(s_1^i)$, looking at line 6 of Algorithm 2, we can see that at step 0 of HISTORIES(P, τ_i), history $h = (s_0^i, Obs(s_0), a^{i_0}, s_1^i, Obs(s_1))$ is created. At each step $j > 0$, since $s_j^i \in comp(s_j^i)$ and $s_{j+1}^i \in comp(s_{j+1}^i)$, looking at lines 12–13 of Algorithm 2, we can see that the extension

$$h' = (s_0^i, Obs(s_0), a_0^i, s_1^i, Obs(s_1) \ldots, s^j, Obs(s_j^i), a_j^i, s_{j+1}^i, Obs(s_{j+1}^i))$$

of h is such that $h' \in H_j$ (line 15 of Algorithm 2). For τ_i, let h_i denote the history

$$h_i = (s_0^i, Obs(s_0), a_0^i, s_1^i, Obs(s_1) \ldots, s_{|S|}^i, Obs(s_{|S|}^i), a_{|S|}^i, s_n^i, Obs(s_n^i)).$$

As h_i is one of the histories created at the last step of the iteration in HISTORIES(P, τ_i), h_i is a member of the output of Algorithm 2 (line 21). This process takes place for each $\tau_i \in \Sigma$ when HISTORIES(P, τ_i) is executed, so the output of each call of HISTORIES(P, τ_k) includes the history h_i, for $1 \leq i \leq k$. Note that all h_i start in the same state s_0. Moreover, since Σ includes all observation traces with $2^{2|P|}$ actions, if $\mathcal{D}'$ has $2^{|P|}$ states, each state in $\mathcal{D}'$ is reached by one such observation trace. i.e. for each transition $(s, a, t) \in T'$, there is some observation trace $\tau_i \in \Sigma$ is of the form $(\ldots, Obs'(s), a, Obs'(t), \ldots)$. Hence, D^{h_i} has $(s, a, t) \in T^{h_i}$, $Obs^{h_i}(s) = Obs'(s)$ and $Obs^{h_i}(t) = Obs'(t)$. Thus, by construction, $\mathcal{D}^{\{h_1,\ldots,h_n\}}$ will then contain all and only the states, transitions and state observations of $\mathcal{D}'$, which means that $\mathcal{D}^{\{h_1,\ldots,h_n\}} = \mathcal{D}'$. And as $\mathcal{D}'$ is deterministic (since trace equivalence is defined for deterministic domains), this means that $\mathcal{D}^{\{h_1,\ldots,h_n\}}$ is in the output of DOMAINS(P, Σ). □

Combining Lemmas 11.2 and 11.1, we then immediately get that DOMAINS(P, Σ) returns the set of domains over P that are trace equivalent to $\mathcal{D}$.

Theorem 11.2 *Consider the learner that on input P (set of propositions) and Σ (set of observation traces) returns the synchronous composition of the domains computed by running* DOMAINS(P, Σ) *(Algorithm 3). This learner outputs a domain that is behaviourally correct with respect to implicit knowledge about $\mathcal{D}$.*

Proof From Lemma 11.2, we know that DOMAINS(P, Σ) returns the set of deterministic domains over P that are bisimilar to $\mathcal{D}$, $\{\mathcal{D}' \mid \mathcal{D} \underline{\leftrightarrow} \mathcal{D}'\}$. The learner then takes this set of domains and returns their synchronous composition. Since, by definition, the behavioural equivalence domain induced by $\mathcal{D}$ is the synchronous composition of $\{\mathcal{D}' \mid \mathcal{D} \underline{\leftrightarrow} \mathcal{D}'\}$, the output of the learner is exactly the behavioural equivalence domain induced by $\mathcal{D}$. □

11.7 Related Work

This work builds upon the framework by Bolander and Gierasimczuk (2015, 2017), where two basic learnability criteria for actions were studied: finite identifiability (conclusively inferring a representation of the correct action in finite time) and identifiability in the limit (inconclusive convergence to a representation of the right action). It has been shown that deterministic actions are finitely identifiable, while arbitrary (non-deterministic) actions are only identifiable in the limit, in the fully observable setting. Moreover, the paper presents exact learning algorithms for deterministic actions that produce DEL event models as output. Our work continues this line of research, since it aims at exact learning, it bounds the size of the required input, and it uses DEL as the action representation language (at least for the first of our learners). The main difference is that the present work considers learning in partially, rather than fully observable, domains.

Our techniques are inspired by the tools of the theory of inductive inference (see, e.g., Jain et al., 2012; Osherson et al., 1997). The condition of behaviourally correct learning is closely related to the classical behaviourally correct learning of recursive functions (Barzdinš, Barzdinš 1974, see also Case and Smith, 1983). The capacity to extract non-explicit knowledge, created by requiring the learned structures to be deterministic, bears close resemblance to the increased power of learning when transitioning from the more general recursive language (set) learning (Gold, 1967) to recursive function learning (for a comparison of the two in the context of BC-learning see Jain et al., 2004). The present paper is by no means the first to transfer techniques from inductive inference to the domain of DEL. A link was introduced in Gierasimczuk (2009a, 2009b), where it was shown that finite identification (Lange & Zeugmann, 1992; Mukouchi, 1992) can be modelled in Public Announcement Logic (Plaza, 1989), and that the elimination process of learning by erasing (Lange et al., 1996) can be seen as iterated upgrade of dynamic doxastic logic (van Benthem, 2007). The revival of finite identification resulted in designing new types of learners, such as preset learners and fastest learners, and gave new insights into the complexity of obtaining definite finite tell tales (DFTTs, also used our present paper) (Gierasimczuk & de Jongh, 2010, 2012). Some of those results were later used to investigate properties of finite identification from complete data (de Jongh & Vargas-Sandoval, 2019; Sandoval, 2020). The more general approach of identifiability in the limit and its connections to doxastic upgrades allowed comparing the learning power of various belief revision methods (Baltag et al., 2019), and gave topological characterisations of learnability (Baltag et al., 2015) followed by an introduction of a dynamic logic for learning theory (DLLT, see Baltag et al., 2019).

There is a rich literature on learning symbolic action models from experience. Research on action learning began in the late 1980s and early 1990s, with systems such as LIVE (Shen & Simon, 1989), EXPO (Gil, 1992) and OBSERVER (Wang, 1996), which learned actions represented as STRIPS-like rules in fully observable domains. Since then, the literature has grown steadily, including some relatively recent overview papers (Arora et al., 2018; Jimenez et al., 2012). Recent work can be compared along several dimensions, such as: the type of observations available to the learner (full, partial, or noisy); the type of actions learned (conditional, unconditional, stochastic, etc.); the methods used (inductive logic programming, neural networks, satisfiability techniques, etc.); and the learning guarantees provided by the learning algorithm (approximate or exact learning).

We discuss those works that are most directly related to the results of this paper.

In recent years, several works have appeared that can learn action descriptions in partially observable environments (Aineto et al., 2019; Amir & Chang, 2008; Cresswell et al., 2013; McCluskey et al., 2009; Molineaux & Aha, 2014; Mourao et al., 2012; Shahaf et al., 2006; Yang et al., 2005, 2007; Zhuo et al., 2009, 2010). In these works, partial observability is induced by selecting at random $n < |P|$ propositional symbols to observe, for each state in the learning input. Each observation of a state s in the learning input is subjected to this process independently, so what is observed about s each time it is visited can be different. In other words, partial observability is unsystematic and observations can be thought of as random subsamples of the full state observation. No attempt is therefore made to *learn* the observation function, as it is just modelled as a random corruption process. The treatment of partial observability in these papers is thus very different from ours. We assume that there is a domain-specific observation function, which is in fact deterministic, and consider the problem of learning *both* the transition and observation functions. We don't only want to learn as much as possible about the underlying transition system; we also want to learn about our own observational limits (and hence the observational limits of other agents in the same state).

Besides treating partial observability differently, most of these works aim at approximate learning: their algorithms are experimentally evaluated, based on an error function, and typically offer no upper bounds on estimation errors. In contrast, we have learning goals that are exact in nature, and prove that the goals are guaranteed to be reached given certain inputs. Amongst learners for partially observable domains, work of Amir et al. (2008), Shahaf et al. (2006) does present exact algorithms for identifying the effects and preconditions of deterministic actions in partially observable domains. The algorithms take an observation trace as input and return a set of deterministic action models that could have led to those observations. The output of the algorithms is exact in the sense of producing all and only those action models that could have led to these observations. However, as observations are random, no attempt is made to learn the observation function, and no bound is given in terms of the number of observations needed to reach their learning goal. Our work differs from theirs in three respects: learning about the observation function, representing actions using DEL, and characterising and bounding the number of observations needed to reach our learning goals.[4]

11.8 Final Remarks and Future Work

We conclude by reflecting on the assumptions made in this paper and exploring some avenues for future research.

Multi-agent learning We have assumed in this paper that the environment involves a single agent. As we mentioned in Sect. 11.2, the ultimate goal of the

[4] We require sound and complete sets of observations for explicit learning, and sound and complete sets of observation traces for implicit learning.

line of research introduced here is to be able to generalise to the multi-agent case, where a learner might end up learning not only what an action does and what is observed, but also what other agents will observe about such an action, and the knowledge or beliefs they will adopt as a result. Learning even just what is *explicitly* known by others is hard, since it requires knowing what they are directly observing and how they are observing it. While an agent knows what it observes, it often doesn't know exactly what others observe. Sometimes common knowledge of what each agent observes can be achieved in a collaborative setting by a process such as *joint attention* (Moore et al., 2014), or by communicating what you observe. But in a non-cooperative setting or one in which communication is imperfect, the problem becomes highly non-trivial. Learning what others can observe, and what they believe or know as a result, is crucial for Theory of Mind reasoning (Premack & Woodruff, 1978) and epistemic planning (Bolander & Andersen, 2011) in unknown domains, and thus remains a key goal for future work.

Situated proactive learning We have assumed that the learner for implicit knowledge has access to all possible observation traces from the initial state s_0. These traces may have been generated by an expert agent or teacher that knows how to traverse the state space to produce such traces. For a situated, proactive learner, i.e. a learner that has to gather such traces starting at s_0, the task of generating the traces is non-trivial and in some cases impossible. If the graph of the transition function is strongly connected (so that it is possible to get from every state to every other state through a sequence of actions), then all such observation traces can be generated from the initial state in one run through the graph, if the agent has a way of recognising the initial state each time it is revisited (it might not, due to the observational limitations). If the agent does not have a way of recognising the initial state, it is less clear how it would be able to explore the unknown graph of the transition function and reach a point in which it's certain that all possible traces have been produced. In some cases, doing so is impossible, e.g. the agent might get stuck in a 'loopy' state from which all outgoing edges are loops. A solution to this might be to allow restarts as in reinforcement learning.

The question of how to collect all observation traces as a situated proactive learner is closely related to the problem of *exploring an unknown graph* (Albers & Henzinger, 2000; Deng & Papadimitriou, 1999; Panaite & Pelc, 1999).

Relaxing domain assumptions We have focused on deterministic domains, in which actions are always applicable and every state is reachable from the initial state s_0. In some cases, these assumptions may not be the most natural. Dropping some of them and generalising the learners in the paper is therefore a possible direction for future research. In some cases, e.g. in the learning algorithm for implicit knowledge, it may be possible to deal with non-deterministic domains by dropping the requirement that only deterministic domains are produced via Algorithms 2 and 3. The algorithms would then produce a larger set of domains matching the observation traces, which includes non-deterministic ones.

Computational complexity We have bounded the *sample complexity* of our learning algorithms. The sample complexity of a learning algorithm is the size of the input required in order to achieve its learning goal. In the case of explicit knowledge, we have bounded the sample complexity of LEARNER(P, A, σ) by the number of transitions in the system (i.e. the size of a sound and complete set of observations). In the case of implicit knowledge, we have bounded the sample complexity of the learner of Theorem 11.2 by the number of observation traces with $2^{2|P|}$ actions starting from s_0, i.e. the size of a sound and complete set of observation traces). On the other hand, we have left time and space complexity issues as future work. These complexities will of course depend on the implementation details. For example, the pseudo-code in Algorithm 2 should not be implemented by creating the histories explicitly. There is a lot of repeated structure in histories, which can be avoided. Note that simply implementing the set of histories as paths in a tree, as depicted in Fig. 11.11, would already yield an improvement, over storing each history separately. This tree implementation is straightforward; more advanced and space-efficient ones may be possible, e.g. using action models, or some other compact encoding of domains.

A more compact representation of the learning output The algorithm presented for implicit knowledge produces the behavioural equivalence domain as output. This domain can of course be very large, as its state space has size $O(2^{|P|})$. We leave for future work the task of learning the behavioural equivalence domain in a representation that is possibly more compact, such as a collection of DEL event models. This would then match the output we provided in the case of explicit knowledge. Ideally, we would compute one set of DEL event models that capture both implicit and explicit knowledge via two distinct indistinguishability relations. This would allow learners to reason about both their implicit and explicitly knowledge and how they are interrelated. An even more compact representation could possibly be obtained with first-order DEL (FODEL) action schemas such as those of Liberman et al. (2020), or with the succinct event models of Charrier and Schwarzentruber (2017). Since FODEL action schemas are represented with variables from first-order logic, learning them would require a non-trivial extension of existing methods. In a domain represented with first-order logic, the agent could possibly be observing ground atoms. It would then have to *generalise* from these ground observation to achieve the level of abstraction characteristic of action schemas. Perhaps a procedure inspired in *least general generalisation* (Plotkin, 1970), widely used for generalising first-order clauses in Inductive Logic Programming, could be adapted for this. Several algorithms for learning less expressive action models, such as those of the Planning Domain Definition Language (PDDL), also perform some type of generalisation, and could provide inspiration.

Acknowledgements Nina Gierasimczuk's research was funded by the Polish National Science Centre Grant 2015/19/B/HS1/03292.

Appendix: Special Acknowledgement by Nina Gierasimczuk

As I arrived in Amsterdam in 2006 to continue my studies at the Institute for Logic, Language and Computation, my interest in theoretical notions of learnability was met with little enthusiasm, at the time Game Theory was far more popular. In my Master thesis (written under the supervision of the late Prof. Marcin Mostowski, (Gierasimczuk, 2007)) I studied learnability of formal languages, with a focus on Dana Angluin's L^* algorithm for regular languages (Angluin, 1987) and its extensions. The underlying notions of learnability: identifiability in the limit and finite identifiability (Gold, 1967) were a natural continuation of my studies, a transition toward more fundamental concepts and higher levels of abstraction. Identifiability in the limit (the concept central to limiting recursion, Gold, 1965) had been characterized by Dana Angluin in 1980 (Angluin, 1980). Her focus was on indexed families of recursive languages, i.e., families $(L_i)_{i \in \mathbb{N}}$ for which there is a total recursive function $f : \mathbb{N} \times \mathbb{N} \to \{0, 1\}$ such that for every $j \in \mathbb{N}$, $L_j = \{w \in \mathbb{N} \mid f(j, w) = 1\}$. The characterization of identifiability in the limit of such families relies on the elegant concept of *tell-tale sets*. I knew and admired this result, but the field was already mature and I could hardly imagine a new significant contribution. And then I received the first ever e-mail from my new mentor (and soon to become PhD adviser) Prof. Dick de Jongh. In the attachment there was the pdf of *Angluin's Theorem for Indexed Families of Recursively Enumerable Sets and Applications* by de Jongh and Kanazawa (1996). The paper extended the original characterisation to recursively enumerable sets. What a relief and validation that was! Getting *more applied* was indeed not the only way to proceed, *more abstract* was valuable too. And so I started my studies with Dick, later entering the stage of joint research, focusing on theoretical models of learning, searching for new results in unlikely places. For us, such a place turned out to be finite identifiability—the simplest kind of learnability—finding the identity of a concept in finite time and with certainty. A characterization similar to the one of Angluin was already given in 1992 by Mukouchi (1992) and, independently, by Lange and Zeugmann (1992). To me, characterizing a mathematical concept is not unlike raising tomb, both as an honour (the concept had been important enough to be characterized), but also as an epistemic closure (the concept is now well-understood and, *ipso facto*, less interesting). But sometimes resurrections happen, and dissections bring about interesting insights. That happened in my joint work with Dick—we found and isolated a new kind of finite identifiability: *fastest learning* (Gierasimczuk & de Jongh, 2013).

The classical perspective on learning, through the lenses of identifiability (both finite and in the limit), gives a symbolic account that is especially well-aligned with some of the classical methods of GOFAI. This alignment gave rise to my collaboration with Thomas Bolander at the Technical University of Denmark. We have combined our respective interests in epistemic planning and exact learning and investigated learnability of action models in Dynamic Epistemic Logic (Bolander & Gierasimczuk, 2015, 2018). One of the continuations is our joint work (together with Andrés Occhipinti Liberman) published in this volume. Here, we focus on learnability with partial observations: agents' access to information is limited. Such scenarios motivate non-trivial uses of communication in learning. Interestingly, a similar topic had come up much earlier, in a conversation with Dick, during one of our working sessions in Science Park 904 in Amsterdam. I distinctly remember we played around with a concept of learning in which two agents had access to mutually disjoint sets of parameters of an unknown concept. These ideas, together with myriads of others, were never realized. Fortunately, the inspiration remained.

Working with Dick was often puzzling. Our research sessions were long and quiet. In the beginning, these prolonged silences made me unsure and slightly uncomfortable. They resembled an awkward elevator ride with a stranger. Over time I learned to cherish this method—thoughts running freely, disciplined by the purpose of the meeting, an occasional comment, a brief exchange of thoughts, a writing on the blackboard. Sometimes it felt as if we were not, in fact, alone. A strange third presence could be felt. Dick, being an intuitionist, would probably disagree with ascribing objective existence to mathematical entities, but to me it was as if we were silently tuning in to a distant, barely accessible broadcast from the Mathematical Beyond.

Dick was a great adviser: demanding, but kind. His authority gave me an intellectual shelter where I could play with my glass beads, uninterrupted. In the face of scientific drawbacks he always advised patience, in the face of academic obstacles—skepticism. He always quietly supported my quests and ambitions. With Dick there is no hyperbolic language, what I'd hear from him was always moderate: 'That's good', 'That's wrong', 'That's sufficient', 'That's important'. He criticized my early proofs: 'Why not do it directly? You have a strange predilection for proofs by contradiction, that's so baroque.' As an aspiring logician I remember feeling proud, what I heard is that now I get to have *a style*.

I am very grateful for being able to call myself Dick's former student and humbled to be one of his academic descendants. I am very thankful to my co-authors, who kindly agreed to offer our joint work for this beautiful collection that bears witness to Dick's varied interests, important contributions, and his astonishing scientific influence.

References

Aineto, D., Celorrio, S. J., & Onaindia, E. (2019). Learning action models with minimal observability. *Artificial Intelligence, 275*, 104–137.

Albers, S., & Henzinger, M. R. (2000). Exploring unknown environments. *SIAM Journal on Computing, 29*(4), 1164–1188.

Amir, E., & Chang, A. (2008). Learning partially observable deterministic action models. *Journal of Artificial Intelligence Research, 33*, 349–402.

Angluin, D. (1980). Inductive inference of formal languages from positive data. *Information and Control, 45*(2), 117–135.

Angluin, D. (1987). Learning regular sets from queries and counterexamples. *Information and Computation, 75*(2), 87–106.

Arora, A., et al. (2018). A review of learning planning action models. *Knowledge Engineering Review, 33*.

Baltag, A., & Moss, L. S. (2004). Logics for epistemic programs. *Synthese, 139*(2), 165–224.
Baltag, A., et al. (2019). A dynamic logic for learning theory. *Journal of Logical and Algebraic Methods in Programming, 109*, 100485. ISSN: 2352-2208.
Baltag, A., Gierasimczuk, N., & Smets, S. (2015). On the solvability of inductive problems: A study in epistemic topolog. In R. Ramanujam (Ed.), *Proceedings Fifteenth Conference on Theoretical Aspects of Rationality and Knowledge, TARK 2015* (Vol. 215, pp. 81–98). Carnegie Mellon University. EPTCS.
Baltag, A., Gierasimczuk, N., & Smets, S. (2019). Truth-tracking by belief revision. *Studia Logica, 107*(5), 917–947.
Barzdinš, J. (1974). Two theorems on the limiting synthesis of functions. In *Theory of algorithms and programs* (Vol. 1, pp. 82–88). In Russian. Latvian State University.
Blackburn, P., de Rijke, M., & de Venema, Y. (2001). *Modal logic* (Vol. 53). Cambridge Tracts in Theoretical Computer Science. Cambridge University Press.
Bolander, T., & Andersen, M. B. (2011). Epistemic planning for single-and multi-agent systems. *Journal of Applied Non-Classical Logics, 21*(1), 9–34.
Bolander, T., & Gierasimczuk, N. (2015). Learning actions models: Qualitative approach. In *LORI* (pp. 40–52).
Bolander, T., & Gierasimczuk, N. (2015). Learning actions models: Qualitative approach. In W. van der Hoek, W. H. Holliday & W. Wang (Eds.), *Logic, Rationality, and interaction—5th international workshop, LORI, Proceedings*, volume 9394 of Lecture notes in computer science (pp. 40–52). Springer, 2015.
Bolander, T., & Gierasimczuk, N. (2017). Learning to act: qualitative learning of deterministic action models. *Journal of Logic and Computation*
Bolander, T., & Gierasimczuk, N. (2018). Learning to act: qualitative learning of deterministic action models. *Journal of Logic and Computation, 28*(2), 337–365.
Case, J., & Smith, C. (1983). Comparison of identification criteria for machine inductive inference. *Theoretical Computer Science, 25*(2), 193–220. ISSN: 0304-3975.
Cassandras, C. G., & Lafortune, S. (2009). *Introduction to discrete event systems*. Springer Scince & Business Media.
Charrier, T., & Schwarzentruber, F. (2017). A succinct language for dynamic epistemic logic. In: *Proceedings of the 16th Conference on Autonomous Agents and Multi-agent Systems. AAMAS '17* (pp. 123–131). International Foundation for Autonomous Agents and Multi-agent Systems.
Chartrand, Gary. (2012). *A first course in graph theory*. Dover Publications.
Cresswell, S. N., McCluskey, T. L., & West, M. M. (2013). Acquiring planning domain models using LOCM. *The Knowledge Engineering Review, 28*(2), 195–213.
de Jongh, D., & Vargas-Sandoval, A. L. (2019). Finite Identification with Positive and with complete data. In Silva, A., et al. (Eds.), *Language, logic, and computation* (pp. 42–63). Springer. ISBN: 978-3-662-59565-7.
de Jongh, D., & Kanazawa, M. (1996). Angluin's theorem for indexed families of r.e. sets and applications. In A. Blum & M. J. Kearns (Eds.), *Proceedings of the Ninth Annual Conference on Computational Learning Theory, COLT 1996* (pp. 193–204). ACM.
Deng, X., & Papadimitriou, C. H. (1999). Exploring an unknown graph. *Journal of Graph Theory, 32*(3), 265–297.
Dissing, L., & Bolander, T. (2020). Implementing theory of mind on a robot using dynamic epistemic logic. In *Proceedings of the International Joint Conference on Artificial Intelligence (IJCAI)*.
Fikes, R. E., & Nilsson, N. J. (1971). STRIPS: A new approach to the application of theorem proving to problem solving. *Artificial intelligence,2*(3–4), 189–208.
Geffner, H., & Bonet, B. (2013). A concise introduction to models and methods for automated planning. *Synthesis Lectures on Artificial Intelligence and Machine Learning, 8*(1), 1–141.
Gierasimczuk, Nina, & de Jongh, Dick. (2010). On the minimality of definite tell-tale sets in finite identification of languages. In D. Grossi, L. Kurzen, & F. R. Velazquez-Quesada (Eds.), *2009 yearbook of logic and interactive rationality (LIRA) seminar*. ILLC.

Gierasimczuk, N. (2007). The problem of learning the semantics of quantifiers. In *TBiLLC'05: 6th International Tbilisi Symposium on Logic, Language, and Computation. Revised Selected Papers*, volume 4363 of Lecture notes in artificial intelligence (pp. 117–126). Springer.

Gierasimczuk, N. (2009a). Bridging learning theory and dynamic epistemic logic. *Synthese, 169*(2), 371–384.

Gierasimczuk, N. (2009b). Learning by erasing in dynamic epistemic logic. In Dediu, A. H., Ionescu, A. M., & Martin-Vide, C. (Eds.), *LATA'09: Proceedings of 3rd International Conference on Language and Automata Theory and Applications* (Vol. 5457, pp. 362–373). Lecture notes in computer science. Springer.

Gierasimczuk, N., & de Jongh, D. (2012). On the complexity of conclusive update. *The Computer Journal, 56*(3), 365–377.

Gierasimczuk, N., & Jongh de Jongh, D. (2013). On the complexity of conclusive update. *The Computer Journal, 56*(3), 365–377.

Gil, Y. (1992). *Acquiring domain knowledge for planning by experimentation*. Technical report. Carnegie Mellon University.

Gold, E. M. (1967). Language identification in the limit. *Information and control, 10*(5), 447– 474.

Gold, E. M. (1965). Limiting recursion. *Journal of Symbolic Logic, 30*(1), 28–48.

Gold, E. M. (1967). Language identification in the limit. *Information and Control, 10*, 447–474.

Gorrieri, R., Versari, C. (2015). *Introduction to concurrency theory: Transition systems and CCS*. Springer.

Jain, S., et al. (1999). *Systems that learn*. MIT Press.

Jain, S., Stephan, F., & Terwijn, S. A. (2004). Counting extensional differences in BClearning. *Information and Computation, 188*(1), 127–142.

Jimenez, S., et al. (2012). A review of machine learning for automated planning. *The Knowledge Engineering Review,27*, 433–467.

Lange, S., & Zeugmann, T. (1992). Types of monotonic language learning and their characterization. In *COLT* (pp. 377–390).

Lange, S., & Zeugmann, T. (1992). Types of monotonic language learning and their characterization. In *Proceedings of the Fifth Annual Workshop on Computational Learning Theory* (pp. 377–390). ACM.

Lange, S., Wiehagen, R., & Zeugmann, T. (1996). Learning by erasing. In S. Arikawa & A. Sharma (Eds.), *ALT* (Vol. 1160, pp. 228–241). Lecture notes in computer science. Springer. ISBN: 3-540-61863-5.

Liberman, A. O., Achen, A., & Rendsvig, R. K. (2020). Dynamic term-modal logics for first-order epistemic planning. *Artificial Intelligence, 286*, 103305.

McCluskey, T. L., et al. (2009). Action knowledge acquisition with opmaker2. In *International conference on agents and artificial intelligence* (pp. 137–150). Springer.

Molineaux, M., & Aha, D. W. (2014). Learning unknown event models. In *AAAI* (pp. 395– 401).

Moore, C., Dunham, P. J., & Dunham, P. (2014). *Joint attention: Its origins and role in development*. Psychology Press.

Mourao, K., et al. (2012). Learning STRIPS operators from noisy and incomplete observations. In *Proceedings of the Twenty-Eighth Conference on Uncertainty in Artificial Intelligence* (pp. 614–623).

Mukouchi, Y. (1992). Characterization of finite identification. In *International workshop on analogical and inductive inference* (pp. 260–267). Springer.

Mukouchi, Y. (1992). Characterization of finite identification. In K. Jantke (Ed.), *AII'92: Proceedings of the International Workshop on Analogical and Inductive Inference*, volume 642 of Lecture notes in computer science (pp. 260–267). Springer.

Osherson, D., et al. (1997). Formal learning theory. In J. van Benthem & A. ter Meulen (Eds.), *Handbook of logic and language* (pp. 737–775). MIT Press.

Panaite, P., & Pelc, A. (1999). Exploring Unknown Undirected Graphs. *Journal of Algorithms, 33*(2), 281–295.

Plaza, J. (1989). Logics of public communications. In M.L. Emrich et al. (Eds.), *Proceedings of the 4th International Symposium on Methodologies for Intelligent Systems* (pp. 201–216).

Plotkin, G. D. (1970). A note on inductive generalization. *Machine Intelligence, 5*(1), 153– 163.

Premack, D., & Woodruff, G. (1978). Does the chimpanzee have a theory of mind. *Behavioral and Brain Sciences, 1*(4), 515–526.

Sandoval, A. L. V. (2020). On the path to the truth: Logical and computational aspects of learning. Ph.D. thesis. Universiteit van Amsterdam.

Sangiorgi, D. (2011). *Introduction to Bisimulation and Coinduction*. Cambridge University Press.

Shahaf, D., Chang, A., & Amir, E. (2006). Learning partially observable action models: Efficient algorithms. In *Proceedings of the National Conference on Artificial Intelligence* (Vol. 21, p. 920). AAAI Press, MIT Press.

Shen, W.-M., & Simon, H. A. (1989). Rule creation and rule learning through environmental exploration. In *IJCAI* (pp. 675–680).

van Benthem, J. (2007). Dynamic logic for belief revision. *Journal of Applied Non-Classical Logics, 2*, 129–155.

van Ditmarsch, H., & Kooi, B. (2008). Semantic results for ontic and epistemic change. In G. Bonanno, W. van der Hoek, & M. Wooldridge (Eds.), *Logic and the foundation of game and decision theory (LOFT 7)* (Vol. 3, pp. 87–117). Texts in logic and games. Amsterdam University Press.

Walsh, T. J., & Littman, M. L. (2008) Efficient learning of action schemas and web-service descriptions. In *AAAI* (Vol. 8, pp. 714–719).

Wang, X. (1996). Learning planning operators by observation and practice. Ph.D. thesis. Carnegie Mellon University.

Yang, Q., Wu, K., & Jiang, Y. (2005). Learning actions models from plan examples with incomplete knowledge. In *ICAPS* (pp. 241–250).

Yang, Q., Wu, K., & Jiang, Y. (2007). Learning action models from plan examples using weighted MAX-SAT. *Artificial Intelligence, 171*(2–3), 107–143.

Zhuo, H. H., et al. (2009). Learning HTN method preconditions and action models from partial observations. In *IJCAI* (pp. 1804–1810).

Zhuo, H. H., et al. (2010). Learning complex action models with quantifiers and logical implications. *Artificial Intelligence, 174*(18), 1540–1569.

Chapter 12
Axiomatizing Origami Planes

Lev Beklemishev, Anna Dmitrieva, and Johann A. Makowsky

Abstract We provide a variant of an axiomatization of elementary geometry based on logical axioms in the spirit of Huzita–Justin axioms for the origami constructions. We isolate the fragments corresponding to natural classes of origami constructions such as Pythagorean, Euclidean, and full origami constructions. The set of origami constructible points for each of the classes of constructions provides the minimal model of the corresponding set of logical axioms. Our axiomatizations are based on Wu's axioms for orthogonal geometry and some modifications of Huzita–Justin axioms. We work out bi-interpretations between these logical theories and theories of fields as described in Makowsky (2018). Using a theorem of Ziegler (1982) which implies that the first order theory of Vieta fields is undecidable, we conclude that the first order theory of our axiomatization of origami is also undecidable.

Keywords Origami · Geometry · Axiom · Interpretation · Undecidability

12.1 Introduction

The ancient art of paper folding, known in Japan and all over the world as *origami*, has a sufficiently long tradition in mathematics. The pioneering book by Sundara Rao (1917) on the science of paper folding attracted attention by Felix Klein. Adolf Hurwitz dedicated a few pages of his diaries to paper folding constructions such as the construction of the golden ratio and of the regular pentagon, see Friedman (2018). The book by Friedman is also an excellent source on the history of mathematical origami.

L. Beklemishev (✉)
Steklov Mathematical Institute of RAS, Moscow, Russia
e-mail: lbekl@yandex.ru

National Research University Higher School of Economics, Moscow, Russia

A. Dmitrieva
University of East Anglia, Norwich, United Kingdom

J. A. Makowsky
Technion – Israel Institute of Technology, Haifa, Israel

N. Bezhanishvili et al. (eds.), *Dick de Jongh on Intuitionistic and Provability Logics*, Outstanding Contributions to Logic 28, https://doi.org/10.1007/978-3-031-47921-2_12

Early books on origami science, such as Young and Young (1905), considered paper folding mainly as recreational mathematics and as a means of introducing geometry to children.

Mathematical origami has been advanced by Margherita Piazzola Beloch in the 1930s. Even though her work was published in Italian at the time of Mussolini and remained essentially unnoticed, she found several key ideas that were rediscovered only much later, see Beloch (1934). In particular, she introduced a new operation, the so-called *Beloch fold*, using which she showed a construction of a segment of length $\sqrt[3]{2}$ and more generally showed how to construct the roots of equations of degrees 3 and 4.

The interest towards mathematical origami reemerged in the 1980s through the work of several enthusiasts such as Humiaki Huzita, who organized the successful *First international meeting on origami science and technology*, Huzita (1989b). Among various mathematical and computational questions studied in relation to origami in modern literature are whether a given crease pattern can be flat-folded, computational complexity of folding problems, and many others (see e.g. Demaine and O'Rourke, (2007)).

The general problem of origami design is, given a three-dimensional shape, to find a sequence of folds to create an origami approximating that shape (if possible) from a square sheet of paper. Great practical advances in this problem have been achieved by Robert J. Lang who developed an algorithm and a program that helped to create folding patterns for certain basic shapes. These basic shapes are subsequently relatively easy to refine into the required artistic images. Using his algorithms Lang designed origami models of very complicated and realistic nature, Lang (2003).

Apart from the visual art, origami science has impressive practical applications in the design of unfoldable structures, for example, satellite solar batteries, telescope mirrors, and even coronary bypasses.

One of the basic mathematical questions about origami, highly relevant for this paper, is origami constructibility: which lengths can be constructed by foldings from a square sheet of paper or, more generally, from a given set of initial points on the plane? This problem is similar to the classical one about compass and ruler constructions and has been studied from the very beginnings of mathematical origami. To systematically investigate this problem and to prove results on non-constructibility, there was a need to formulate basic rules of the game: a finite set of operations to which one can reduce any complex folding.

Justin (1989) and Huzita (1989a) identified a list of six such operations (H-1), ..., (H-6), later called the Huzita–Justin or Huzita–Hatori axioms. (The seventh operation proposed by Justin and Hatori was later shown to be reducible to the other ones.)

(H-1): Given two points P_1 and P_2, one can make a fold that passes through both of them.

(H-2): Given two points P_1 and P_2, one can make a fold that places P_1 onto P_2.

(H-3): Given two lines ℓ_1 and ℓ_2, one can make a fold that places ℓ_1 onto ℓ_2.

(H-4): Given a point P and a line ℓ, one can make a fold orthogonal to ℓ that passes through P.

(H-5): Given two points P_1 and P_2 and a line ℓ, one can make a fold that places P_1 onto ℓ and passes through P_2.

(H-6): Given two points P_1 and P_2 and two lines ℓ_1 and ℓ_2, one can make a fold that places P_1 onto ℓ_1 and P_2 onto ℓ_2.

(H-7): Given a point P and two lines ℓ_1, ℓ_2 one can make a fold orthogonal to ℓ_1 that places P onto ℓ_2.

We note that (H-6) is essentially the Beloch fold mentioned earlier. Some of the expositions also take the points in (H-1) and (H-2) to be distinct and then demand the lines obtained in (H-1), (H-2) and (H-4) to be unique. This obviously does not affect origami constructibility. Therefore, for the reasons of simplicity, we describe the rules without the requirements of uniqueness.

Robert et al. (2009) showed that (H-1)–(H-7) can be characterized as all the operations that define a unique fold by alignment of points and finite line segments. Using these operations elegant solutions of the two classical problems—the doubling of the cube and the angle trisection—were found.

Alperin (2000) characterized the classes of points (corresponding to certain subfields of $\mathbb{C}$) constructible using foldings defined by natural subsets of Huzita–Justin operations. Our paper uses the ideas of Alperin in the part that deals with the interpretations between geometric theories and the theories of the corresponding classes of fields.

As mentioned above, Huzita–Justin axioms were not meant to be understood as axioms in a logical sense but rather as specifying a (not necessarily deterministic or always defined) set of operations generating the origami constructible points on the plane. The distinction between logical axioms and the operational approach was clear in the early XXth century works on origami mathematics (see Chap. 5 of Friedman (2018), especially the discussion on p. 285). In fact, it was considered a weakness of the origami approach compared to axiomatic geometry in the tradition of Hilbert. The aim of this paper is to connect the two views and to adapt the principles of origami constructions to serve as logical axioms of planar geometry.

In Makowsky (2018, 2019) a proof was outlined of the statement that the first order theory of origami planes was undecidable. Although the proof strategy is feasible, the exact definition of the first order theory of origami planes was left imprecise. In particular, the role of the betweenness relation was overlooked. The purpose of this paper is to provide a precise definition of the first order theory of origami planes and to establish its properties.

The second aim of the paper is to work out mutual first order interpretations between our logical theories of origami and certain classes of fields, as suggested in Makowsky (2019). Ziegler's theorem then shows the algorithmic undecidability of the first order theories of these structures.

Even though these results are not really surprising, there are many details to take care of in their accurate proofs. The proofs rely on certain background from several areas: the development of axiomatic orthogonal geometry, the construction of coordinatization in the context of sufficiently weak axioms of geometry, and the logical techniques of first order interpretations. Since we deal with the questions of first order definability and interpretability, we need to be careful about the details of the constructions usually presented in a less formal way. Therefore, our paper has a long introductory part where we present these background results. We assume that the reader has some basic knowledge of elementary algebra, analytic geometry and first order logic, including the notion of interpretation.

The plan of the paper is as follows. In Sect. 12.2 we present background on orthogonal geometry using the book by Wu (1984) as our main reference source. We introduce the notion of metric Wu plane which is basic for all further developments. In Sect. 12.3 we present background on interpretations. In Sect. 12.4 we present the details of the coordinatization of Wu planes. This is an expanded version of the corresponding section of Makowsky (2019). In Sect. 12.5 we prove the result on the bi-interpretability between the classes of metric Wu planes and of Pythagorean fields. In Sect. 12.6 we state a general undecidability result for geometric theories of metric Wu planes based on Ziegler's theorem.

In Sect. 12.7 we introduce appropriate logical versions of the Huzita–Justin axioms. We also show that metric Wu planes satisfy the first four Huzita–Justin axioms. In Sect. 12.8 we discuss the role of order axioms, introduce ordered metric Wu planes and their relation to ordered Pythagorean fields. In Sect. 12.9 we introduce Euclidean and Vieta fields and prove our main Theorems 12.7 and 12.8. Finally, in Sect. 12.10 we discuss the remaining open questions.

12.2 Background on Axiomatic Geometry

In this section we present necessary background on axiomatic geometry and begin the discussion how to adapt the principles of origami constructions to serve as logical axioms of planar geometry. There are two issues here: what are the basic predicates, in which the logical analogs of the axioms (H-1), ..., (H-6) can be formulated, and what are the additional axioms which govern the basic predicates.

Looking at the usual statements of (H-1), ..., (H-6) we see that these principles use the primitive notions of points, lines (folds), incidence (*"a fold ℓ passes through a point P"*), which are basic for all standard axiomatizations of elementary geometry. In addition, the predicate *"a fold along ℓ identifies points P and Q"* and possibly some others are used.

The notion of a fold along a given line is similar to that of reflection with respect to a line. There are axiomatizations of geometry based on reflection as a basic notion (see Bachmann, 1973). However, in this paper we find it convenient to rely on the well-developed axiomatizations of geometry based on the notion of orthogonality of lines.

The predicate $\ell \perp m$ *"lines ℓ and m are orthogonal"* is sufficiently natural from the point of view of origami constructions: orthogonality can be tested by folding a paper along one of the lines. Moreover, we will see later that the majority of principles corresponding to Huzita–Justin axioms are easily expressible using incidence and orthogonality. Thus, we take Wu's axiomatization of orthogonal plane geometry as basic (Wu, 1984). We use this book as a reference source for careful proofs of many technical statements that we will need.

As we will see, for some of the logical versions of Huzita–Justin axioms, namely (H-3), (H-5) and (H-6), the notion of *betweenness* also plays a role. We are going to analyze the situation in Sect. 12.8. We define the predicate $Be(P_1, P_2, P_3)$ to hold whenever P_2 belongs to the line segment P_1P_3 and all three points are distinct. This predicate is then added to the vocabulary in order to state the full axiomatization of origami geometry.

Our axiomatization uses two basic sorts of variables, denoting lines and points. We use $P, P_1, \ldots, P_i$, and more liberally, upper case letters, to denote points and $\ell, \ell_1, \ldots, \ell_i$, and more liberally, lower case letters, to denote lines.

We use the following basic relations:

(i) the *incidence* relation $P \in \ell$ between points and lines,
(ii) the *orthogonality* relation $\ell_1 \perp \ell_2$ between two lines,
(iii) and the *betweenness* (or *order*) relation between three points P_1, P_2, P_3 denoted by $Be(P_1, P_2, P_3)$.

The two-sorted language can be considered as a notational variant of a single-sorted one, as the set of lines and the set of points are definable using the incidence relation: ℓ is a line iff $\exists P(P \in \ell)$, and P is a point iff $\exists \ell(P \in \ell)$. This allows us to apply various standard notions and results for (one-sorted) first order logic in our context.

We denote by τ_{wu} the vocabulary corresponding to (i)-(ii) and by τ_{o-wu} the vocabulary corresponding to (i)-(iii). By $\mathrm{FOL}(\tau_{wu})$ and $\mathrm{FOL}(\tau_{o-wu})$ we denote the corresponding sets of first order formulas. $\mathrm{FOL}_{\in}$ will denote the set of formulas in the language of the incidence predicate.

Next we turn to the axioms of two-dimensional orthogonal geometry as presented in Wu (1984). The axioms are subdivided into several groups.

Hilbert's axioms of incidence

(I-1): For any two distinct points A, B there is a unique line ℓ with $A \in \ell$ and $B \in \ell$.
(I-2): Every line contains at least two distinct points.
(I-3): There exist three distinct points A, B, C such that no line ℓ contains all of them.

Hilbert's (sharper) axiom of parallels

(ParAx): Let ℓ be any line and A a point not on ℓ. Then there exists one and only one line determined by ℓ and A that passes through A and does not intersect ℓ.

Axiom schema of infinity and Desargues' axioms

(InfLines): Given distinct non-collinear A, B, C and ℓ with $A \in \ell$, $B, C \notin \ell$ we construct a line ℓ_1 going through C and parallel to AB and define A_1 as the intersection of ℓ_1 and ℓ. Inductively, we define ℓ_n as a line going through C and parallel to $A_n B$ and define A_{n+1} as its intersection with ℓ. Then all the A_i are distinct.

(De-1): If the three pairs of the corresponding sides of two triangles ABC and $A'B'C'$ are all parallel to each other, i.e., $AB \parallel A'B'$, $AC \parallel A'C'$, $BC \parallel B'C'$, then the three lines AA', BB', CC' joining the corresponding vertices of these two triangles are either parallel to each other or concurrent.

(De-2): If two pairs of the corresponding sides of two triangles ABC and $A'B'C'$ are parallel to each other, say $AB \parallel A'B'$, $AC \parallel A'C'$, and the three lines joining the corresponding vertices are distinct yet either concurrent or parallel to each other, then the third pair of the corresponding sides are also parallel to each other, i.e., $BC \parallel B'C'$.

Definition 12.1 A $\tau_{\in}$ structure Π is a *Desarguesian plane* if it satisfies (I-1, I-2, I-3), the axiom schema of infinity (InfLines), (ParAx) and the two axioms of Desargues (De-1) and (De-2).

In order to introduce the orthogonality axioms in the plane, we consider a new relation of orthogonality $\perp$ and the language τ_{wu} consisting of $\in$ and $\perp$.

Orthogonality axioms

(O-1): $\ell_1 \perp \ell_2$ iff $\ell_2 \perp \ell_1$.

(O-2): Given O and ℓ_1, there exists exactly one line ℓ_2 with $\ell_1 \perp \ell_2$ and $O \in \ell_2$.

(O-3): If $\ell_1 \perp \ell_2$ and $\ell_1 \perp \ell_3$ then $\ell_2 \parallel \ell_3$ or $\ell_2 = \ell_3$.

(O-4): For every O there is an ℓ with $O \in \ell$ and $\ell \not\perp \ell$.

(O-5): The three heights of a triangle intersect in one point.

Concerning Axiom (O-4) we remark that lines ℓ such that $\ell \perp \ell$ are called *isotropic*.

Caveat: Without any axioms of order, the axioms of metric Wu planes do not exclude the existence of isotropic lines. Since our analysis concerns the role of the axioms of order in the statements of origami principles, we do not want to assume that there are no isotropic lines outright.

Definition 12.2 A τ_{wu} structure Π is an *orthogonal Wu plane* if it is a Desarguesian plane satisfying orthogonality axioms (O-1, O-2, O-3, O-4, O-5).

Axiom of symmetric axis. We assume that we are working in an orthogonal Wu plane. In order to formulate the next axiom, we follow Wu (1984, p. 22, Definition 3) to define the relation of being a symmetric point using only the Incidence relation.

For two arbitrary points $A \neq B$ on a line ℓ, take an arbitrary $E \notin \ell$ and construct a line ℓ' parallel to ℓ such that $E \in \ell'$. Let D be the intersection of ℓ' and a line going through B parallel to AE. Then $ABDE$ is a parallelogram. Finally, construct C as the intersection of ℓ and the line going through D and parallel to EB. Then, due to Desargues' axioms, point C is independent of the choice of E. We say that C is the *symmetric point of A with respect to B*. In addition, for any point P we say that P is a symmetric point of P with respect to P.

Next we define a notion of a midpoint, following Wu (1984, p. 23, Definition 4). Let A, B be two points on a line ℓ. Draw through A a line ℓ' distinct from ℓ and take thereon a point M' distinct from A. Construct the symmetric point B' of A with respect to M' and draw through M' a line $M'M \parallel B'B$, meeting ℓ at M. Then M is independent of the choice of ℓ' and M', and is called the *midpoint* of A and B. Additionally, we define the midpoint of two coincident points to be the point itself.

Now following Wu (1984, p. 75), we use the relation of orthogonality to define symmetric axis. For any pair A, B of two distinct points, let the unique line through the midpoint of A and B and perpendicular to the line AB be the *perpendicular bisector of* A, B. Clearly, if AB is an isotropic line, then its perpendicular bisector is AB itself.

Let the perpendicular bisector of A, B be ℓ. We call A the *symmetric point of* B *with respect to* ℓ or ℓ *the symmetric axis of* A, B. Any point A on ℓ is said to be a symmetric point of itself with respect to ℓ. (Whenever ℓ is isotropic, it is the symmetric axis for any two points A, B on ℓ.) We denote this relation as $Sym(A, \ell, B)$. It will be important in our treatment of origami geometry.

Let ℓ_1 be any line and ℓ be a non-isotropic line. By Wu (1984, p. 76, Property 1), the points symmetric to the points from ℓ_1 with respect to ℓ are also lying on a unique line, say ℓ_2. In this case we call ℓ_2 the *symmetric line of* ℓ_1 *with respect to* ℓ, or ℓ a *symmetric axis of* ℓ_1 *and* ℓ_2.

Finally, we can add the following axiom to the list.

(AxSymAx): Any two intersecting non-isotropic lines have a symmetric axis.

Now we are ready to give

Definition 12.3 A τ_{wu} structure Π is a *metric Wu plane* if it is an orthogonal Wu plane satisfying (AxSymAx).

Following Wu, we use the word 'metric' here, since assuming the axiom of symmetric axes gives us some metric properties, in particular the notion of congruence.

12.3 Background on Interpretations

In this section we briefly recall the notions of translation scheme, interpretation and bi-interpretability of first order theories, following the terminology and notations of Makowsky (2019). More background on interpretations can be found, for example, in Shoenfield (1987), Hodges (1993), Friedman and Visser (2014), Makowsky (2004). Particular interpretations that we consider here are many-dimensional relative interpretations with parameters.

Definition 12.4 Let K and L be first-order relational signatures, n a positive integer, $\mathbf{p}$ a list of variables. An (n-dimensional) *translation scheme* Σ with definable parameters $\mathbf{p}$ from L to K is specified by three items:

- a formula $\delta_\Sigma(x_0, \ldots, x_{n-1}, \mathbf{p})$ of signature K,
- for each relational symbol $R(y_0, \ldots, y_{m-1})$ of L, a formula $R_\Sigma(\mathbf{x}_0, \ldots, \mathbf{x}_{m-1}, \mathbf{p})$ of K in which the $\mathbf{x}_i$ are disjoint n-tuples of distinct variables,
- a formula $\psi_\Sigma(\mathbf{p})$ of K.

Intuitively, δ_Σ denotes the domain of Σ; R_Σ denote the relations of Σ and $\psi_\Sigma(\mathbf{p})$ denotes the admissible range of parameters of Σ.

In case L is multi-sorted we may use a δ_Σ for each sort. Then the defining formulas for different sorts can also have different dimensions.

Any translation scheme Σ from L to K naturally defines a map from L-formulas to K-formulas by making it commute with the propositional connectives and relativizing the quantifiers to the domain specified by δ_Σ. We denote this map by $\phi \mapsto \phi^\Sigma$.

Definition 12.5 Let T be a K-theory and S be an L-theory. A translation scheme Σ from L to K is an *interpretation of S in T* if

1. T proves $\exists \mathbf{p}\, \psi_\Sigma(\mathbf{p})$ and $\forall \mathbf{p}\, (\psi_\Sigma(\mathbf{p}) \to \exists x_0 \ldots \exists x_{n-1} \delta_\Sigma(x_0, \ldots, x_{n-1}, \mathbf{p}))$.
2. For any L-sentence ϕ,

$$S \vdash \phi \;\Rightarrow\; T \vdash \forall \mathbf{p}\, (\psi_\Sigma(\mathbf{p}) \to \phi^\Sigma).$$

Let Σ be an interpretation of S in T. Let M be a model of T, and let $\mathbf{m}$ be a tuple of elements of M such that $M \models \psi_\Sigma(\mathbf{m})$. The interpretation Σ naturally defines an internal model $\Sigma^*(M, \mathbf{m})$ of S within M. The domain of $\Sigma^*(M, \mathbf{m})$ is the quotient of the set $D := \{\mathbf{x} \in M : M \models \delta_\Sigma(\mathbf{x}, \mathbf{m})\}$ by the equivalence relation defined on D by the interpreted equality relation. Formulas R_Σ then define the evaluation of the signature of S in $\Sigma^*(M, \mathbf{m})$. In fact, $\Sigma^*(M, \mathbf{m})$ will be a model of S, since for all $\alpha \in S$ we have $M \models \alpha^\Sigma(\mathbf{m})$.

We can often ignore the dependence of $\Sigma^*(M, \mathbf{m})$ on the choice of parameters $\mathbf{m}$ and will denote it by $\Sigma^*(M)$. Next we introduce the notion of bi-interpretability of theories (and of the corresponding classes of models).

Definition 12.6 Let T be a K-theory and S be an L-theory. Suppose there are interpretations Σ of S in T and Ξ of T in S. Then, for each model M of T, $\Xi^*(\Sigma^*(M))$ is a model of T and, for each model N of S, $\Sigma^*(\Xi^*(N))$ is a model of S. We demand that models in these pairs are isomorphic and, moreover, definably isomorphic, in the following sense.

Suppose there are formulas α in K and β in L, such that for any model M of T, α defines in M an isomorphism between M and $\Xi^*(\Sigma^*(M))$, for any admissible choice of parameters, and, for any model N of S, β defines in N an isomorphism between B and $\Sigma^*(\Xi^*(N))$, for any admissible choice of parameters. Then we say that T and S are *bi-interpretable*.

Bi-interpretation is a rather strong form of equivalence of theories, in particular it reduces the decision problem for one theory to another.[1]

[1] There also are well-known weaker conditions under which interpretations preserve decidability of theories. For example, it is sufficient to require that each model of S is isomorphic to a model of the form $\Sigma^*(M)$ for some model M of T, see Lemma 4.1 in Makowsky (2017).

Proposition 12.1 *If S and T are bi-interpretable and S is decidable, then so is T.*

Rather than providing a proof we will give concrete examples how this works below.

12.4 Background on Coordinatization

Definition 12.7 Pythagorean field is a field for which every sum of two squares is a square:

$$\forall x, y\, \exists z\; x^2 + y^2 = z^2$$

We would like to describe mutual interpretations between the classes of metric Wu planes and of Pythagorean fields.

Given a field $\mathcal{F}$ one can define a plane Π in a standard manner via Cartesian coordinates. We denote such a Π as $PP^*_{wu}(\mathcal{F})$, following (Makowsky, 2019). It is rather obvious that the two sorts (points and lines) and the predicates of incidence and orthogonality will be definable in the language of fields, which in fact yields a first-order (two-dimensional) interpretation. Notice that this interpretation is parameter-free.

On the other hand, given a plane Π we can define a field $\mathcal{F}$ in it by a classical construction known as 'coordinatization'. This construction was explicitly introduced by M. Hall in Hall (1943). M. Hall credits von Staudt (1857), Hilbert (1971) for the original idea. A good exposition can be found in Blumenthal (1980), Szmielew (1983).

Since coordinatization is more complicated than the opposite map, it is not immediately obvious that it yields a first-order interpretation. In fact, the amount of work needed to see it is considerable. The purpose of this section is to give a logically-minded reader enough details, without writing out all the first-order formulas explicitly. We follow here Ivanov (2016), which contains a particularly nice exposition of this construction.

Let Π be a stucture satisfying (I-1, I-2, I-3) and (ParAx), with two distinguished intersecting lines ℓ_0, m_0 in Π. Let O be the point of intersection of the lines ℓ_0 and m_0. Take any point Z_0 such that $Z_0 \notin \ell_0 \cup m_0$. All the constructions below will depend on a particular choice of parameters ℓ_0, m_0, and Z_0 in this configuration.

Lemma 12.1 *There is a formula bij$(X, Y, \ell_0, m_0, Z_0) \in \mathrm{FOL}_{\in}$ which, for every choice of ℓ_0, m_0 and Z_0 as above, defines a bijection between the points of ℓ_0 and of m_0.*

Proof Let d be the line going through the points O and Z_0. Take any $X \in \ell_0$ and let $h(X)$ be the point at the intersection of d and the line m_X parallel to m_0 containing

X. Let $y(X) \in m_0$ be the point at the intersection of m_0 and the line ℓ_X parallel to ℓ_0 and containing $h(X)$. Clearly $f : \ell_0 \to m_0$ given by $f(X) = y(X)$ is a bijection and is FOL definable by some formula $bij(X, Y, \ell_0, m_0, Z_0)$. □

We are going to define a structure $RF^*_{field}(\Pi)$ (this notation is once again taken from Makowsky (2019)) whose universe is denoted as K, which we take to be $\{A : A \in \ell_0\}$. Thinking of ℓ_0 and m_0 as the axes of a *coordinate system* we can identify the points of Π with elements of K^2 as follows.

Let P be a point of Π. The projection of a point P onto ℓ_0 is defined by the point $X \in \ell_0$ which is the intersection of ℓ_0 with the line m_P parallel to m_0 and $P \in m_P$. After analogously projecting a point P onto m_0, we get a point $Y \in m_0$. Then the coordinates of P is a pair $(X, f^{-1}(Y)) \in K^2$.

We define 0 and 1 in K by saying that the point O has coordinates $(0, 0)$ and the point Z_0 has coordinates $(1, 1)$. Coordinates and elements of K will also be denoted by lower case letters. It should be clear from the context whether lower case letters denote lines or elements of K.

Next we define the *slope* $sl(\ell) \in K \cup \{\infty\}$ of a line ℓ in Π. If ℓ is parallel to ℓ_0, its slope is 0 and it is called a *horizontal* line. If ℓ is parallel to m_0, its slope is ∞ and it is called a *vertical* line. For ℓ not vertical, let ℓ_1 be the line parallel to ℓ and passing through 0. Let $(1, a)$ be the coordinates of the intersection of ℓ_1 with the vertical line ℓ_2 passing through $(1, 0)$. Then we define the slope by $sl(\ell) = a \in K$.

This shows:

Lemma 12.2 *There is a first order formula* $slope(\ell, A, \ell_0, m_0, Z_0) \in \mathrm{FOL}_\in$ *which expresses* $sl(\ell) = A$*, for any choice of parameters* ℓ_0, m_0, Z_0*. There is also a first order formula* $slope_\infty(\ell, m_0) \in \mathrm{FOL}_\in$ *expressing* $sl(\ell) = \infty$.

Lemma 12.3 *(i) Two lines* ℓ, ℓ_1 *have the same slope,* $sl(\ell) = sl(\ell_1)$ *iff they are parallel.*
(ii) For the line d, defined in Lemma 12.1, we have $sl(d) = 1$ *(because* $(1, 1) \in d$*).*

We now define a ternary operation $T : K^3 \to K$ on the set $K = \{A : A \in \ell_0\}$. We think of $T(a, x, b) = \langle ax + b\rangle$ as the result of multiplying a with x and then adding b. But we yet have to define multiplication and addition.

Let $a, b, x \in K$. Let ℓ be the unique line with $sl(\ell) = a \neq \infty$ intersecting the line m_0 at the point P_1 with coordinates $P_1 = (0, b)$. Let $\ell_1 = \{(x, z) \in K^2 : z \in K\}$. The line ℓ intersects ℓ_1 at a unique point, say $P_2 = (x, y)$. We set $T(a, x, b) = y$.

Lemma 12.4 *There is a formula* $Ter(a, x, b, y, \ell_0, m_0, Z_0) \in \mathrm{FOL}_\in$*, where* a, b, x, y *range over coordinates and* ℓ_0, m_0, Z_0 *are parameters of lines and points, which expresses that* $T(a, x, b) = y$.

Lemma 12.5 *The ternary operation* $T(a, x, b)$ *has the following properties and interpretations:*

(T-1): $T(1, x, 0) = T(x, 1, 0) = x$.
$T(1, x, 0) = x$ means that the auxiliary line $d = \{(x, x) \in K^2 : x \in K\}$ *is a*

line with $sl(d) = 1$.
$T(x, 1, 0) = x$ *means that the slope of the line* ℓ *passing through* $(0, 0)$ *and* $(1, x)$ *is given by* $sl(\ell) = x$.

(T-2): $T(a, 0, b) = T(0, a, b) = b$.
The equation $T(a, 0, b) = b$ *means that the line* ℓ *defined by* $T(a, x, b) = y$ *intersects* m_0 *at* $(0, b)$.
The equation $T(0, a, b) = b$ *means that the horizontal line* ℓ_1 passing through $(0, b)$ consists of the points $\{(a, b) \in K^2 : a \in K\}$.

(T-3): *For all* $a, x, y \in K$ *there is a unique* $b \in K$ *such that* $T(a, x, b) = y$.
This means that for every slope s *different from* ∞ *there is a unique line* ℓ *with* $sl(\ell) = s$ *passing through* (x, y).

(T-4): *For every* $a, a', b, b' \in K$ *and* $a \neq a'$ *the equation* $T(a, x, b) = T(a'x, b')$ *has a unique solution* $x \in K$.
This means that two lines ℓ_1 *and* ℓ_2 *with different slopes not equal to* ∞ *intersect at a unique point* P.

(T-5): *For every* $x, y, x', y' \in K$ *and* $x \neq x'$ *there is a unique pair* $a, b \in K$ *such that* $T(a, x, b) = y$ *and* $T(a, x', b) = y'$.
This means that any two points P_1, P_2 *not on the same vertical line belong to a unique line* ℓ *with slope different from* ∞.

A structure $\langle K, T_K, 0, 1\rangle$ with a ternary operation T_K and $0, 1 \in K$ satisfying (T-1)–(T-5) is called a *planar ternary ring*, in short PTR. We also define addition $add_T(a, b, c)$ by $T(a, 1, b) = c$ and multiplication $mult_T(a, x, c)$ by $T(a, x, 0) = c$.

Following Makowsky (2019), we denote the structure $(K; add_T, mult_T, 0, 1)$ as $RF^*_{field}(\Pi)$. It is shown in Hilbert (2013), that if Π is a Desarguesian plane, then $RF^*_{field}(\Pi)$ is a skew-field (a field, in which the commutativity of the multiplication is not assumed) of characteristic 0. Moreover, as proved in Wu (1984, p. 42, Theorem 1), for any such Π and any two choices of ℓ_0, m_0, Z_0, there is an isomorphism between the two obtained skew-fields.

12.5 Metric Wu Planes and Pythagorean Fields

The following theorem is stated in a somewhat weaker form in Makowsky (2019), however it is based on the previous classical results, in particular, by Hall (1943) and Wu (1984).

Theorem 12.1 *(i) Let* $\mathcal{F}$ *be a Pythagorean field of characteristic* 0. *Then* $PP^*_{wu}(\mathcal{F})$ *is a metric Wu plane.*

(ii) Let Π *be a metric Wu plane. Then* $RF^*_{field}(\Pi)$ *is a Pythagorean field of characteristic* 0.

(iii) $RF^*_{field}(PP^*_{wu}(\mathcal{F}))$ *is isomorphic to* $\mathcal{F}$.

(iv) $PP^*_{wu}(RF^*_{field}(\Pi))$ *is isomorphic to* Π.

(v) *The isomorphisms in (iii) and (iv) are definable, that is, form a bi-interpretation between the classes of metric Wu planes and of Pythagorean fields of characteristic* 0.

Proof (i) Axioms (I-1, I-2, I-3) and (ParAx) are shown in Hartshorne (2000, Proposition 14.1). The infinity axiom holds, since $\mathcal{F}$ has characteristic 0. Considering Desargues' axioms, Proposition 14.4 in Hartshorne (2000) shows that Pappus theorem holds in a plane defined over a field. Then by Hessenberg's theorem, Wu (1984, p. 67), Desargues' axioms also hold.

We naturally define lines $a_0x + b_0y + c_0 = 0$ and $a_1x + b_1y + c_1 = 0$ to be orthogonal if $a_0a_1 + b_0b_1 = 0$. Then the axiom (O-1) holds by commutativity of multiplication. The axioms (O-2) and (O-3) hold since we are able to solve systems of linear equations.

Considering axiom (O-5), let (a_1, b_1), (a_2, b_2) and (a_3, b_3) be the coordinates of a triangle. Then a line going through (a_1, b_1) and perpendicular to the opposite side is defined by an equation

$$\frac{x - a_1}{b_3 - b_2} + \frac{y - b_1}{a_3 - a_2} = 0.$$

Then since linear equations are solvable in a Pythagorean field of characteristic 0, we can write equations for all three heights and see that the orthocenter (x, y) exists. Therefore, (O-5) holds.

We know that $1 \neq 0$ and hence any line of the form $x = c$ is non-isotropic. Then for any point (x_0, y_0) there is a non-isotropic line $x = x_0$ passing through it and (O-4) holds.

Finally, suppose an angle is formed by two non-isotropic lines given by $l_1x + m_1y + n_1 = 0$ and $l_2x + m_2y + n_2 = 0$. Then the internal and external bisectors of the angle are given by the two equations

$$\frac{l_1x + m_1y + n_1}{\sqrt{l_1^2 + m_1^2}} = \pm\frac{l_2x + m_2y + n_2}{\sqrt{l_2^2 + m_2^2}}.$$

Since $\mathcal{F}$ is Pythagorean, the roots exist and are not equal to 0, because the lines are non-isotropic. Therefore, bisectors exist and (AxSymAx) holds.

(ii) This statement is extensively discussed in Wu (1984). As mentioned above, Hilbert showed in *Grundlagen der Geometrie* that $RF^*_{field}(\Pi)$ forms a skew-field of characteristic 0. Wu first proves in Wu (1984, Sect. 2.1) that Linear Pascalian axiom (a version of Pappus theorem) is sufficient to obtain the commutativity of multiplication and then in Wu (1984, p. 72) shows that Linear Pascalian axiom holds in any metric Wu plane.

Finally, to conclude that $RF^*_{field}(\Pi)$ is Pythagorean, we refer to the Pythagorean Theorem (known as 'kou-ku theorem' Wu (1984)) proved in Wu (1984, p. 97).

(iii) It is shown in Wu (1984, p. 42, Theorem 1), that any choice of parameters ℓ_0, m_0 and Z_0 in the definition of $RF^*_{field}(\Pi)$ gives us isomorphic fields. Wu provides an explicit construction of such an isomorphism and we shortly describe it here.

Let ℓ_0, m_0, Z_0 and ℓ_0', m_0', Z_0' be two choices of parameters and let O, I and O', I' be the corresponding zero and one points. First consider the case when $O = O'$ but $\ell_0 \neq \ell_0'$. Then as shown by Wu the parallel projection with respect to the line II' defines the required isomorphism. Now consider the case $\ell_0 \parallel \ell_0'$ and $OO' \parallel II'$. Then the parallel projection with respect to the line OO' defines the required isomorphism. Finally, observe that any other case can be reduced to those two by taking compositions of isomorphisms. For example, in a general case of ℓ_0 and ℓ_0' being neither coincident nor parallel and their intersection point not equal to O or O', construct a line ℓ_0'' through O parallel to ℓ_0'. Then take I'' to be a point on ℓ_0'' such that $O'O \parallel I'I''$. The line ℓ_0'' and points O, I'' define a field by taking $m_0'' \perp \ell_0''$, $O \in m_0''$ and Z_0'' defined by I''. Now using the first case obtain an isomorphism between fields of ℓ_0 and ℓ_0'' and using the second case an isomorphism between fields of ℓ_0'' and ℓ_0'. Then the required isomorphism is their composition.

Clearly, if we choose ℓ_0, m_0 to be the axes of $PP^*_{wu}(\mathcal{F})$ and $Z_0 = (1, 1)$, then the constructed field will be isomorphic to $\mathcal{F}$. Hence, it will be isomorphic to $RF^*_{field}(PP^*_{wu}(\mathcal{F}))$ for every choice of ℓ_0, m_0, Z_0.

This result can also be found in Hall (1943, Theorem 5.9).

(iv) Let Π be a metric Wu plane and let $\mathcal{F} = RF^*_{field}(\Pi)$. In order to establish an isomorphism between $PP^*_{wu}(\mathcal{F})$ and Π we need to define two maps: a map for points and a map for lines.

Points of $PP^*_{wu}(\mathcal{F})$ are pairs $(x, y) \in \mathcal{F}^2$. We recall that the universe of $\mathcal{F}$ is the set of points incident to the line ℓ_0, and that there is a definable bijection f between the points of ℓ_0 and m_0 (the coordinate axes in Π, parameters of the considered interpretation). Hence, given (x, y) we can define two auxiliary lines: m_x going through $x \in \ell_0$ and parallel to m_0, and ℓ_y, going through $f(y) \in m_0$ and parallel to ℓ_0. Let A be the intersection of m_x and ℓ_y. We map (x, y) to A, and it is clear that A has coordinates (x, y). Thus, we have described a (definable) bijection between the sets of points of $PP^*_{wu}(\mathcal{F})$ and Π.

Lines of $PP^*_{wu}(\mathcal{F})$ can be specified by equations $ax + by + c = 0$ where not all of a, b, c are 0. Thus, a line is interpreted by a triple $(a, b, c) \in \mathcal{F}^3$. Two lines are defined to be equal if (a, b, c) and (a', b', c') are proportional. A point (x, y) is incident to a line (a, b, c) if $ax + by + c = 0$.

Each line in $PP^*_{wu}(\mathcal{F})$ is equal to a line defined by the equation $y = ax + b$ or to a vertical line $x = c$. We construct the corresponding line in Π by drawing, in the first case, a line through the point $(0, b)$ with the slope a, and in the second case a vertical line (parallel to m_0) through $(c, 0)$. This maps lines in $PP^*_{wu}(\mathcal{F})$ to lines in Π and is clearly a (definable) bijection preserving the incidence relation.

Concerning the orthogonality relation, we may assume that the coordinate axes ℓ_0 and m_0 in Π are selected to be orthogonal. (By Wu, the field $\mathcal{F}$ does not depend on the choice of parameters, up to isomorphism.) We can define lines (a, b, c) and (a', b', c') in $PP^*_{wu}(\mathcal{F})$ to be orthogonal iff $aa' + bb' = 0$. Then the usual arguments show that this agrees with the orthogonality of the corresponding lines in Π.

(v) In the proof of (iii) we use the isomorphisms established in Wu (1984, p. 42, Theorem 1). The construction of these isomorphisms mostly involves drawing various parallel lines and is clearly definable. The isomorphism between $\mathcal{F}$

and $RF^*_{field}(PP^*_{wu}(\mathcal{F}))$ where ℓ_0, m_0 are the axes of $PP^*_{wu}(\mathcal{F})$ and $Z_0 = (1, 1)$, is defined by a map $x \mapsto (x, 0)$, which is first-order definable. The construction given in the proof of (iv) explicitly provides us with a definable isomorphism. □

12.6 Undecidability

To establish the undecidability of the theory of Pythagorean fields we refer to the following theorem of Ziegler (1982), Beeson (1982).

Theorem 12.2 *Let T be a finite subtheory of the theory of the field of reals* $(\mathbb{R}; +, \times)$. *Then*

(i) T is undecidable;
(ii) The same holds for the extension of T by the axioms stating that the characteristic of the field is 0.

Although the second part is not mentioned as a result in Ziegler (1982), Beeson (1982), it easily follows from Ziegler's proof.

Corollary 12.1 *The theories of Pythagorean fields and of Pythagorean fields of characteristic* 0 *are undecidable.*

Using the bi-interpretability of Pythagorean fields and metric Wu planes we obtain the undecidability of the theory of metric Wu planes. In fact, we prove a more general theorem establishing the undecidability of a sufficiently wide class of geometric theories. As a preparation for its proof, we define syntactic translations between formulas in the language of fields and formulas in the language τ_{wu}.

Consider any formula ϕ in the language of fields. Using the formulas *add* and *mult* from the construction of $RF^*_{field}(\Pi)$ to interpret addition and multiplication, we obtain a formula $\phi^{wu}(\ell_0, m_0, Z_0)$ in the language τ_{wu}, where ℓ_0, m_0, Z_0 are the parameters (free variables) of the formula.

Let $Par(\ell_0, m_0, Z_0)$ denote the formula stating that ℓ_0 and m_0 are lines intersecting in exactly one point and that Z_0 is not incident with either ℓ_0 or m_0. These conditions definably specify the admissible values of the parameters. Then, for any metric Wu plane Π,

$$RF^*_{field}(\Pi) \models \phi \iff \Pi \models \forall \ell_0, m_0, Z_0\ (Par(\ell_0, m_0, Z_0) \rightarrow \phi^{wu}). \qquad (12.1)$$

Similarly, for the other interpretation, consider any formula ϕ in the language of τ_{wu}. Using the formulas from the construction of PP^*_{wu} to interpret the two sorts of variables, equality, incidence and orthogonality, we obtain a formula ϕ^{field} in the language of fields. Then, for any field $\mathcal{F}$,

$$PP^*_{wu}(\mathcal{F}) \models \phi \iff \mathcal{F} \models \phi^{field}. \qquad (12.2)$$

Now we are ready to state a geometric version of Ziegler's Theorem.

Let $\Pi_{\mathbb{R}}$ denote the real plane $PP^*_{wu}(\mathbb{R})$. Let WU denote the first order theory of metric Wu planes and let PF denote the first order theory of Pythagorean fields of characteristic 0.

Theorem 12.3 *Let T be a finite set of axioms in the vocabulary τ_{wu} such that $\Pi_{\mathbb{R}} \models T$. Then $T \cup \mathrm{WU}$ is undecidable.*

Proof Let $T' = \{\phi^{field} \mid \phi \in T\}$. Then by Ziegler's theorem, $T' \cup \mathrm{PF}$ is undecidable.

We want to prove that

$$T' \cup \mathrm{PF} \models \phi \iff T \cup \mathrm{WU} \models \forall \ell_0, m_0, Z_0\, (Par(\ell_0, m_0, Z_0) \to \phi^{wu}). \quad (12.3)$$

Then, since the translation $(\cdot)^{wu}$ is computable, this provides a computable reduction of $T' \cup \mathrm{PF}$ to $T \cup \mathrm{WU}$ and proves that the latter is undecidable.

To prove (12.3), suppose $T \cup \mathrm{WU} \models \forall \ell_0, m_0, Z_0\, (Par(\ell_0, m_0, Z_0) \to \phi^{wu})$. Take any $\mathcal{F} \models T' \cup \mathrm{PF}$. Then by Theorem 12.1, $PP^*_{wu}(\mathcal{F})$ is a metric Wu plane and, using (12.2), we have $PP^*_{wu}(\mathcal{F}) \models T \cup \mathrm{WU}$. Hence,

$$PP^*_{wu}(\mathcal{F}) \models \forall \ell_0, m_0, Z_0\, (Par(\ell_0, m_0, Z_0) \to \phi^{wu}).$$

Then, by (12.1), $RF^*_{field}(PP^*_{wu}(\mathcal{F})) \models \phi$ and, since $\mathcal{F} \cong RF^*_{field}(PP^*_{wu}(\mathcal{F}))$, we obtain $T' \cup \mathrm{PF} \models \phi$.

Suppose $T' \cup \mathrm{PF} \models \phi$. Consider any $\Pi \models T \cup \mathrm{WU}$. Let $\mathcal{F} = RF^*_{field}(\Pi)$. By Theorem 12.1, $\mathcal{F}$ is a Pythagorean field of characteristic 0 and Π is isomorphic to $PP^*_{wu}(\mathcal{F})$. Then, using (12.2), we obtain $\mathcal{F} \models T' \cup \mathrm{PF}$. Therefore, $\mathcal{F} \models \phi$ and by (12.1), $\Pi \models \forall \ell_0, m_0, Z_0\, (Par(\ell_0, m_0, Z_0) \to \phi^{wu})$. It follows that $T \cup \mathrm{WU} \models \forall \ell_0, m_0, Z_0\, (Par(\ell_0, m_0, Z_0) \to \phi^{wu})$.

This completes the proof of (12.3) and thereby of Theorem 12.3. □

Corollary 12.2 *The theory of metric Wu planes is undecidable.*

12.7 Logical Huzita–Justin Axioms

Looking at the Huzita–Justin axioms it appears as if one could axiomatize origami geometry in the language of incidence and orthogonality only, taking as the underlying geometry a metric Wu plane. However, it turns out that there are multiple reasons for choosing as the underlying geometry an *ordered* metric Wu plane.

Firstly, the Huzita–Justin axiom (H-3), which says that for any two lines l_1, l_2 there is a fold which places l_1 on l_2, is only true if l_1, l_2 are non-isotropic (not orthogonal to themselves). Our formulation of (H*3) takes this into account.

Secondly, Axiom (H-5) states that, given two points P_1 and P_2 and a line ℓ, one should be able to construct a fold that places P_1 onto ℓ and passes through P_2.

However, such a fold only exists provided P_1 is closer to the given line ℓ than P_2. This condition needs to be expressible in the language, which we achieve by introducing the betweenness relation, as formulated in (H*5). Other possibilities are discussed at the end of the paper.

We say that a metric Wu plane is orderable if it can be equipped with a ternary relation $Be(A, B, C)$ which satisfies the Hilbertian axioms of betweenness. In Proposition 12.4 we prove that a metric Wu plane is orderable iff it has no isotropic lines.

A reasonable axiomatization of origami geometry can be obtained from ordered metric Wu planes by adding a finite set of axioms. Metric Wu planes already satisfy Huzita–Justin axioms (H-1), (H-2), (H*3), (H-4) where (H*3) is our modification of (H-3), see Proposition 12.2. An *ordered origami plane* is an ordered metric Wu plane which satisfies also our modified axioms (H*5) and (H*6), see Sect. 12.9. This deviates from the definition given in Makowsky (2018, 2019) where an ordered origami plane is defined in an inconsistent way.

The ordered metric Wu planes are bi-interpretable with ordered Pythagorean fields, whose first-order theory is undecidable by Ziegler's theorem. Similarly, we obtain that the first order theory of our axiomatization of origami geometry is also undecidable.

Huzita–Justin axioms were not meant to be axioms in the logical sense, but rather rules of folding. Yet, one can try to naively formulate them in the language τ_{wu} by treating the requirement to construct an object (satisfying given conditions) by a classical existential statement. Huzita–Justin axioms are naturally stated using the relation $Sym(P_1, \ell, P_2)$ "*points P_1 and P_2 are symmetric with respect to line ℓ*" defined in Sect. 12.2. Then one obtains the following versions of Huzita–Justin axioms.

(H-1): Given two points P_1 and P_2, one can make a fold that passes through both of them:

$$\forall P_1, P_2 \, \exists \ell \, (P_1 \in \ell \wedge P_2 \in \ell).$$

(H-2): Given two points P_1 and P_2, one can make a fold that places P_1 onto P_2:

$$\forall P_1, P_2 \, \exists \ell \, Sym(P_1, \ell, P_2).$$

(H-3): Given two lines ℓ_1 and ℓ_2, one can make a fold that places ℓ_1 onto ℓ_2:

$$\forall \ell_1, \ell_2 \, \exists k \, \forall P_1 \, \left(P_1 \in \ell_1 \rightarrow \exists P_2 \, (P_2 \in \ell_2 \wedge Sym(P_1, k, P_2))\right).$$

(H-4): Given a point P and a line ℓ, one can make a fold orthogonal to ℓ that passes through P:

$$\forall P, \ell \, \exists k \, (P \in k \wedge \ell \perp k).$$

(H-5): Given two points P_1 and P_2 and a line ℓ_1, one can make a fold that places P_1 onto ℓ_1 and passes through P_2:

$$\forall P_1, P_2, \ell_1 \exists \ell_2 \, (P_2 \in \ell_2 \wedge \exists P_3 \, (Sym(P_1, \ell_2, P_3) \wedge P_3 \in \ell_1)).$$

(H-6): Given two points P_1 and P_2 and two lines ℓ_1 and ℓ_2, one can make a fold that places P_1 onto ℓ_1 and P_2 onto ℓ_2:

$$\begin{aligned} \forall P_1, P_2, \ell_1, \ell_2 \exists \ell_3 \, (\exists Q_1 \, (Sym(P_1, \ell_3, Q_1) \wedge Q_1 \in \ell_1) \wedge \\ \wedge \exists Q_2 \, (Sym(P_2, \ell_3, Q_2) \wedge Q_2 \in \ell_2)) \, . \end{aligned}$$

Since the original Huzita–Justin axioms talk only about the possibility of the existence of folds, Axioms (H-5) and (H-6) formulated above do not hold in a real plane. The exceptional configurations, where a considered fold does not exist, have to be described explicitly. In order to fix that we are going to amend them with the most obvious conditions under which the lines would indeed exist. On the other hand, the formalizations of Axioms (H-1), (H-2), (H-3), (H-4) obviously hold in the real plane and, as explained below, almost hold in any metric Wu plane.

We would like to state that metric Wu planes satisfy the first four origami axioms (H-1), (H-2), (H-3), (H-4). However, since there could exist non-isotropic lines in a metric Wu plane, they do not necessarily satisfy (H-3). Thus, we modify this axiom.

(H*3): Given two non-isotropic lines ℓ_1 and ℓ_2, there is a fold (line) that places ℓ_1 onto ℓ_2.

Proposition 12.2 *Every metric Wu plane satisfies the origami axioms (H-1), (H-2), (H*3), (H-4).*

Proof (H-1) is equivalent to (I-1) and (H-4) is equivalent to (O-2). To prove (H-2) we use the construction from Wu (1984, p. 75).

(AxSymAx) is an analogue of (H*3) for intersecting lines, so we may only consider the case of parallel lines. Take ℓ_1 and ℓ_2 parallel to each other. Take any point $P_1 \in \ell_1$ and drop a perpendicular from P_1 on ℓ_2. Let the intersecting point be P_2. Then we claim that the perpendicular bisector of $P_1 P_2$ is the line we need. □

12.8 Ordered Metric Wu Planes and Pythagorean Fields

We have established a correspondence between Pythagorean fields and metric Wu planes. Following Alperin (2000), the next step would be a correspondence between Euclidean fields, defined below, and planes satisfying some analogue of (H-5).

Definition 12.8 A *Euclidean field* is a formally real Pythagorean field such that every element is either a square or the opposite of a square:

$$\forall x \exists y \, (x = y^2 \vee -x = y^2).$$

As we will discuss in the next section, Euclidean fields are always uniquely ordered, therefore we want our plane to be in some sense "ordered" as well. One way to do so would be to take as an axiom that there are no isotropic lines. Then the corresponding field would be formally real and hence orderable (see below).

We take a different approach and follow Wu (1984). As the concept 'lying between' is not definable in a metric Wu plane, we introduce a new relation of Betweenness and the axioms that describe it. We interpret $Be(P_1, P_2, P_3)$ as *three distinct points are on the same line and P_2 is between P_1 and P_3*. Note, that if three points are not all distinct, none of them lie between the others. Let τ_{o-wu} be the signature consisting of $\in$, $\perp$ and Be.

Axioms of betweenness

(B-1): Let A, B, C be three distinct points on a line. If B lies between A and C, then B also lies between C and A.

(B-2): For any two distinct points A and C on a line, there always exists another point B which lies between A and C, and another point D such that C lies between A and D.

(B-3): Given any three distinct points A, B, C on a line, one and only one of the following three cases holds: B lies between A and C, A lies between B and C, and C lies between A and B.

(B-4): (Pasch) Assume the points A, B, C and ℓ in general position, i.e. the three points are not on one line, none of the points is on ℓ. Suppose there is a point D at which ℓ and the line AB intersect. If $Be(A, D, B)$, then there is $D' \in \ell$ with $Be(A, D', C)$ or $Be(B, D', C)$.

Definition 12.9 A τ_{o-wu} structure Π is an ordered metric Wu plane if it is a metric Wu plane satisfying axioms of betweenness (B-1), (B-2), (B-3), (B-4).

Proposition 12.3 *Every ordered metric Wu plane satisfies (H-3) and hence the origami axioms (H-1), (H-2), (H-3), (H-4).*

Proof There are no isotropic lines in ordered metric Wu planes as proven in Wu (1984, p. 107, Theorem 3). □

If $\mathcal{F}$ is an ordered field, we define the relation of betweenness on $PP^*_{wu}(\mathcal{F})$ in the standard way. If Π is an ordered metric Wu plane, we follow Wu (1984, p. 105) to define an order on $RF^*_{field}(\Pi)$. Using the result, Wu (1984, p. 103, Separation property 1), we separate all points on ℓ_0 distinct from 0 into two parts. Then 0 lies between A, B when A, B lie in different parts, and 0 does not lie between A, B when A, B lie in the same part. We define those numbers in $RF^*_{field}(\Pi)$ whose corresponding points lie in the same part as 1 to be positive numbers and those whose corresponding points lie in the other part to be negative numbers. Then we can say that $a < b$ whenever $b - a$ is a positive number.

Theorem 12.4 *(i) Let $\mathcal{F}$ be an ordered Pythagorean field. Then $PP^*_{wu}(\mathcal{F})$ is an ordered metric Wu plane.*

(ii) Let Π *be an ordered metric Wu plane. Then* $RF^*_{field}(\Pi)$ *is an ordered Pythagorean field.*
(iii) $RF^*_{field}(PP^*_{wu}(\mathcal{F}))$ *is isomorphic to* $\mathcal{F}$*.*
(iv) $PP^*_{wu}(RF^*_{field}(\Pi))$ *is isomorphic to* Π*.*
(v) The isomorphisms in (iii) and (iv) are definable, that is, form a bi-interpretation between the classes of ordered metric Wu planes and of ordered Pythagorean fields.

Proof (i) Using properties of ordered fields, it is easy to check that $PP^*_{wu}(\mathcal{F})$ satisfies the axioms of betweenness (B-1, B-2, B-3, B-4).

(ii) We only need to show that $RF^*_{field}(\Pi)$ is an ordered field. This is proved on Wu (1984, p. 105, Theorem 1).

(iii) By Theorem 12.1 it is sufficient to check that the relation of order is preserved. As discussed on Wu (1984, p. 105), if we take different ℓ_0, m_0, Z_0 in the construction of $RF^*_{field}(\Pi)$, the canonical isomorphism between the obtained fields will preserve order. If we choose ℓ_0, m_0 to be the axes of $PP^*_{wu}(\mathcal{F})$ and $Z_0 = (1, 1)$, then the field is clearly isomorphic to $\mathcal{F}$. It follows that $\mathcal{F}$ is isomorphic to $RF^*_{field}(PP^*_{wu}(\mathcal{F}))$ for any choice of parameters.

(iv) Let $\mathcal{F}$ denote the field $RF^*_{field}(\Pi)$. By Theorem 12.1 it is sufficient to check that the betweenness relation is preserved under the isomorphism of metric Wu planes Π and $PP^*_{wu}(\mathcal{F})$.

Since collinearity is preserved, it is sufficient to consider the betweenness relation for points on the same line. Suppose three points A, B, C on a line ℓ in Π are given. Assume ℓ is not vertical and consider the intersections A', B', C' of ℓ_0 axis and lines parallel to m_0 and going through A, B, C respectively. By Corollary 1 in (1984, p. 104), $Be(A, B, C)$ holds in an ordered metric Wu plane if and only if $Be(A', B', C')$ does. On the other hand, the interpretation of betweenness in $PP^*_{wu}(\mathcal{F})$ for points of the coordinate axis is the same as that in Π. This shows the claim in the case ℓ is not vertical.

If ℓ is vertical, we consider the projections of A, B, C on the m_0 axis and the corresponding points on ℓ_0 via the bijection f. By the same principle, f preserves the betweenness on respective coordinate axes, hence $Be(A, B, C)$ holds in Π iff it holds in $PP^*_{wu}(\mathcal{F})$.

(v) We use the same isomorphisms as in Theorem 12.1, which are already showed to be definable. □

After establishing Theorem 12.4, we also want to obtain an ordered analogue of Theorem 12.3. For this purpose, we need an ordered version of Theorem 12.2. Although in Ziegler (1982), Beeson (1982) only theories in the language of fields are concerned, the proof of Ziegler's Theorem essentially consists of constructing a model of T, in which the ring of integers is interpretable. Then it is easy to see that the proof still holds for the case of ordered fields, which gives us the following result.

Theorem 12.5 *Let* T *be a finite subtheory of the theory of the ordered field of reals* $(\mathbb{R}; +, \times, \leq)$*. Then*

(i) *T is undecidable;*
(ii) *The same holds for the extension of T by the axioms stating that the characteristic of the field is* 0.

Then using Theorem 12.4 and the same technique as in Theorem 12.3, we obtain an ordered version of the geometrical Ziegler's theorem.

Let $\Pi_{\mathbb{R}}$ denote the real plane $PP^*_{wu}(\mathbb{R})$. Let OWU denote the first order theory of ordered metric Wu planes.

Theorem 12.6 *Let T be a finite set of axioms in the vocabulary τ_{o-wu} such that $\Pi_{\mathbb{R}} \models T$. Then $T \cup$ OWU is undecidable.*

Next we consider the question of orderability. Recall that a field is orderable (or *formally real*) if -1 is not a sum of squares. For Pythagorean fields this is equivalent to saying that -1 is not a square. The statement that there are no isotropic lines plays a similar role for metric Wu planes.

Proposition 12.4 *A metric Wu plane Π is orderable if and only if there are no isotropic lines in Π.*

Proof In one direction, we have already mentioned a theorem of Wu that ordered metric Wu planes have no isotropic lines. In the other direction, we assume a Π without isotropic lines is given and consider as a coordinate system a pair of orthogonal lines and the corresponding field $\mathcal{F} = RF^*_{field}(\Pi)$. By Theorem 12.1 (iv) $PP^*_{wu}(\mathcal{F})$ is isomorphic to Π.

We claim that $\mathcal{F}$ is formally real. Assume the contrary, then $d^2 + 1 = 0$ in $\mathcal{F}$, for some d. Then the line defined by the points $(1, 0)$ and $(0, d)$ (as well as the parallel line given by the equation $dx + y = 0$) is isotropic. Since $\mathcal{F}$ is formally real, $PP^*_{wu}(\mathcal{F})$ is orderable, but it is isomorphic to Π. □

12.9 Euclidean and Vieta Fields and Origami Axioms

As already mentioned in the previous section, our goal is to establish a correspondence between Euclidean fields and planes satisfying some amended version of (H-5). Recall that a Euclidean field is a formally real Pythagorean field such that every element is either a square or the opposite of a square.

The non-zero squares of a Euclidean field constitute a positive cone, hence (see Becker, 1974) Euclidean fields admit a unique ordering:

$$x \leq y \iff \exists z\ (x + z^2 = y).$$

Since the ordering is definable, one often considers Euclidean fields as ordered fields. An ordered field is Euclidean if and only if each positive element in it is a square.

Proposition 12.5 *The first order theory of Euclidean fields is undecidable.*

Proof Ziegler's Theorem (Theorem 12.2). □

Next we formulate our amended version of Axiom (H-5) in which we add an appropriate precondition for the constructed fold to exist. Below we use the notion *a point A is closer to a line ℓ than to a point B*, $Closer(A, \ell, B)$, which can be formulated in the language τ_{o-wu} by saying that there exist points $H \in \ell$ and B' and a line $m \ni A$ such that $Sym(H, m, B')$, while $Be(A, B', B)$ or $B = B'$.

(H*5): Given two points P_1 and P_2 and a line ℓ_1, if P_2 is closer to ℓ_1 than to P_1, then there is a fold (line) that places P_1 onto ℓ_1 and passes through P_2.

Definition 12.10 A τ_{o-wu} structure Π is a *Euclidean ordered metric Wu plane* if it is an ordered metric Wu plane satisfying (H*5).

Theorem 12.7 *(i) Let $\mathcal{F}$ be a Euclidean field. Then $PP^*_{wu}(\mathcal{F})$ is a Euclidean ordered metric Wu plane.*
(ii) Let Π be a Euclidean ordered metric Wu plane. Then $RF^*_{field}(\Pi)$ is a Euclidean field.
(iii) Furthermore, $RF^*_{field}(PP^*_{wu}(\mathcal{F}))$ is isomorphic to $\mathcal{F}$.
(iv) $PP^*_{wu}(RF^*_{field}(\Pi))$ is isomorphic to Π.
(v) The isomorphisms in (iii) and (iv) are definable, that is, form a bi-interpretation between the classes of Euclidean ordered metric Wu planes and of Euclidean fields.

Proof (i) In order to show (H*5) it is enough to prove that a circle intersects a line whenever the radius is smaller than the distance between the center and the line. In $PP^*_{wu}(\mathcal{F})$ that is equivalent to solving a quadratic equation. Therefore, if $\mathcal{F}$ is Euclidean, (H*5) holds.
(ii) Consider any positive $s \in RF^*_{field}(\Pi)$ and show that the square root of s exists. Without loss of generality we assume $s > 1$, since otherwise we can find a square root of s^{-1}. Let $P_1 = (0, s)$, $P_2 = (0, \frac{s-1}{2})$ and let ℓ_0 be the x axis. Then by (H*5) we can find a point $P_3 = (x, y)$, such that $P_3 \in \ell_0$ and $|P_1P_2| = |P_3P_2|$. The first condition gives us $y = 0$. The second one, if we calculate the squares of distances, means $\left(\frac{s-1}{2}\right)^2 + x^2 = \left(\frac{s+1}{2}\right)^2$, which is equivalent to $x^2 = s$.
(iii)–(v) We use Theorem 12.4, since every Euclidean field is a Pythagorean field and every Euclidean ordered metric Wu plane is an ordered metric Wu plane. □

Then by Theorem 12.6, we obtain:

Corollary 12.3 *The theory of Euclidean ordered Wu planes is undecidable.*

Finally, we would like to find a correct version of (H*6) and to establish its correspondence with Vieta fields as defined below.

Definition 12.11 A *Vieta field* is a Euclidean field in which every element is a cube:

$$\forall x\ \exists y\ y^3 = x.$$

It follows from Cardano formula that any cubic polynomial over a Vieta field has at least one root.

Proposition 12.6 *The first order theory of Vieta fields is undecidable.*

Proof Ziegler's Theorem (Theorem 12.2). □

The following version of (H-6) is inspired by Ghourabi et al. (2012, Proposition 6).

(H*6): Given two points P_1 and P_2 and two lines ℓ_1 and ℓ_2, if $P_1 \notin \ell_1$, $P_2 \notin \ell_2$, ℓ_1 and ℓ_2 are not parallel and points are distinct or lines are distinct, then there is a fold (line) that places P_1 onto ℓ_1 and P_2 onto ℓ_2.

Definition 12.12 Π is an *ordered Wu origami plane* if it is an ordered metric Wu plane which also satisfies (H*5) and (H*6).

Theorem 12.8 *(i) Let $\mathcal{F}$ be a Vieta field. Then $PP^*_{wu}(\mathcal{F})$ is an ordered Wu origami plane.*
(ii) Let Π be an ordered Wu origami plane. Then $RF^*_{field}(\Pi)$ is a Vieta field.
(iii) Furthermore, $RF^*_{field}(PP^*_{wu}(\mathcal{F}))$ is isomorphic to $\mathcal{F}$.
(iv) $PP^*_{wu}(RF^*_{field}(\Pi))$ is isomorphic to Π.
(v) The isomorphisms in (iii) and (iv) are definable, that is, form a bi-interpretation between the classes of ordered Wu origami planes and of Vieta fields.

Proof (i) It sufficies to show that (H*6) holds in $PP^*_{wu}(\mathcal{F})$. We use Ghourabi et al., (2012, Proposition 6) to conclude that the conditions we chose are sufficient for the existence of a fold. Although the original result was proven specifically for the real plane, we note that it essentially uses only the Vieta property of $\mathbb{R}$ and therefore holds for any plane over a Vieta field.
(ii) Take any $r \in RF^*_{field}(\Pi)$. Let $P_1 = (-1, 0)$, $P_2 = (0, -r)$, ℓ_1 be the line $x = 1$ and ℓ_2 the line $y = r$. Then by (H*6) we can find a line ℓ. Drop perpendiculars from P_1 and P_2 on ℓ and let the constructed points be H_1 and H_2. Then since a reflection in ℓ maps P_1 onto ℓ_1 and P_2 onto ℓ_2, the coordinates of H_1 and H_2 have to be of the form $(0, s)$ and $(t, 0)$ respectively. Lines P_1H_1 and P_2H_2 have to be parallel, therefore $t = \frac{r}{s}$. Finally, H_1H_2 is perpendicular to P_1H_1, which means $\frac{r}{s} - s^2 = 0$. Hence, $s^3 = r$.
(iii)–(v) Once again these follow from Theorem 12.4, since every Vieta field is a Pythagorean field and every ordered Wu origami plane is an ordered metric Wu plane. □

Corollary 12.4 *The theory of ordered Wu origami planes is undecidable.*

12.10 Discussion

We have axiomatized the classes of orthogonal Wu planes using versions of origami axioms and established their bi-interpretations with the first order theories of fields (corresponding to the classes of Pythagorean, Euclidean and Vieta fields). A few natural questions concerning the axiomatization of geometry via origami constructions were left open.

One such question concerns the choice of the considered language. Although the orthogonality of lines is natural from the point of view of origami constructions—orthogonality can be tested just by a single fold—we see that the Huzita–Justin axioms are easily formulated using the notion of symmetry of two points w.r.t. a line $Sym(P, \ell, Q)$. It would be natural to consider this notion as basic and orthogonality as definable. The *Sym* predicate behaves well provided the metric Wu plane is orderable, that is, has no isotropic lines.

Problem 12.1 Find a natural axiomatization of orderable metric Wu planes in terms of $\in$ and *Sym*.

A similar question can be asked about betweenness. We have based our axiomatization on the standard Hilbertian axioms for betweenness. One can, however, consider as basic the relation $Closer(A, \ell, B)$ which holds if A is closer to line ℓ than to B. It is the relation that was used in the statement of (H*5).

Problem 12.2 Find a natural axiomatization of the class of ordered metric Wu planes in terms of $\in$, $\perp$ and *Closer*. In particular, this requires that there is a first-order formula that works as a definition of betweenness in each structure satisfying these axioms.

Another question concerns the definability of betweenness in Euclidean metric Wu planes. Recall that in Euclidean fields the ordering is definable. This suggests that there is an axiomatization of Euclidean ordered metric Wu planes in the language τ_{wu} only. In fact, one such axiomatization based on the so-called *Euclidean axiom of betweenness* is well-known (Degen & Profke, 1976, p. 149). This axiom defines the betweenness relation $Be(A, B, C)$ by stating that A, B, C are collinear and there exists a point D such that $DA \perp DC$ and $DB \perp AC$. Then, a metric Wu plane is Euclidean ordered if and only if it has no isotropic lines and the above relation Be satisfies the usual axioms of betweenness.

It would be interesting to know if the Euclidean axiom of betweenness can be replaced by its alternative suggested by the origami axiom (H-5). First, *define* $Closer(A, \ell, B)$ by saying that there is a fold m that goes through A and places B on ℓ. Second, define $Be(A, B, C)$ by saying that A, B, C are collinear and there is a line $\ell \ni B$ such that $\ell \perp AC$, $Closer(A, \ell, C)$ and $Closer(C, \ell, A)$. State (some of) the betweenness axioms for the defined relation. Though this approach may work, there are a number of details to be worked out here that we leave for a future study.

Problem 12.3 Find an axiomatization of Euclidean orderable Wu planes in the language τ_{wu} that would be natural from the point of view of origami.

Yet another interesting direction of study is to develop a constructive version of origami geometry as a logical theory based on intuitionistic logic, in the spirit of the work of Beeson (2015). In such a theory existential statements would yield actual origami constructions rather than just be classically true.

Problem 12.4 Develop a constructive version of origami geometry.

12.11 Special Acknowledgements

This paper is dedicated to Professor Dick de Jongh. The subject of the paper, the logical analysis of origami geometry, has little to do with Dick's own work. Even so, we hope that Dick will like its almost recreational character and the use of logical notions such as interpretations.

Dick's influence on the first author of the paper was quite a formative one. Dick was one of the first logicians who established contacts with the colleagues in Moscow and Tbilisi at the end of the Soviet Union. Over the years, his involvement ranged from arranging humanitarian aid and donations of equipment to organizing workshops, lectures and mutual visits. The close ties between our communities today are consequences of these selfless efforts. Thirty years ago, still a PhD student, Lev came to Amsterdam for the first time for a month's visit, and it was Dick who was his host and guide to all things new, both in academia and in the western culture. Since then, Dick remains for Lev a constant source of support and advice.

Recently, Dick has also functioned as the academic mentor of the second author during her studies at the Master of Logic programme in Amsterdam. Being new to academia and foreign culture, just as Lev used to be, Anna deeply appreciated all the help and support that Dick provided.

Acknowledgements The work of Lev Beklemishev and Anna Dmitrieva was supported by the Academic Fund Program at the National Research University Higher School of Economics (HSE) in 2019–2020 (grant No. 19-04-050) and by the Russian Academic Excellence Project "5–100".

References

Alperin, R. C. (2000). A mathematical theory of origami constructions and numbers. *The New York Journal of Mathematics, 6*(119), 133.

Bachmann, F. (1973). *Aufbau der Geometrie aus dem Spiegelungsbegriff*. Springer.

Becker, E. (1974). Euklidische Körper und euklidische Hüllen von Körpern. *Journal für die reine und angewandte Mathematik, 1974*(268–269), 41–52.

Beeson, M. (1982). *Some undecidable field theories*. Translation of Ziegler. Available at www.michaelbeeson.com/research/papers/Ziegler.pdf

Beeson, M. (2015). A constructive version of Tarski's geometry. *Annals of Pure and Applied Logic, 166*(11), 1199–1273.

Beloch, M. P. (1934). Alcune applicazioni del metodo del ripiegamento della carta di sundara-row. *Atti dell'Acc. di Scienze, Mediche. Nat Mat Ferrara Serie II,11*, 186–189.

Blumenthal, L. M. (1980). *A modern view of geometry*. Courier Corporation.

Degen, W., & Profke, L. (1976). *Grundlagen der affinen und euklidischen Geometrie*. Teubner.

Demaine, E. D., & O'Rourke, J. (2007). *Geometric folding algorithms*. Cambridge University Press.

Friedman, M. (2018). *A history of folding in mathematics: Mathematizing the margins*. Birkhäuser.

Friedman, H., & Visser, A. (2014). When bi-interpretability implies synonymy. *Logic Group Preprint Series, 320*.

Ghourabi, F., Kasem, A., & Kaliszyk, C. (2012). Algebraic analysis of Huzita's origami operations and their extensions. In *International workshop on automated deduction in geometry* (pp. 143–160). Springer.

Hall, M. (1943). Projective planes. *Transactions of the American Mathematical Society, 54*(2), 229–277.

Hartshorne, R. (2000). *Geometry: Euclid and beyond*. Springer.

Hilbert, D. (1971). *Foundations of Geometry, Second Edition, translated from the Tenth Edition, revised and enlarged by Dr Paul Bernays*. The Open Court Publishing Company.

Hilbert, D. (2013). *Grundlagen der Geometrie*. Springer.

Hodges, W. (1993). *Model theory, volume 42 of Encyclopedia of mathematics and its applications*. Cambridge University Press.

Huzita, H. (1989a). Axiomatic development of origami geometry (pp. 143–158).

Huzita, H. (Ed.). (1989b). *Proceedings of the 1st International Meeting of Origami, Science and Technology*. Comune di Ferrara and Centro Origami Diffusion.

Ivanov, N. V. (2016). Affine planes, ternary rings, and examples of non-desarguesian planes. arXiv:1604.04945

Justin, J. (1989). Résolution par le pliage de équation du troisieme degré et applications géométriques. In *Proceedings of the First International Meeting of Origami Science and Technology* (pp. 251–261).

Lang, R. J. (2003). *Origami design secrets: Mathematical methods for an ancient art*. A. K. Peters.

Lang, R. J., & Alperin, R. C. (2009). One-, two-, and multi-fold origami axioms. In *Origami4: Fourth international meeting of origami science, mathematics, and education* (pp. 383–406).

Makowsky, J. A. (2017). Can one design a geometry engine? on the (un) decidability of affine Euclidean geometries. arXiv:1712.07474

Makowsky, J. A. (2018). The undecidability of orthogonal and origami geometries. In *International workshop on logic, language, information, and computation* (pp. 250–270). Springer.

Makowsky, J. A. (2004). Algorithmic uses of the Feferman-Vaught theorem. *Annals of Pure and Applied Logic, 126*(1–3), 159–213.

Makowsky, J. A. (2019). Can one design a geometry engine? *Annals of Mathematics and Artificial Intelligence, 85*(2–4), 259–291.

Shoenfield, J. R. (1967). *Mathematical logic*. Addison-Wesley Publishing Company.

Sundara Rao, T. (1917). In W. Beman & D. Smith (Eds.), *Geometric exercises in paper folding*. The Open Court Publishing Company.

Szmielew, W. (1983). *From affine to Euclidean geometry, an axiomatic approach*. Polish Scientific Publishers (Warszawa-Poland) and D. Reidel Publishing Company (Dordrecht-Holland).

von Staudt, K. G. C. (1857). *Beiträge zur Geometrie der Lage* (Vol. 2). F. Korn.

Wu, W.-T. (1984). *Mechanical theorem proving in geometries*. Springer (Original in Chinese, 1984).

Young, G. C., & Young, W. H. (1905). *The first book of geometry*. Chelsea Publishing Company.

Ziegler, M. (1982). Einige unentscheidbare Körpertheorien. In V. Strassen E. Engeler, H. Läuchli (Eds.), *Logic and Algorithmic, an international symposium held in honour of E. Specker* (pp. 381–392). L'enseignement mathématique.

Chapter 13
The Complete Bibliography of Dick de Jongh

Dick de Jongh

1. de Jongh, D. (1964). Onderzoekingen over de intuitionistische propositielogica. Master's thesis, Universiteit van Amsterdam. https://drive.google.com/file/d/14drwxJqi31FXHEfmuXAZb5QhpAcKEr7t/view?usp=sharings
2. de Jongh, D., & Troelstra, A. S. (1966). On the connection of partially ordered sets with some pseudo-Boolean algebras. *Indagationes Mathematicae,* (3).
3. de Jongh, D. (1968). Essay on I-valuations. In Braffort & van Scheepen (Eds.), *Automation in language translation and theorem proving: Some applications of mathematical logic* (pp. 263–267). European Commission.
4. de Jongh, D. (1968). *Investigations on the Intuitionistic Propositional Calculus*, Doctoral thesis, University of Wisconsin. https://eprints.illc.uva.nl/id/eprint/1837/2/HDS-05-Dick-de-Jongh.text.pdf
5. de Jongh, D., & Gabbay, D. M. (1969). A sequence of decidable finitely axiomatizable intermediate logics with the disjunction property. *Technical Reports*, Vol. 35, U.S. Office of Naval Research Information Systems Branch.
6. de Jongh, D. (1969). *The maximality of the intuitionistic predicate calculus with respect to Heyting's arithmetic*. Manuscript. https://drive.google.com/file/d/1y6aZDa1dRafVU-Z305eHACfG2taTGvgg/view?usp=sharing
7. de Jongh, D. (1970). A characterization of the intuitionistic propositional calculus. In *Intuitionism and Proof Theory, Proceedings of the Buffalo 1968 Conference*. North-Holland Publishing Company.
8. de Jongh, D., & Beth, E. W. (1971). *A survey of modern logic* (translation Dutch-English, in cooperation with S. de Jongh-Kearl). Reidel Publishing Company.
9. de Jongh, D., Ehrenfeucht, D. A., Geiser, J., & Gordon, C. E. (1971) Semantics for non-iterated local observation. Manuscript.
10. de Jongh, D. (1973). Formulas of one propositional variable in intuitionistic arithmetic. In *Reports 73–03*. Mathematisch Instituut, Roetersstraat 15.

D. de Jongh (✉)
Institute for Logic, Language and Computation, University of Amsterdam, Amsterdam, The Netherlands
e-mail: d.h.j.dejongh@uva.nl

N. Bezhanishvili et al. (eds.), *Dick de Jongh on Intuitionistic and Provability Logics*, Outstanding Contributions to Logic 28, https://doi.org/10.1007/978-3-031-47921-2_13

11. de Jongh, D., & Gabbay, D. M. (1974) A sequence of decidable intermediate logics with the disjunction property. *The Journal of Symbolic Logic, 39*, 67–78.
12. de Jongh, D., & Smoryński, C. (1976). Kripke models and the intuitionistic theory of species. *Annals of Mathematical Logic, 9*, 157–186.
13. de Jongh, D., & Parikh, R. (1977) Well-partial orderings and hierarchies. In *Proceedings of the Koninklijke Akademie van Wetenschappen, series A* (Vol. 80).
14. de Jongh, D. (1980) A class of intuitionistic connectives. In J. Barwise, H. J. Keisler & K. Kunen (Eds.), *The Kleene symposium* (pp. 103–111). North Holland.
15. de Jongh, D., van Benthem, J., Groenendijk, J., Stokhof, M., & Verkuyl, H. (1982). (under the name: Gamut) *Logica, Taal en Betekenis, Boek1, Inleiding in de Logica*. Aula Paperback no. 76, Spectrum.
16. de Jongh, D., van Benthem, J., Groenendijk, J., Stokhof, M., Verkuyl, E. H. (1982). (under the name: Gamut) *Logica, Taal en Betekenis, Boek 2, Intensionele Logica en Logische Grammatica*. Aula Paperbacks no. 77, Spectrum.
17. de Jongh, D. (1982). Formulas of one propositional variable in intuitionistic arithmetic. In T. en van Dalen (Ed.), *Brouwer centenary symposium* (pp. 51–64). North-Holland.
18. de Jongh, D., & Bowen, K. A. (1986). Some complete logics for branched time part I. *ITLI-1986-05*, ILLC, University of Amsterdam.
19. de Jongh, D., Groenendijk, J., & Stokhof, M. (Eds.). (1987). Foundations of pragmatics and lexical semantics. *Groningen-Amsterdam Studies in Semantics (GRASS)* no. 7, Foris Publications.
20. de Jongh, D., Groenendijk, J., & Stokhof, M. (Eds.). (1987). The theory of generalized quantifiers and the theory of discourse representation. *Groningen-Amsterdam Studies in Semantics (GRASS)* no. 8, Foris Publications.
21. de Jongh, D. (1987). A simplification of a completeness proof of Guaspari and Solovay. *Studia Logica, 46*, 193–203.
22. de Jongh, D., & Montagna, F. (1987). Generic generalized Rosser fixed points. *Studia Logica, 46*, 187–192.
23. de Jongh, D., & Montagna, F. (1988). Provable fixed points. *Zeitschrift für Mathematische Logik und Grundlagen der Mathematik, Bd. 35*, 229–250.
24. de Jongh, D., & Montagna, F. (1989). Much shorter proofs. *Zeitschrift für Mathematische Logik und Grundlagen der Mathematik, Bd. 35*, 247–260.
25. de Jongh, D., & Veltman, F. (1990). Provability logics for relative interpretability. In P. Petkov (Ed.), *Mathematical Logic, Proceedings of the Heyting 1988 Summer School in Varna* (pp. 31–42). Plenum Press.
26. de Jongh, D., & Visser, A. (1991). Explicit fixed points in interpretability logic. *Studia Logica, 50*, 39–50.
27. de Jongh, D., & Jumelet M., & Montagna, F. (1991). On the proof of Solovay's theorem. *Studia Logica, 50*, 51–70.

28. de Jongh, D., & Montagna, F. (1991). Rosser orderings and free variables. *Studia Logica, 50*, 71–80.
29. de Jongh, D., & van Benthem, J. F. A. K. Groenendijk, J., Stokhof, M., & Verkuyl, H. (1991). (under the name: Gamut) *Logic, language and meaning, Books 1 and 2*. Chicago University Press (reviewed and enlarged translation of 11 and 12).
30. de Jongh, D., & Hendriks, A., Renardel de Lavalette, G. R. (1993). Computations in fragments of intuitionistic propositional logic. *Journal of Automated Reasoning, 7*, 537–561.
31. de Jongh, D., & Visser, A. (Eds.). (1993). *Special Issue of Annals of Pure and Applied Logic, Devoted to the PIA symposium* (Vol. 61, pp. 1–2).
32. de Jongh, D., & Chagrova, L. (1995). The decidability of dependency in intuitionistic propositional logic. *Journal of Symbolic Logic, 60*, 498–504.
33. de Jongh, D., & Visser, A., van Benthem, J. F. A. K., & Renardel de Lavallette, G. R. (1995). NNIL, a study in intuitionistic propositional logic. In A. Ponse, M. de Rijke & Y. Venema (Eds.), *Modal logic and process algebra, a bisimulation perspective* (pp. 289–326). CSLI Lecture Notes no. 53.
34. de Jongh, D., & Hendriks, A. (1996). Finitely generated Magari algebras and arithmetic. In A. Ursini & P. Aglianò (Eds.), *Logic and algebra* (pp. 137–160). Marcel Dekker, Inc.
35. de Jongh, D., & Kanazawa, M. (1996). Angluin's theorem for indexed families of r.e. sets and applications. In *Proceedings of the Nineth Annual Conference on Computational Learning Theory* (pp. 193–204). The Association for Computing Machinery.
36. de Jongh, D., & Visser, A. (1996). Embeddings of Heyting algebras. In W. Hodges, M. Hyland, C. Steinhorn & J. Truss (Eds.), *Logic: From foundations to applications*, European Logic Colloquium (pp. 187–214). Clarendon Press.
37. de Jongh, D., & Visser, A. (1996). Bisimulations, model descriptions and propositional quantifiers. In *Eighth ESSLLI summer school*. Czech Republic.
38. de Jongh, D., Osherson, D., Martin, E., & Weinstein, S. (1997). Formal learning theory. In van Benthem, J. F. A. K., Ter Meulen, A. (Eds.), *Handbook of logic and language* (pp. 737–775). Elsevier Science Publishers.
39. de Jongh, D., & Japaridze, G. (1998). The logic of provability. In S. Buss (Ed.), *Handbook of proof theory* (pp. 475–546). North-Holland/Elsevier.
40. de Jongh, D., & Pianigiani, D. (1998). Solution of a problem of David Guaspari. In E. Orlowska (Ed.), *Logic at work, essays dedicated to the memory of Helena Rasiowa* (pp. 246–254). Physica-verlag, Springer.
41. de Jongh, D., & van Ulsen, P. (1999). Beth's non-classical valuations. In *Un Logicien Consciencieux, La Philosophie de Evert Willem Beth* (Vol. 3, pp. 279–302). Philosophiae Scientiae.
42. de Jongh, D., & Veltman, F. (1999). The modal completeness of ILW. In J. Gerbrandy, M. Marx, M. de Rijke & Y. Venema (Eds.), *Essays dedicated to Johan van Benthem on his 50th Birthday*, CD-ROM JFAK, Vossiuspers Amsterdam University Press.

43. de Jongh, D., & Areces, C., & Hoogland, E. (2000). Interpolation, definability and fixed points in interpretability logics. In M. Zakharyaschev, K. Segerberg, M. de Rijke & H. Wansing (Eds.), *Advances in modal logic*. Lecture Notes (Vol. 2, pp. 53–76). Stanford, Ca, CSLI.
44. de Jongh, D. (2000) Learning and grammar. In J. F. Quesada, F. J. Salguero & A. Nepomuceno (Eds.), *Proceedings of the First Workshop on Logic and Language*, 2000–2001 in ILL-PR (pp. 139–146). Instituto de Lógica, Lenguaje e Información, Un. de Sevilla, Editorial Kronos, Sevilla.
45. de Jongh, D. (2003). Intuicionismo. In D. M. J. Braquinho (Ed.), *Enciclopédia de Termes Lógico-Filosóficos* (pp. 413–417). Gradiva.
46. de Jongh, D., & Hendriks, A. (2003). Characterization of strongly equivalent logic programs in intermediate logics. *The Journal for the Theory and Practice of Logic programming (TPLP), 3*, 259–270 (2003)
47. de Jongh, D. (2004). Volledigheid en onvolledigheid. Afscheidscollege. *ILLC-prepublications* PP-2004-30, Universiteit van Amsterdam. https://eprints.illc.uva.nl/id/eprint/141/1/PP-2006-48.text.pdf
48. de Jongh, D., Veltman, F., & Verbrugge, R. (2004). Completeness by construction for tense logics of linear time. In F. Veltman (Ed.), *Liber Amicorum for Dick de Jongh*. Institute of Logic, Language and Computation, Amsterdam. https://staff.fnwi.uva.nl/f.j.m.m.veltman/papers/FVeltman_CompbyConstr.pdf
49. de Jongh, D., Iemhoff, R., & Zhou, C. (2005). Properties of intuitionistic provability and preservativity. *Logic Journal of the IGPL, 13*(6), 615–636.
50. de Jongh, D. (2006). The Incompleteness Theorems, their content and their meaning. *Philosophical Trends*. https://eprints.illc.uva.nl/id/eprint/223/1/PP-2006-48.text.pdf
51. de Jongh, D., & Bezhanishvili, N. (2006). Intuitionistic logic. *Technical Report* PP-2006-26, ILLC, Universiteit van Amsterdam. https://eprints.illc.uva.nl/id/eprint/200/1/PP-2006-25.text.pdf
52. de Jongh, D. (2006). Intuicionismo, Ediciones Universidad de Salamanca Azafea. Rev. filos. 8, 53–69A.
53. de Jongh, D., & Liu, F. (2006). Optimality, belief and preference. In S. Artemov & R. Parikh (Eds.), *Proceedings of the Workshop on Rationality and Knowledge*. ESSLLI.
54. de Jongh, D., & Bezhanishvili, G., Bezhanishvili, N. (2008). The Kuznetsov-Gerciu and Rieger-Nishimura logics: The boundaries of the finite model property. *Logic and Logical Philosophy, 17*, 73–110.
55. de Jongh, D., & Gencer, Ç. (2009). Unifiability in extensions of K4. *Logic Journal of IGPL, 17*(2), 159–172.
56. de Jongh, D., Bílková, M., & Joosten, J. (2009). Interpretability in PRA. *Annals of Pure and Applied Logic, 161*(2), 128–138.
57. de Jongh, D., & Liu, F. (2009). Preference, priorities and belief. In T. Grüne-Yanoff, S. O. Hansson (Eds.), *Preference change*. Springer.
58. de Jongh, D., & Ghosh, S. (2009). Comparing strengths of beliefs explicitly. In *Proceedings of the 2nd Multi-agent Logics, Languages, and Organizations Federated, MALLOW 2009, CEUR Workshop Proceedings* (Vol. 494).

59. de Jongh, D., & Yang, F. (2011). Jankov's theorems for intermediate logics in the setting of universal models. In N. Bezhanishivili et al. (Eds.), *TbiLLC 2009*, LNAI 661, par. 8 (pp. 53–76).
60. de Jongh, D., Visser, A., & Verbrugge, R. (2011). Intermediate logics and the de Jongh property. *Archive for Mathematical Logic, 50*, 197–213.
61. de Jongh, D., Renardel de Lavalette, G. R., & Hendriks, A. (2012). Intuitionistic implication without disjunction. *Journal of Logic and Computation, 22*(3), 375–404.
62. de Jongh, D., & Bezhanishivili, N. (2012). Extendible formulas in two variables in intuitionistic logic. *Studia Logica, 100*, 61–89.
63. de Jongh, D., & Ghosh, S. (2013). Comparing strengths of beliefs explicitly. *Logic Journal of the IGPL, 21*, 488–514.
64. de Jongh, D., & Gierasimczuk, N. (2013). On the complexity of conclusive update. *The Computer Journal, 56*(3), 365–377.
65. de Jongh, D., Gierasimczuk, N., & Hendricks, V. F. (2014). Logic and learning. In A. Baltag & S. Smets (Eds.), *Johan van Benthem on logic and information dynamics, volume 5 of Outstanding Contributions to Logic* (pp. 267–288). Springer.
66. de Jongh, D., Colacito, A., & Vargas Sandoval, A. (2017). Subminimal negation. *Soft Computer, 21*, 165–174.
67. de Jongh, D., & Zhao, Z. (2017). Positive formulas in intuitionistic and minimal logic. In *10th international Tbilisi symposium on logic, language, and computation, TbiLLC, Gudauri 2015, volume 10148 of Lecture notes in computer science*, (pp. 229–250). Springer.
68. de Jongh, D., Bezhanishvili, N., Tzimoulis, A., & Zhao, Z. (2017). Universal models for the positive fragment of intuitionistic logic. In H. H. Hansen (Ed.), *11th international Tbilisi symposium on logic and language, and computation, TbiLLC 2015 revised selected papers*. Lecture notes in computer science (Vol. 10148, pp. 229–250).
69. de Jongh, D., & Maleki, F. S. (2017). Subintuitionistic Logics with Kripke Semantics. In *11th international Tbilisi symposium on logic, language, and computation, TbiLLC 2015*, LNCS, Volume 10148 (pp. 333–354). Springer.
70. de Jongh, D., & Maleki, F. S. (2017). Weak subintuitionistic logics. *Logic Journal of the IGPL, 25*(2), 214–231.
71. de Jongh, D., & Bezhanishvili, N. (2018). Stable formulas in intuitionistic logic. *Notre Dame Journal of Formal Logic, 59*(3), 307–324.
72. de Jongh, D., & Maleki, F. S. (2018). Subintuitionistic logics and the implications they prove. *Indagationes Mathematicae, 6*(3), 1525–1545.
73. de Jongh, D., & Maleki, F. S. (2019). Two neighborhood semantics for subintuitionistic logics. In *12th international Tbilisi symposium on logic, language, and computation, TbiLLC 2017*, LNCS, Volume 11456 (pp. 64–85). Springer.
74. de Jongh, D., Bezhanishvili, N., & Colacito, A. (2019). A study of subminimal logics of negation and their modal companions. In *Post-proceedings of TbiLLC 2017*, Lecture notes in computer science (Vol. 11456, pp. 21–41). Springer.

75. de Jongh, D., & Vargas-Sandoval, A. L. (2019). Finite identification with positive and with complete data. In *Post-proceedings of TbiLLC 2017*, Lecture notes in computer science (Vol. 11456, pp. 42–63). Springer.
76. de Jongh, D., & Maleki, F. S. (2021). Binary modal companions for subintuitionistic logics. *Mathematics, logic, and their philosophies: Essays in Honour of Mohammad Ardeshir* (pp. 35–52). Springer.
77. de Jongh, D., Ilin, J., & Yang, F. (2021). NNIL formulas revisited: universal models and finite model property. *Journal of Logic and Computation, 31*(2), 573–596.
78. de Jongh, D., de Almeida Borges, A., Joosten, J., & Visser, A. (2022). The strictly positive fragment of provability logic for Heyting arithmetic. Manuscript.
79. de Jongh, D., & Maleki, F. S. (2023). Binary modal logic and unary modal logic. *Logic Journal of the IGPL*, 2023; jzac083, https://doi.org/10.1093/jigpal/jzac083.

The manufacturer's authorised representative in the EU is Springer Nature Customer Service Centre GmbH, Europaplatz 3, 69115 Heidelberg, Germany. If you have any concerns regarding our products, please contact ProductSafety@springernature.com

Printed and bound by CPI Group (UK) Ltd, Croydon, CR0 4YY

28/09/2026

02224065-0003